PRACTICAL GUIDE TO FINITE ELEMENTS

MECHANICAL ENGINEERING

A Series of Textbooks and Reference Books

Editor

L. L. Faulkner

Columbus Division, Battelle Memorial Institute
and Department of Mechanical Engineering
The Ohio State University
Columbus, Ohio

1. *Spring Designer's Handbook*, Harold Carlson
2. *Computer-Aided Graphics and Design*, Daniel L. Ryan
3. *Lubrication Fundamentals*, J. George Wills
4. *Solar Engineering for Domestic Buildings*, William A. Himmelman
5. *Applied Engineering Mechanics: Statics and Dynamics*, G. Boothroyd and C. Poli
6. *Centrifugal Pump Clinic*, Igor J. Karassik
7. *Computer-Aided Kinetics for Machine Design*, Daniel L. Ryan
8. *Plastics Products Design Handbook, Part A: Materials and Components; Part B: Processes and Design for Processes*, edited by Edward Miller
9. *Turbomachinery: Basic Theory and Applications*, Earl Logan, Jr.
10. *Vibrations of Shells and Plates*, Werner Soedel
11. *Flat and Corrugated Diaphragm Design Handbook*, Mario Di Giovanni
12. *Practical Stress Analysis in Engineering Design*, Alexander Blake
13. *An Introduction to the Design and Behavior of Bolted Joints*, John H. Bickford
14. *Optimal Engineering Design: Principles and Applications*, James N. Siddall
15. *Spring Manufacturing Handbook*, Harold Carlson
16. *Industrial Noise Control: Fundamentals and Applications*, edited by Lewis H. Bell
17. *Gears and Their Vibration: A Basic Approach to Understanding Gear Noise*, J. Derek Smith
18. *Chains for Power Transmission and Material Handling: Design and Applications Handbook*, American Chain Association
19. *Corrosion and Corrosion Protection Handbook*, edited by Philip A. Schweitzer
20. *Gear Drive Systems: Design and Application*, Peter Lynwander
21. *Controlling In-Plant Airborne Contaminants: Systems Design and Calculations*, John D. Constance
22. *CAD/CAM Systems Planning and Implementation*, Charles S. Knox
23. *Probabilistic Engineering Design: Principles and Applications*, James N. Siddall

24. *Traction Drives: Selection and Application*, Frederick W. Heilich III and Eugene E. Shube
25. *Finite Element Methods: An Introduction*, Ronald L. Huston and Chris E. Passerello
26. *Mechanical Fastening of Plastics: An Engineering Handbook*, Brayton Lincoln, Kenneth J. Gomes, and James F. Braden
27. *Lubrication in Practice: Second Edition*, edited by W. S. Robertson
28. *Principles of Automated Drafting*, Daniel L. Ryan
29. *Practical Seal Design*, edited by Leonard J. Martini
30. *Engineering Documentation for CAD/CAM Applications*, Charles S. Knox
31. *Design Dimensioning with Computer Graphics Applications*, Jerome C. Lange
32. *Mechanism Analysis: Simplified Graphical and Analytical Techniques*, Lyndon O. Barton
33. *CAD/CAM Systems: Justification, Implementation, Productivity Measurement*, Edward J. Preston, George W. Crawford, and Mark E. Coticchia
34. *Steam Plant Calculations Manual*, V. Ganapathy
35. *Design Assurance for Engineers and Managers*, John A. Burgess
36. *Heat Transfer Fluids and Systems for Process and Energy Applications*, Jasbir Singh
37. *Potential Flows: Computer Graphic Solutions*, Robert H. Kirchhoff
38. *Computer-Aided Graphics and Design: Second Edition*, Daniel L. Ryan
39. *Electronically Controlled Proportional Valves: Selection and Application*, Michael J. Tonyan, edited by Tobi Goldoftas
40. *Pressure Gauge Handbook*, AMETEK, U.S. Gauge Division, edited by Philip W. Harland
41. *Fabric Filtration for Combustion Sources: Fundamentals and Basic Technology*, R. P. Donovan
42. *Design of Mechanical Joints*, Alexander Blake
43. *CAD/CAM Dictionary*, Edward J. Preston, George W. Crawford, and Mark E. Coticchia
44. *Machinery Adhesives for Locking, Retaining, and Sealing*, Girard S. Haviland
45. *Couplings and Joints: Design, Selection, and Application*, Jon R. Mancuso
46. *Shaft Alignment Handbook*, John Piotrowski
47. *BASIC Programs for Steam Plant Engineers: Boilers, Combustion, Fluid Flow, and Heat Transfer*, V. Ganapathy
48. *Solving Mechanical Design Problems with Computer Graphics*, Jerome C. Lange
49. *Plastics Gearing: Selection and Application*, Clifford E. Adams
50. *Clutches and Brakes: Design and Selection*, William C. Orthwein
51. *Transducers in Mechanical and Electronic Design*, Harry L. Trietley
52. *Metallurgical Applications of Shock-Wave and High-Strain-Rate Phenomena*, edited by Lawrence E. Murr, Karl P. Staudhammer, and Marc A. Meyers
53. *Magnesium Products Design*, Robert S. Busk
54. *How to Integrate CAD/CAM Systems: Management and Technology*, William D. Engelke

55. *Cam Design and Manufacture: Second Edition*; with cam design software for the IBM PC and compatibles, disk included, Preben W. Jensen
56. *Solid-State AC Motor Controls: Selection and Application*, Sylvester Campbell
57. *Fundamentals of Robotics*, David D. Ardayfio
58. *Belt Selection and Application for Engineers*, edited by Wallace D. Erickson
59. *Developing Three-Dimensional CAD Software with the IBM PC*, C. Stan Wei
60. *Organizing Data for CIM Applications*, Charles S. Knox, with contributions by Thomas C. Boos, Ross S. Culverhouse, and Paul F. Muchnicki
61. *Computer-Aided Simulation in Railway Dynamics*, by Rao V. Dukkipati and Joseph R. Amyot
62. *Fiber-Reinforced Composites: Materials, Manufacturing, and Design*, P. K. Mallick
63. *Photoelectric Sensors and Controls Selection and Application*, Scott M. Juds
64. *Finite Element Analysis with Personal Computers*, Edward R. Champion, Jr., and J. Michael Ensminger
65. *Ultrasonics: Fundamentals, Technology, Applications: Second Edition, Revised and Expanded*, Dale Ensminger
66. *Applied Finite Element Modeling: Practical Problem Solving for Engineers*, Jeffrey M. Steele
67. *Measurement and Instrumentation in Engineering: Principles and Basic Laboratory Experiments*, Francis S. Tse and Ivan E. Morse
68. *Centrifugal Pump Clinic: Second Edition, Revised and Expanded*, Igor J. Karassik
69. *Practical Stress Analysis in Engineering Design: Second Edition, Revised and Expanded*, Alexander Blake
70. *An Introduction to the Design and Behavior of Bolted Joints: Second Edition, Revised and Expanded*, John H. Bickford
71. *High Vacuum Technology: A Practical Guide*, Marsbed H. Hablanian
72. *Pressure Sensors: Selection and Application*, Duane Tandeske
73. *Zinc Handbook: Properties, Processing, and Use in Design*, Frank Porter
74. *Thermal Fatigue of Metals*, Andrzej Weronski and Tadeusz Hejwowski
75. *Classical and Modern Mechanisms for Engineers and Inventors*, Preben W. Jensen
76. *Handbook of Electronic Package Design*, edited by Michael Pecht
77. *Shock-Wave and High-Strain-Rate Phenomena in Materials*, edited by Marc A. Meyers, Lawrence E. Murr, and Karl P. Staudhammer
78. *Industrial Refrigeration: Principles, Design and Applications*, P. C. Koelet
79. *Applied Combustion*, Eugene L. Keating
80. *Engine Oils and Automotive Lubrication*, edited by Wilfried J. Bartz
81. *Mechanism Analysis: Simplified and Graphical Techniques, Second Edition, Revised and Expanded*, Lyndon O. Barton
82. *Fundamental Fluid Mechanics for the Practicing Engineer*, James W. Murdock
83. *Fiber-Reinforced Composites: Materials, Manufacturing, and Design, Second Edition, Revised and Expanded*, P. K. Mallick

84. *Numerical Methods for Engineering Applications*, Edward R. Champion, Jr.
85. *Turbomachinery: Basic Theory and Applications, Second Edition, Revised and Expanded*, Earl Logan, Jr.
86. *Vibrations of Shells and Plates: Second Edition, Revised and Expanded*, Werner Soedel
87. *Steam Plant Calculations Manual: Second Edition, Revised and Expanded*, V. Ganapathy
88. *Industrial Noise Control: Fundamentals and Applications, Second Edition, Revised and Expanded*, Lewis H. Bell and Douglas H. Bell
89. *Finite Elements: Their Design and Performance*, Richard H. MacNeal
90. *Mechanical Properties of Polymers and Composites: Second Edition, Revised and Expanded*, Lawrence E. Nielsen and Robert F. Landel
91. *Mechanical Wear Prediction and Prevention*, Raymond G. Bayer
92. *Mechanical Power Transmission Components*, edited by David W. South and Jon R. Mancuso
93. *Handbook of Turbomachinery*, edited by Earl Logan, Jr.
94. *Engineering Documentation Control Practices and Procedures*, Ray E. Monahan
95. *Refractory Linings Thermomechanical Design and Applications*, Charles A. Schacht
96. *Geometric Dimensioning and Tolerancing: Applications and Techniques for Use in Design, Manufacturing, and Inspection*, James D. Meadows
97. *An Introduction to the Design and Behavior of Bolted Joints: Third Edition, Revised and Expanded*, John H. Bickford
98. *Shaft Alignment Handbook: Second Edition, Revised and Expanded*, John Piotrowski
99. *Computer-Aided Design of Polymer-Matrix Composite Structures*, edited by Suong Van Hoa
100. *Friction Science and Technology*, Peter J. Blau
101. *Introduction to Plastics and Composites: Mechanical Properties and Engineering Applications*, Edward Miller
102. *Practical Fracture Mechanics in Design*, Alexander Blake
103. *Pump Characteristics and Applications*, Michael W. Volk
104. *Optical Principles and Technology for Engineers*, James E. Stewart
105. *Optimizing the Shape of Mechanical Elements and Structures*, A. A. Seireg and Jorge Rodriguez
106. *Kinematics and Dynamics of Machinery*, Vladimír Stejskal and Michael Valášek
107. *Shaft Seals for Dynamic Applications*, Les Horve
108. *Reliability-Based Mechanical Design*, edited by Thomas A. Cruse
109. *Mechanical Fastening, Joining, and Assembly*, James A. Speck
110. *Turbomachinery Fluid Dynamics and Heat Transfer*, edited by Chunill Hah
111. *High-Vacuum Technology: A Practical Guide, Second Edition, Revised and Expanded*, Marsbed H. Hablanian
112. *Geometric Dimensioning and Tolerancing: Workbook and Answerbook*, James D. Meadows
113. *Handbook of Materials Selection for Engineering Applications*, edited by G. T. Murray

114. *Handbook of Thermoplastic Piping System Design*, Thomas Sixsmith and Reinhard Hanselka
115. *Practical Guide to Finite Elements: A Solid Mechanics Approach*, Steven M. Lepi

Additional Volumes in Preparation

Applied Computational Fluid Dynamics, edited by Vijay K. Garg

Friction and Lubrication in Mechanical Design, A. A. Seireg

Machining of Ceramics and Composites, edited by Said Jahanmir and M. Ramulu

Fluid Sealing Technology, Heinz Konrad Muller and Bernard Nau

Heat Exchange Design Handbook, T. Kuppan

Mechanical Engineering Software

Spring Design with an IBM PC, Al Dietrich

Mechanical Design Failure Analysis: With Failure Analysis System Software for the IBM PC, David G. Ullman

PRACTICAL GUIDE TO FINITE ELEMENTS

A SOLID MECHANICS APPROACH

STEVEN M. LEPI
Visteon
Dearborn, Michigan

MARCEL DEKKER, INC.   NEW YORK • BASEL • HONG KONG

Library of Congress Cataloging-in-Publication Data

Lepi, Steven M.
Practical guide to finite elements : a solid mechanics approach /
Steven M. Lepi.
p. cm. -- (Mechanical engineering ; 115)
Includes bibliographical references and index.
ISBN 0-8247-0075-9
1. Finite element method. I. Series: Mechanical engineering
(Marcel Dekker, Inc.) ; 115
TA347.F5L44 1998
620.1'05'0151535--dc21 97-52825
CIP

This book is printed on acid-free paper.

Cover photo credit: Karin Tuttle, Ford Motor Company.

Headquarters
Marcel Dekker, Inc.
270 Madison Avenue, New York, NY 10016
tel: 212-696-9000; fax: 212-685-4540

Eastern Hemisphere Distribution
Marcel Dekker AG
Hutgasse 4, Postfach 812, CH-4001 Basel, Switzerland
tel: 44-61-261-8482; fax: 44-61-261-8896

World Wide Web
http://www.dekker.com

The publisher offers discounts on this book when ordered in bulk quantities. For more information, write to Special Sales/Professional Marketing at the headquarters address above.

Current printing (last digit):
10 9 8 7 6 5 4 3 2 1

PRINTED IN THE UNITED STATES OF AMERICA

To Kim

The eternal spring of true friendship
flows forth precious memories of time past
and the promise of kindred spirit
in the years ahead.

Preface

This book originated more than a decade ago, as a short paper that attempted to provide a brief overview of how finite element techniques can be used to solve basic solid mechanics problems. My former engineering manager thought that it would be helpful if I could provide a brief explanation of the method, drawing upon my academic study in engineering, my experience as a finite element analyst, and additional experience as a part-time calculus instructor. One of the first things I did was to meet with several managers to decide whom the potential audience was, and at what technical level the paper should be written. After many conversations, some of which lapsed into exchanges redolent of the comic strip *Dilbert*, it became apparent that the level of engineering knowledge in industry varies considerably with the particular type of industry, an individual's job classification, and time elapsed from contact with academia. (Some of the more interesting suggestions were that I should "not mention integrals, because they scare people away," and "don't use matrices—nobody understands matrices"; other suggestions were less intelligible.)

There are many fine books on the topic of finite element analysis, several of which are referenced in this work. However, the majority of them are written on a level that appears to be somewhat beyond the comprehension of engineers who have been away from the academic world for more than three or so years. (I make this rough estimate of "brain drain" based upon my experience teaching finite element analysis to engineers at both my present place of employment and at a local university.) This book was written for the practicing engineer, and might also be used as a primer for an introductory college level course on finite element analysis. It does not replace the standard, comprehensive, college level finite element textbooks that are currently available.

Chapter 1 prepares the reader for the topics that are covered in the remainder of the book. Use of energy methods to determine displacement, the Ritz method, and stress-strain basics are presented in a cursory fashion. As throughout the rest of the book, sources of more in-depth information are provided. Chapter 1 concludes with an overview of the analysis process in Section 1.6, and reinforces the need for engineering judgment in specifying a mechanical idealization.

Chapter 2 introduces finite element basics using a one-dimensional rod element. The global stiffness matrix, convergence, finite element equilibrium, and matrix operations are considered. A quick look at sources of fundamental errors associated with finite elements is given in Section 2.8 to conclude the chapter. In Chapter 3, the one-dimensional finite element equations are expanded to allow the solution of problems in two and three dimensions.

Chapter 4 presents a handful of elements that are used in two- and three-dimensional problems of solid mechanics. These elements, all non-parametric, use the minimum number of nodes necessary to define the particular element type under consideration. Assumptions associated with Euler-Bernoulli beams and Kirchhoff plates are listed. In Chapter 5, parametric elements are introduced, along with the associated techniques of numerical integration. Chapter 6 considers how surface tractions and body loads are approximated using consistent nodal forces, and concludes with a brief exposure to some elementary means of imposing essential boundary conditions on the global system of equilibrium equations. Chapter 7 concludes the text with hints regarding the practical application of the finite element method, such as what type of elements can be used for a given idealization, element distortion, and some common pitfalls.

I did not write this book for those who are "afraid of integrals and don't understand matrices." The study of finite element analysis is complicated by the fact that it draws upon many areas of technical study, such as physics, calculus, numerical methods, computer programming, and at least one engineering field of study, such as solid mechanics. To gain an appreciation for the method, one requires either a fundamental recall of the technical fields of study just mentioned, or the desire to review them in some depth. This book targets engineering professionals who either currently have, or at one time had, a good understanding of the requisite body of knowledge. Most individuals in industry who read this book will likely need to augment their reading with periodic reviews of the references mentioned within. It is presumed that those currently pursuing advanced engineering degrees would find the pace of the reading too slow, while engineering professionals in industry for 3–5 years may find the pace about right.

Although many years have passed since my high school days, I must first express my appreciation to Mr. Rinna (Melvindale High School), whose kindness, character, and patience I shall not forget. Professor Barna Szabó (Washington University) reviewed the manuscript and provided detailed comments, suggestions, and guidance, for which I am very grateful. I would like to thank Mr. Jack Collins (Ohio State University) for his review of topics in Chapter 7 related to mechanical failure, and also thank Mr. Leroy Sturges (Iowa State University) for his input on topics related to structural mechanics. I am indebted to my manager, Paul Nicastri (Visteon), for providing general encouragement and assistance with copyright issues. Mary Sheen (the poet) provided much needed editing of the earlier versions of this work. Finally, thanks to my good friend of many years, Mr. Robert Rossi (Visteon), whose critical review and editing abilities proved invaluable.

Steven M. Lepi
slepi@ford.com

Contents

Terminology Used in This Text

Brick Element
A hexagonal (six-sided) volume element, typically having eight or 20 nodes.

Code
A term used to describe software written to solve finite element problems. In the past, the "code" was typically a separate software package from the finite element pre-processor.

Closed-Form Solution
A solution found using a single differential equation that has an elementary anti-derivative.

Complete Polynomial
If a polynomial in one variable is complete to order n, it must contain all terms less than order n, including a constant term. The polynomial below is complete to order two:

$$U(X) = a_1 + a_2 X + a_3 X^2$$

Constraint
A mathematical relationship imposed upon a displacement variable. When a constraint involves more than one finite element node, the term *multipoint constraint* (MPC) is used.

Deformation
The change in any physical dimension of a structure due to external loads or reactions.

Displacement
The change in the spatial coordinates of a point (or points), either by translational or rotational motion, as measured with respect to some initial reference position.

DOF
Degree of freedom—if a point is allowed to move along only one coordinate axis, the point has one degree of freedom.

Equivalent Nodal Forces
Distributed loads in a finite element model are characterized by equivalent nodal forces.

Equivalent Nodal Forces, Kinematic Equivalence
Equivalent nodal forces may also be computed by assuming that, in the absence of rigid body modes, the nodes of a finite element displace with regard to the amount of stiffness associated with the particular nodes. Forces that account for the deformable nature of a finite element are termed *kinematically equivalent nodal forces*. When the kinematically equivalent nodal forces are computed using the element's shape functions, the nodal forces are termed *consistent nodal forces*.

Equivalent Nodal Forces, Static Equivalence
Equivalent nodal forces can be computed by assuming that the finite element model is a rigid body. Then, using summation of forces and summation of moments, static equivalent forces at each node are determined. Forces computed in this manner are termed *statically equivalent nodal forces*. However, these forces are not unique in that an infinite number of combinations could produce the same resultant. Furthermore, these forces need not produce the correct finite element results because the finite element model is a deformable body, not a rigid body. See below.

Functional
A single variable functional takes the form of an integral equation containing one independent variable, a function of the independent variable, and any number of derivatives of the function. In the functional below, x is the independent variable, u the dependent variable, while F denotes a function of the independent variable, derivatives, and the dependent variable.

$$I(u) = \int F\left(u, \frac{du}{dx}, \frac{d^2u}{dx^2}, \ldots \frac{d^nu}{dx^n}, x\right) dx$$

Hooke's Law
Specifies a linear relationship between external forces applied to an elastic structure and the resulting displacement—can be expressed in terms of stress and strain as well as force and displacement.

Hybrid Element
A finite element that uses a combination of stress and displacement assumptions within an element instead of the commonly used displacement assumption.

Idealization
The process of making engineering assumptions to convert a physical problem into a mathematical model.

Ill-Conditioning of Stiffness Matrix (solving $KD=F$)
A situation where the values in the stiffness matrix are such that the solution for the displacement vector is extremely sensitive to computer round-off error.

Isotropic
A material that is isotropic has the same material properties in all directions.

Jacobian
The Jacobian can be considered a "scaling factor" that scales distances measured in two separate coordinate systems. The inverse of the Jacobian is used in finite element analysis to relate derivatives in one coordinate system to derivatives in another.

Line Element
A one-dimensional element, e.g., 2-node line for beam bending.

Linear Element
An element that uses a linear displacement assumption is called a linear element. One must be careful with this definition. Consider an element that is used in applications where strain is defined in terms of first order derivatives. If an element is known to have a linear displacement assumption, one would logically assume that strain, being the first derivative of displacement, must be constant. This is not always the case, however, as explained in Chapter 5.

Linear Finite Element Solution
Using a linear solution, it is assumed that the stiffness matrix terms and the direction of the load vectors remain constant.

Linear Theory in Solid Mechanics
Used in reference to the presumption of small (infinitesimal) deformations and strains in solid mechanics problems.

Mechanism
Occurs in finite elements that have some deficiency in their formulation. Sometimes the deficiency is intentionally designed into a finite element to improve other aspects of the element's performance. A mechanism is characterized by patterns of deformation in a finite element that result in the absence of strain energy, while the actual structure would experience strain energy.

Mesh (Finite Element Mesh)
When using finite element methods to solve structural problems, the term "mesh" is used to describe the assemblage of finite elements that describes the structure being analyzed.

Multipoint Constraint
See constraint.

Normal Stress
Normal stress on a given surface is due to resultant forces acting normal to the surface; normal stress causes deformation within a structure. See also shear stress.

Patch Test
A method of testing finite elements to determine if they perform acceptably under less than ideal conditions.

Pivot
During Gaussian elimination, each equation in the active stiffness matrix is divided by its respective pivot. Pivots that are very small or negative typically suggest a problem with the finite element model.

Poisson's Ratio
In a uniaxially loaded bar, Poisson's ratio is the quotient of the lateral strain divided by the axial strain.

Positive Definite
When solving the finite element matrix equations, a positive definite matrix suggests that the structure being modeled is both stable and adequately restrained. This condition is characterized by all positive pivots during Gaussian elimination. The Eigenvalues of a positive definite system are all greater than zero.

Post-Processor
Describes software used to evaluate the results of a finite element model, typically using color contour graphing routines, etc.

Pre-Processor
Describes software used to perform various tasks associated with finite element analysis, for instance, generating finite elements, applying loads and boundary conditions, and performing model checks. Some finite element software is marketed with pre-processing, solver, and post-processing capability all in one suite.

Quad Element
A four-edged (quadrilateral) surface element; 4-, 8-, and 9-node versions are common.

Reactions
Forces (and/or moments) at restrained points in a finite element model that balance the externally applied loads. (Reactions also balance body loads, such as gravity.)

Restraint
A degree of freedom at a node in a finite element model is restrained when a zero value is assigned to it. In structural mechanics, a restraint is analogous to connecting the structure to "ground."

Shape Function
Solving the finite element matrix equations generates values of displacement at the nodes of each element. One use of the shape function is to interpolate displacement at locations within the element from the nodal values of displacement.

Shear Stress
Causes elastic deformation within a structure without changing the structure's volume.

Solver
A finite element solver is a software routine that solves the matrix equations which evolve from the finite element formulation of a problem. Some solvers can solve linear problems only, while others solve linear and non-linear problems. Solvers may incorporate other software routines in addition to the solver capability, for instance, a matrix optimization routine.

Surface Element
A two-dimensional element, e.g., 4-node quadrilateral for plane stress.

Surface Traction
A system of forces acting upon the surface of a body (or structure) that induces stress internal to the body.

Tet Element
A tetrahedral (four-sided) volume element; 4- and 10-node versions are common.

Tri Element
A triangular surface element; 3- and 6-node versions are common.

Volume Element
A three-dimensional element, e.g., 10-node tetrahedral for three-dimensional stress, or an 8-node, hexahedral brick element.

Young's Modulus
Quantifies the stiffness of a material, typically in association with an applied tensile load. Young's modulus for a given material is often computed by generating a force versus deflection curve of a tensile test specimen. The value for Young's modulus is determined by observing the linear portion of the curve, and computing the *slope*.

Symbols Used in This Text

A — Cross sectional area

$\underline{\underline{A}}$ — Matrix of nodal coordinates, used to establish interpolation functions and surface area

D — Displacement

$\underline{\underline{D}}$ — Vector of nodal displacements for an entire structure: The global displacement vector

$^{e}\underline{\underline{D}}$ — Vector of nodal displacements for Element e

E — Elastic modulus, Young's modulus

$\underline{\underline{F}}$ — Vector of nodal forces for an entire structure: The global force vector

$^{e}\underline{\underline{F}}$ — Vector of nodal forces for Element e, composed of concentrated, surface, body, initial stress, and initial strain contributions

K — Spring constant

$\underline{\underline{K}}$ — Stiffness matrix for an entire structure: The global stiffness matrix

$^{e}\underline{\underline{K}}$ — Stiffness matrix for a single element

$\underline{\underline{N}}$ — Shape function matrix

P — Magnitude of a concentrated load

r, s, t — Coordinate directions in parametric space, not necessarily orthogonal

U, V, W — Displacement in X-, Y-, and Z-coordinate directions (global)

u, v, w — Displacement in x-, y-, and z-coordinate directions (local)

X, Y, Z — Orthogonal coordinate directions in global space

x, y, z — Orthogonal coordinate directions in local space

${}^{e}U_{a}$	Nodal displacement, for Element *e*, Node *a*, in the *X*-coordinate direction
${}^{e}\tilde{U}(X)$	A one-dimensional interpolation function for displacement in the *X*-coordinate direction, for Element *e*
$\sim$	Tilde symbol is used to represent an approximate value of a continuous variable
$\equiv$	The triple bar is used in place of the words "is defined as"
${}^{e}\Pi$	Total potential energy for Element *e*
γ	Shear strain
ε	Normal strain
τ	Stress, both normal and shear

Matrix Notation

$$\underline{\underline{A}} = \begin{Bmatrix} A_1 \\ A_2 \\ A_3 \end{Bmatrix} \qquad \underline{\underline{B}} = \begin{bmatrix} B_{11} & B_{12} \\ B_{21} & B_{22} \end{bmatrix} \qquad \underline{\underline{C}} = \begin{bmatrix} C_{11} & C_{12} & C_{13} \end{bmatrix}$$

A double underscore is used to denote column and row vectors, in addition to square matrices.

$$^{e}\underline{\underline{D}} = \begin{Bmatrix} ^{e}U_a \\ ^{e}V_a \\ ^{e}U_b \\ ^{e}V_b \\ \vdots \end{Bmatrix}$$

Vector of nodal displacements for Element e.

$$\underline{\underline{\breve{D}}} = \begin{Bmatrix} \tilde{U}(X,Y,Z) \\ \tilde{V}(X,Y,Z) \\ \tilde{W}(X,Y,Z) \end{Bmatrix}$$

Vector of displacement interpolation functions.

$$^{e}\underline{\underline{K}}\,^{e}\underline{\underline{D}} = {}^{e}\underline{\underline{F}}$$

Finite element equilibrium equations for Element e.

$$^{e}\underline{\underline{F}} = {}^{e}\underline{\underline{F^{c}}} + {}^{e}\underline{\underline{F^{S}}} + {}^{e}\underline{\underline{F^{\beta}}}$$

The nodal force vector for each element is composed of loads from concentrated, surface, and body contributions.

$$\underline{\underline{F}} = \underline{\underline{F^{c}}} + \sum_{e=1}^{e=nel} {}^{e}\underline{\underline{F^{s}}} + \sum_{e=1}^{e=nel} {}^{e}\underline{\underline{F^{\beta}}} \qquad nel \equiv number\ of\ elements$$

The global force vector composed of the concentrated load vector, plus the summation of surface and body contributions from each element. Although not discussed in this text, the force vector can contain contributions from initial stress and initial strain.

$$\underline{\underline{K}}\,\underline{\underline{D}} = \underline{\underline{F}}$$

Finite element equilibrium equations for an entire structure.

Chapter 1

Finite Element Analysis: Background Concepts

"If he [the teacher] is indeed wise he does not bid you enter the house of his wisdom but rather leads you to the threshold of your own mind."

—Kahlil Gibran

1.1 Introduction

Finite element analysis (FEA) is used in a variety of engineering applications, including solid mechanics, fluid mechanics, heat transfer, and acoustics. In solid mechanics, FEA can be used to determine the maximum stress in a structure that is subjected to external forces. In heat transfer problems, the objective may be to determine the peak temperature in a structure loaded by an internal heat source, while a typical fluid mechanics problem might consider the exit velocity of a fluid, given entrance and exit pressures.

Describing how FEA is applied in each of these disciplines becomes quite complicated, requiring an abstract mathematical approach, while restricting the discussion to a single discipline limits the mathematical abstraction. This guide considers only simple solid mechanics problems, with the hope of conveying basic FEA concepts while employing "limited" mathematics. Although only solid mechanics problems are considered in this guide, one should be aware that FEA is used in various fields of study, with FEA software often marketed as "general purpose."

Why use a solid mechanics approach to explain FEA? One reason is that the development of finite element techniques is strongly linked to the field of stress and structural analysis, and continues to be well utilized in this area. Another reason for using a solid mechanics approach is that individuals with various technical backgrounds often have an intuitive feel for basic solid mechanics problems, such as a concentrated load acting upon a simple rod, while having more difficulty, perhaps, with heat transfer and fluid mechanics problems. In addition, finite element analysis, when applied to solid mechanics problems, may be explained in terms of the *total potential energy principle*, as described later in this chapter.

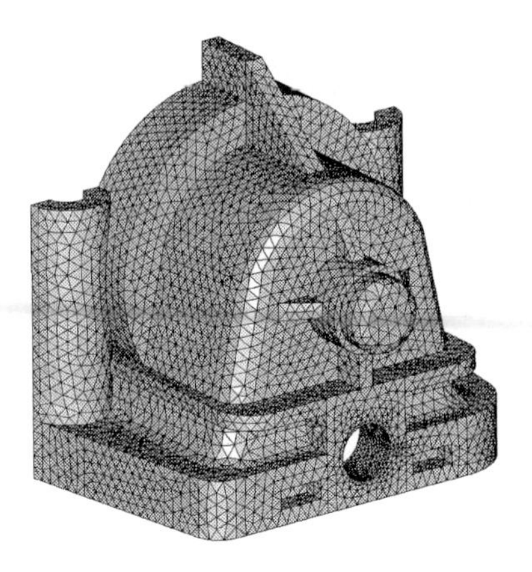

Automotive Starter Motor Housing
using Tetrahedral Elements
(Courtesy of R. Miller, Visteon)

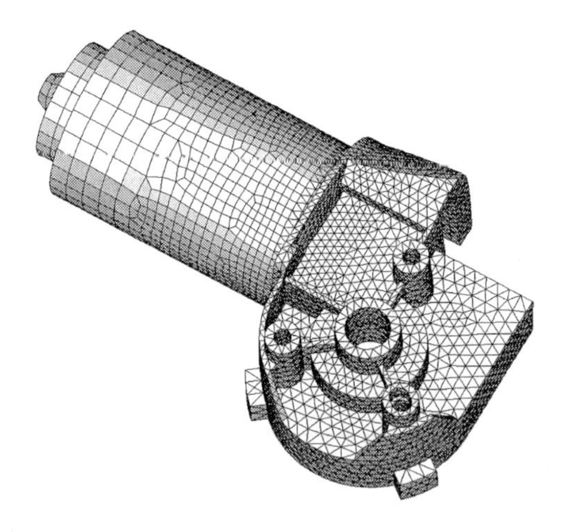

Windshield Wiper Motor Assembly
Tetrahedral and Quadrilateral Elements
(Courtesy of E. Riedy, SDRC)

Using the total potential energy principle allows finite element basics to be explained without explicitly employing *techniques of variational calculus* which increase the level of mathematical abstraction.

1.2 What Is Finite Element Analysis?

Key Concept: Finite element analysis is a numerical method that generates approximate solutions to engineering problems which are often posed in terms of differential equations. The method *partitions* a structure into simply shaped portions called *finite elements*, generates an approximate solution for the *variable of interest* within each element, then combines the approximate solutions. The assemblage of solutions describes the variable of interest for the entire structure.

The Variable of Interest
When performing engineering analysis, differential equations are routinely encountered. Ideally, a single differential equation would characterize the problem being analyzed, and this single differential equation would be solved by closed-form methods to obtain a continuous expression for the *variable of interest*. The term *closed-form* is primarily used in this text in reference to an engineering analysis problem that is solved using a single differential equation which has an elementary anti-derivative; a table of elementary anti-derivatives ("integrals") is generally included in most calculus books.

While the "ideal" analysis problem would be expressed in terms of a single differential equation having a closed-form solution, solid mechanics problems are often posed in terms of two- and three-dimensional space. To obtain a solution to a two- or three-dimensional problem, assuming that a solution exists, several differential equations are typically required to describe *variables* of interest. The basic concepts of the finite element method described in the first few chapters of this text will be explained using problems described in one-dimensional space, using a single differential equation.

In many structural analysis problems the variable of interest is displacement within an elastic body which is subjected to external loads. Once displacement is known, strain may be computed, and then stress, perhaps using Hooke's law.

Closed-Form Solution Often Difficult to Obtain
In many structural problems of practical interest, even those posed in one-dimensional space, it is often difficult (or even impossible) to generate a closed-form solution. If difficulties are encountered while attempting a closed-form solution, finite element analysis can often provide an approximate solution; in some simple cases, the closed-form and finite element solutions are identical.

Closed-Form Solution to a Differential Equation

Consider the steering link assembly shown in Figure 1.1, illustrative of those found in automotive suspension systems. The socket portion on the left end of the assembly is fixed (restrained), while a distributed load is applied to the right end.

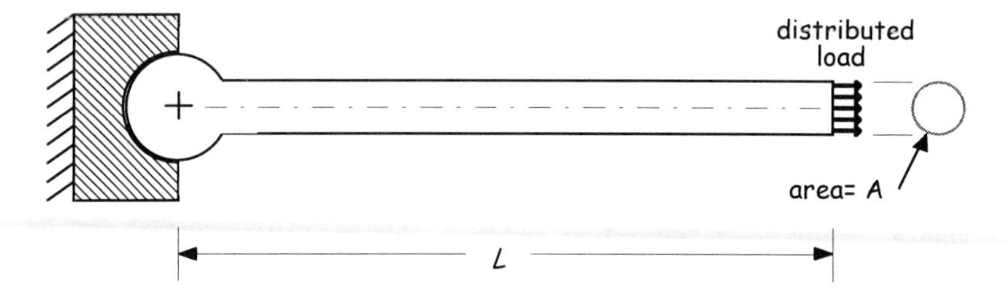

Figure 1.1 Steering Link Assembly

An analyst would like to evaluate whether excessive axial displacement will occur when a distributed load is imposed upon the right end of the link. The analyst might choose to ignore the fact that the ball portion has a larger cross section than the rest of the link, and thus simplify the problem by considering a rod of uniform cross sectional area (A), as illustrated in Figure 1.2.

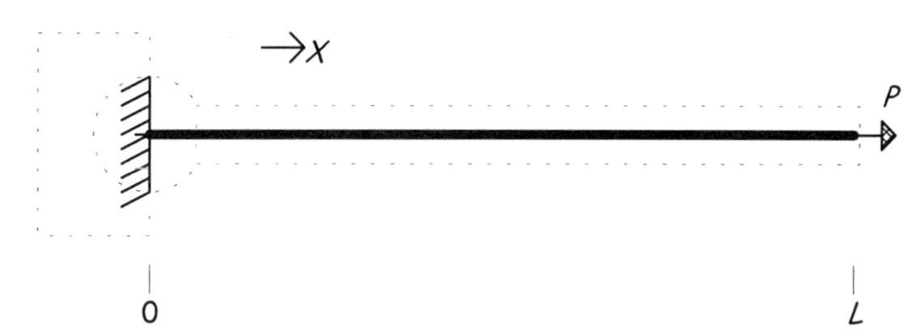

Figure 1.2 Idealized Link

The sketch of the *idealized* structure above reveals that the link, now considered as a uniform rod, is restrained on the left end while an equivalent, concentrated load P replaces the original distributed load acting upon the right end. Poisson effects, the true nature of how the ball is restrained in the socket, and localized effects at the ball/shaft interface will be ignored in this model. In actual practice, the tightness of fit at the joint, contact area between the ball and socket, etc., would need to be considered to achieve an accurate prediction of axial displacement.

The concept of Poisson's ratio, mentioned above, is very important in structural analysis. The analyst should be familiar with the Poisson effect since it often has a significant impact on the nature of deformation in elastic bodies. A good explanation of the Poisson effect is found in Juvinal [1].

Idealization Is Used to Simplify Analysis Problems

Using idealization techniques associated with solid mechanics problems, the actual three-dimensional geometry of the link assembly in Figure 1.1 can be represented by a one-dimensional rod. The sketch of the idealized rod, shown in Figure 1.2, is to scale in terms of the length but the cross sectional area is described by the mathematical constant A. Thus, in this case, the idealization process has allowed a three-dimensional body to be represented by a one-dimensional analogue.

The idealization above is employed to aid in defining and solving a differential equation. When using FEA, the same (or similar) types of idealization techniques are employed. Some basic mechanical idealization techniques will be considered in Section 1.6. Here at the outset, however, the reader is duly warned that the *effectiveness of the finite element method hinges upon the proper use of idealization.* This point cannot be overemphasized. Without some level of idealization, analysis problems in general, and finite element analysis problems specifically, cannot be solved in a practical fashion. Attempting to capture the effects of every geometric, material, and mathematical detail of an actual structure is virtually impossible. One soon realizes that the capability of the finite element method (and other analysis methods) is quickly exceeded without the use of idealization and the simplification that follows. A word of caution, though: while typical solid mechanics problems must be simplified, oversimplification often leads to analysis results of dubious value. In this regard, the words of Albert Einstein come to mind: "Everything should be made as simple as possible, but not simpler."

In practice, a sound understanding of engineering principles allows the analyst to make assumptions that lead to substantial simplification without overlooking the key attributes of the problem. The topic of engineering assumptions, and the associated simplification that can be achieved through idealization, will be considered in Section 1.6, and elsewhere throughout the text.

The Governing Differential Equation

Back to the problem of displacement in the steering link. Since the distributed load acting upon the right end of the link is uniform, its resultant force passes through the centroid of all resisting sections such that a state of *uniaxial stress* exists. Again, the effects near the socket end of the link will be discounted in this example.

The axial displacement of a uniform rod in a state of uniaxial stress is characterized by the governing differential equation:

$$\frac{d}{dX}\left(EA\frac{dU}{dX}\right)=0 \qquad (0\le X\le L) \tag{1.2.1}$$

The differential equation above was derived using the principle of static force equilibrium—see Appendix A for details. The variables in (1.2.1) are the modulus of elasticity E, cross sectional area A, axial displacement U, and the distance along the length of the rod, X. Equation 1.2.1 is valid for loads that do not stress the material beyond the presumed linear-elastic range.

In solving the differential equation given by (1.2.1), one arbitrary constant is introduced into the solution each time the differential equation is integrated. Since (1.2.1) is a second order differential equation, two integration steps are required, thus two arbitrary constants will be introduced into the solution. A solution containing arbitrary constants actually represents a family of solutions. To obtain a *unique* solution to a differential equation, boundary conditions must be invoked to eliminate the arbitrary constants. In this particular problem, the boundary conditions are:

$$U(0)=0 \qquad (A)\left(E\frac{dU}{dX}\right)\bigg|_{X=L}=P \tag{1.2.2}$$

The first boundary condition above indicates that displacement at the left end of the rod is zero while the second boundary condition suggests that at the right end of the rod, the product of uniaxial stress and the cross sectional area is equal to the applied load. (Recall that uniaxial stress is expressed as $\tau_{uni} = E\, dU/dX$).

Solving the Governing Differential Equation

The closed-form solution for $U(X)$ is found by twice integrating (1.2.1) and subsequently applying the boundary conditions given by (1.2.2). To begin, each side of (1.2.1) is multiplied by dX:

$$d\left(EA\frac{dU}{dX}\right)=0$$

Both sides of the preceding equation are integrated:

$$\int d\left(EA\frac{dU}{dX}\right) = \int 0$$

Performing the integration and adding an arbitrary constant:

$$EA\frac{dU}{dX} = c_1 \tag{1.2.3}$$

Manipulating (1.2.3), then integrating again:

$$\int dU = \int\left(\frac{c_1}{EA}\right)dX \qquad (0 \le X \le L) \tag{1.2.4}$$

Noting that the variables that form the integrand of (1.2.4) are constant along the total length of the rod, an expression for *U(X)* is generated by closed-form integration:

$$U(X) = \frac{c_1 X}{EA} + c_2 \qquad (0 \le X \le L) \tag{1.2.5}$$

Equation 1.2.5 represents the closed-form solution for displacement in the rod as a function of the distance along the rod's longitudinal axis, *X*. The arbitrary constants in (1.2.5), c_1 and c_2, can be eliminated by applying the boundary conditions given by (1.2.2):

$$U(0) = 0 \qquad \left(EA\frac{dU}{dX}\right)\bigg|_{X=L} = P$$

Applying the preceding boundary conditions to (1.2.5), the arbitrary constants are eliminated and displacement for this problem is uniquely defined as:

$$U(X) = \frac{PX}{EA} \qquad (0 \le X \le L) \tag{1.2.6}$$

Displacement at any point along the axis of the rod may now be computed, assuming that *P*, *A*, and *E* are known. This completes the example of using a closed-form method to generate an expression for displacement.

With Slight Change, a Closed-Form Solution Is No Longer Possible
If all problems were as simple as the one shown on the previous page, there would be little need for FEA. Consider a slightly different problem that illustrates how quickly a simple problem becomes more difficult.

To save weight, the design engineer suggests that the cross section of the link nearest the load be decreased to one-half the original size. Once again, an expression for displacement as a function of X is required. The distributed load is increased in proportion to the decrease in cross sectional area, such that the product of the new distributed load, acting upon the reduced cross sectional area, is again equal to P, the same force as before.

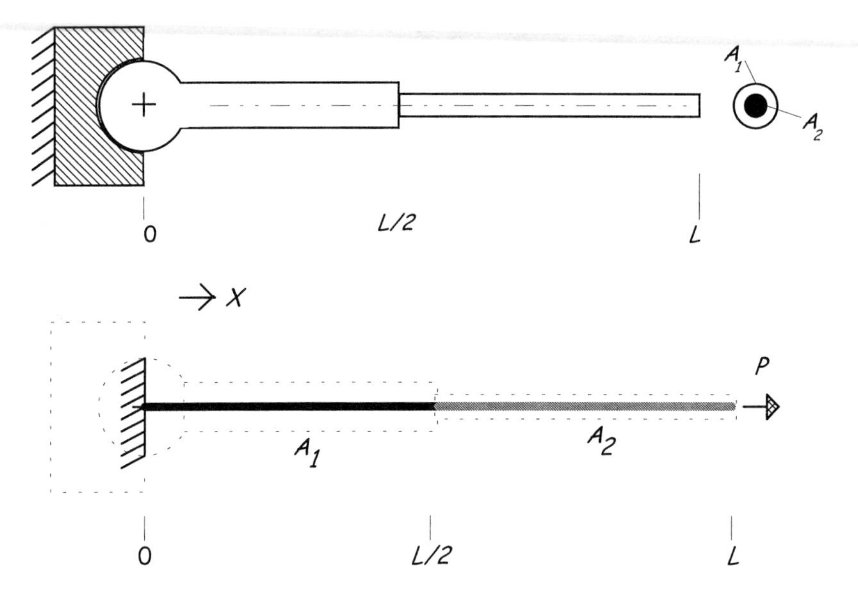

Figure 1.3 Modified Steering Link and Associated Idealization

The governing differential equation given by (1.2.1) is applied and, following the same steps as before, (1.2.4) is again obtained:

$$\int dU = \int \left(\frac{c_1}{EA} \right) dX \qquad (0 \leq X \leq L)$$

Now, what expression should be used for A in the integral above, since neither A_1 nor A_2 from Figure 1.3 are valid along the entire length of the rod? In this problem, a single differential equation is no longer sufficient to provide an expression for $U(X)$ along the entire length of the link.

An Approximate Solution for Displacement

An approximate solution for displacement in the modified link of Figure 1.3 may be obtained by partitioning the link into two portions, with a function for displacement generated on each. Used together, the two functions form an approximate solution for the entire structure.

To begin, the link is partitioned into two portions of length $L/2$. Although there are several ways to produce an expression for displacement on each portion, assume that integral equations are established on each, as illustrated in Figure 1.4.

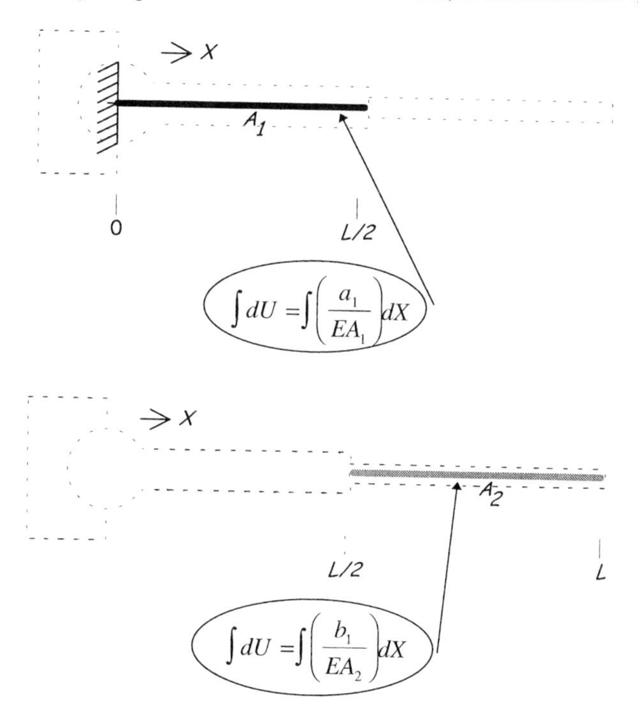

Figure 1.4 Modified Link with Two Partitions

When the two integrals above are solved for $U(X)$, two solutions are generated, each representing displacement on a portion of the rod:

$$^{1}U(X) = \frac{a_1 X}{EA_1} + a_2 \qquad \left(0 \le X \le \frac{L}{2}\right) \tag{1.2.7}$$

$$^{2}U(X) = \frac{b_1 X}{EA_2} + b_2 \qquad \left(\frac{L}{2} \le X \le L\right) \tag{1.2.8}$$

Assuming that displacement is zero at the left end of the rod, and continuous at the interface of the two partitions, the four arbitrary constants in (1.2.7) and (1.2.8) can be eliminated, if one further assumes that the axial force is equal to P anywhere along the length of the rod. Used together, the two equations above can be considered as an approximate solution for the entire rod.

Approximate Solutions—Some Derivatives Are Not Continuous

The *piecewise* solution above is an approximation. Even though the individual solutions (1.2.7) and (1.2.8) represent a closed-form solution to the governing differential equation for a *portion* of the rod, the assembled solution does not represent a closed-form solution for the *entire* rod. Why?

Consider that a discontinuity in slope occurs at $L/2$, the point where the cross section changes from A_1 to A_2. In other words, the derivative dU/dX is not continuous from the large cross section to the smaller. Mathematically speaking, this translates into discontinuous strain at $L/2$, suggesting the existence of a three-dimensional state of stress. However, recall that the individual solutions were derived from differential equations which initially assumed a uniaxial stress state. Hence, an error is introduced, because the solutions cannot account for three-dimensional stress.

Approximate solutions can simplify problems by ignoring continuity requirements on certain derivatives of the dependent variable. Although the error caused by ignoring continuity requirements was discounted in the case of the steering link, the analyst should be aware of the ramifications of doing so. If the objective were to analyze stress in the steering link, instead of simply examining axial displacement, ignoring displacement continuity would not be advisable, since the stress at the discontinuity is (theoretically) infinite. The relaxation of slope continuity requirements is also a characteristic of the finite element method, and will be considered again.

Other Variables in the Integrand May Also Be Discontinuous
The preceding problem illustrated that, using a single differential equation, a step change in cross sectional area presents some difficulty in obtaining a closed-form solution for displacement. A structure with an abrupt change in a material property, such as the elastic modulus, presents the same type of difficulty as a step change in geometry. To illustrate, consider a steering link that is composed of two materials, as illustrated in Figure 1.5. As before, the differential equation of equilibrium leads to the same integral expression as (1.2.4):

$$\int dU = \int \left(\frac{c_1}{EA} \right) dX \qquad (0 \le X \le L) \tag{1.2.9}$$

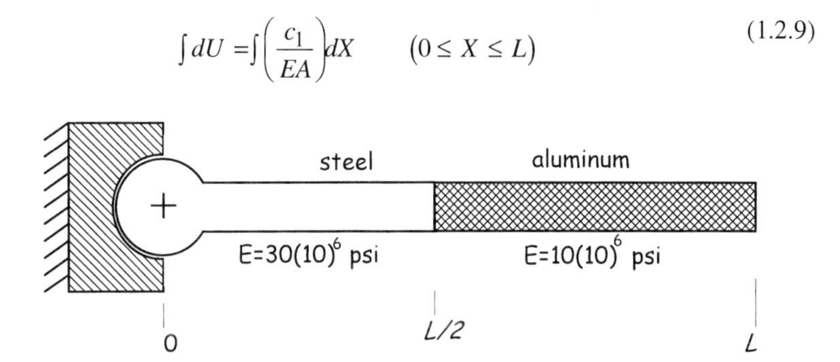

Figure 1.5 Steering Link with Two Materials

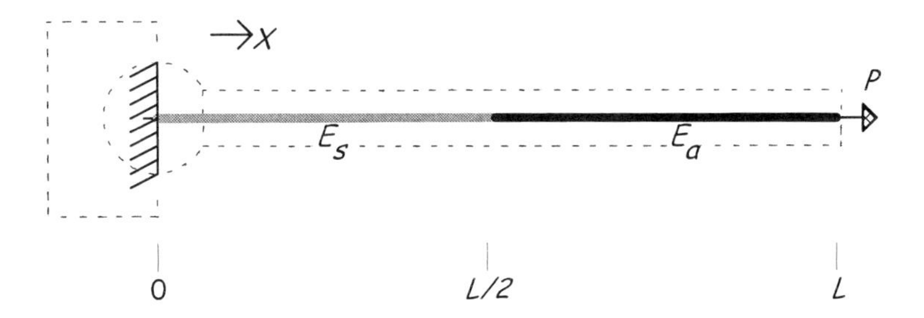

Figure 1.6 Idealized Link

One of the variables in the integral given by (1.2.9) is discontinuous. In this case, the elastic modulus term (E) makes a step change at $L/2$ (Figure 1.5). Similarly, an abrupt change in loading can cause the same type of problem as an abrupt change in modulus. For example, if an additional concentrated load were applied at $L/2$, the arbitrary constant c_1 in (1.2.9) would be discontinuous at that point.

So far, it is apparent that when discontinuities in material properties, geometry, or loads occur, solving a differential equation becomes difficult.

Approximate Solution When Terms in Integrand Are Difficult to Integrate
As shown in the previous examples, an approximate solution is helpful when discontinuities are present. Approximate solutions are also helpful when variables are continuous but difficult to describe over the entire domain of the structure. Consider the problem of displacement in a tapered shaft with a uniaxial load, illustrated in Figure 1.7.

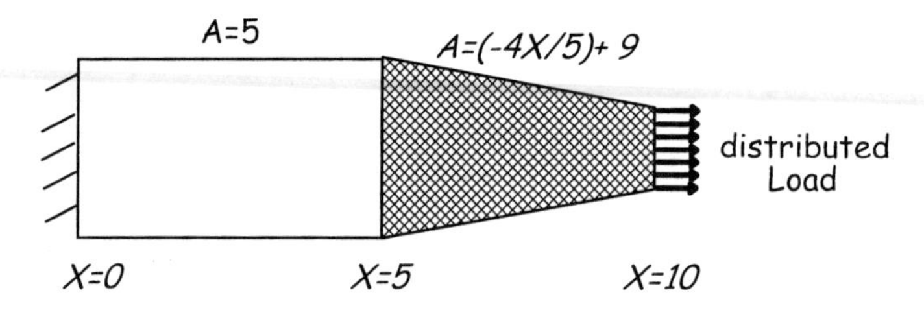

Figure 1.7 Tapered Shaft

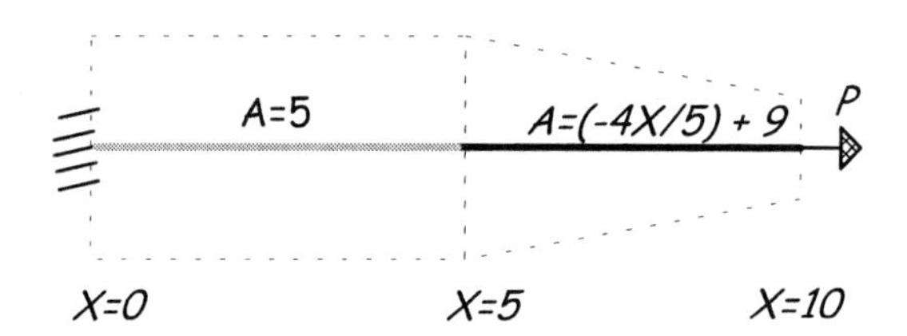

Figure 1.8 Idealization for Tapered Shaft

In Figure 1.8, the cross sectional area term is seen to be continuous at $X=5$ but what simple, integrable mathematical expression for A is valid along the entire length of the rod? An approximate solution can be employed.

To solve this problem, the tapered rod is partitioned into two portions, integrals established and unique expressions for A used in each (Figure 1.9).

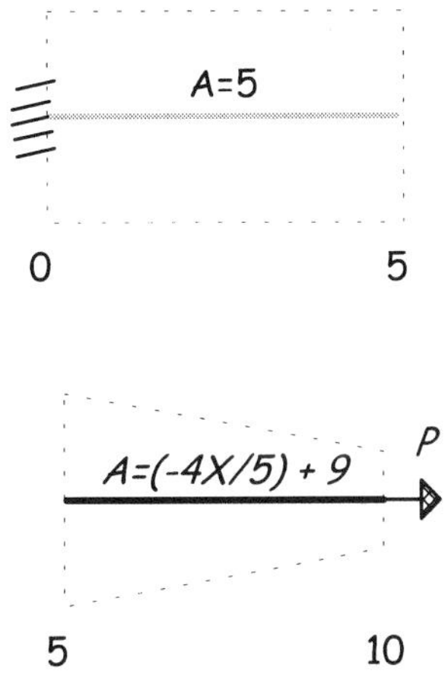

A=5

0 5

$$\int dU = \int \frac{a_1}{E(5)} dX \qquad (0 \le X \le 5)$$

A=(-4X/5) + 9

P

5 10

$$\int dU = \int \frac{b_1}{E\left(-\frac{4X}{5}+9\right)} dX \qquad (5 \le X \le 10)$$

Figure 1.9 Approximate Solution for Tapered Shaft

The two equations above can be integrated to obtain expressions for displacement. Together, the solutions represent an approximate, piecewise solution for displacement in the tapered shaft.

Notice that the partitions used in the examples above *are not finite elements*—finite elements have not been introduced yet—only the concepts of idealization, partitioning, and approximate solutions have been considered thus far. Finite elements, which make use of the partitioning and approximation concepts, will be introduced in Section 1.6.

What Is Finite Element Analysis?

Finite Element Analysis (FEA) is a numerical method, typically used to find approximate solutions to complex, real-world engineering problems posed in terms of differential equations. FEA utilizes partitioning concepts, similar to those illustrated in the examples above. However, there are several differences between the concepts illustrated above and the finite element method. These differences will be examined in the chapters that follow. For now, consider some significant characteristics of the finite element method:

- Using FEA, a structure is partitioned in a manner analogous to the examples of the steering link and the tapered shaft, with a solution for the variable of interest generated for each portion. However, using the finite element method, the simply shaped partitions that evolve from the idealization are defined by points (*nodes*) on the boundary of the partition (*element).* In the simplest case, a solution for one variable of interest is generated within the boundary of each element.

- The finite element method uses numerical methods, not closed-form solutions, and creates a system of linear, algebraic equations which is easily solved using matrix methods and a digital computer.

- FEA uses a *functional*, not a *differential* equation, as the underlying basis for establishing an approximate solution.

- A functional is an integral equation and, in solid mechanics problems, represents the total potential energy of the structure and the external loads. Functionals will be considered further in Section 1.3.

Displacement Based FEA

In solid mechanics problems, the state of stress and strain is often a concern and, since these metrics can be calculated using a function of displacement, displacement is often the variable of interest. The *displacement based* finite element method generates an approximate solution for displacement, then typically uses displacement to calculate approximate values of stress and strain.

Since the displacement based approach is currently the most common, the term *finite element method* is typically used without the modifier *displacement based.* There are other means of developing a finite element solution. For instance, a hybrid method that uses both an assumed displacement field along with an assumed stress field can be employed. Only the displacement based approach will be discussed in this text; see Pian [2] and [3] for discussion on other approaches.

Why Use FEA for Stress and Structural Analysis?

Although finding the solution to a problem posed in terms of a differential equation is possible for simply shaped structures such as uniform rods, beams, plates, etc., attempting to find a closed-form solution for a typical structure encountered in engineering practice can be difficult. As discussed earlier in this section, three factors affect the ability to obtain a closed-form solution:

- The use of several different materials within the same structure
- Complicated or discontinuous geometry
- Complicated loading (or complicated boundary conditions, in general)

Why use FEA? In many practical applications, one or more of the three factors above are present, and a closed-form solution is generally not practical; in such cases, FEA is often employed to obtain an approximate solution. The approximate solution is generated in the form of a field variable (displacement in solid mechanics) and this variable is typically used to calculate approximate values for other engineering metrics, such as stress and strain.

Even when the three factors above are not present, a closed-form solution for the differential equations of equilibrium in two- and three-dimensional solid mechanics problems is not generally possible, since a *system* of differential equations needs to be solved. In multi-dimensional problems, the finite element method may therefore be thought of as a means of creating a system of linear, algebraic equations that approximates a system of differential equations.

Section 1.3 considers how the principle of total potential energy is employed to obtain equilibrium displacement in elastic structures subjected to static loads.

1.3 Energy and Displacement

Key Concept: The finite element method uses a *functional* instead of a differential equation to solve engineering analysis problems. When used to solve solid mechanics problems, the functional represents the total potential *energy* of the system.

The approximate solutions considered in the text so far have been produced using a differential equation. Differential equations were used because they are familiar to anyone who has completed an undergraduate engineering degree. However, energy principles are often used to solve solid mechanics problems. A brief introduction to energy principles is given below.

Minimum Energy and Equilibrium Displacement

One way to explain how energy can be used to calculate displacement is to examine a 1 Kg ball released at the top of a bowl, as illustrated in Figure 1.10. The principle of minimum potential energy suggests that after dissipating its kinetic energy, the ball will come to rest at a position where its potential energy is minimum. The potential energy of the ball is a function of both the force associated with the ball, and its height:

$$Potential\ Energy = (Force)(Height) \qquad (a)$$

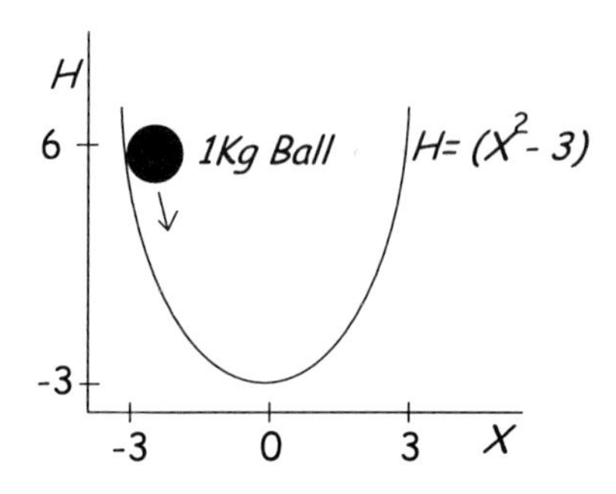

Figure 1.10 Potential Energy

Assuming that the force acting upon the ball is a function of gravitational acceleration:

$$Force = (mass)(acceleration) = (1Kg)\left(9.8\frac{m}{s^2}\right) = 9.8N \qquad (b)$$

In this example, the height of the ball is characterized by the function:

$$Height = (X^2 - 3) \qquad (c)$$

Potential energy can be expressed using Equations *a*, *b*, and *c* above:

$$Potential\ Energy = (9.8)(X^2 - 3)$$

The questions is, "at what position in the bowl will the ball come to rest?" The answer can be found by minimizing the potential energy function with respect to the ball's position, *X*. Recall that to find the minimum value of a single variable function, the derivative of the function is generated, and then set to zero (Fisher [4]):

$$\frac{d}{dX}(Potential\ Energy) = \frac{d}{dX}9.8(X^2 - 3) = 0$$

or:

$$X = 0$$

When $X=0$, the potential energy is at a minimum, and this value of X represents the equilibrium positionof the ball in the bowl. One can see from the graph in Figure 1.10 that for any other value of X, the height, and therefore potential energy, is greater than at $X=0$. This suggests that the ball, at any point other than $X=0$, would be inclined to move, seeking a state of lower potential.

The next example will consider a slightly more complex case of using energy to find equilibrium displacement, namely, the total potential of a spring-mass system, illustrated in Figure 1.11.

The Principle of Total Potential Energy

The case illustrated in Figure 1.10 is quite simple in that the phenomenon is characterized by considering only two factors: the force acting on the ball and the distance of the ball from a reference point. More complex cases are encountered in the field of solid mechanics. For instance, observe in Figure 1.11 that the external load, together with the spring, is considered as a structural *system*.

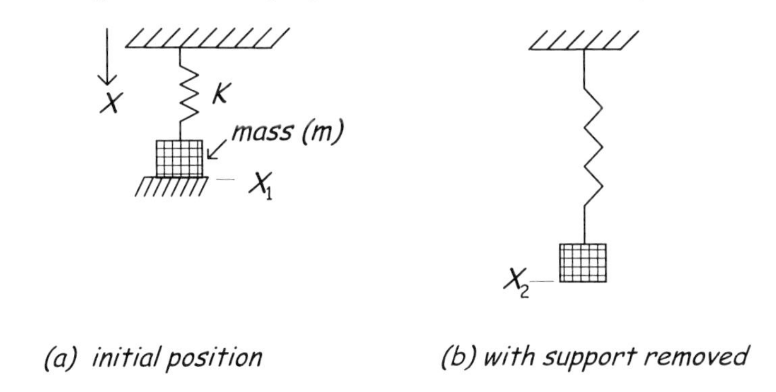

Figure 1.11 Energy in a Mass-Spring System

The load in Figure 1.11 is induced by gravity acting upon the mass. Within the system, the combination of the elastic energy of the spring and the work potential of the external load is termed the *total potential energy* (TPE).

For a broad class of solid mechanics problems, a mathematical expression for the total potential energy can be established, and equilibrium displacement determined, by applying the principle of minimum total potential energy. Consider the following statement of the principle of minimum total potential energy, as applied to structural systems:

> *Of all possible displaced configurations that a loaded structure can assume, the displacement that satisfies the essential boundary conditions, and minimizes the total potential energy of the system, is the equilibrium displacement.*

The concept of minimum total potential energy, as applied to the finite element method, is considered in Zienkiewicz [5]. Essential boundary conditions, playing a fundamental role in engineering analysis, will be introduced in the next section. For now, consider how TPE can be used to determine the equilibrium displacement in the spring-mass system depicted in Figure 1.11. As shown, a mass is connected to ground by a linear spring, and supported at X_1; gravity acts upon the mass to provide potential energy. When the support is removed the mass begins to travel and, assuming that its kinetic energy is dissipated, comes to rest at its static equilibrium position, X_2. In this example, displacement is defined as the distance between the initial location of the mass and its final position, while the force on the mass is generated by gravitational acceleration, g:

$$U \equiv X_2 - X_1 \qquad F = mg$$

It can be shown that the total potential energy of the system *at rest* is given as:

$$TPE = \frac{1}{2}KU^2 - FU \tag{1.3.1}$$

The first term in (1.3.1) represents the elastic energy of the spring, while the second term represents the work potential of the load. As the mass moves lower, its potential to do work decreases linearly with displacement, while the energy in the spring increases quadratically. The summation of both effects, Equation 1.3.1, is plotted in Figure 1.12. Analogous to the ball in the bowl, the equilibrium displacement of the mass in Figure 1.11 can be determined by setting the first derivative of TPE, (1.3.1), equal to zero, then solving for U. Hence, taking the first derivative of (1.3.1):

$$\frac{d}{dU}(TPE) = \frac{d}{dU}\left(\frac{1}{2}KU^2 - FU\right) = KU - F$$

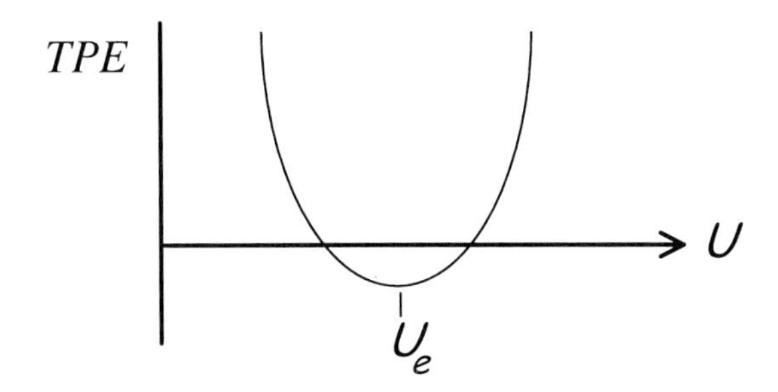

Figure 1.12 Total Potential Energy as a Function of Displacement

Setting the equation above to zero and rearranging:

$$KU = F \qquad \therefore U_e = \frac{F}{K}$$

The equation above is the equilibrium equation for a linear spring. For a single degree of freedom system, the equilibrium equation indicates that static equilibrium displacement is described by a single value, quantified by the magnitude of the external load and spring constant.

Now consider a slightly more complex structural system, the steering link illustrated in the previous examples. The steering link is a continuous system, having the characteristic that variables of interest, such as displacement, are described as *continuous variables*, as opposed to point values. In other words, in dealing with a continuous system, one is concerned with a variable as function, as opposed to a discrete value defined at a single point.

It was shown that the principle of total potential energy can be used to find equilibrium displacement in discrete systems. The same principle can also be used to find equilibrium displacement in a continuous body, such as the steering link. Again, the link will be idealized as a rod restrained at one end, as shown in Figure 1.13. A continuous function, not simply a single value, is required to describe equilibrium displacement in the rod, since there exist an infinite number of points on the rod, each point having a distinct value of displacement at equilibrium.

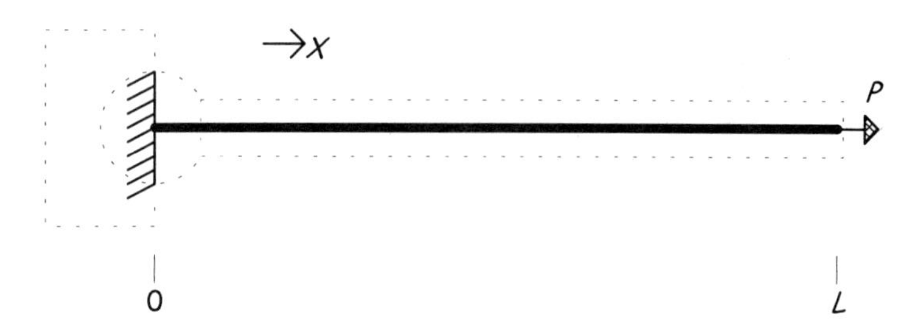

Figure 1.13 Rod Idealization for the Steering Link Problem

Functionals

The fact that equilibrium displacement in a rod is represented by a continuous function renders a more complex expression for TPE, as compared to the discrete mass-spring system; observe (1.3.2):

$$TPE = \frac{1}{2}\int_0^L EA\left(\frac{dU}{dX}\right)^2 dX - P\,U(L) \tag{1.3.2}$$

As with the discrete mass-spring system, the first term in the expression for total potential energy represents elastic energy, while the second term represents the work potential of the external load. One might recognize that the derivative of displacement, dU/dX, represents normal strain in a uniaxially loaded rod, hence, the integral portion of (1.3.2) represents the elastic *strain energy* of the rod. Notice that the expression for total potential energy given by (1.3.2) is no longer a function of a single variable but is now expressed as a *functional.* A functional takes the form of an integral equation, and a *single variable* functional is characterized by:

- A single, independent variable
- A function of the independent variable
- One (or more) derivatives of the function

For the functional expressed by (1.3.2), X is the independent variable, $U(X)$ the function, and dU/dX the derivative; no higher order derivatives appear in this particular functional.

Now, recall that to compute equilibrium displacement, total potential energy must be minimized. In the discrete mass-spring system example, the objective was to minimize a *function*, (1.3.1), with respect to a single variable. For a continuous system, minimization of a *functional* is required. In the case of the rod, the objective is to minimize (1.3.2) *with respect to a function, U(X).* The problem of minimizing a functional is somewhat more complex than the differential calculus problem of minimizing a function. To obtain a closed-form solution for this type of problem, techniques of *variational calculus* are employed.

In minimization problems of single variable differential calculus, the question is, "What *single value* of the independent variable, when introduced into the given function, will yield a local minimum of the function?" In an analogous variational calculus problem, the question becomes "What *function* of the independent variable, when introduced into the given functional, will yield a minimum value of the functional?"

A classic variational calculus problem examines the displacement in a cable that is strung between two points, as illustrated Figure 1.14 (Mathews [6]).

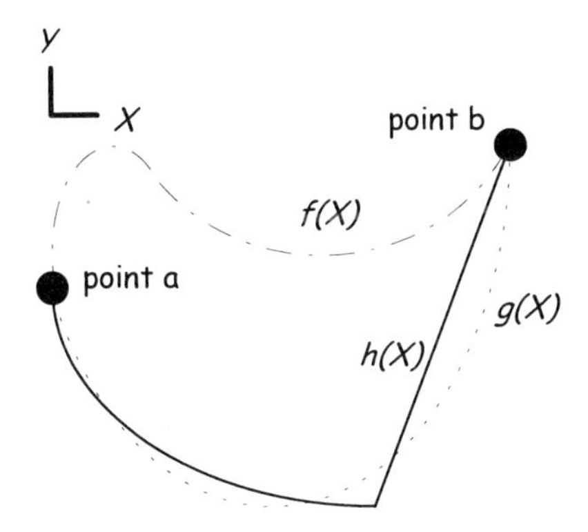

Figure 1.14 Flexible Cable Connecting Two Points

Assuming that gravity acts in the negative *Y*-coordinate direction, which function in Figure 1.14 describes how the cable actually connects Points *A* and B? While three candidate functions are shown, there exist an infinite number of possibilities. Techniques of variational calculus provide a closed-form means of generating the function that describes how the cable actually joins the two points.

Many problems from the fields of engineering and physics may be stated in terms of a differential equation or in an equivalent variational form. Variational forms are considered in Appendix B. The variational form may offer some advantages over the differential form. For example, a broad range of solid mechanics problems can be expressed in a variational form that is equivalent to an expression for TPE. As such, the variational form may offer a more intuitive feel for the problem, as opposed to using a differential equation that does not represent a tangible quantity. One "disadvantage" of the variational form, as compared to the differential form, is that it can introduce a potential for imprecision by the nature of the functional. The imprecision in solid mechanics problems is based upon the fact that the functional contains a *weak form* of the problem statement, with derivatives of lower order than those used in the differential form. As a result, if one were to examine derivatives of the solution generated from a weak formulation, one would find that they can exhibit discontinuities at certain points. However, the disadvantage of the weak form may be overlooked in light of the fact using a single differential equation, one may not be able to obtain *any* solution. In other words, the weak form approach amounts to trading derivative continuity in return for an *approximate* solution. The discontinuity that the weak form can produce is the same type that was illustrated by using equations (1.2.7) and (1.2.8) to express displacement in the modified steering link in Figure 1.3.

To obtain a closed-form solution (assuming that one exists) for the function that minimizes the total potential energy functional of a given problem, variational calculus is required. However, using the Ritz method and differential calculus, one can also obtain an *approximate* solution for a function that minimizes the functional.

1.4 The Ritz Method and Minimum Energy

Key Concept: Total potential energy in a continuous structure can be expressed in terms of a functional. Through the process of applying the principle of minimum total potential energy to the functional, a minimizing function representing the equilibrium displacement within the structure is established.

To obtain a closed-form solution for a minimizing function, variational calculus is needed. However, a minimizing function that *approximately* represents equilibrium displacement may be found using the Ritz Method. Some fundamental aspects of the Ritz approach are considered in this section.

The Ritz Method—Finding an Approximate Expression for Displacement

Why should the Ritz Method be used to obtain an approximate solution if one can obtain a closed-form solution using variational calculus? There are two reasons. Firstly, a closed-form, variational calculus solution only applies to one unique problem. Thus, using a closed-form solution, every unique problem requires a different solution procedure. Secondly, in some cases, a closed-form solution may simply not exist. By analogy, closed-form integration of a single differential equation was not possible in the example of the modified steering link or the tapered shaft.

In contrast to closed-form variational techniques that apply to one particular problem, an approximate solution technique like the Ritz method can be *applied to a wide range of variational calculus problems.* An approximate solution can also be used when a closed-form solution does not exist. To illustrate the difference between closed-form and approximate solutions to the problem of minimizing a functional, consider the analogy of closed-form integration versus numerical integration. Referring to Figure 1.15, assume that computation of the area bounded by the X-axis and the function $f(X)=X^3$, in the interval $(0 \leq X \leq 1)$, is required.

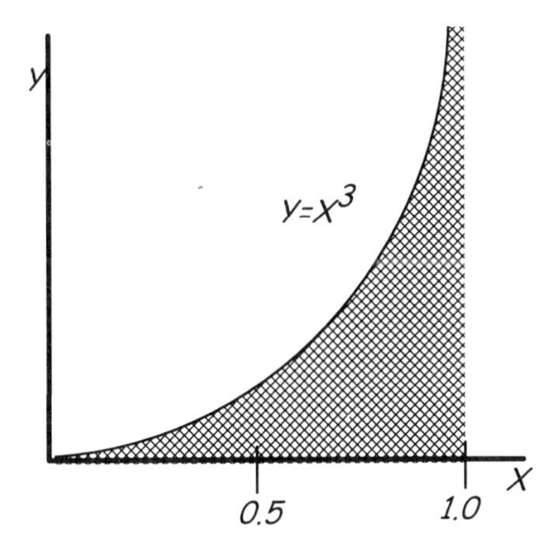

Figure 1.15 Area Underneath Curve

The closed-form solution is simply:

$$Area = \int_0^1 X^3 dX = \left. \frac{X^4}{4} \right|_0^1 = 0.250$$

The integrand in the expression above, X^3, has a known anti-derivative, and the closed-form solution is found by applying the Fundamental Theorem of Calculus. However, the following integral has no closed-form solution, since no simple anti-derivative exists:

$$Area = \int_0^1 exp(x^2)\, dX$$

Although no closed-form solution exists for the integral of $exp(x^2)$, a numerical solution could be invoked to yield an approximate solution.

Advantages of Numerical Methods

Although a closed-form solution is possible for only one of the integrals above, numerical integration can be used to provide an approximate solution to either. To illustrate the principle of numerical integration, the curve from Figure 1.15 is considered again in Figure 1.16.

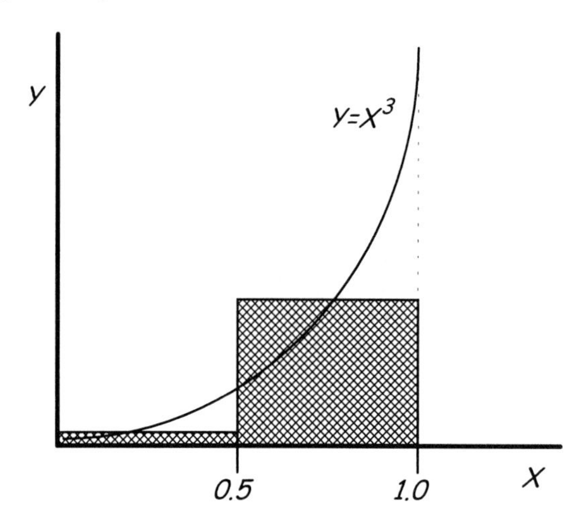

Figure 1.16 Numerical Integration

One numerical integration technique uses rectangles to approximate the area associated with partitions on the domain. The area of each rectangle is calculated, then the contributions from all the individual areas are summed to approximate the actual, total area. Using two rectangles (as shown), with the height of each rectangle determined by the value of the function at the midpoint of the rectangle, the computed area is 0.219; note that the exact solution was shown to be 0.250. Using more, smaller rectangles, the numerical solution will better represent the exact value. The same numerical procedure could be used for any integral of this type, even for integrals that have no closed-form solution.

The preceding example of numerical integration suggests three important concepts:

- Approximate, numerical methods can be used when closed-form solutions cannot

- Regardless of the particular problem, the same numerical procedure can be followed to obtain a solution

- An assemblage of simple geometric shapes (rectangles in this case) can be employed to approximate a complicated shape, such as the area under the curve of a higher order polynomial

It will be shown in Chapter 2 that the three concepts above apply to the finite element method as well, since the finite element method is also a numerical method that uses a partitioning concept. Numerical methods lend themselves to computer applications, since the same procedure, hence the same programming, can be used regardless of the specifics of the problem. In other words, regardless of whether one wishes to integrate the function $Y=X^3$ or a more complicated function, such as $Y=exp(x^2)$, the same numerical method is used.

Convergence

One additional analogy between numerical integration and the finite element method can be drawn. When using numerical integration, the more partitions that are used, the better the approximation for the integral becomes. That is, as the partitions become finer, the approximate solution *converges* to the exact solution. In the limit, as number of partitions approaches infinity, the approximation is expected to be essentially equal to the exact solution. The same principle applies to the finite element method, as will be shown in Chapter 2.

Ritz Method—An Approximate Function to Minimize a Functional

Assume that the problem of finding a function which describes equilibrium displacement in a continuous structure has been cast into variational form, rendering a total potential energy functional. An approximate function for equilibrium displacement, via the Ritz method, will be considered instead of a closed-form solution.

To apply the Ritz method, four steps are followed:

1. A *form* for displacement, in terms of unknown parameters, is assumed.
2. The assumed form is introduced into the functional.
3. Differentiation and integration are performed, *rendering a function.*
4. The resulting function is minimized using differential calculus.

Example 1.1—Ritz Method for Steering Link
Using the Ritz method, the *variational calculus* problem of minimizing a functional is replaced by the less complicated, approximate, *differential calculus* problem of minimizing a function. For example, consider again the steering link problem, illustrated once more in Figure 1.17.

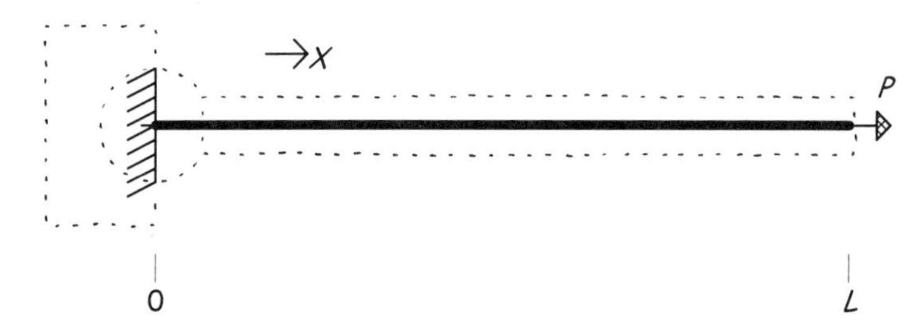

Figure 1.17 Idealization for the Steering Link Problem

The link is idealized as a rod, and the total potential energy functional is expressed as:

$$TPE = \frac{1}{2}\int_0^L EA\left(\frac{dU}{dX}\right)^2 dX - P\,U(L) \tag{1.4.1}$$

The functional in (1.4.1) can be established by examining the strain energy in a differential size slice of a rod-like structure, or by manipulating the governing differential equation; both techniques are briefly mentioned in Appendix B. If a functional cannot be established, the *method of weighted residuals* may provide another means of approaching solid mechanics problems using partitioning techniques. Indeed, the weighted residual method is actually more robust than the finite element method but may be less intuitive, in the context of solid mechanics problems. Weighted residual methods are used in preference to the finite element method in the finite element textbook by Burnett [7, p. 88].

Returning to the problem of the link stated in functional form, the displacement boundary condition associated with (1.4.1) is the same as that used for the differential equation:

$$U(0) = 0 \tag{1.4.2}$$

Equation 1.4.2 indicates that displacement at the left end of the rod is fixed. The first step of the Ritz method is now applied.

Step 1—Assume a Displacement Function in Terms of Unknown Parameters
In this case a linear polynomial is chosen for the assumed form, with the tilde overscript denoting that the displacement assumption is approximate:

$$\tilde{U}(X) = a_1 + a_2 X \tag{1.4.3}$$

In displacement based solid mechanics problems, the assumed form is called the displacement assumption. Polynomials are often used for the assumed form because they are easy to differentiate and integrate and are also "well behaved," such that the function does not exhibit asymptotic behavior, for instance. Why use a linear polynomial? There exist certain guidelines for choosing displacement assumptions, and these guidelines will be covered in more depth in Chapter 4. For now, consider that a one-dimensional displacement assumption must meet four requirements; the displacement assumption must:

1. Be sufficiently differentiable
2. Be "complete in terms"
3. Have linearly independent terms
4. Satisfy the essential boundary conditions

Sufficiently differentiable means that the displacement assumption must be of high enough order so that when introduced into the functional, the derivative of the displacement assumption is not zero. Hence, in this problem, one could not choose a constant for the displacement assumption. Complete in terms means that if a quadratic displacement assumption is chosen, the constant and linear terms must also be present. In regard to Ritz Requirement 3, the stipulation of "linearly independent terms" requires that no term is merely a linear combination of the other terms.

The last requirement for the Ritz displacement assumption is that the displacement assumption must meet the *essential* boundary conditions. Whether a boundary condition is essential or not is determined by examining the functional for the given problem. Essential boundary conditions for structural problems consist of boundary conditions placed upon displacement and *p–1 derivatives* of displacement, where *p* is the highest order derivative in the functional. Examining (1.4.1), note that the highest order derivative (the only derivative in this case) is unity, such that *p–1* is zero, meaning that derivatives of order zero (i.e., no derivatives) are essential.

In contrast, the functional for other types of problems may contain second order derivatives, such that *p–1=1*. For example, the classical mechanics of materials approach to beam bending can be cast into variational form, employing a functional with a *second order* derivative. In such a case *p–1=1*, meaning that displacement and *first order derivatives* of displacement are essential. Figure 1.18 attempts to illustrate the physical significance of essential boundary conditions, as they apply to the aforementioned beam bending problem.

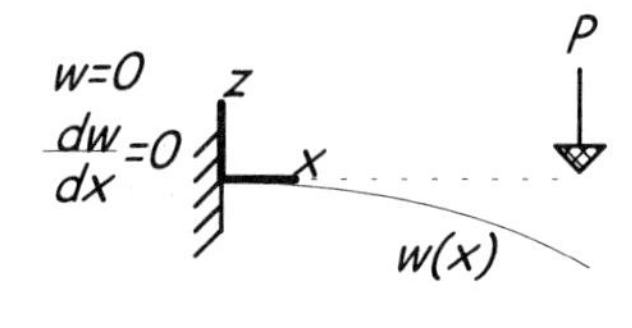

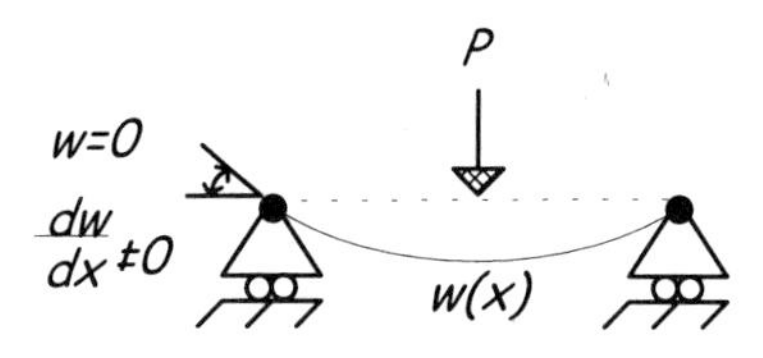

Figure 1.18 Cantilever and Simply Supported Beams

Notice that kinematic restraints of a *cantilever* beam, illustrated at the top of Figure 1.18, are characterized not only by zero displacement at the end of the beam but also by the fact that the derivative of displacement (the slope) is also zero. In the *simply supported* beam, however, the slope must *not* be constrained to zero, if the true nature of the end support is to be properly characterized. The key point is that in some problems, such as beam bending, it is *essential* that both displacement and slope[1] restraints be specified to yield a solution that represents the actual response of the structure. In other problems, a derivative does not

[1] Alternately, angle of rotation may be used instead of a derivative.

represent a kinematic boundary condition, and no constraint is placed upon its value.

In the current problem of the rod, the order of the derivative in the functional reveals that $p-1=0$, hence, derivatives need not be treated as essential boundary conditions. Only one type of essential boundary condition exists in this problem, namely, the boundary condition placed upon the displacement function itself, Equation 1.4.2. The displacement assumption must therefore meet the requirement:

$$\tilde{U}(0) = a_1 + a_2 0 = 0 \qquad \therefore a_1 = 0$$

Using $a_1=0$ in (1.4.3), the displacement assumption that meets the essential boundary condition for this problem is expressed as:

$$\tilde{U}(X) = a_2 X \tag{1.4.4}$$

Step 2—Introduce the Assumed Form into the Functional
Since (1.4.4) meets the essential boundary condition, it can be introduced into the functional given by (1.4.1):

$$\tilde{\Pi} = \frac{1}{2}\int_0^L EA\left(\frac{d}{dX}a_2 X\right)^2 dX - P\, a_2 L \tag{1.4.5}$$

The symbol $\tilde{\Pi}$ is now used to represent approximate, total potential energy, because the algebra from this point on becomes cumbersome using the alternate expression, "TPE".

Step 3—Differentiation and Integration Are Performed
Performing the differentiation indicated in (1.4.5):

$$\tilde{\Pi} = \frac{1}{2}\int_0^L EA(a_2)^2\, dX - P\, a_2 L$$

The required integration is performed, taking note that E, A, and a_2 are independent of X:

$$\tilde{\Pi} = \frac{1}{2} EA(a_2)^2 L - P\, a_2 L \tag{1.4.6}$$

During the process of introducing the displacement assumption into the functional and performing differentiation and integration, the expression for total potential energy is transformed from a functional expression, (1.4.5), into a simple function, (1.4.6). Hence, the Ritz method allows a variational calculus problem to be transformed into an approximate, differential calculus problem.

Step 4—The Resulting Total Potential Energy Function Is Minimized

The expression for total potential energy given by (1.4.6) is minimized with respect to the parameters. In this case the function contains only one parameter but in general, any number of a_i's (greater than zero) may appear in the expression. Minimization would then consist of taking partial derivatives with respect to each a_i and setting each resulting equation to zero. In the current problem, (1.4.6) is minimized with respect to only one parameter, a_2:

$$\frac{\partial}{\partial a_2}\left(\tilde{\Pi}\right)=\frac{\partial}{\partial a_2}\left(\frac{1}{2}EA(a_2)^2 L - P\, a_2 L\right)=EA(a_2)L - PL$$

Setting the derivative of the equation above to zero and solving for a_2:

$$a_2 = \frac{P}{EA}$$

Substituting a_2 into (1.4.4):

$$\tilde{U}(X) = a_2 X = \frac{PX}{EA} \tag{1.4.7}$$

In this case, the Ritz method has rendered the same results as the closed-form solution to the governing differential equation, (1.2.6). When the displacement assumption and the solution to the differential equation have the same form, the Ritz procedure and the closed-form solution can produce identical results. For the rod idealization (Figure 1.17) the closed-form solution to the governing differential equation is a linear polynomial. Since a linear polynomial was chosen as the displacement assumption in (1.4.3), the Ritz method and the closed-form solution have produced the same answer.

The Ritz displacement assumption need not be *exactly* the same as the closed-form solution, it only needs to *contain* the correct solution. In this example, the displacement assumption could have been linear, quadratic, or higher order, and the closed-form solution would still have been obtained, since the closed-form solution (a linear polynomial) is a subset of a higher order polynomial.

Although polynomials are often used, the Ritz displacement assumption is actually stated in more general terms. For one-dimensional problems the Ritz displacement assumption can be expressed as:

$$\widetilde{U}(x) = f_0 + a_1 f_1(x) + a_2 f_2(x) + \ldots + a_n f_n(x) \tag{1.4.8}$$

The $f_i(x)$ denote Ritz *trial functions,* which allow the displacement assumption to meet the essential boundary conditions of the particular problem, while the a_i are the adjustable parameters, already discussed. The constant term, f_0, is added to account for the possibility of a non-homogeneous boundary condition. The Ritz method can also be utilized for solid mechanics problems in two- and three-dimensional space. In the three-dimensional case, each trial function could be a function of three spatial coordinates, x, y, and z, for instance.

The Finite Element Method—Partitioning a Structure

The finite element method, when used to solve solid mechanics problems, is essentially the same as the Ritz method illustrated above. The finite element method establishes a functional, specifies displacement assumptions, and performs a minimization process. However, the finite element method first partitions a structure into finite sized elements and then employs individual polynomial displacement assumptions that are *valid within the domain of their respective elements.* The individual displacement assumptions are subjected to certain requirements, but the requirements are less rigorous than those that would be imposed upon the solution to the differential form of the problem.

Summary—FEA in Solid Mechanics Problems

Since strain and stress can be calculated with displacement, one popular approach to stress and structural analysis is to consider displacement as the *variable of interest.* If a problem is simple enough, a single differential equation may be used to generate an expression for displacement. In more complicated situations, an approximate solution may be necessary.

Although it may be possible to create an approximate solution for displacement by piecing together differential equation solutions, it has been shown that the Ritz method is very effective in a wide range of solid mechanics problems. As will be discussed in Chapter 2, the finite element method and the Ritz method are essentially the same. One difference is that, using the finite element method, the structure is partitioned into finite sized elements and a solution is generated within the domain of each. Mathematically, this is equivalent to using the Ritz method with a very specific set of Ritz trial functions. The term "finite element" is used to distinguish finite sized elements (elements of measurable size) from the infinitesimally small, differential elements that are considered from the study of differential equations. [2]

The displacement based finite element method eventually produces a system of linear algebraic equations, which, when solved, leads to a continuous expression for displacement within the domain of each element. Used together, the individual expressions from each element describe displacement for the entire structure.

1.5 Stress-Strain Basics

Key Concept: When applied to solid mechanics problems, the finite element method is often used to determine the state of stress and strain within a loaded structure. This section reviews some basic stress-strain concepts.

The purpose of reviewing stress and strain basics is to provide a basis for further discussion of the finite element concepts presented in the remainder of the text. The review is not intended to aid the individual in performing stress or structural analysis. The material is presented as a *review*. It presumes that the reader has at one time considered the topics in sufficient detail. In-depth discussion on stress, strain, and mechanics of materials are found in references such as Higdon [8] and Timoshenko [9].

Normal Stress

Consider the *normal stress* in a bar loaded under uniaxial tension as illustrated in Figure 1.19.

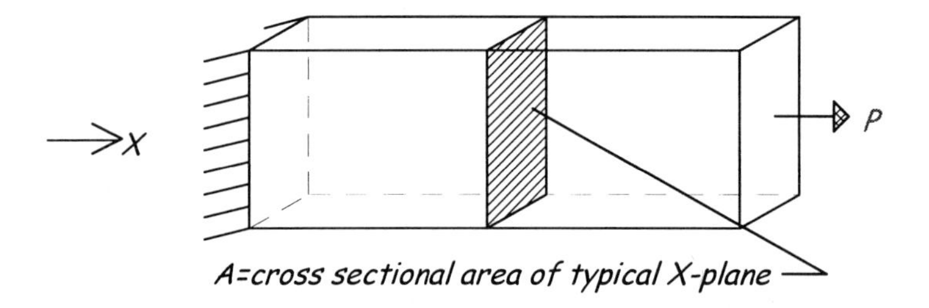

Figure 1.19 Bar Under Uniaxial Load, Typical X-plane

[2] Roughly speaking, a slice of measurable width is *finite*, while a slice immeasurably small (approaching zero) is *infinitesimal*.

Restraints are imposed on the left face of the bar to prevent displacement, while a load is applied to the right face, such that the resultant force (P) passes through the centroids of all resisting cross sections. Assume that the bar is composed of a typical engineering material, low carbon steel for instance, and that the applied load does not stress the material beyond the linear elastic region.

Figure 1.19 suggests that a plane cut through a structure is defined by the coordinate axis that is orthogonal to it, hence, any slice of the bar that is orthogonal to the X-axis is considered an X-plane. For a bar in uniaxial tension, the normal stress on a typical X-plane, in the X- coordinate direction, is given as:

$$S_{XX} = \frac{P}{A_0} \tag{1.5.1}$$

The first subscript associated with the S variable in (1.5.1) refers to the *plane* upon which the stress is acting, while the second subscript refers to the *direction.* Equation 1.5.1 is valid for all X-planes except those near the restrained face, where Poisson effects induce a three-dimensional state of stress. (Recall that loading a typical engineering material in one direction induces strain in other directions as well.) Note also that the subscript associated with the A variable in (1.5.1) refers to the *original* cross sectional area of the bar.

When the material in the bar behaves in an elastic fashion, the Poisson effect causes a slight reduction in the cross sectional area. The difference between the original cross section and that under load is very small when the bar is composed of a typical engineering material and loaded in the elastic range. Hence, the original cross sectional area dimension, A_0, is typically used in the expression for normal stress without significant error. When A_0 is used, the stress computed by (1.5.1) is termed engineering normal stress, or simply *engineering stress.*

Shear Stress

When analyzing the strength of a structure, it is often helpful to examine both normal stress and *shear stress*. Figure 1.20 illustrates shear and normal forces on a plane oriented at arbitrary angle ϕ with respect to the bottom of the bar originally shown in Figure 1.19. Equations 1.5.2 and 1.5.3 below define the normal stress, and the shear stress, respectively, on the N plane. As before, the first subscript refers to the plane, the second the direction in which the stress is acting:

$$S_{NN} = \frac{F_N}{B_0} \tag{1.5.2}$$

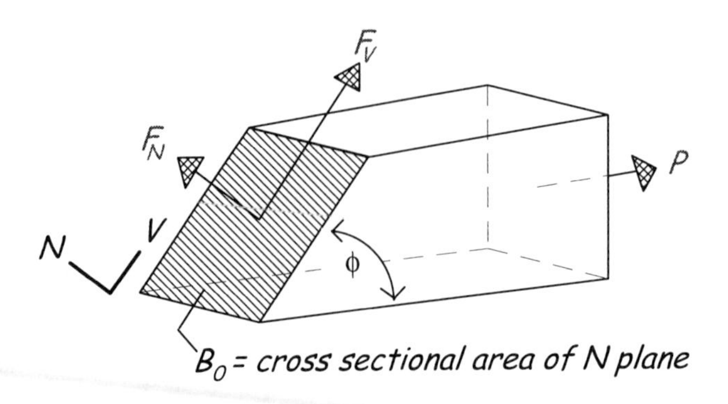

Figure 1.20 Shear and Normal Forces on Arbitrary Plane of Restrained Bar

$$S_{NV} = \frac{F_V}{B_0} \tag{1.5.3}$$

Notice that stress is a second order *tensor* metric, defined in terms of magnitude, direction, and *plane*.

In the examples of stress given so far, it is tacitly assumed that the stress is the same anywhere on a given plane. However, stress on a given plane can actually vary from location to location, depending upon the geometry of the structure, the external loads, and the restraints. Hence, *stress at a point* also needs to be considered, as illustrated in Figure 1.21.

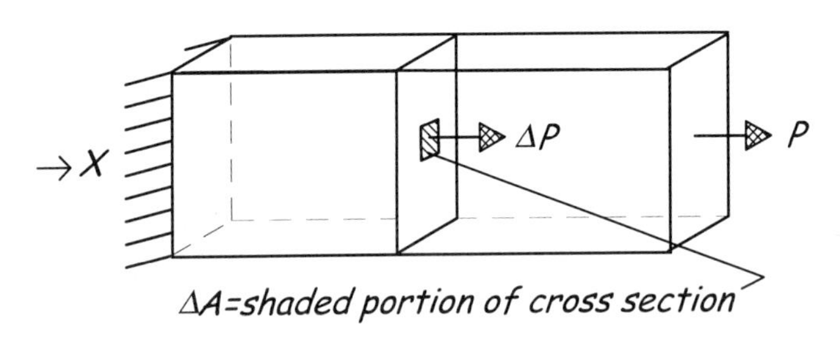

Figure 1.21 Rectangular Bar Loaded Under Uniaxial Load

Stress at a Point

Notice in Figure 1.21 that a portion of the load, ΔP, is distributed on a portion of the cross sectional area, ΔA. As one considers increasingly smaller portions of the cross section of the bar, smaller amounts of the total load are associated with each portion. That is, as ΔA decreases in size and finally becomes a point, the concept of *stress at a point* is realized as the limiting value:

$$\tau_{XX} = \lim_{\Delta A \to 0} \frac{\Delta P}{\Delta A} = \frac{dP}{dA} \tag{1.5.4}$$

The concept of stress at a point is applicable to both normal and shear stress, on both the major axis planes (X, Y, and Z planes) and arbitrary planes, N.

General State of Stress

In general, the state of stress within a structure may be defined in terms of a differential size element and nine stress components, as illustrated in Figure 1.22.

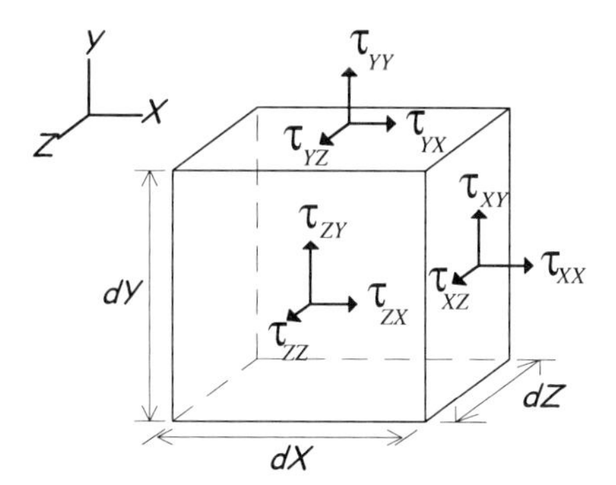

Figure 1.22 Components of Stress

Although nine stress components are shown in Figure 1.22, only six are actually required to describe a general state of stress, since shear stresses are related in the following manner:

$$\tau_{YX} = \tau_{XY} \qquad \tau_{ZX} = \tau_{XZ} \qquad \tau_{ZY} = \tau_{YZ}$$

The state of strain is also an important concern and is considered on the following page.

Normal Strain

Consider a bar of original length L_0, loaded in uniaxial tension, as shown in Figure 1.23. With the load applied, the bar stretches a distance of ΔL.

Somewhat similar to the example of engineering stress, engineering normal strain, or simply *engineering strain*, is defined in terms of the original length of the bar, L_0, and the amount of stretch the bar experiences:

$$e_{XX} = \frac{stretch}{original\ length} = \frac{\Delta L}{L_0} \tag{1.5.5}$$

As with engineering stress, engineering strain is applicable when both displacement and strain are small.

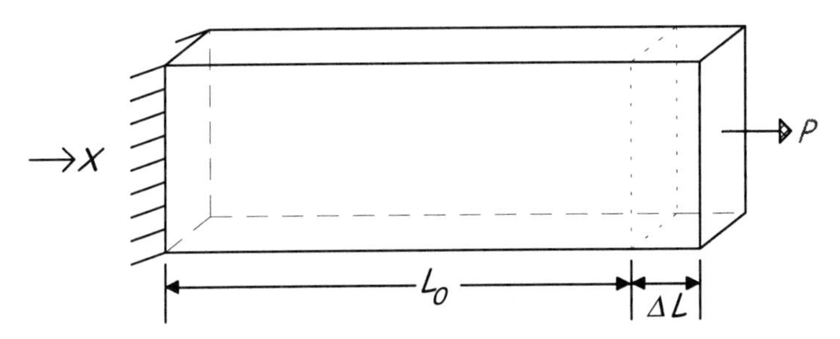

Figure 1.23 Strain in Bar Under Uniaxial Load

Strain at a Point

Figure 1.24 considers a slice of an unloaded bar, with the shaded region depicting a typical slice. Points *a* and *b* define the arbitrary, finite length of the slice ΔX, where $\Delta X = X_b - X_a$.

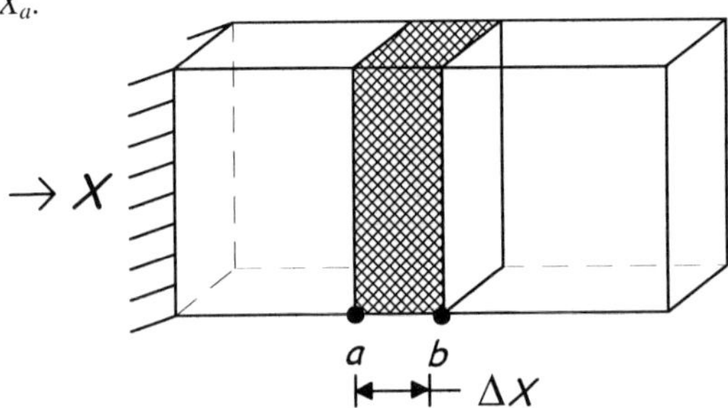

Figure 1.24 A Slice of an Unloaded Bar

A load is subsequently applied to the bar in Figure 1.24, such that a state of uniaxial stress develops, as illustrated in Figure 1.25.

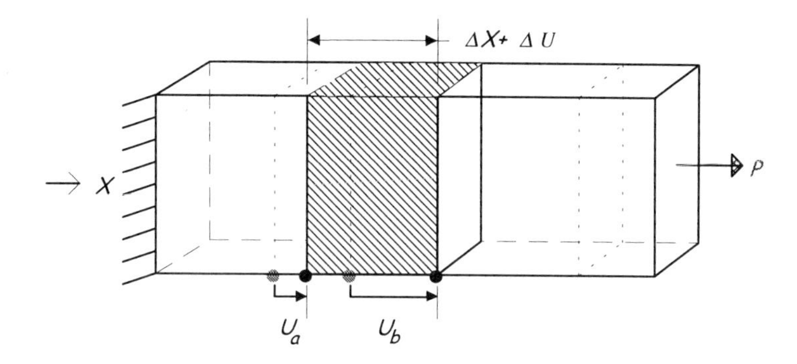

Figure 1.25 Stretching of Bar with Load Applied

Point a displaces an amount U_a, and point b displaces U_b. Defining the change in displacement as $\Delta U=U_b-U_a$, observe that, under load, the slice stretches to a new length of $\Delta X+\Delta U$. If ΔX is considered as the original length of the slice, and ΔU the amount of stretch, engineering strain can be computed as:

$$e_{XX} = \frac{stretch}{original\ length} = \frac{\Delta U}{\Delta X}$$

Analogous to normal stress at a point, the normal strain at a point can also be thought of using the limit concept. In this case normal strain can be defined as the amount of stretch in a slice, as thinner and thinner slices are considered, until the thickness of the slice approaches zero:

$$\varepsilon_{XX} = \lim_{\Delta X \to 0} \frac{\Delta U}{\Delta X} = \frac{\partial U}{\partial X} \tag{1.5.6}$$

If a differentiable function for axial displacement in terms of the X variable exists, (1.5.6) states that the normal strain in the X direction can be determined by taking the first derivative of the function with respect to X. In a similar manner, it can be shown that if V and W are displacements in the Y and Z directions, normal strains in the Y and Z directions are computed as:

$$\varepsilon_{YY} = \frac{\partial V}{\partial Y} \qquad \varepsilon_{ZZ} = \frac{\partial W}{\partial Z} \tag{1.5.7}$$

Partial derivatives are used to defined strain because in the general case, displacement may be a function of more than one variable. Equations for shear strain, stated below, may be developed by examining the angular change at the corners of the differential cube shown in Figure 1.22; see Higdon [8] for a good explanation of shear strains.

$$\gamma_{XY} = \frac{\partial U}{\partial Y} + \frac{\partial V}{\partial X} \qquad \gamma_{YZ} = \frac{\partial V}{\partial Z} + \frac{\partial W}{\partial Y} \qquad \gamma_{ZX} = \frac{\partial W}{\partial X} + \frac{\partial U}{\partial Z}$$

The symbol γ is used to denote shear strains. The symbol ε, *when associated with shear strains,* indicates that the value of shear strain is a component of the symmetric strain tensor. For example, considering shear strain in the X–Y plane, the relationship between the two shear strain measures is:

$$\varepsilon_{XY} = \frac{1}{2}\gamma_{XY} = \frac{1}{2}\left(\frac{\partial U}{\partial Y} + \frac{\partial V}{\partial X}\right)$$

The tensor values for shear strain will not be of concern in this text; see Ford [10, p. 177], for a discussion of strain tensors.

The equations for strain given above are applicable for "small" strains and displacements. For large strains, a more precise definition of strain requires that second order strains be included, as shown in (1.5.8). The strain measures in (1.5.8) are sometimes called Green-Lagrange strain; Cook [11, p. 437].[3]

When strains are small, the squared terms in Green's strain can be neglected, such that the resulting terms are equal to those given by (1.5.6) and (1.5.7). The question is, "What is small strain?" Ford [10, p. 121], suggests that normal strains less than 1 percent can be considered small, while Bathe [12, p. 303] suggests normal strains less than 4 percent can be considered small. Shear strains less than 0.001 radian are small according to Malvern [13, p. 121]. The small strain assumption is valid for many well designed structures that are constructed of typical engineering metals.

[3] The terms listed are the non-tensor equivalents of a strain tensor derived from the Green deformation tensor. Malvern [13, p. 158]

$$
\begin{aligned}
\varepsilon_{XX} &= \frac{\partial U}{\partial X} + \frac{1}{2}\left(\left(\frac{\partial U}{\partial X}\right)^2 + \left(\frac{\partial V}{\partial X}\right)^2 + \left(\frac{\partial W}{\partial X}\right)^2\right) \\
\varepsilon_{YY} &= \frac{dV}{dY} + \frac{1}{2}\left(\left(\frac{\partial U}{\partial Y}\right)^2 + \left(\frac{\partial V}{\partial Y}\right)^2 + \left(\frac{\partial W}{\partial Y}\right)^2\right) \\
\varepsilon_{ZZ} &= \frac{\partial W}{\partial Z} + \frac{1}{2}\left(\left(\frac{\partial U}{\partial Z}\right)^2 + \left(\frac{\partial V}{\partial Z}\right)^2 + \left(\frac{\partial W}{\partial Z}\right)^2\right) \\
\gamma_{XY} &= \frac{\partial U}{\partial Y} + \frac{\partial V}{\partial X} + \left(\frac{\partial U}{\partial X}\frac{\partial U}{\partial Y} + \frac{\partial V}{\partial X}\frac{\partial V}{\partial Y} + \frac{\partial W}{\partial X}\frac{\partial W}{\partial Y}\right) \\
\gamma_{YZ} &= \frac{\partial W}{\partial Y} + \frac{\partial V}{\partial Z} + \left(\frac{\partial U}{\partial Y}\frac{\partial U}{\partial Z} + \frac{\partial V}{\partial Y}\frac{\partial V}{\partial Z} + \frac{\partial W}{\partial Y}\frac{\partial W}{\partial Z}\right) \\
\gamma_{ZX} &= \frac{\partial U}{\partial Z} + \frac{\partial W}{\partial X} + \left(\frac{\partial U}{\partial Z}\frac{\partial U}{\partial X} + \frac{\partial V}{\partial Z}\frac{\partial V}{\partial X} + \frac{\partial W}{\partial Z}\frac{\partial W}{\partial X}\right)
\end{aligned}
\tag{1.5.8}
$$

It can be shown that if the material is such a structure is not strained beyond the elastic limit, the strains are not likely to exceed 1 percent. Sherman [14] compares the range of elastic strains that steel, aluminum, and plastic plate structures would be expected to operate under. In this text, it will be assumed that strain components are defined without Green's strain and the associated second order derivatives.

Hooke's Law for a Body Under Uniaxial Tension

From experimental study, Robert Hooke (1635–1703) determined that when bodies composed of springy [elastic] material are stretched by an external force, they exhibit a linear relationship between the applied force and the resulting displacement. The linear relationship between the force and displacement is called Hooke's Law, and forms the foundation of elasticity theory. (See Timoshenko [9], Timoshenko [15, p. 17], and Higdon [8, p. 92] for more on Hooke's law and elasticity theory.) For a body consisting of a homogeneous, isotropic, linear-elastic material, and loaded such that a state of uniaxial stress exists, Hooke's Law may be expressed as:

$$
\tau_{uni} = E\varepsilon_{XX} \tag{1.5.9}
$$

Equation 1.5.9 states that uniaxial stress is equal to the normal strain in the X direction multiplied by a constant, E, known as the elastic modulus (or Young's modulus). Using (1.5.9), it is tacitly assumed that the material is not loaded beyond its elastic limit, and that the load is applied in the X-coordinate direction.

The Uniaxial Tensile Test and Elastic Modulus

Equation 1.5.9 shows that a value for elastic modulus is needed to compute uniaxial stress. How is the elastic modulus in tension determined for a given material? Typically, a test specimen for a given material is fabricated and mounted in a machine that applies an increasing amount of tensile load to the specimen. As the load increases, an electronic gauge attached to the specimen continually records displacement, while a load cell on the machine records the tensile force (P) applied to the specimen. Using the recorded force data, along with the value of the specimen's original cross sectional area, engineering stress may be calculated for various values of force, using (1.5.1):

$$S_{XX} = \frac{P}{A_0}$$

Engineering strain may then be calculated using displacement values from the tensile test, the original length of the specimen, and (1.5.5): [4]

$$e_{XX} = \frac{stretch}{original\ length} = \frac{\Delta L}{L_0}$$

The engineering stress and strain values, calculated using data from the tensile test, are typically plotted in a manner similar to that illustrated in Figure 1.26; a ductile metal that experiences strain-hardening, such as low-carbon steel, could produce a curve somewhat similar to that shown. Other examples are given in Higdon [8] and Collins [16, p. 102, 278].

Using the engineering values of stress and strain, one can compute an approximate elastic modulus, using the equation:

$$E = \frac{S_{XX}}{e_{XX}} \tag{1.5.10}$$

Standard practice is to employ the modulus computed by (1.5.10) in the relationship given by (1.5.9). However, we cheat a little in doing so, since (1.5.9) expresses a relationship between *true* stress and *true* strain while the modulus computed by (1.5.10) uses engineering stress and strain. As it turns out, for many

[4] In a tensile specimen, the original length (L_0) actually refers to a specific portion of the specimen, i.e., the "gauge length." A gauge length of 2 inches is common for many ASTM tensile test specimens. The displacement (ΔL) is measured within the gauge length portion of the specimen.

engineering materials of practical interest, the error in using engineering stress and strain to compute elastic modulus is insignificant if the material remains elastic and is not stressed beyond its proportional limit.

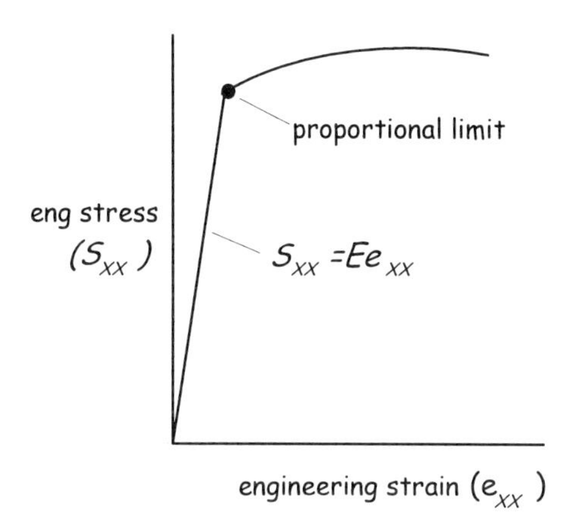

Figure 1.26 Idealized Engineering Stress vs. Engineering Strain

Hooke's law may also be employed for shear stress:

$$\tau_{XY} = G\gamma_{XY} \tag{1.5.11}$$

The constant G in the equation above is defined as the modulus of rigidity or *shear modulus*. The shear modulus may be experimentally determined in a manner somewhat analogous to the procedure used to establish the elastic modulus. However, a torsional load, as opposed to a tensile load, is applied to determine the shear modulus. It can be shown that the shear modulus for many typical engineering materials is related to Poisson's ratio and the elastic modulus by the equation:

$$G = \frac{E}{2(1+\upsilon)} \tag{1.5.12}$$

Using Displacement to Calculate Stress

Stress can be computed using derivatives of an appropriate displacement function. For instance, uniaxial stress can be expressed using the derivative of the function that represents axial displacement. To obtain this relationship, the derivative expression for normal strain, (1.5.6), is substituted into (1.5.9):

$$\tau_{uni} = E\frac{\partial U}{\partial X} \tag{1.5.13}$$

If displacement as a function of X is known, (1.5.13) can be used to calculate uniaxial stress in a loaded rod, as illustrated in the following example.

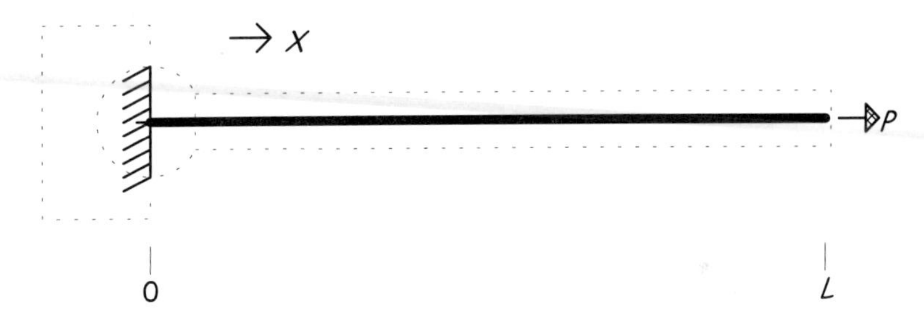

Figure 1.27 Idealized Link: Using Displacement to Calculate Stress

For this problem, the following information is given:

$$E = 30(10)^6\ psi \quad U(X) = 4X(10)^{-4}in \quad A = \frac{1}{6}in^2 \quad P = 2000lb_f$$

The objective is to calculate uniaxial stress using displacement. Assume that *U(X)* is known, perhaps from solving a differential equation, and Young's modulus is known for this particular material to be 30 $(10)^6$. Substituting the modulus and the known expression for displacement into (1.5.13):

$$\tau_{uni} = 30(10)^6 \frac{\partial}{\partial X}\left[4X(10)^{-4}\right]$$

Performing the operations in the preceding equation:

$$\tau_{uni} = 12000psi \tag{1.5.14}$$

Recall that in the linear, elastic range, the engineering stress is essentially equal to the true stress, such that the answer given by (1.5.14) can be checked using (1.5.1):

$$\tau_{uni} \approx S_{XX} = \frac{P}{A_0} = \frac{2000 lb_f}{\frac{1}{6} in^2} = 12000 psi$$

Why Use Displacement to Calculate Stress?

Since stress can be calculated by simply dividing the applied load by the original cross sectional area, why use displacement to calculate stress? The reason is that only in the case of uniaxial stress (for small strain and displacement) can such a simple calculation be made. In general, using elasticity theory, normal stress in any one coordinate direction is dependent upon forces in orthogonal directions. For a general three-dimensional state of stress, Hook's law is expressed as:

$$\begin{aligned}
\tau_{XX} &= \frac{E}{(1+\nu)(1-2\nu)}\left[(1-\nu)\varepsilon_{XX} + \nu(\varepsilon_{YY} + \varepsilon_{ZZ})\right] \\
\tau_{YY} &= \frac{E}{(1+\nu)(1-2\nu)}\left[(1-\nu)\varepsilon_{YY} + \nu(\varepsilon_{XX} + \varepsilon_{ZZ})\right] \\
\tau_{ZZ} &= \frac{E}{(1+\nu)(1-2\nu)}\left[(1-\nu)\varepsilon_{ZZ} + \nu(\varepsilon_{XX} + \varepsilon_{YY})\right] \\
\tau_{XY} &= G\gamma_{XY} = \frac{E}{2(1+\nu)}\gamma_{XY} \\
\tau_{YZ} &= G\gamma_{YZ} = \frac{E}{2(1+\nu)}\gamma_{YZ} \\
\tau_{ZX} &= G\gamma_{ZX} = \frac{E}{2(1+\nu)}\gamma_{ZX}
\end{aligned} \qquad (1.5.15)$$

Normal stress in any direction is a function of three strain components, Young's modulus, and Poisson's ratio, ν. Note that the three equations for normal stress do not depend upon the shear strains; see Higdon [8, p. 100] for details concerning normal and shear stress.

Equations (1.5.15) are utilized when analyzing problems from the Theory Of Elasticity point of view. An alternative method of stress analysis uses a "mechanics of materials approach"; both approaches are discussed in Ugural [17, p. 65]. One difference between the two approaches is that the mechanics of materials approach does not enforce strain compatibility, nor does the mechanics of materials approach account for stress concentrations. A key point illustrated in this section is that calculating uniaxial stress is simply a matter of differentiating a continuous function of displacement then multiplying by an elastic constant. It was shown how the elastic constant can be determined; the question is, how are continuous functions for displacement generated? As shown, a function of displacement may be established using a single, homogeneous, linear, differential equation, if the problem is simple enough. When problems become more complex, other methods of obtaining expressions for displacement, such as the finite element method, may be employed. A basic mathematical explanation of the finite element method will be given in Chapter 2.

1. 6 The Analysis Process

Key Concept: The process of mechanical engineering analysis, whether using sophisticated computer software or closed-form calculations, often involves several fundamental steps, such as properly posing a problem in engineering terms, generating a mechanical idealization, and solving the resulting differential equations.

Fundamental Steps in Engineering Analysis
Figure 1.28 illustrates how a typical engineering analysis project might progress.

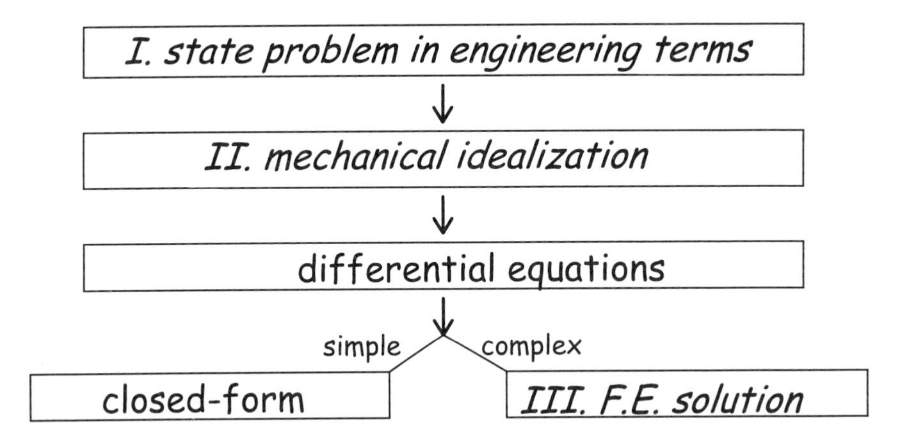

Figure 1.28 Flow of a Typical Analysis Project

In general, analyzing stress and structural problems often requires posing a problem in engineering terms, developing a mechanical idealization, and solving the differential equations that evolve from the idealization. If the differential equations are simple enough, they can be solved by closed-form methods; for more complicated situations, a numerical solution, perhaps using finite elements, is required. Bear in mind that when using finite element software, the equations that describe the response of the structure are invisible to the user.

This section considers the three major topics denoted in Figure 1.28.

I. State Analysis Problem in Engineering Terms

Generating a clear problem statement (or engineering hypothesis) is often a difficult task. The question "Will this structure fail in service?" does not constitute a problem stated in engineering terms, and posing an analysis problem as such does little good, since typical analysis software will not flag the analyst with a statement such as "THIS STRUCTURE IS GOING TO FAIL." Consider a better-posed problem statement:

> Assuming that the structure under investigation is in a state of plane stress, what is the maximum von Mises stress when the structure is subjected to a static, distributed load of 1000 psi on one end, while the opposing end is fully restrained? The material is assumed to be homogeneous and isotropic; a linear relationship between stress and strain is presumed in the range of the loads that will be applied.

Notice that the ability to properly pose a problem relies upon an understanding of loads, stress metrics, boundary conditions, and material behavior. The problem statement above focuses upon the maximum value of von Mises stress that develops in the structure with the given loading and restraints. It also clearly states the purpose of the analysis, and the metric that will be used to evaluate the strength of the structure.

The scope of an engineering analysis must often be narrow and well defined in order to be efficient and effective. Using finite element analysis, one cannot typically build a model and then "see what it will tell you." Considerable thought is needed to properly pose an analysis problem, conceive a model that will yield *relevant* results, and use the results to effect an engineering decision.

II. Mechanical Idealization

The term *idealization*, as used in this text, may be loosely defined as the process of making engineering assumptions to convert a physical problem into mathematical terms. The idealization process renders an equation (or system of equations) that is solved by either closed-form methods or a numerical method, such as the finite element method. To further investigate mechanical idealization, consider six categories of assumptions and simplifications associated with the process of mechanical idealization in solid mechanics problems:

- Linearity
- Boundary condition assumptions
- Stress-strain assumptions
- Geometric simplification
- Material assumptions
- Loading assumptions

Idealization often allows the essence of an intractable problem to be expressed in simplified mathematical form such that analysis can be performed. Without idealization, analyzing many structural problems would be impractical due to difficulties in replicating complex geometry, storing vast amounts of data, and implementing complicated mathematical principles to simulate many subtle responses that would not significantly affect an engineering design decision. Note that idealization is independent of the finite element method; it is used to initiate problems that may be solved by either closed-form methods or numerical means.

It is often said that finite element analysis simply "does not work," or is "not applicable to this type of problem." Indeed, there are many engineering analysis problems that do not lend themselves to finite element solutions. However, it is the author's experience that in many instances where the finite element method "does not work," or "fails to produce accurate results," the underlying problem is not with finite element capability but *with the quality and applicability of the engineering assumptions that were applied in the idealization process*. Again, the quality of a finite element analysis hinges upon a proper idealization, and good engineering judgment is required to consistently generate proper idealizations.

Mechanical Idealization—Linearity

There exist two major categories of idealizations used in association with the solid mechanics problems, namely, idealizations based upon the presumption of linearity and those that account for non-linear behavior. At the outset, it is important for the analyst to identify which type of idealization is applicable to a particular problem, since there are major differences in the way each type is approached. Many finite element analysts use different software programs depending upon whether a problem requires a linear or non-linear solution.

Linear, Static Solution: The fundamental assumption that allows a linear solution is that an *incremental change in the magnitude of the external load renders a proportional change in displacement*, and this relationship remains constant for the range of loads that will be applied. For example, if the magnitude of the load is increased by a factor of two, the resulting displacement within the structure will increase by the same factor. Generally speaking, for this to occur both the stiffness of the structure and the direction of the external forces must remain constant.

To illustrate some basic concepts related to linearity, consider a simple spring-force system having a spring constant K and an applied load F. The spring constant can be thought of as a proportionality factor between the applied load F and the resulting displacement D:

$$KD = F \tag{1.6.1}$$

The problem of computing displacement in the spring-force system described by (1.6.1) is a linear problem as long as the spring constant K remains constant, and the direction of the load vector does not deviate, as the magnitude of the external load changes.

To compare linear and non-linear behavior, consider the spring on the left hand side of Figure 1.29, where a small force is incrementally applied until a maximum magnitude of P_1 is reached.

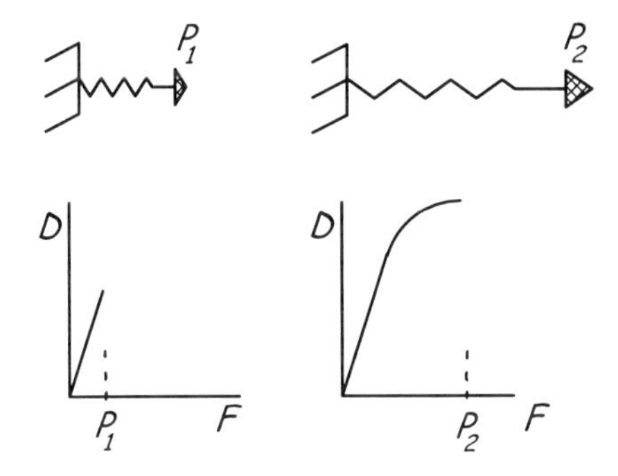

Figure 1.29 Linear and Non-Linear Problems

The relationship between the applied force and the resulting displacement remains linear, as suggested by the graph of displacement versus force on the left hand side. Now, imagine that the force acting upon the spring is increased to such an extent that the spring is stretched beyond its elastic limit, as shown on the right hand side. If it is assumed that the spring becomes less stiff after its elastic limit is reached, the relationship between force and displacement might appear as illustrated in the graph on the right. The cause of the non-linear behavior in this instance is due to the changing stiffness of the spring.

In will be shown in Chapter 2 that the finite element method produces an equation that is analogous to (1.6.1), except that the expression represents a system of equations, where the scalar proportionality factor K is replaced with a square matrix. In finite element problems, this square matrix, called the K-matrix, represents stiffness, just like the spring constant. And, just like the spring problem above, the K-matrix must remain constant if the finite element problem is to be considered linear. The question then is "What determines if the K-matrix is constant?" Three major factors control whether or not the K-matrix is constant:

A. The elastic modulus for a given material
B. *Geometric stiffness*
C. *Contact* boundary conditions

A. Stiffness and Elastic Modulus: The elastic modulus affects the finite element stiffness matrix. For a simple analogy, consider the stiffness of a rod with original cross sectional area A_o, loaded in uniaxial tension. Assume that the material in the rod behaves in a manner similar to the material illustrated in Figure 1.26. Using the relationship between stress and strain given in that figure, and the associated equation for engineering strain, (1.5.5), uniaxial stress may be expressed in terms of the elastic modulus and engineering strain:

$$S_{XX} = Ee_{XX} \qquad e_{XX} = \frac{\Delta L}{L_o} \qquad \therefore S_{XX} = E\frac{\Delta L}{L_o} \tag{1.6.2}$$

Uniaxial stress may also be expressed in terms of the applied force magnitude P and the cross sectional area, as in (1.5.1):

$$S_{XX} = \frac{P}{A_o} \tag{1.6.3}$$

Substituting the right hand side of (1.6.3) for the left hand side of (1.6.2), and rearranging into "$KD=F$" form:

$$\underset{\substack{\uparrow \\ K}}{\left(\frac{EA_0}{L_0}\right)} \Delta L = P \tag{1.6.4}$$

The variables within the parentheses in (1.6.4) constitute the stiffness of the rod. Note that the stiffness is directly proportional to the elastic modulus and cross sectional area, and inversely proportional to the length. It will be shown in Chapter 2 that the equation given by (1.6.4) is the same equation that evolves from a simple 2-node line element for rod applications. We can now examine the variables affecting the stiffness of the rod, and how non-linear behavior is introduced.

When displacement (and the associated strain) in the rod is small, E in (1.6.4) is constant; at higher levels of strain, E is no longer constant, as illustrated by the changing slope of the curve in Figure 1.26. When E is not constant throughout the entire range of loading, the stiffness in (1.6.4) is not constant, violating the fundamental requirement for linearity. To use a *linear* finite element model, the analyst must assume that a structure is loaded such that E remains constant. An analogous argument can be advanced for two- and three-dimensional problems.

B. Stiffness and Geometry: *Geometric stiffness* is a term used to describe the situation where changes in a structure's geometry under load affects the proportionality factor between force and displacement. Geometric stiffness controls the phenomenon of structural collapse and stress stiffening; collapse and its associated strain energy will be detailed in Chapter 2.

To understand collapse, consider a structure that is well restrained and subjected to an external compressive load that is incrementally increasing. Assume that initial incremental increases in the external load result in proportional increases in displacement. When further increases in the external load bring about disproportionately large increases in displacement, infinitely large displacements become possible (theoretically speaking). The reason for the disproportionate change in displacement is that the structure, now deformed relative to its original unloaded configuration, no longer has the same stiffness. The change in stiffness, along with the presence of an external load of specific magnitude and direction, causes the structure to collapse.

How is collapse analyzed? Using non-linear finite element techniques, the collapse problem may be simulated by incrementally increasing the external load and evaluating the stiffness matrix after each increment. A significant decrease in any of the terms of the stiffness matrix signals that collapse is possible. Although collapse is a non-linear problem, *some* problems of collapse may be approximated by bifurcation buckling, which requires only a linear solution. Cook [11, p. 432] recommends using non-linear techniques for most problems associated with collapse. It is noted that the collapse load predicted using bifurcation buckling analysis can differ substantially from both experimentally determined loads and loads predicted by non-linear analysis methods. (The collapse load is called the *critical load* in bifurcation buckling analysis.)

A phenomenon similar to bifurcation buckling is that of snap-through. When snap-through occurs, the structure passes from a stable condition to an unstable one, and then back to a stable configuration, which is different from the initial configuration of the structure. An example of snap-through analysis is given by Keene [18].

Another way in which geometry affects stiffness is when gross deformation of a loaded body occurs. Recall that in structures fabricated from typical engineering metals, deformation under load is very small relative to the dimensions of the structure, hence, the final, loaded geometry of the structure is nearly the same as the initial configuration. In the case of a uniaxial tensile specimen loaded such that stress is below the proportional limit, this *small deformation* assumption allows stress to be computed by dividing the external load by the *original* cross sectional area, as given by (1.6.3).

Notice that the stiffness computed by (1.6.4) is also a function of the cross sectional area. If the cross sectional dimension of a uniaxial tensile specimen under load is nearly the same as when unloaded, the stiffness (K) calculated using (1.6.4) is typically accurate enough. However, a substantial decrease in cross section occurs when the phenomenon of *necking* takes place, as in the case where ductile test specimens are subjected to uniaxial tensile testing; see Crandall [19, p. 323] and Ling [20]. As the cross section decreases, (1.6.4) predicts a decrease in stiffness, again indicating non-linear behavior.

If an analyst were modeling the behavior of a ductile, steel tensile specimen, incrementally loaded to the point of ultimate fracture, two sources of non-linearity would be present: Material non-linearity due to the changing relationship between stress and strain beyond the proportional limit, and geometric non-linearity, due to large changes cross sectional area when the specimen begins to neck. In such a case, the modulus, cross sectional area, and length in (1.6.4) cannot be considered constant, and a linear analysis is no longer sufficient.

There is one additional manner in which geometry can affect stiffness. The stiffness of a structure depends upon its orientation with respect to the external loads. For example, a long slender beam may be quite stiff in response to an axial load but quite flexible in response to a load that bends the beam. Hence, it is important that the geometric orientation of a structure remain constant while loads are applied. However, when large deformation of a structure occurs, the orientation of the structure can change with respect to the structure's initial coordinate system. If this happens to a beam, for instance, terms in the stiffness matrix that were originally computed to resist bending loads may now correspond to axial loads. A non-linear analysis is required to account for such changes in stiffness due to geometric orientation.

C. Stiffness and Boundary Conditions: A major source of non-linearity occurs when a boundary condition is a function of displacement. For instance, consider a cantilever beam that is subjected to an incrementally increasing tip load as a function of time, illustrated in Figure 1.30. Suppose when the load reaches a value of P_c, the beam contacts a rigid body, and further displacement at the point of contact is prohibited. In such a case, a contact boundary condition has occurred, and the stiffness of the structure is affected. In the case illustrated, the effective length of the beam becomes smaller when contact occurs, resulting in a stiffer beam, as reflected by the smaller slope in the graph of displacement versus load shown in Figure 1.30.

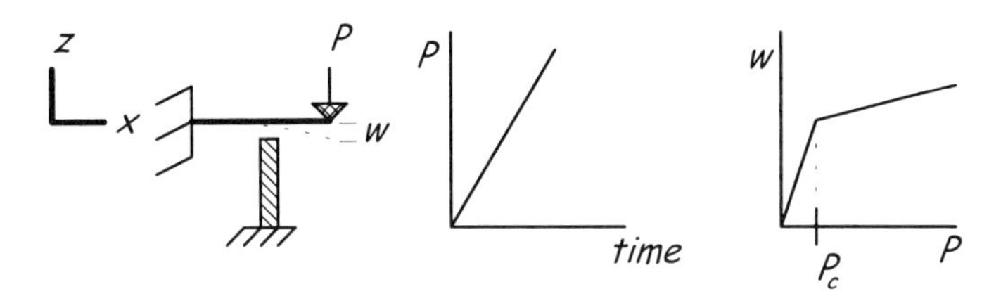

Figure 1.30 Contact Boundary Condition

Recall again that both the stiffness and the direction of the external forces must remain constant if linear analysis techniques are to be employed. There are cases where it is possible for the direction of the resultant of the applied load to change with large deformation. For example, consider a pressure load applied to one surface of a hollow, thin walled cube. Assume that the cube is very elastic, and as the pressure increases considerable deformation occurs, causing the surface to become concave. The forces due to the pressure acting upon the surface must then change direction, since the pressure is no longer projected onto a flat plane. Using linear finite element solution procedures, no account is made of the changing direction of the force. A non-linear method that considers changing (updated) geometry, and the associated change in forces resulting from surface pressure, is required to correctly simulate this phenomenon. The term *follower force* is used to describe a force that changes with deformation of the structure.

In some cases, all sources of non-linearity discussed previously may exist simultaneously. Non-linear analysis is a complex subject, and further discussion is beyond the scope of this text. For topics in non-linear finite element analysis, see Bathe [12]; a list of various sources of non-linearity is given on page 302.

Mechanical Idealization—Boundary Conditions

Essential boundary conditions determine how an idealized structure will resist rigid body motion when external loads are applied. It is the resistance to rigid body motion, disregarding inertial and/or thermal effects, that induces strain and stress in a structure. The idealization process often begins with a sketch of the structure under consideration, depicting essential boundary conditions, perhaps making use of the symbols illustrated in Figure 1.31.

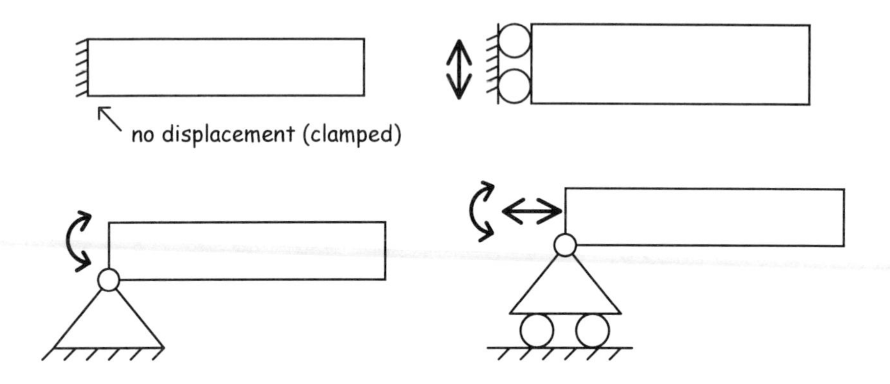

Figure 1.31 Essential Boundary Conditions Symbols

Mechanical Idealization—Stress and Strain Assumptions

Using finite element methods, many typical engineering problems are simply beyond the limits of practicality without the proper use of stress (or strain) assumptions; details related to stress-strain assumptions are found in Ugural [17]. In the following, eight types of assumptions will be considered:

A. Uniaxial stress
B. Plane stress
C. Plane strain
D. Axisymmetric solid (Solid body of revolution)
E. Euler-Bernoulli beam stress
F. Kirchhoff plate bending stress
G. Three-dimensional stress
H. Shell stress

The purpose of using a stress-strain assumption is to reduce mathematical complexity while simplifying the geometry required to represent a given structure. Proper choice of a stress-strain assumption can provide increased accuracy with less computational expense. For example, three-dimensional bodies can often be represented by two-dimensional geometry, or perhaps even one-dimensional geometry, as in the case of the steering link problem illustrated in Figure 1.2.

In the examples of stress and strain assumptions that follow, the typical coordinate system will have either two orthogonal directions, X and Y, or three, X, Y, and Z; the respective displacements are then U, V, and W. It is further assumed, in all the examples, that *stress, strain, and displacement in a loaded structure are referenced to the coordinate system that was defined in the original, unloaded, un-deformed structure*. This assumption is adequate when strains and displacements are small, since the loaded, deformed configuration will appear to be essentially the same as the un-deformed, unloaded structure. However, this assumption is not valid in cases where large strains occur, for instance, when a ductile, uniaxial test specimen begins to neck.

A. Uniaxial Stress: The least complex stress assumption, *uniaxial stress*, has already been considered in the analysis of the steering link. In the (idealized) pinned truss members of a suspension bridge, illustrated in Figure 1.32, uniaxial stress might be an appropriate assumption.

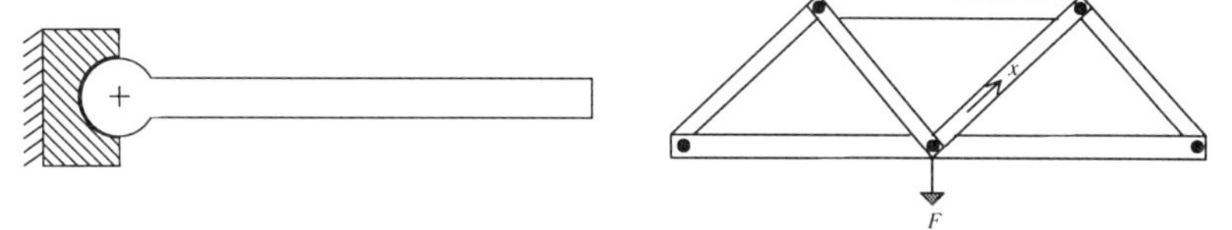

Figure 1.32 Uniaxial Stress

When using a uniaxial stress idealization, the cross sectional geometry is presumed uniform, and characterized by a mathematical constant, such that a line may used to represent an idealized member in the state of uniaxial stress.

B. Plane Stress: The presumption of *plane stress* may be applied when all out-of-plane *stress* components are either zero or small enough to be considered insignificant. For example, plane stress may be assumed in the plate under tensile load, as illustrated in Figure 1.33. It is presumed that *displacement through the thickness is not restrained, i.e.,* displacement in the Z direction is non-zero.

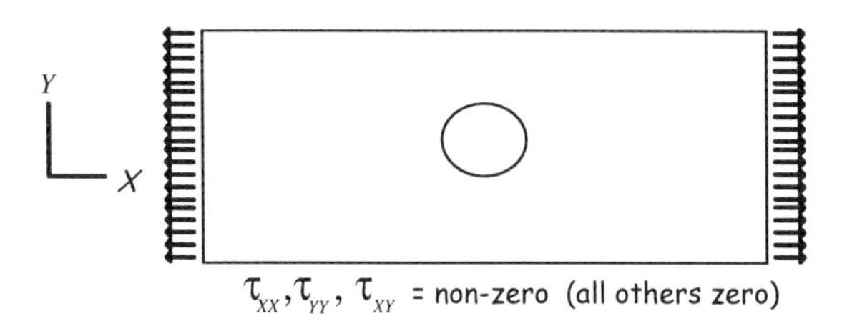

Figure 1.33 Plane Stress in a Plate with Hole

Recall that due to Poisson effects, stretching the plate with loads in the *X* and *Y* directions can cause displacement in the Z direction as well. Restraining displacement in Z would induce stress in the thickness direction of the plate, and a state of plane stress would no longer exist. In addition, when a plate becomes thick, some of the assumptions associated with the plane stress condition are no longer valid.[5] Using a plane stress idealization, the thickness is assumed to be uniform and described by a mathematical constant, therefore, a surface may be used to represent this type of idealized structure.

C. Plane Strain: When a structure is thick in one coordinate direction, and the strain and displacement components in this coordinate direction are presumed zero, a state of plane strain may exist. Plane strain is presumed to exist in a typical slice of a long dam with water pressure acting upon one surface, as illustrated in Figure 1.34

A distinguishing characteristic of plane strain is that *displacement through the thickness is restrained, and therefore equal to zero.* Restraining displacement in the *Z* coordinate direction (through the thickness) induces normal stress in *Z*, which is opposite of the situation that occurs with plane stress. Using a plane strain idealization, the cross section is assumed to be uniform and infinitely thick.

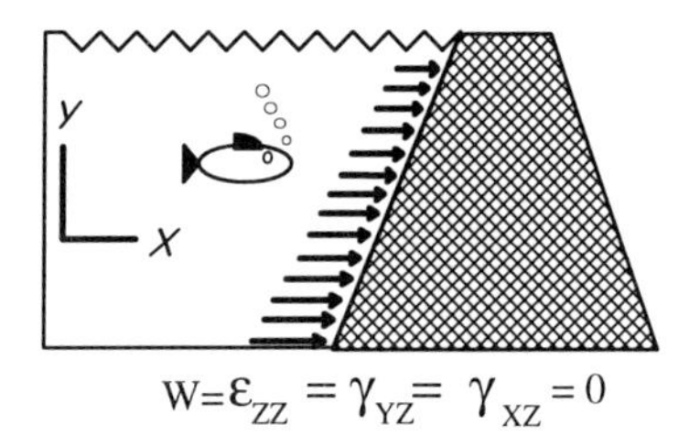

$$W=\varepsilon_{ZZ}=\gamma_{YZ}=\gamma_{XZ}=0$$

Figure 1.34 Plane Strain in a Cross Section of a Water Retention Dam

---

[5] Plane stress requires the following:

1. $\tau_{ZZ}=\tau_{XZ}=\tau_{YZ}=0$
2. τ_{XX}, τ_{YY}, and τ_{XY} independent of *Z*

It can be shown that the *stress function* that satisfies both the differential equations of equilibrium and the strain compatibility equations contains the term $\left(\frac{-\nu}{2(\nu+1)}\Theta_0 Z^2\right)$, violating Requirement 2 above (Timoshenko [9, p. 276]). However, if Z (the thickness) is small, the error is small; if Poisson's ratio is zero, no error exists. The discussion of plane stress and strain mentioned here is taken from a macroscopic point of view, i.e., from the perspective of the entire structure. The same topic can be considered from a microscopic viewpoint, as is done when fracture mechanics is considered; Gross [21].

Therefore, a surface may be used to represent an idealized structure in a state of plane strain, and no mathematical constant for the thickness is required.

D. Axisymmetric Stress: When both the geometry and loading of a structure are symmetrical about a structure's axis of revolution, an *axisymmetric solid of revolution* idealization may be applicable, as illustrated in Figure 1.35. Here, the coordinate system consists of three components: the radial direction R, the circumferential or *hoop* direction θ, and axial direction, Z. The corresponding displacements are U, V, and W, respectively. Notice that by revolving the generating surface 360 degrees about the Z-axis, a *solid of revolution* is created. A surface is therefore sufficient to represent an axisymmetric *solid* idealization.

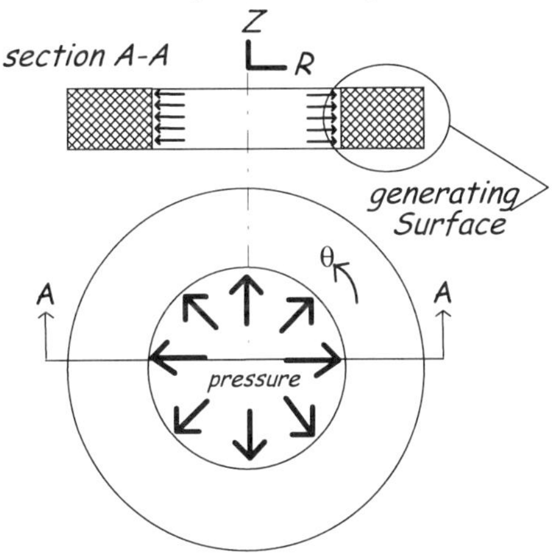

Figure 1.35 An Axisymmetric Ring

There are other types of axisymmetric assumptions, such as the *axisymmetric thin shell* assumption. If the ring in Figure 1.35 were very thin, such as a wedding band, an axisymmetric thin shell idealization might be used, in which case a *line* parallel to, and revolved about, the Z-axis would characterize the geometry. The axisymmetric solid idealization (without torsion) assumes that displacement in the hoop direction is zero, along with shear stresses $\tau_{R\theta}$ and $\tau_{Z\theta}$. Refer to Timoshenko [9] for a discussion on axisymmetric stress.

E. Beam Stress: Another common stress assumption is that of elementary (Euler-Bernoulli) *beam stress* that occurs in *narrow*, *slender* members due to the presence of bending loads. Beam geometry is characterized by *one* dimension much greater than the other two, for example, when the axial length of the beam illustrated in Figure 1.36 is much greater than the other two directions.

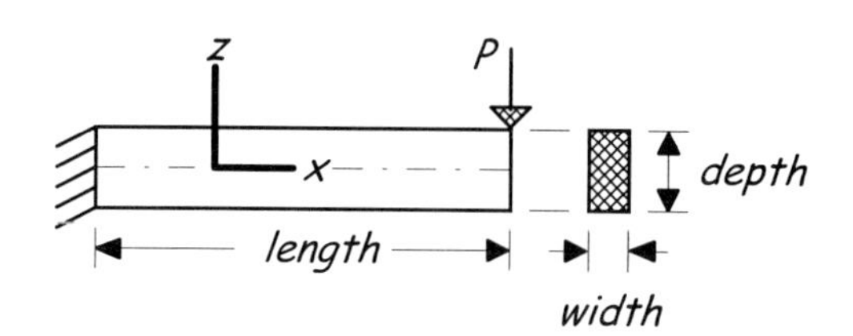

Figure 1.36 Elementary Beam Stress

Beam slenderness is characterized by the ratio of the length divided by the depth:

$$l/d \text{ ratio} = \frac{length}{depth}$$

Beam narrowness is characterized by the ratio of depth to width:

$$d/w \text{ ratio} = \frac{depth}{width}$$

When beams are slender ($l/d \geq 10$), and narrow ($d/w \geq 2$), the Euler-Bernoulli beam formulation may be applicable. For non-slender (deep) and/or wide beams, an alternate formulation should be employed to obtain suitable accuracy. Details of assumptions consistent with slender (Euler-Bernoulli) beams will be discussed in Chapter 4.

F. Plate Bending Stress: A somewhat more complex stress assumption is that of *Kirchhoff plate bending stress*, which occurs in thin, flat structures subjected to loads normal to the midsurface, like the bracket illustrated in Figure 1.37. As previously mentioned, Euler-Bernoulli beams are slender members, i.e., one dimension is much larger than the other two. Kirchhoff plates are *thin*, with *two* dimensions of the structure significantly larger than the third. While normal stress in the x direction (τ_{xx}) is the only significant stress component in the Euler-Bernoulli beam, bending stress in the Kirchhoff plate is presumed to occur in both x and y directions, i.e., τ_{xx}, τ_{yy}. In-plane shear stress, τ_{xy}, is also presumed non-zero. Hence, the Kirchhoff plate bending stress assumption is somewhat more involved than the Euler-Bernoulli beam assumption.

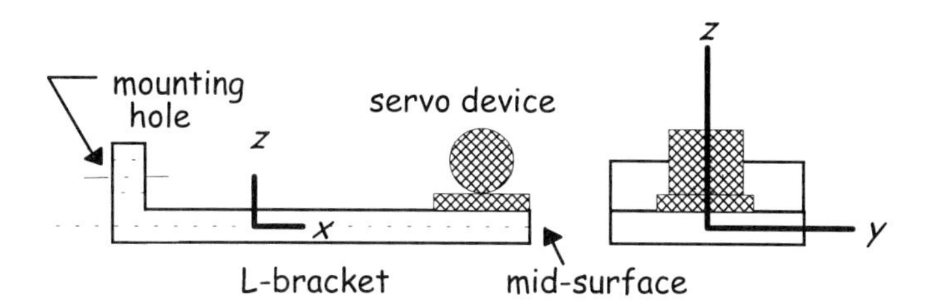

Figure 1.37 L-bracket with Servo Attached, Thin Plate

Like the Euler-Bernoulli beam, a Kirchhoff plate assumes that the external load is not carried by membrane deformation, meaning that in-plane forces are presumed zero at the mid-surface. However, when *membrane stretching* occurs, by deforming a plate into a doubly curved surface for instance, in-plane forces are induced at the mid-surface. In-plane tensile forces at the mid-surface increase the effective transverse stiffness of the plate, somewhat in the same way that increasing tension on a cable allows a weight to be supported with a decreased amount of deflection, as illustrated in Figure 1.38.

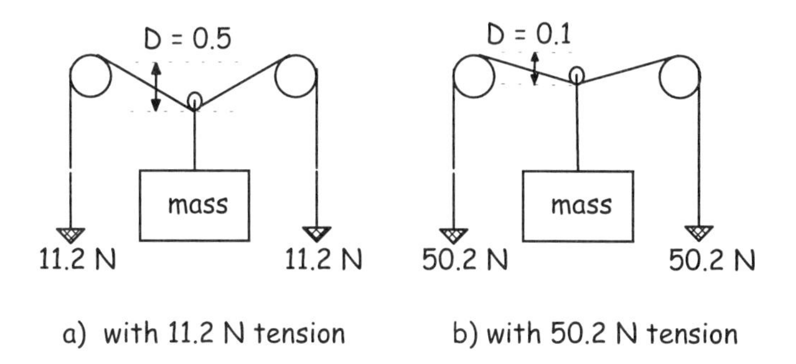

Figure 1.38 Membrane Forces Affect Transverse Stiffness

Assume that the mass in the figure exerts a 10 N force in the vertical direction. With an 11.2 N tension in the cable, the deflection is 0.5 m, hence the "stiffness" is:

$$K = \frac{F}{D} = \frac{10}{0.5} = 20$$

When the tension in the cable is increased to 50.2 N, the deflection is reduced to 0.1 m, and the associated stiffness becomes:

$$K = \frac{F}{D} = \frac{10}{0.1} = 100$$

The sketch in Figure 1.38 suggests that tensile, membrane forces can increase the effective *transverse stiffness* of a structure, while compressive loads will have the opposite affect. In plates, tensile membrane forces appear in the form of in-plane, extensional forces that are non-zero at the mid-surface of the plate. When modeling a flat plate using the Kirchhoff plate bending assumption, no account is made for changes in transverse stiffness due to in-plane forces. In addition, no account is taken of transverse shear. The analyst should be aware of the limitations of the Kirchhoff plate bending assumption. Details of the Kirchhoff plate assumptions will be discussed in Chapter 4.

G. Three-Dimensional Stress: When using the assumption of three-dimensional stress, three-dimensional (volume) geometry is required to represent the idealization, and a substantial increase in time is required for model creation. In addition, three-dimensional models consume vast amounts of computer resources in terms of hard disk storage and computational time. For example, consider both a square surface meshed with N elements per side and a cube with N elements per side. Cook [11, p. 574] shows that a rough estimate for the increase in computational expense in going from a surface to a volume idealization is N^3. Thus, a volume with ten elements per side is 1000 times more computationally expensive than a surface with ten elements per side. Therefore, a three-dimensional stress assumption is often used as a last resort, when no other stress assumption can be invoked. The seasoned analyst will typically consider all stress-strain assumptions that could lead to substantial reduction in geometry creation, and if none are applicable, use a three-dimensional stress assumption.

H. Shell Stress: Shell theory is somewhat similar to plate theory, in that it considers thin structures. However, while plate theory is limited to conditions where in-plane forces at the mid-surface are zero (or small), shell theory attempts to couple the effects of in-plane forces with bending stiffness in structures that are thin and *curved.* Shell theory, significantly more complex than plate theory, is beyond the scope of this guide; see Timoshenko [22], Bull [23], and Cook [11] for more information on structural shells. As with plates, a surface representation is used for shell idealizations.

Stress-strain assumptions can be based upon elasticity theory, using some form of generalized Hooke's law, or upon a mechanics of materials approach. The Euler-Bernoulli beam idealization, for example, is based upon a mechanics of materials approach, while equations applicable to plane stress evolve from

Hooke's law. Both approaches allow significant reduction in a problem's complexity by making limiting assumptions about the geometry, loading, and material response of a structure. The mechanics of materials approach is often applied to bodies that are designed to perform a particular *structural* function; beams, plates, and shells, are bodies that would typically be considered *structural members*.

A finite element that is developed to characterize the response of a structural member is called a *structural element*. Beams, plates, and shells are all considered structural elements, with the beam being the least complex. In contrast, a *continuum element* is designed to characterize the response of a body whose stress-strain behavior is based upon some form of Hooke's law. Elements used for uni-axial stress, plane stress, plane strain, axisymmetric stress, and three-dimensional stress are continuum elements.

Mechanical Idealization—Geometric Simplification

As shown, the process of mechanical idealization utilizes assumptions regarding linearity, boundary conditions, and stress-strain relationships to simplify the analysis process. The fourth component of a mechanical idealization to be considered is geometric simplification. The necessity of geometric simplification in analysis, particularly when using finite element methods, cannot be overstated. Without a substantial reduction in the amount of geometric detail, the capability of the finite element method is quickly exceeded. Even if an idealization requires the assumption of three-dimensional stress, along with the associated three-dimensional (volume) geometry, some geometric details of a structure are typically eliminated to allow practical application of finite element procedures.

There exist three idealization techniques that are commonly used to reduce the level of geometric complexity. As shown, a stress or strain assumption can allow a three-dimensional problem to be described using one- or two-dimensional geometry. For example, the three-dimensional steering link was reduced to a one-dimensional idealization, allowing a significant reduction in the amount of geometry construction necessary to represent the structure. The second way to reduce geometric complexity is to ignore all geometric details that do not significantly impact the state of stress in a structure. For instance, removing fillets and chamfers that are not in the load path. A third means of reducing geometric complexity is to employ symmetry arguments, such as mirror symmetry. The use of symmetry typically requires both symmetrical geometry and symmetrical loading.

Consider a plate of uniform thickness, t, under tensile load, as illustrated in Figure 1.39.

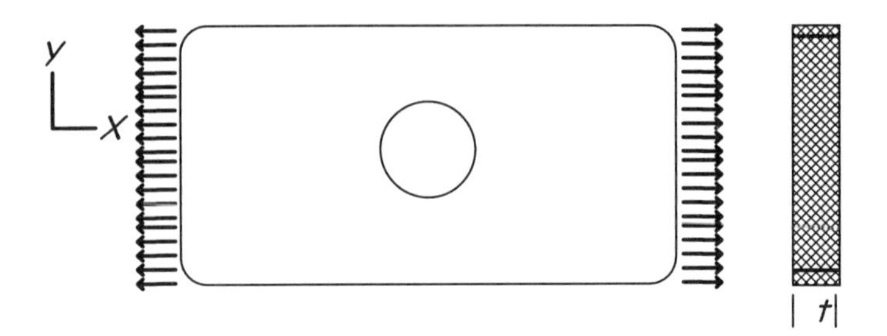

Figure 1.39 Plate Under Uniform Tensile Load

The objective is to reduce the level of geometric complexity required to characterize stress in the plate. To achieve the desired reduction, the plate is first assumed to be in a state of plane stress. Recall that a surface can be used to represent a plane stress idealization. Since both the loading on the plate and the geometry exhibit mirror symmetry, further reduction in geometrical detail can be realized by employing a quarter-surface representation, illustrated in Figure 1.40.

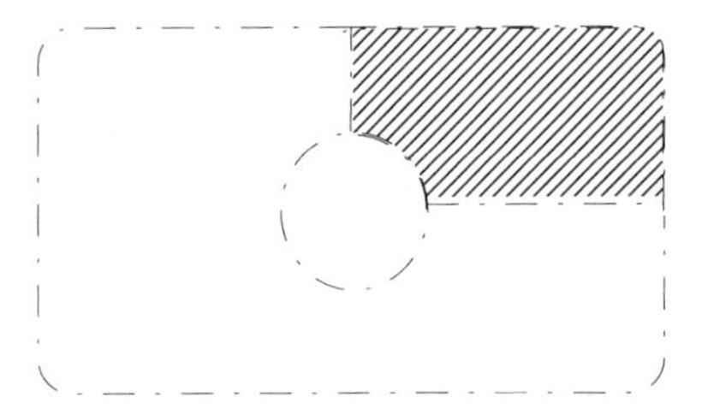

Figure 1.40 One-Quarter Surface, Plane Stress Idealization

Using the idealization depicted, a substantial amount of time will be saved in terms of both creating geometry to represent the actual part, and solving the equations necessary to obtain a solution for displacement within the plate.

Depending upon geometry and loading, mirror symmetry may also be used with other stress-strain assumptions, including three-dimensional stress. Other types of symmetry arguments, such as cyclic or repetitive symmetry, can also be employed. See Meyer [24, p. 249], Dewhirst [25], and [26] for further examples of symmetry and *anti-symmetry*.

Mechanical Idealization—Material Behavior
In developing an idealization, the analyst must consider the type of material that the structure is constructed of, and how the material is expected to respond when loaded. For instance, structures composed of typical engineering metals are presumed to exhibit linear material response if the external loads are small enough so that stress remains below the proportional limit. In conjunction with the assumption of linear material response, elasticity is also often presumed, such that removing the external load will result in the structure returning to its original configuration.

In other instances, a material may be elastic while the stress-strain relationship is non-linear, as in the case of natural rubber. Somewhat conversely, a material may be described as perfectly *plastic*. This type of behavior may be exhibited when a ductile, steel workpiece undergoes large plastic deformation during metal forming operations. In such a case, the elastic response of the material is completely neglected. In still other cases, the presumption of *elastic-plastic* material response may be necessary, in which case linear elastic response is presumed upon initial loading while non-linear, elastic-plastic response is considered when loads cause the material to be stressed beyond the elastic limit; Crandall [19, p. 277].

Other assumptions, such as homogeneous and isotropic material properties, are typically presumed for common engineering materials used under typical conditions. There are cases where the typical presumptions are no longer valid, and these cases may be associated with:

- Temperature effects
- Loading rate effects
- Time dependency (creep, stress relaxation)
- "Non-standard" material (plastic, natural rubber, non-isotropic metals)
- Repeated loading

Since common engineering materials are neither *completely* homogeneous, elastic, nor isotropic, some degree of idealization regarding material behavior is always required.

Mechanical Idealization—Loading Assumptions
Some assumption must be made regarding the type of loading that is applied to a structure. In reality, most loads are not static but are presumed so if the rate of loading is sufficiently low. In addition, interpretation of stress and strain results is influenced by whether or not the actual load is monotonic, repeated, or cyclic.

Loads that are confined to a relatively small portion of a structure are often considered to be concentrated or "point loads." Applying this assumption to the classical mechanics of materials idealization of a beam is standard procedure, while applying this same assumption to a longitudinal section of a beam that is analyzed using a elasticity theory approach results in a stress concentration. To understand the effects of such loading, the analyst should be aware of the mechanics of materials approach to stress and structural analysis. Attention should be paid to how it differs from the theory of elasticity approach, and how the finite element method idealizes both of these approaches. In addition to concentrated loads, loads may be spread out over a very large surface area. If so, these loads may be characterized by *distributed loads*.

III. Finite Element Procedures

The finite element method can provide an approximate solution to the differential equations that evolve from the process of mechanical idealization. The four finite element procedures below comprise a large portion of an analysis project:

- Create mesh boundary and the associated elements
- Input model data (mat props, element props, loads, restraints)
- Invoke finite element solver
- Evaluate results

The first three procedures above are concerned with finite element model preparation or *pre-processing*, while the task of evaluating results is termed *post-processing*.

Finite Element Procedures—Creating Mesh Boundaries and Mesh

As mentioned, lines, surfaces, and volumes are used to describe the geometry that evolves from the process of mechanical idealization; finite elements are constructed in the same form: line elements, surface elements, and volume elements. Finite elements are specified by points or *nodes* on the boundary of the element, as illustrated in Figure 1.41. Both the coordinate location of the nodes, along with *element connectivity*, are necessary to describe a finite element. The element connectivity prescribes how the nodes should be connected to form the element. For the 4-node surface element in Figure 1.41, four sets of *X*- and *Y*-coordinates, along with the element connectivity *a,b,c,d,* are needed to specify the element. It is standard convention to state element connectivity in a counter-clockwise manner.

The assemblage of elements that is used to represent a structure is called a finite element *mesh*. To produce a mesh, present day finite element pre-processors typically employ two separate types of geometric entities: *mesh boundary* geometry, and the geometry of the finite elements (the mesh). A mesh boundary may be thought of as the envelope in which finite elements are generated.

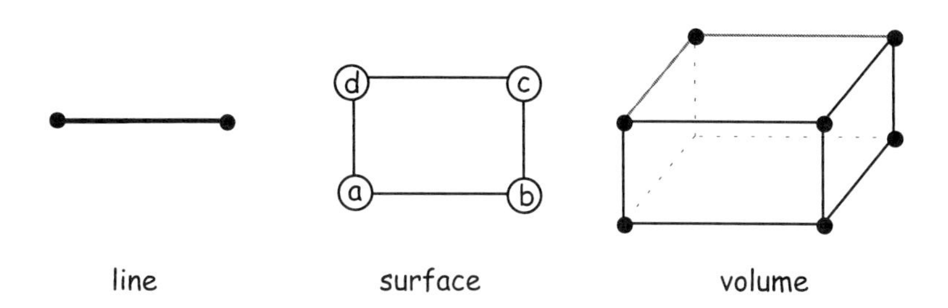

Figure 1.41 Line, Surface, and Volume Elements

The choice of mesh boundary geometry (line, surface, or volume) corresponds to the type of element that will be created within the boundary. Once the mesh boundary is created, finite elements can be generated within it, "automatically." Some details of mesh creation will be discussed in Chapter 7.

Advantage of Using a Mesh Boundary: One advantage of creating a mesh boundary instead of individual elements is that the mesh boundary typically represents more of the structure than any one individual element, reducing the task of specifying the geometry of a structure. In addition, the geometric entities that are used to create mesh boundary geometry are more robust than the geometry of typical finite elements. For example, the curved edge of a mesh boundary may be represented by a third (or higher) order polynomial, while finite elements on the same edge might only be able to employ straight line segments. Hence, it typically takes far fewer geometric entities to define a structure using mesh boundary geometric entities than it does using finite elements. The restrictive nature of element boundary shape is discussed in Chapter 4.

In Figure 1.42, the one-dimensional mesh boundary for the steering link of the previous examples is shown. Since only one element is used, there is not much need for a mesh boundary because the geometry of the element can be described as easily as the boundary. However, if 100 elements were needed within the same boundary, it would be much easier to describe a mesh boundary, using just two points, and then use *automatic meshing* to generate and record the data for 100 elements within the boundary. Another advantage of using a mesh boundary is that the analyst can experiment with the effects of changing the type and number of elements without having to re-specify the geometry of the structure each time.

In more elaborate finite element models, different types of elements are used within the same model. For example, surface elements may be combined with volume elements *if special precautions are taken.*[6] Combining different element types will be considered again in Chapter 7.

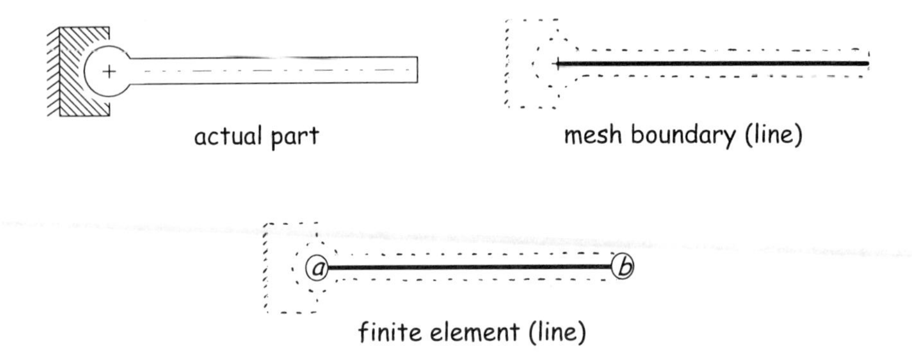

Figure 1.42 Finite Element Created Within a Mesh Boundary

It is noted that when different types of elements are used within the same model, then differing types of mesh boundaries would also be present. The need for mesh boundaries is more pronounced in two- and three-dimensional idealizations. More examples of mesh boundaries will follow.

Mesh Boundary Geometry and CAD Geometry: If one creates accurate CAD geometry, is it simply a matter of meshing the geometry to produce a finite element model? There have been many articles that would lead one to believe that CAD geometry is the same as analysis geometry, and that it is only a matter of time before CAD and analysis are "integrated"; Porter [27].

However, the fact is that CAD geometry is typically quite different from the geometry used for finite element mesh boundaries. It appears that the confusion lies in the fact that many companies, and individuals, view FEA as simply an extension of solid modeling. This issue is clearly expressed by Rodamaker [28]:

There is a general tendency, primarily among companies whose primary product is computer aided drafting systems, to visualize finite element analysis

[6] Care must be taken when combining elements of different types, since some element types may not be compatible with others. The analyst must consult suitable documentation to avoid serious errors.

> *as a sub-topic of solid modeling. Their view is that a solid model is created and converted to a finite element model using an automated meshing technique and the analysis is run. This is a serious over trivialization of the finite element process.*

Furthermore, since some individuals believe that finite element models are equivalent to CAD drawings, they deduce that drafting people should create FEA models, since they are responsible for CAD; see Robinson [29].

Why is finite element mesh boundary geometry not the same as solid modeling (CAD) geometry? Two reasons will be offered, although there are others. Firstly, it should be apparent from the reading so far that finite element analysis in solid mechanics problems is based upon *mechanical idealization*. Hence, the geometry of the elements and the associated mesh boundaries are often highly idealized, using two-, and in some cases, one-dimensional geometry. As such, the actual geometry of the structure may bear little resemblance to the geometry of the finite element idealization; Lepi [30].

Secondly, the *goal* of creating CAD geometry is usually diametrically opposed to the goal of creating FEA geometry. A designer creates CAD geometry to render an unambiguous representation of a structure, generally with the intention of using the geometry to provide a plan from which a physical structure can be built. In other words, the designer's goal is to produce an extremely accurate computerized drawing, pattern, or sketch to be used as a plan for fabrication. In contrast, the analyst wishes to produce the minimum amount of geometry necessary to allow him to analyze (investigate) the nature of a structure. Ideally, the analyst would create no geometry at all, and simply analyze a structure using a closed-form solution to a differential equation.

The above discussion is not intended to suggest that there exists little connection between the CAD designer and the analyst. On the contrary, a team approach to analysis may indeed be best. Using such an approach, an analyst specifies the mesh boundary geometry needed for a given idealization. If the mesh boundary geometry is complex, the analyst may enlist the help of the designer to build the mesh boundary, since designers typically possess more advanced geometry creation skills. If both the designer and analyst use the same software, then the mesh boundary geometry can be easily transferred from the designer to the analyst, without any loss of data. The analyst then can create a mesh, or various meshes, and proceed to analyze the structure. See Lepi [30] for more on the topic of the team approach to analysis.

Finite Element Procedures—Input Model Data

After a mesh boundary and the associated elements are created, the data necessary to perform an analysis is specified. The following data is typically needed to solve a simple structural mechanics problem:

- Material properties (elastic modulus, Poisson's ratio)
- Element properties (cross sectional area, or thickness)
- External loads (concentrated loads, distributed loads, etc.)
- Displacement restraints

If volume elements are used, element properties such as cross sectional area and thickness are not required since the geometry of the element is totally specified by the location of the element's nodes. However, for beam and plate elements, the nodal locations only describe one or two dimensions of the element, hence a cross sectional area or thickness property must be defined. A check of the model should be performed before submitting it to the solver to ensure that data has been specified for element properties, materials, loads, and restraints. During the model check, the quality of the mesh should also be checked to ensure that the geometry of the elements is not overly distorted such that poor accuracy results. Element distortion is discussed in Chapter 7.

Finite Element Procedures—Invoke Solver

The last step in the pre-processing phase is to choose a solution routine, then invoke the calculation phase of the analysis. A *linear, static* analysis is the most simple solution routine, similar to the problems that are introduced in undergraduate mechanics of materials course work. For example, a straight, slender, uniform steel beam, subjected to a static bending load that produces small deflections and small strains, may be considered a linear, static problem.

Finite Element Procedures—Evaluate Results

After the finite element solver is invoked and the mathematical computations have been completed, the results (stress, strain, displacement) need to be evaluated. The results are often evaluated graphically, using color contour plots to distinguish between varying the levels of stress, strain, or displacement, within a given structure.

Elementary Applications for Finite Elements

The rest of this section considers a few examples of how idealization and finite element procedures are applied in elementary applications. *The examples given are not intended to show how to solve certain engineering problems* but rather illustrate how the principles of idealization and meshing are employed in FEA. The details of these models will be considered in Chapter 7.

Although the steering link and the non-uniform shaft from earlier examples were illustrated in one-dimensional space, finite element models are typically constructed in either two- or three-dimensional space. The spatial coordinates X and Y, and the related translational displacements U and V, characterize a problem in two-dimensional space; X, Y, and Z, along with translational displacements U, V, and W are used for continuum elements in three dimensions. (As noted earlier, an exception to this occurs when using axisymmetric solids, where an R, θ, Z coordinate system is employed.)

In addition to three translational displacements, *structural* elements also employ up to three *rotational displacements*. Hence, an element can have as many as six displacement components per node, or six nodal *degrees of freedom* (DOF). In the illustrations that follow, restraints at the nodes of finite elements are denoted by the symbol:

U, V

The symbol above indicates that displacement in the X and Y coordinate directions, i.e., U and V, are fixed.

How Many Elements to Use?

In all of the application examples that follow, the number of elements chosen is not representative of the number that would be needed for suitable accuracy. There are several factors that determine how many elements are required for a given idealization. The degree of accuracy of the solution is one factor that determines the number of elements needed; this topic will be covered in Chapter 2. Another factor, the placement of external loads and boundary conditions, may dictate where element boundaries are drawn. Thirdly, the shape of the elements determines how many elements to use. If a mesh is generated with elements that are *poorly shaped*, the analyst may try using a greater number of elements to improve the element quality. The topic of element quality will be discussed in Chapter 7.

Steering Link Assembly in Two-Dimensional Space

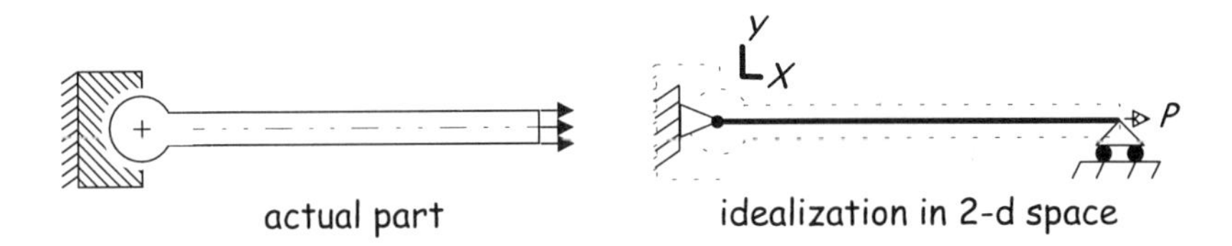

Idealization	
Stress-strain:	Uniaxial stress
Geometric:	One-dimensional geometry
Material:	Linear, elastic, isotropic, homogeneous
Loads:	Concentrated load replaces distributed
Boundary conditions:	Left end is pinned joint, right end is a slider

U, V ① ② F_{X2} V

line mesh boundary

element (2-node line)

Finite Element Model	
Mesh boundary:	One-dimensional
Element:	2-node line element for truss applications in 2-D space
Element Property:	Cross sectional area
Nodal D.O.F.:	Translations U and V at each node
Load:	$F_{X2}=P$ *(concentrated load, in X-direction, at Node 2)*
Restraints:	U, $V=0$ at Node 1, $V=0$ at Node 2
Solution procedure:	Linear, static

The line element for truss applications is used either in two-dimensional or three-dimensional space, depending upon how the problem is posed, and what type of idealization is used. The two-dimensional truss element assumes that the joints are pinned, and no bending is supported, therefore, vertical displacement must be restrained at the right end, as illustrated in the figures above.

Structural Plate with Hole in Two-Dimensional Space

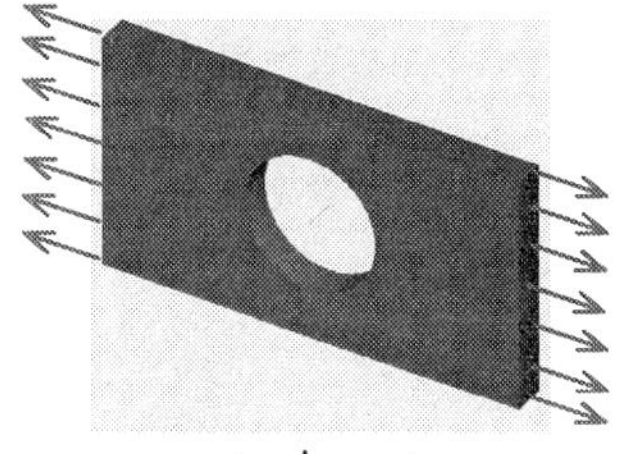

actual part

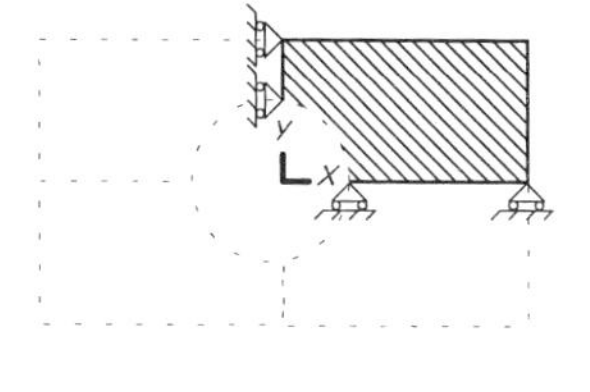

1/4 surface mesh boundary

Idealization	
Stress-strain:	Plane stress
Geometric:	Two-dimensional geometry, 1/4 symmetry
Material:	Linear, elastic, isotropic, homogeneous
Loads:	Distributed load on right edge
Boundary conditions:	Left edge restrained *U*, bottom edge restrained *V*

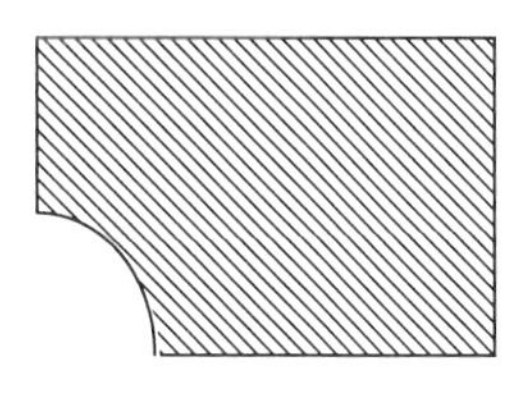

surface mesh boundary

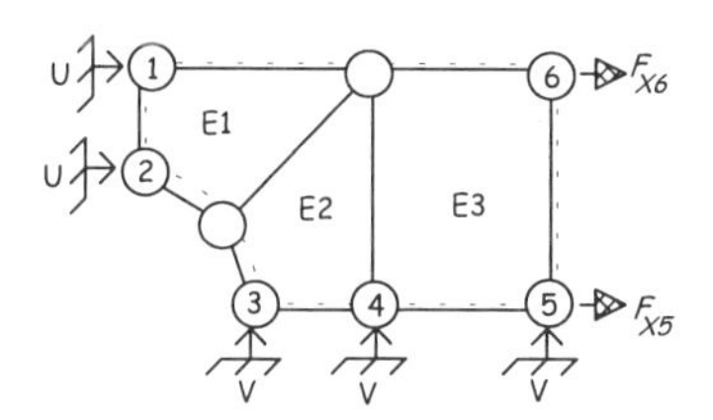

finite element mesh

Finite Element Model	
Mesh boundary:	Two-dimensional (surface)
Element:	4-node, surface element for plane stress applications
Element Property:	Thickness
Nodal D.O.F.:	Translations *U* and *V* at each node
Load:	Equivalent nodal force, *X*-direction, Nodes 5 and 6
Restraints:	*U=0,* Nodes 1 and 2, *V=0,* Nodes 3, 4, and 5
Solution procedure:	Linear, static

The boundary conditions invoked in the finite element model above are called symmetrical boundary conditions, and they simulate the effect of having the full geometry. Symmetrical boundary conditions will be considered in Chapter 7.

Ring Under Internal Pressure

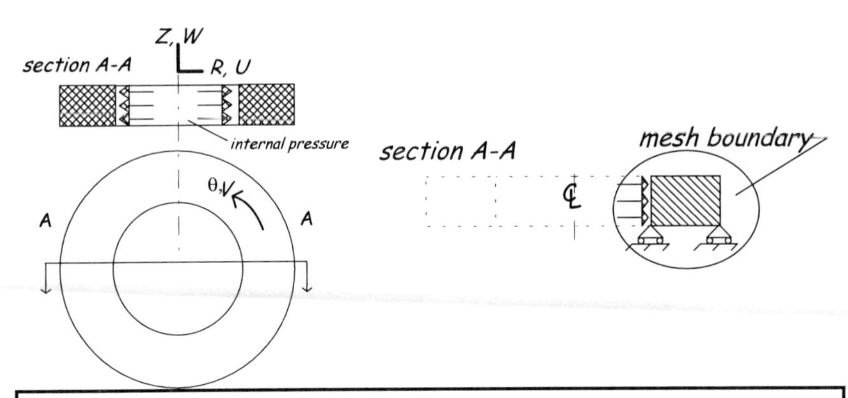

Idealization	
Stress-strain:	Axisymmetric stress
Geometric:	Two-dimensional geometry
Material:	Linear, elastic, isotropic, homogeneous
Loads:	Distributed load on inner surface
Boundary conditions:	Axial motion is restrained

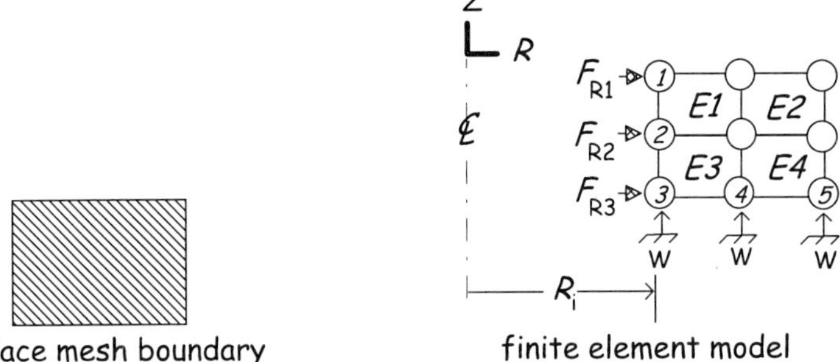

Finite Element Model	
Mesh boundary:	Two-dimensional
Element:	4-node, surface element for axisymmetric solids in 2-D
Element Property:	None
Nodal D.O.F.:	Translations *U* and *W*, in the *R* and *Z* directions, respectively
Load:	F_R: Equivalent nodal forces to characterize distributed load
Restraints:	*W*=0 at Node 3, 4, and 5
Solution procedure:	Linear, static

Structural Beam in Two-Dimensional Space

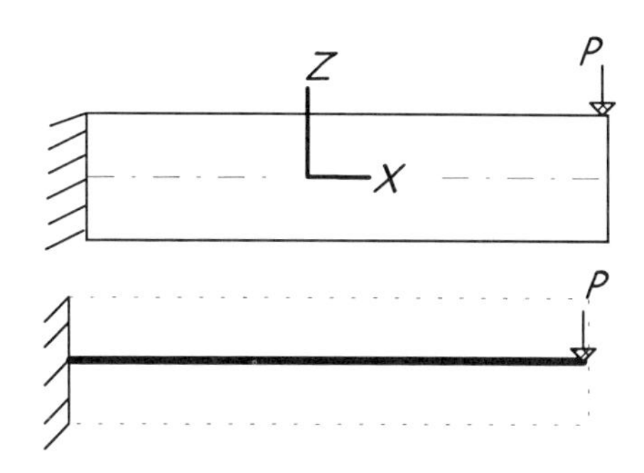

Idealization	
Stress-strain:	Euler-Bernoulli beam stress
Geometric:	One-dimensional geometry
Material:	Linear, elastic, isotropic, homogeneous
Loads:	Concentrated load at end of beam
Boundary conditions:	Left end of beam clamped

F_{Z2}

W, R_y ① ②

mesh boundary

finite element model

Finite Element Model	
Mesh boundary:	One-dimensional (line)
Element:	2-node, line element for Euler-Bernoulli beams in 2-D space
Element Property:	Cross sectional area, 2nd moment of cross sectional area
Nodal D.O.F.:	Translations W, and rotation represented by dW/dX
Load:	F_{Z2}: Nodal force
Restraints:	$W = dW/dX = 0$ at Node 1
Solution procedure:	Linear, static

Like truss elements, beams are used in either two-dimensional or three-dimensional space.

Structural Plate in Three-Dimensional Space

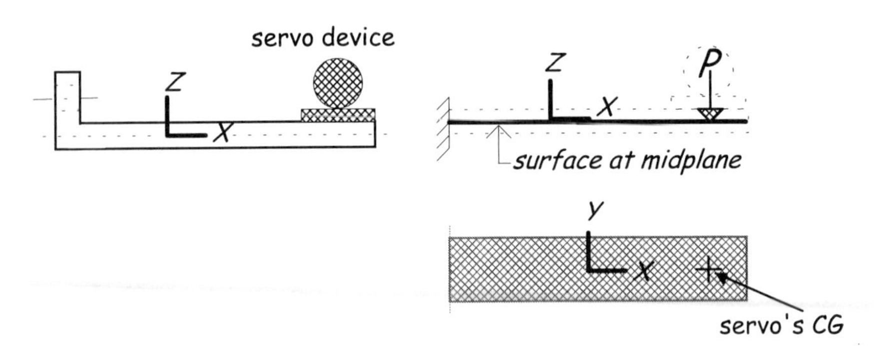

Idealization	
Stress-strain:	Kirchhoff plate bending stress
Geometric:	Two-dimensional
Material:	Linear, elastic, isotropic, homogeneous
Loads:	Servo weight approximated by a concentrated load
Boundary conditions:	Left end of plate clamped

Finite Element Model	
Mesh boundary:	Two-dimensional (surface)
Element:	4-node, surface element in three-dimensional space
Element Property:	Thickness
Nodal D.O.F.:	Three translations, *U*, *V*, and *W*, and three rotations
Load:	F_{Z3}, F_{Z4}, F_{Z5}, F_{Z6}: Nodal forces to approximate load
Restraints:	All six DOF's restrained at Nodes 1 and 2.
Solution procedure:	Linear, static

Element 3 is proportioned such that applying equal loads at Nodes 3, 4, 5, and 6 in the *Z* coordinate direction yields a resultant equal to *P*.

References

1. Juvinall, R.C., *Stress, Strain, And Strength*, McGraw-Hill Book Company, N.Y., 1967
2. Pian, T.H.H., Tong, P., "Basis of Finite Element Methods for Solid Continua", International Journal of Numerical Methods in Engineering, Vol. 1, No. 1, p. 26, 1969
3. Pian, T.H.H., "Derivation of Element Stiffness Matrices by Assumed Stress Functions," AIAA Journal, Vol. 2, No. 7, pp. 1333–1336, 1964
4. Fisher, R.C., Ziebur, A.D., *Calculus and Analytic Geometry*, Prentice-Hall, Inc., Englewood Cliffs, N.J., 1975
5. Zienkiewicz, O.C., *The Finite Element Method in Structural and Continuum Mechanics*, p. 18, McGraw-Hill, N.Y., 1970
6. Mathews, R., Walker, R.L., *Mathematical Methods Of Physics*, W.A. Benjamin, Inc., N.Y., 1965
7. Burnett, D.S., *Finite Element Analysis, Addison-Wesley, Reading*, MA, 1987
8. Higdon, A., *Mechanics Of Materials*, John Wiley & Sons, N.Y., 1985
9. Timoshenko, S.P., *Theory Of Elasticity*, McGraw-Hill, N.Y., Re-issued 1987
10. Ford, Sir Hugh, *Advanced Mechanics Of Materials*, 2nd ed., Ellis Horwood LYT, Chichester, 1977
11. Cook, R.D., *Concepts And Applications Of Finite Element Analysis*, John Wiley & Sons, N.Y., 1989
12. Bathe, K., *Finite Element Procedures In Engineering Analysis*, Prentice-Hall, Englewood Cliffs, N.J., 1982
13. Malvern, L.E., *Introduction To The Mechanics Of A Continuous Medium*, Prentice-Hall, Englewood Cliffs, N.J., 1969
14. Sherman, K.C., Bankert, R.J., Nimmer, R.P., "Engineering Performance Parameter Studies for Thermoplastic, Structural Panels," Proceedings of ANTEC, pp. 640–644, 1989
15. Timoshenko, S.P., *History Of Strength Of Materials*, Dover Publications, N.Y., 1983
16. Collins, J.A., *Failure of Materials in Mechanical Design*, John Wiley & Sons, New York, 1981
17. Ugural, A.C., Fenster, S.K., *Advanced Strength and Applied Elasticity, Elsevier*, N.Y., 1987
18. Keene, R., "Successful Snap-through," BenchMark, NAFEMS, Glasgow, U.K., p. 26–28, September, 1992
19. Crandall, S.H., *An Introduction to the Mechanics of Solids*, second edition, McGraw-Hill Book Company, 1972
20. Ling, Y., "Uniaxial True Stress-Strain after Necking," AMP Journal of Technology, Vol 5., p. 37–48, AMP, Inc.

21. Gross, T.S., "Micromechanisms of Montonic and Cyclic Crack Growth," in ASM Handbook, Volume 19, pp. 42–48, 1996
22. Timoshenko, S.P., *Theory Of Plates And Shells*, McGraw-Hill, Inc., N.Y., reissued 1987
23. Bull, J.W., *Finite Element Applications To Thin-Walled Structures*, Elsevier, London, 1990
24. Meyer, C., ed., *Finite Element Idealization*, American Society of Civil Engineers, N.Y., 1987
25. Dewhirst, D.L., "Exploiting Symmetry in Finite Element Models," Project No. AJ445, Ford Motor Company, Dearborn, 1996
26. Anonymous, "Practical Finite Element Modeling And Techniques Using MSC/NASTRAN," NA/1/000/PMSN, The MacNeal-Schwendler Corporation, LosAngeles, 1990
27. Porter, S., "Tightening The Link Between Design and Analysis," in Computer Graphics World, pp. S1–S14, August, 1994
28. Roadamaker, M.C., "Integrating Modal Testing and Finite Element Methods," Sound and Vibration, pp. 4–5, January, 1989
29. Robinson, J., "FEA training is a Must for engineers," in Machine Design, p. 146, 7 March 1994
30. Lepi, S., Reading, M., "A Team Approach to Engineering Analysis: Technical and Efficiency Issues," Autofact '94 Conference Proceedings, Section 13, SME, Dearborn, 1994

Chapter 2

Finite Element Concepts In One-Dimensional Space

"...a nation where they will not be judged by the color of their skin but by the content of their character."

—M.L. King, Jr.

2.1 Review

Key Concept: Commercial finite element software is typically designed to solve engineering analysis problems which are posed in two- or three-dimensional space. However, the mathematics required for analysis in multi-dimensional space can be quite extensive, therefore, Chapter 2 uses *one-dimensional examples* to introduce finite element concepts. Before beginning the discussion of finite element concepts, a brief review of Chapter 1 highlights is given.

In Search of Displacement

In solid mechanics problems, establishing displacement as a function of a loaded structure's spatial coordinates is valuable, since strain and stress can be calculated using derivatives of displacement and Hooke's law. In some simple cases, a closed-form solution for displacement may be generated using a single differential equation which is based upon static equilibrium of forces. However, as mentioned in Chapter 1, closed-form solutions are not always possible due to:

- Discontinuous or complex structural geometry
- The use of several materials within the same structure
- Complex loading and boundary conditions

Multi-Dimensional Stress Requires a System of Differential Equations

One additional reason why closed-form solutions are not always possible: when the stress state is two- or three-dimensional, a system of differential equations needs to be solved to obtain functions that describe equilibrium displacement.

Observe the differential equations of equilibrium for a three-dimensional state of stress, given by (2.1.1).

$$\frac{\partial \tau_{XX}}{\partial X}+\frac{\partial \tau_{YX}}{\partial X}+\frac{\partial \tau_{ZX}}{\partial X}=F_{\beta X}$$
$$\frac{\partial \tau_{XY}}{\partial Y}+\frac{\partial \tau_{YY}}{\partial Y}+\frac{\partial \tau_{ZY}}{\partial Y}=F_{\beta Y} \qquad (2.1.1)$$
$$\frac{\partial \tau_{XZ}}{\partial Z}+\frac{\partial \tau_{YZ}}{\partial Z}+\frac{\partial \tau_{ZZ}}{\partial Z}=F_{\beta Z}$$

The force on the right hand side of each equation in (2.1.1) represents a body force, with units of force per unit volume. Body forces are generated by gravitational or centrifugal loading, for example.

Using the elasticity theory approach to stress analysis, each of the three normal stress variables in (2.1.1) is a function of three normal strain components, as shown by the generalized Hooke's law equations, (1.5.15):

$$\tau_{XX}=\frac{E}{(1+v)(1-2v)}\left[(1-v)\varepsilon_{XX}+v\left(\varepsilon_{YY}+\varepsilon_{ZZ}\right)\right]$$
$$\tau_{YY}=\frac{E}{(1+v)(1-2v)}\left[(1-v)\varepsilon_{YY}+v\left(\varepsilon_{XX}+\varepsilon_{ZZ}\right)\right]$$
$$\tau_{ZZ}=\frac{E}{(1+v)(1-2v)}\left[(1-v)\varepsilon_{ZZ}+v\left(\varepsilon_{XX}+\varepsilon_{YY}\right)\right]$$

Substituting the equations above into (2.1.1), and recalling that strain can be expressed in terms of first order derivatives, it is apparent that a *system of second order differential equations* is required to establish equilibrium displacements using the elasticity approach. Hence, when solving analysis problems using (2.1.1) and Hooke's law, something other than a closed-form solution will generally be required because it is not practical (or even possible in most cases) to generate a closed-form solution to a system of differential equations subjected to various boundary conditions. In addition to solving the differential equations, strain compatibility is required when using the theory of elasticity approach; see Ugural [1, pp. 65–66] for more on solving the differential equations of equilibrium and strain compatibility.

It will be shown that for two- and three-dimensional stress problems, the finite element method serves to replace a system of differential equations with an approximate system of linear, algebraic equations. The system of linear equations is then solved using matrix methods and a digital computer.

Energy Method, not Differential Equations

It was shown in Chapter 1 that an expression for equilibrium displacement can be established by minimizing an expression for total potential energy. An energy approach is advantageous because it can:

- Lend an intuitive feel to solid mechanics problems
- Yield a system of equations that can be solved in an efficient manner
- Provide a formal proof of convergence

Having proof of convergence assures that the approximation for the variable of interest can be an increasingly better representation, given that certain rules are followed. Other methods of approximating variables of interest can be used, such as collocation, least squares, finite differences, etc., but the finite element energy approach has become a very popular means to solve solid mechanics problems for several reasons. For instance, the finite element method generates a *continuous expression* for the field variable, as opposed to a point-by-point approximation generated using other methods, such as the finite difference method. Also, the system of equilibrium equations that is generated by the finite element method has properties that allow it to be solved in an efficient manner. Techniques for solving the finite element equilibrium equations will be discussed in more depth in Section 2.7. Various approximation techniques related to engineering analysis are cited in Bathe [2, p. 102].

Total Potential Energy Functional

In a discrete structural system (i.e., a mass, spring and external force) total potential energy is expressed in terms of a function. Using differential calculus, the total potential energy function can be minimized and in doing so, a value for equilibrium displacement is established. However, in a continuum (e.g., a continuous structure), total potential energy is expressed in terms of a functional, and equilibrium displacement cannot be expressed using only one value; a *continuous function* is required to describe equilibrium displacement in a continuous system.

While the familiar *differential calculus* can be used to minimize a *function* representing total potential energy, *variational calculus* is required to obtain a closed-form solution for displacement by minimizing a *functional.*

A closed-form solution for equilibrium displacement in a continuous system can *sometimes* be found by minimizing a functional using techniques of variational calculus. However, the Ritz method produces a solution for equilibrium displacement using an approximate technique, and it can be applied to a wide range of problems, even when closed-form techniques fail.

Other Types of Functionals
In solid mechanics problems, a functional representing total potential energy can often be established and, by virtue of the minimum energy principle, a function for equilibrium displacement is generated by minimization. Functionals may also be established irrespective of physical principles such as total potential energy. For instance, a functional for a uniaxial stress idealization can be established by manipulating the governing differential equation of equilibrium given by (1.2.1), and this manipulation results in exactly the same functional as developed specifically to represent total potential energy. Indeed, for many structural analysis problems, the same functional is established by either manipulating differential equations or by establishing an expression for total potential energy.

Functionals for other types of engineering analysis (e.g., heat transfer analysis) may also be established by manipulating a differential equation. However, in heat transfer problems, the functional does not represent a physical quantity, such as total potential energy.

Regardless of how a functional is established, or what it represents, the functional is minimized by either variational calculus or an approximate means. Through the process of minimization, an expression for the variable of interest is established. Functionals are considered in more depth in Appendix B.

In some cases it is not possible to cast a problem into functional form. In such circumstances, a *weighted residual method* may often be utilized to obtain an approximate solution. While weighted residual methods are actually more versatile than using a quadratic functional with the Ritz approach, the latter was used in this text because it was thought to be less abstract.

Difficulties with the Classical Approach to the Ritz Method
The Ritz method generates a continuous expression for displacement that is *valid for the entire structure*. Consider the Ritz method as applied to the problem of the modified steering link from Section 1.2, shown again in Figure 2.1.

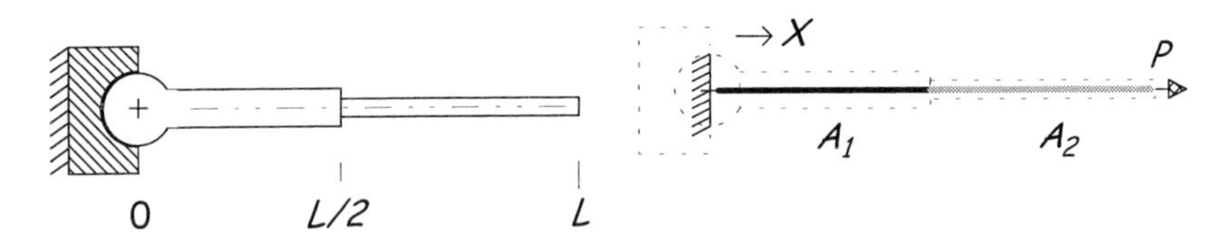

Figure 2.1 Modified Link

Recall the total potential energy functional for a uniaxially loaded rod, (1.3.2):

$$\Pi = \frac{1}{2}\int_0^L EA\left(\frac{dU}{dX}\right)^2 dX - P\,U(L)$$

Attempting to apply the Ritz process to the entire link results in the same problem as using a differential equation on the entire link: the cross sectional area term is not continuous. In light of this, one might suggest that the domain be partitioned, and the total potential energy functional divided, such that integration is performed within the domain of each partition:

$$\Pi = {}^1\Pi + {}^2\Pi = \frac{1}{2}\int_0^{L/2} EA\left(\frac{dU}{dX}\right)^2 dX + \frac{1}{2}\int_{L/2}^{L} EA\left(\frac{dU}{dX}\right)^2 dX + \ldots \qquad (2.1.2)$$

The above is somewhat analogous to using a differential equation on each partition of the structure, as discussed in Section 1.2. It will be shown in the following sections that the finite element method is characterized by partitioning and minimization concepts, using a functional as the mathematical basis for generating equilibrium equations.

Although it was stated that the Ritz method generally considers a structure as a single, continuous domain, in an early paper Courant [3] illustrates the application of the method to a *symmetric portion* of a structure. Then, based upon symmetry arguments, the results are considered valid for the entire structure. This of course differs from the concept of partitioning, since partitions can be defined irrespective of symmetry. It will be shown that the finite element method can be considered a modern application of the Ritz method, using the partitioning concept.

2.2 Finite Element Basics

Key Concept: The finite element method partitions a structure into simply shaped portions (*finite elements*), establishes a functional, and then performs a minimization process. In essence, the finite element method is the same as the Ritz method, but it employs partitioning and specific types of displacement assumptions.

The Finite Element Method Uses Partitioning

The reader is again reminded that the discussion in this text is based upon the presumption that the finite element techniques discussed within are applied to solid mechanics problems, with displacement as the field variable of interest, and the functional representing total potential energy.

The Ritz method establishes a functional for the entire structure under consideration. A displacement assumption is chosen, introduced into the functional, and the functional minimized, generating an approximate expression for equilibrium displacement; the expression is valid for the *entire structure*.

This process is essentially the same as the finite element method. However, using the finite element method, the structure is first partitioned into simply shaped elements which are defined by nodes on the element's boundary. In the simplest case, one solution, in the form of an approximate equilibrium displacement function, is established *for each element.* Displacement for the entire structure is described by the assemblage all of the individual solutions.

Finite Element Interpolation Functions

The finite element method employs a functional and assumes a form for the unknown function of displacement. The displacement assumption is initially expressed in terms of a polynomial with adjustable parameters (a_i's); subsequently, the displacement assumption is typically transformed into an expression in terms of nodal variables, i.e., an *interpolation function.*

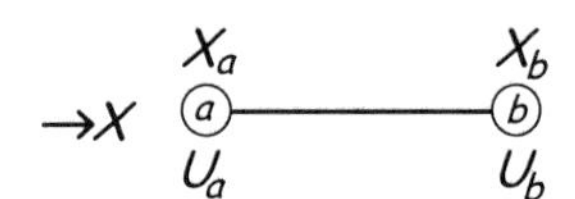

Figure 2.2 A 2-Node Finite Element

A 2-node line element in one-dimensional space is depicted in Figure 2.2, with its associated displacement interpolation function given by (2.2.1), below.

$$\tilde{U}(X)=\left(\frac{X_b-X}{X_b-X_a}\right)U_a+\left(\frac{X-X_a}{X_b-X_a}\right)U_b \qquad \left(X_a \le X \le X_b\right) \qquad (2.2.1)$$

This element can be used for problems that consider members under uniaxial load with displacement in one coordinate direction, such as the steering link problem from Chapter 1. In one-dimensional space, the 2-node line element may be called a bar or *rod element.*

The X variables with subscripts in (2.2.1) are discrete ("point") variables, representing the spatial coordinates at the nodes, while the U variables with subscripts are also discrete variables, representing nodal displacements. The free, continuous variable X (without subscript) can assume any value in the range $X_a \leq X \leq X_b$.

Finite elements typically interpolate displacement within the element's domain from the nodal values. For example, when the X variable in the interpolation function given by (2.2.1) takes on the coordinate value of a particular node, the displacement function is simply equal to the nodal value:

$$\textit{if } X = X_a, \; \tilde{U}(X) = U_a; \quad \textit{if } X = X_b, \; \tilde{U}(X) = U_b$$

At a location other than a node, displacement is interpolated from the nodal values. For example, consider displacement interpolation for the 2-node element in Figure 2.3.

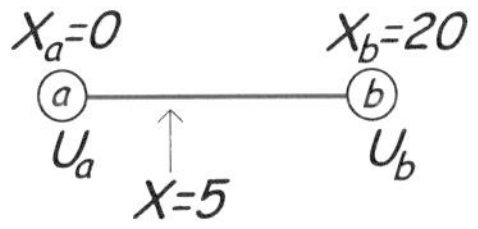

$X_a=0$ $X_b=20$

U_a $X=5$ U_b

$$\tilde{U}(X) = \left(\frac{20 - X}{20}\right) U_a + \left(\frac{X}{20}\right) U_b$$

Figure 2.3 Interpolating Displacement

The element is 20 units long, and a value for displacement at the point $X=5$ is required. The values of the nodal coordinates are substituted into (2.2.1), yielding the interpolation function shown in Figure 2.3. Now, substituting $X=5$ into that interpolation function:

$$\tilde{U}(5) = \left(\frac{3}{4}\right) U_a + \left(\frac{1}{4}\right) U_b$$

The equation above suggests that the displacement at $X=5$ is a weighted average of the nodal values U_a and U_b. Since the position $X=5$ is nearest Node a, the displacement function is weighted more heavily by the displacement at that node. Although linear interpolation was illustrated using the 2-node line element for rod applications above, interpolation functions, both linear and higher order, may also be developed for elements of various shapes (see Chapter 4).

Finite Element Interpolation Function Requirements
Interpolation functions for finite elements are typically designed to meet certain requirements so that when elements are joined together, the mesh exhibits desirable convergence characteristics. An in-depth discussion of the requirements for finite element interpolation functions will be presented in Chapter 4. For now, the objective is to show how an interpolation function for a very simple element is developed.

There are several ways to generate finite element interpolation functions. The interpolation function in (2.2.1) was established using a *mathematical approach.* Using the mathematical approach, a displacement assumption (typically a polynomial) is specified such that it:

1. Has sufficiently differentiable terms
2. Is complete
3. Has linearly independent terms
4. Satisfies essential boundary conditions *of the element*

The requirements above are suitable for one-dimensional problems, and are essentially the same requirements as for the Ritz method, with the exception of Requirement 4. Using the Ritz method, no "elements" are defined, therefore, the displacement assumption is constrained to satisfy the essential boundary conditions of the *structure*.

Essential and Natural Boundary Conditions
Two types of boundary conditions are associated with the finite element formulation of a given analysis problem: *essential* and *natural* boundary conditions. Recall from Section 1.4 that essential boundary conditions are determined by examining the value of p, the highest order derivative found in the total potential energy functional. Also, recall that in solid mechanics problems, essential boundary conditions deal with kinematic constraints, such as the allowable translation and rotation of a structure. Natural boundary conditions deal with external forces and moments which act upon the structure.

Constraining Finite Element Interpolation Functions
It will be shown that the finite element functional for a rod idealization requires the satisfaction of only one type of essential boundary condition, and that boundary condition pertains to displacement only; no derivative constraints are required. For the 2-node rod element, enforcing the two mathematical constraints given by (2.2.2) will satisfy interpolation function Requirement 4, above:

$$\tilde{U}(X_a) = U_a \qquad \tilde{U}(X_b) = U_b \tag{2.2.2}$$

By invoking the displacement constraints specified by (2.2.2), the displacement assumption is transformed into a finite element interpolation function that ensures displacement continuity between adjacent elements. For two- and three-dimensional elements, the essential boundary conditions must be continuous not only at the nodes but also *across the boundaries of adjacent elements* to satisfy Requirement 4. Example 2.1 will illustrate how a simple, one-dimensional displacement assumption is transformed into a finite element interpolation function.

Work Potential Terms in the Finite Element Functional

The finite element method employs a functional that contains at least one unknown function of displacement and has at least two terms: An *integral term* representing strain energy, and a *work potential term*, consisting of forces multiplied by displacements. The work potential term for the finite element functional uses forces and *displacements at each node*; consider below a functional for the rod element of Figure 2.2:

$$\tilde{\Pi} = \frac{1}{2}\int_{X_a}^{X_b} EA\left(\frac{d\tilde{U}}{dX}\right)^2 dX - F_a U_a - F_b U_b \tag{2.2.3}$$

The work potential terms indicate that external forces can be applied at any node. This suggests that the finite element method has more flexibility then the Ritz method when imposing natural boundary conditions. The tilde overscript indicates that the expression in (2.2.3) is approximate, in general.

Applying the Finite Element Method Manually

The finite element method, as typically used, is programmed for use with a digital computer. However, to illustrate basic principles, the following examples employ a manual approach. This can only be done (in a practical fashion) for the least complicated finite element problems.

Example 2.1—A Finite Element Analysis of Steering Link with One Element

The objective of this example is to establish, using just one element, a function of displacement for the steering link problem from Chapter 1. Figure 2.4 illustrates the steering link and the associated rod idealization.

The steering link, idealized as a uniform rod, is composed of steel having an elastic modulus of $30(10)^6$ psi. The cross sectional area is 5 in^2 and an external load P, equal to 30 lb, is applied to the right end. The total potential energy of the rod element, expressed by (2.2.3), will be minimized to establish a function that describes equilibrium displacement in the rod.

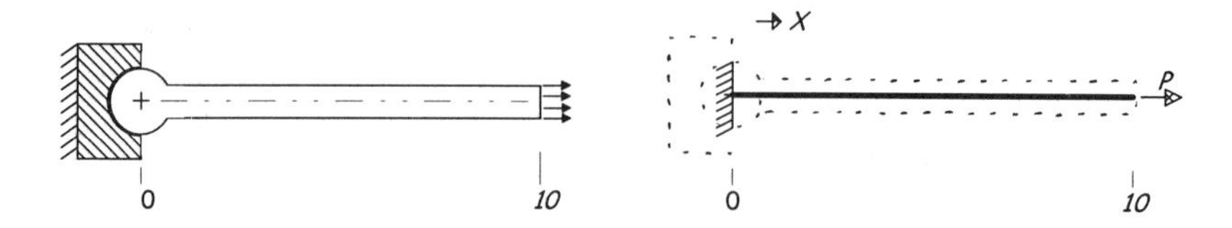

Figure 2.4 Steering Link and Idealization

A Procedure for "Manual" Finite Element Analysis

The following steps will be taken to obtain a finite element solution to the problem posed by Example 2.1:

1. Idealize the problem and define elements
2. Choose a displacement assumption
3. Transform the displacement assumption into an interpolation function
4. Substitute the interpolation function into the functional
5. Integrate, transforming the functional into a function
6. Minimize the function from Step 5, yielding the equilibrium equations
7. Solve the resulting equilibrium equations

Examining the functional for this problem, (2.2.3), note that the unknown function is $U(X)$, therefore, an element will be designed to approximate $U(X)$ within the element by *interpolating* displacement in the X-coordinate direction. Other engineering analysis problems, posed in two or three-dimensional space, may require interpolation functions in more than one coordinate direction. This will be discussed in more detail in Chapter 4. For now, consider in the example below how the most simple of finite elements (a rod element in one-dimensional space) can be developed to solve a problem that employs a uniaxial stress idealization.

Step 1—Idealize the Problem and Define Elements

The steering link model, using one rod finite element, is shown in Figure 2.5.

Referring to Figures 2.4 and 2.5, the essential and force boundary conditions are seen to be:

$$U_a=0 \qquad F_b=P=30\text{lb}_f$$

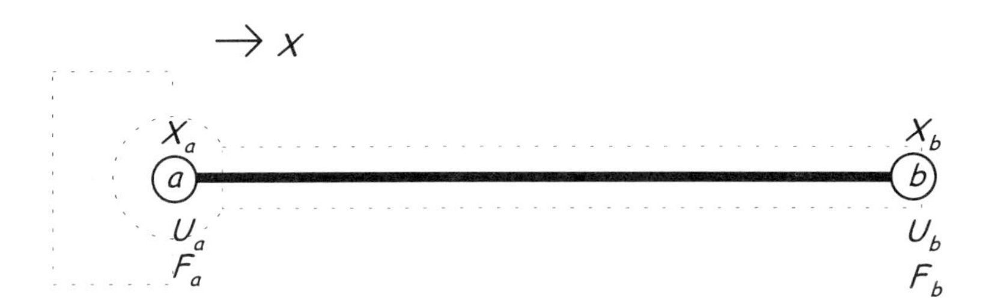

Figure 2.5 Finite Element Model of Link, Using One Element

Step 2—Choose a Displacement Assumption
Polynomial displacement assumptions, well behaved and easy to integrate, are typically chosen for displacement assumptions. The general form for a polynomial displacement assumption in one dimensional finite element problems is expressed as:

$$\tilde{U}(X) = a_1 + a_2 X + a_3 X^2 + \ldots + a_n X^{n-1}$$

The a_i's in a displacement assumption are sometimes referred to as the *generalized coordinates* or, alternately, as the *basis function coefficients*.

Referring to the four requirements for the finite element interpolation functions previously mentioned, note that the displacement assumption must have sufficiently differentiable terms. Using a polynomial displacement assumption, at least a linear polynomial is required in this problem. Note that if only a constant term were used, the integral term of (2.2.3) would become zero when the derivative is computed, yielding a trivial problem. Requirement 2 states that the displacement assumption must be complete, meaning that if a polynomial of order n is chosen, all of the terms less than n must also be included. Requirement 3 specifies that the terms must be linearly independent, such that adding a term to the displacement assumption that is simply a scalar multiple of a previous term is not acceptable. In general, it is not acceptable to add a term that can be expressed as a linear combination of the existing terms.

Can a quadratic (or higher) order displacement assumption be used with the 2-node rod element shown above? Not if the essential boundary conditions of the element are to be enforced. For example, if the displacement assumption is to meet the essential boundary conditions of the element, then the following conditions must hold:

$$\tilde{U}(X_a) = U_a \quad \textit{and} \quad \tilde{U}(X_b) = U_b,$$

With the two essential boundary conditions above, two arbitrary constants can be eliminated. As such, the displacement assumption is limited to a linear polynomial:

$$\tilde{U}(X) = a_1 + a_2 X \tag{2.2.4}$$

If a quadratic displacement assumption is used, three arbitrary constants are present in the displacement assumption:

$$\tilde{U}(X) = a_1 + a_2 X + a_3 X^2$$

For the 2-node element there now exist three unknown parameters but only two equations:

$$\tilde{U}(X_a) = a_1 + a_2 X_a + a_3 X_a{}^2 = U_a$$
$$\tilde{U}(X_b) = a_1 + a_2 X_b + a_3 X_b{}^2 = U_b$$

Since one of the three a_i's cannot be eliminated, it would have to be expressed in terms of something other than the original boundary conditions of the element. A quadratic displacement assumption *could* be used, if another node were added to the element. The additional node, having an additional nodal value of displacement, would provide an additional essential element boundary condition to be specified, providing a means of eliminating the third arbitrary constant.[1]

Based upon this information, a linear displacement assumption will be used in this example:

$$\tilde{U}(X) = a_1 + a_2 X \tag{2.2.5}$$

Step 3—Transform the Displacement Assumption into an Interpolation Function
The displacement assumption is converted into a finite element interpolation function. If done by a purely mathematical approach, this is accomplished by imposing the element boundary conditions as constraints upon the displacement

[1] Some elements, called incompatible elements, use more terms in the displacement assumption than the essential boundary conditions of the element call for. In two- and three-dimensional problems, this (typically) prevents continuity of the essential boundary conditions at the interface of two adjacent elements. Incompatible elements will be mentioned again in Chapter 4.

assumption, (2.2.5):

$$\tilde{U}(X_a) = a_1 + a_2 X_a = U_a$$
$$\tilde{U}(X_b) = a_1 + a_2 X_b = U_b \qquad (2.2.6)$$

There are other means of developing finite element interpolation functions, for instance, simply by "inspection," or by applying a known type of interpolation function. More details on this topic in Chapter 4.

Solving for the a_i's in (2.2.6), substituting back into (2.2.5), and rearranging:

$$\tilde{U}(X) = \left(\frac{X_b - X}{X_b - X_a}\right) U_a + \left(\frac{X - X_a}{X_b - X_a}\right) U_b \qquad (2.2.7)$$

One can test the validity of the interpolation function in (2.2.7) by substituting the coordinate value X_a for the X variable, resulting in U_a. Similarly, when X is assigned the value X_b, the result is U_b. Equation 2.2.7 can be simplified by noting that the length of the element, H, is equal to $X_b - X_a$:

$$H \equiv X_b - X_a \qquad (2.2.8)$$

Using (2.2.8) in (2.2.7):

$$\tilde{U}(X) = \left(\frac{X_b - X}{H}\right) U_a + \left(\frac{X - X_a}{H}\right) U_b \qquad (2.2.9)$$

Step 4—Substitute the Interpolation Function into the Functional

Recalling (2.2.3):

$$\tilde{\Pi} = \frac{1}{2}\int_{X_a}^{X_b} EA\left(\frac{d\tilde{U}}{dX}\right)^2 dX - F_a U_a - F_b U_b$$

Substituting (2.2.9) into (2.2.3):

$$\tilde{\Pi} = \frac{1}{2}\int_{X_a}^{X_b} EA\left(\frac{d}{dX}\left[\left(\frac{X_b - X}{H}\right) U_a + \left(\frac{X - X_a}{H}\right) U_b\right]\right)^2 dX - F_a U_a - F_b U_b \qquad (2.2.10)$$

Differentiating the interpolation function with respect to X and rearranging:

$$\tilde{\Pi} = \frac{1}{2}\int_{X_a}^{X_b} EA\left(\frac{U_b - U_a}{H}\right)^2 dX - F_a U_a - F_b U_b \tag{2.2.11}$$

Step 5—Integrate the Functional
Integrating (2.2.11) and rearranging:

$$\tilde{\Pi} = \frac{1}{2}\left(\frac{EA}{H^2}\right)(U_b - U_a)^2 (X_b - X_a) - F_a U_a - F_b U_b \tag{2.2.12}$$

Upon integrating (2.2.11), a simple *function* of the nodal variables is established, (2.2.12). Using the definition of H from (2.2.8), Equation 2.2.12 can be simplified:

$$\tilde{\Pi} = \frac{1}{2}\left(\frac{EA}{H}\right)(U_b - U_a)^2 - F_a U_a - F_b U_b \tag{2.2.13}$$

Step 6—Minimize TPE, Yielding Finite Element Equilibrium Equations
The function (2.2.13) can now be minimized with respect to each nodal variable, using differential calculus, and the result equated to zero:

$$\frac{\partial \tilde{\Pi}}{\partial U_a} = (2)\frac{1}{2}\left(\frac{EA}{H}\right)(U_b - U_a)(-1) - F_a = 0 \tag{2.2.14}$$

$$\frac{\partial \tilde{\Pi}}{\partial U_b} = (2)\frac{1}{2}\left(\frac{EA}{H}\right)(U_b - U_a)(1) - F_b = 0 \tag{2.2.15}$$

Equations 2.2.14 and 2.2.15 are the *finite element equilibrium equations* (for this problem) and they can be simplified to yield:

$$\left(\frac{EA}{H}\right)(U_a - U_b) = F_a \tag{2.2.16}$$

$$\left(\frac{EA}{H}\right)(-U_a + U_b) = F_b \tag{2.2.17}$$

The equilibrium equations are generally expressed in matrix form:

$$\frac{EA}{H}\begin{bmatrix} 1 & -1 \\ -1 & 1 \end{bmatrix}\begin{Bmatrix} U_a \\ U_b \end{Bmatrix} = \begin{Bmatrix} F_a \\ F_b \end{Bmatrix} \tag{2.2.18}$$

The matrix form of the equilibrium equations will not be utilized at the present time.

Step 7—Solve the Resulting Equilibrium Equations
Recalling the variables mentioned in Step 1, (2.2.17) can be evaluated:

$$\left(\frac{EA}{H}\right)\left(\underset{\substack{\Uparrow \\ 0}}{(-U_a)} + U_b\right) = \underset{\substack{\Uparrow \\ P}}{F_b} \qquad \therefore U_b = \frac{PH}{EA} \tag{2.2.19}$$

With all of the nodal displacement values known, the solution phase of the analysis is complete, and displacement can be uniquely defined for this problem using the interpolation function. Recalling the interpolation function for this problem, (2.2.9):

$$\tilde{U}(X) = \left(\frac{X_b - X}{H}\right)\underset{\substack{\Uparrow \\ 0}}{(U_a)} + \left(\frac{X - \overset{\substack{0 \\ \Downarrow}}{(X_a)}}{H}\right)\underset{\substack{\Uparrow \\ \frac{PH}{EA}}}{(U_b)} = \frac{PX}{EA} \tag{2.2.20}$$

As was the Ritz solution, the finite element solution for this problem is identical to the closed-form solution, and only one element was required to achieve this result. The finite element method can yield the same result as the closed-form solution to the differential equation, using only one element, when the displacement assumption is of the same form as the solution to the differential equation. Substituting the values for P, E, and A into (2.2.20):

$$\tilde{U}(X) = \frac{PX}{EA} = \frac{30X}{30(10)^6(5)} = 2.0(10)^{-7}(X) \tag{2.2.21}$$

Equation 2.2.21 is plotted in Figure 2.6. This completes the finite element analysis of the steering link using one element.

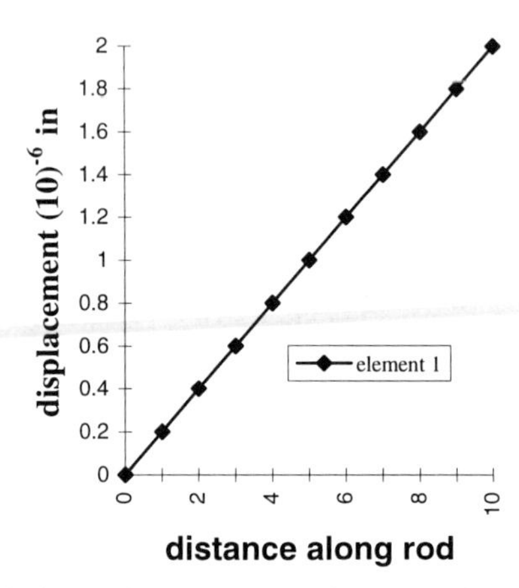

Figure 2.6 Displacement in Rod, Single Element Model

The Finite Element Stiffness Matrix (*K*)

The finite element equilibrium equations are expressed in matrix form by (2.2.18). The equations are analogous to the scalar equation for a linear spring, $KD=F$, where K is the spring constant, D is displacement, and F is the force applied to the spring. These terms are illustrated in Figure 2.7, where X_0 is the initial, relaxed position of the spring and X_f is the final position.

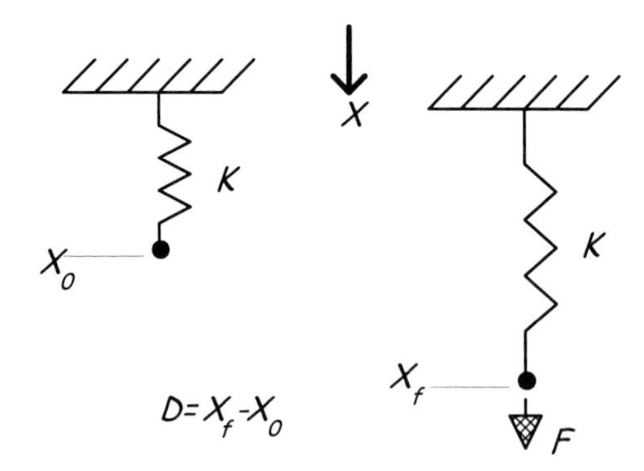

Figure 2.7 Linear Spring System

The finite element equilibrium equations can be considered in $KD=F$ form, however, the terms now represent matrices, not merely scalar values:

$$\underbrace{\frac{EA}{H}\begin{bmatrix} 1 & -1 \\ -1 & 1 \end{bmatrix}}_{\Uparrow \; {}^{e}\underline{\underline{K}}} \underbrace{\begin{Bmatrix} U_a \\ U_b \end{Bmatrix}}_{\Uparrow \; {}^{e}\underline{\underline{D}}} = \underbrace{\begin{Bmatrix} F_a \\ F_b \end{Bmatrix}}_{\Uparrow \; {}^{e}\underline{\underline{F}}} \tag{2.2.22}$$

The preceding subscripts indicate that the variables apply to a single element, not the entire structure. The K-matrix is termed the *stiffness matrix*, and for the simple 2-node rod element, the following are defined:

$${}^{e}\underline{\underline{K}} = \frac{EA}{H}\begin{bmatrix} 1 & -1 \\ -1 & 1 \end{bmatrix} \qquad {}^{e}\underline{\underline{D}} = \begin{Bmatrix} U_a \\ U_b \end{Bmatrix} \qquad {}^{e}\underline{\underline{F}} = \begin{Bmatrix} F_a \\ F_b \end{Bmatrix}$$

When applying FEA to problems with more than one element, it is convenient to use matrix notation. If not familiar with fundamental matrix algebra, one should consult suitable reference; see Section 2.7 for a brief review.

Local and Global Node Numbering; Connectivity Table

In finite element practice, there exist two sets of node numbers: global, and local. In the example above, the local numbers were used; since only one element was present, this was sufficient. However, when using two or more elements, a distinction between local and global numbering must be made.

When a finite element mesh is generated, each node is assigned a *global* element number. Consider a two-element model of the steering link with both global and local node numbering, as depicted in Figure 2.8. Local node numbering specifies the location of a node with respect to the element, while the global numbers are referenced to the assembled model. For example, Node a of a two-node rod element is always on the left end of the element while Node b is on the right. The sequencing for global numbers is not so straight forward. Although the global numbers in Figure 2.8 are sequenced from left to right, the global numbering scheme that yields the greatest computational efficiency in more complex models is generally not so obvious. Global node numbering and associated computational efficiency will be mentioned again in Section 2.7. A *connectivity table* is used to indicate what global node number is assigned to each of the element's local nodes; the connectivity table will be used in the following example.

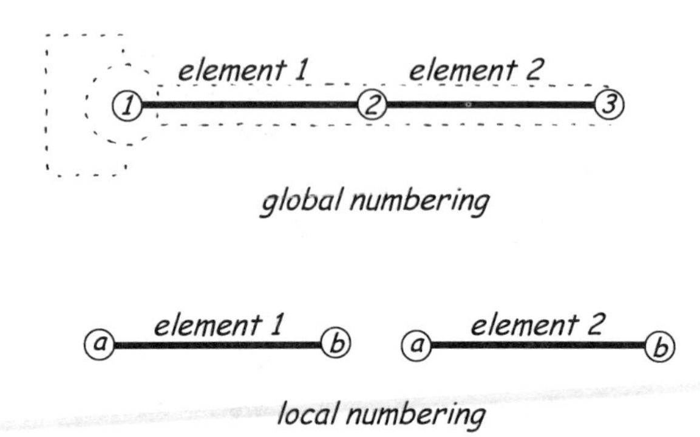

connectivity table

element	node a	node b
1	1	2
2	2	3

Figure 2.8 Connectivity Table for Link Model with Two Elements

Using More Than One Element in a Simple Finite Element Model

The following example uses two finite elements to perform an analysis of the uniaxially loaded steering link.

Example 2.2—A Finite Element Analysis of a Link, Using Two Elements

A finite element analysis using two rod elements for the steering link problem from Example 2.1 is presented, using the same seven steps that were used before. This example introduces the distinction between global and local nodes, and also points out that if one element correctly represents the exact displacement in a structure, adding additional elements offers no beneficial effect.

Step 1—Idealize and Define Elements

The steering link finite element model is shown in Figure 2.9, with two elements of equal length. The connectivity table indicates how the global nodes are connected to form each element.

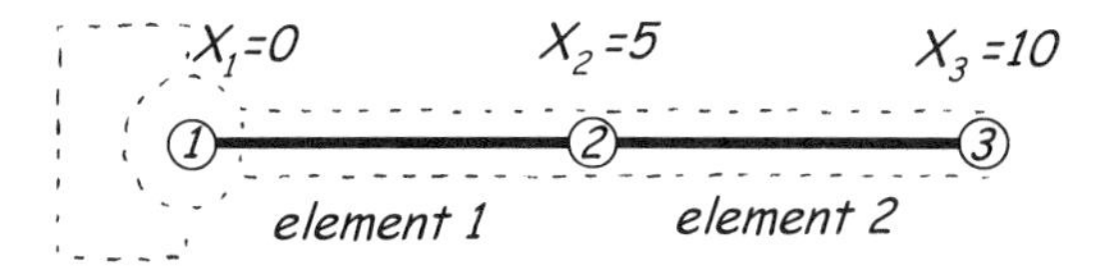

connectivity table

element	node a	node b
1	1	2
2	2	3

Figure 2.9 Model of Link, Two Elements

Step 2—Choose a Displacement Assumption
A linear displacement assumption will be used for each element, with a preceding superscript used to distinguish between Elements 1 and 2.

$$^1\tilde{U}(X) = a_1 + a_2 X$$
$$^2\tilde{U}(X) = b_1 + b_2 X$$

Step 3—Transform Displacement Assumption into Interpolation Function
Repeating the same process as in Example 2.1, two finite element interpolation functions are generated by constraining the displacement assumption to conform to the essential boundary conditions of the element. Notice that we must be more careful describing variables when two elements are used, hence the added superscripts in the interpolation functions below:

$$^1\tilde{U}(X) = \left(\frac{^1X_b - X}{^1H}\right) {^1U_a} + \left(\frac{X - {^1X_a}}{^1H}\right) {^1U_b}$$
$$^2\tilde{U}(X) = \left(\frac{^2X_b - X}{^2H}\right) {^2U_a} + \left(\frac{X - {^2X_a}}{^2H}\right) {^2U_b}$$

Using the connectivity table and variables described in Figure 2.9, the two equations above yield:

$$
\begin{aligned}
{}^{1}\tilde{U}(X) &= \left(\frac{5-X}{5}\right)U_1 + \left(\frac{X}{5}\right)U_2 \\
{}^{2}\tilde{U}(X) &= \left(\frac{10-X}{5}\right)U_2 + \left(\frac{X-5}{5}\right)U_3
\end{aligned}
\qquad (2.2.23)
$$

Step 4—Substitute the Interpolation Function into the Functional

Using two elements, the strain energy portion of the total potential energy functional is divided into two intervals and an additional work potential term is added:

$$
\tilde{\Pi} = {}^{1}\tilde{\Pi} + {}^{2}\tilde{\Pi} = \left(\frac{1}{2}\right)\int_0^5 EA\left(\frac{d\,{}^{1}\tilde{U}}{dX}\right)^2 dX + \left(\frac{1}{2}\right)\int_5^{10} EA\left(\frac{d\,{}^{2}\tilde{U}}{dX}\right)^2 dX - F_1U_1 - F_2U_2 - F_3U_3 \qquad (2.2.24)
$$

For convenience, derivatives of interpolation functions (2.2.23) are computed before substituting into (2.2.24):

$$
\frac{d\,{}^{1}\tilde{U}}{dX} = \left(\frac{U_2-U_1}{5}\right) \qquad \frac{d\,{}^{2}\tilde{U}}{dX} = \left(\frac{U_3-U_2}{5}\right) \qquad (2.2.25)
$$

Substituting (2.2.25) into the expression for total potential energy, (2.2.24):

$$
\tilde{\Pi} = \left(\frac{1}{2}\right)\int_0^5 EA\left(\frac{U_2-U_1}{5}\right)^2 dX + \left(\frac{1}{2}\right)\int_5^{10} EA\left(\frac{U_3-U_2}{5}\right)^2 dX - F_1U_1 - F_2U_2 - F_3U_3 \qquad (2.2.26)
$$

Step 5—Integrate the Functional

Equation 2.2.26 is integrated. All of the variables within the integral are independent of the X variable, and can be removed from the integral to simplify the integration process, yielding:

$$
\tilde{\Pi} = \left(\frac{1}{2}\right)EA\left(\frac{U_2-U_1}{5}\right)^2 (5-0) + \left(\frac{1}{2}\right)\left(\frac{U_3-U_2}{5}\right)^2 (10-5) - F_1U_1 - F_2U_2 - F_3U_3
$$

Step 6—Minimize, Yielding Finite Element Equilibrium Equations
The expression for total potential energy above is simplified, minimized with respect to each nodal displacement, and then equated to zero:

$$\frac{\partial \tilde{\Pi}}{\partial U_1} = \frac{EA}{5}(U_2 - U_1)(-1) - F_1 = 0$$

$$\frac{\partial \tilde{\Pi}}{\partial U_2} = \frac{EA}{5}(U_2 - U_1)(1) + \frac{EA}{5}(U_3 - U_2)(-1) - F_2 = 0$$

$$\frac{\partial \tilde{\Pi}}{\partial U_3} = \frac{EA}{5}(U_3 - U_2)(1) - F_3 = 0$$

Notice that for each nodal DOF, the minimization process renders one equilibrium equation. The equilibrium equations are simplified and arranged as:

$$\frac{EA}{5}(U_1 - U_2 + 0U_3) = F_1 \tag{2.2.27}$$

$$\frac{EA}{5}(-U_1 + 2U_2 - U_3) = F_2 \tag{2.2.28}$$

$$\frac{EA}{5}(0U_1 - U_2 + U_3) = F_3 \tag{2.2.29}$$

To aid in casting the equations above into matrix format, a zero precedes a variable that was not contained in the original equation; the zero acts as a "place-holder" to line up the corresponding entries in each matrix. Equations 2.2.27–2.2.29 can be expressed in matrix form:

$$\frac{EA}{5}\begin{bmatrix} 1 & -1 & 0 \\ -1 & 2 & -1 \\ 0 & -1 & 1 \end{bmatrix}\begin{Bmatrix} U_1 \\ U_2 \\ U_3 \end{Bmatrix} = \begin{Bmatrix} F_1 \\ F_2 \\ F_3 \end{Bmatrix} \tag{2.2.30}$$

It is further noted that no externally applied force (or reaction force) exists at Node 2, hence, $F_2=0$; also, the external force applied at Node 3 is equal to P:

$$\frac{EA}{5}\begin{bmatrix} 1 & -1 & 0 \\ -1 & 2 & -1 \\ 0 & -1 & 1 \end{bmatrix}\begin{Bmatrix} U_1 \\ U_2 \\ U_3 \end{Bmatrix} = \begin{Bmatrix} F_1 \\ 0 \\ P \end{Bmatrix} \tag{2.2.31}$$

The global stiffness for the two-element model is therefore expressed as:

$$\underline{\underline{K}} = \frac{EA}{5}\begin{bmatrix} 1 & -1 & 0 \\ -1 & 2 & -1 \\ 0 & -1 & 1 \end{bmatrix} \tag{2.2.32}$$

Step 7—Solve the Resulting Equilibrium Equations

The system of equations in (2.2.31) cannot be solved until essential boundary conditions are imposed. In practice, the imposition of essential boundary conditions is handled by efficient means, as will be discussed in Chapter 6. For now, consider that for this problem, U_1 is zero. Since the first column of the *K*-matrix is multiplied by U_1, and in this problem U_1 is zero, this has the same effect as if the first column of the stiffness matrix were zero:

$$\frac{EA}{5}\begin{bmatrix} 1 & -1 & 0 \\ -1 & 2 & -1 \\ 0 & -1 & 1 \end{bmatrix}\begin{Bmatrix} 0 \\ U_2 \\ U_3 \end{Bmatrix} = \begin{Bmatrix} F_1 \\ 0 \\ P \end{Bmatrix}$$

or:

$$\frac{EA}{5}\begin{bmatrix} -1 & 0 \\ 2 & -1 \\ -1 & 1 \end{bmatrix}\begin{Bmatrix} U_2 \\ U_3 \end{Bmatrix} = \begin{Bmatrix} F_1 \\ 0 \\ P \end{Bmatrix} \tag{2.2.33}$$

Since the reaction force at Node 1, F_1, is unknown at this time, the top equation does not add any additional information to help solve the system of equations for U_2 and U_3. The top row is therefore removed, resulting in two equations and two unknowns[2]:

$$\frac{EA}{5}\begin{bmatrix} 2 & -1 \\ -1 & 1 \end{bmatrix}\begin{Bmatrix} U_2 \\ U_3 \end{Bmatrix} = \begin{Bmatrix} 0 \\ P \end{Bmatrix}$$

[2] With all three equations, there exist three equations and three unknowns: U_2, U_3, and F_1. Removing the first row results in two equations and two unknowns, U_2 and U_3. Why not remove Row 2, also? Doing so would leave only one equation (the bottom row) but two unknowns, U_2 and U_3.

The simultaneous solution to the system of equations above can be shown to be:

$$U_2 = \frac{5P}{EA} \qquad U_3 = \frac{10P}{EA} \tag{2.2.34}$$

Equations (2.2.34) are substituted into (2.2.23), and, recalling that $U_1=0$, the resulting equations are:

$${}^1\tilde{U}(X) = \frac{PX}{EA} \qquad {}^2\tilde{U}(X) = \frac{PX}{EA} \tag{2.2.35}$$

The known variables are substituted into (2.2.35) and the resulting equations plotted in Figure 2.10. The graph reveals that the displacement functions are the same for both elements.

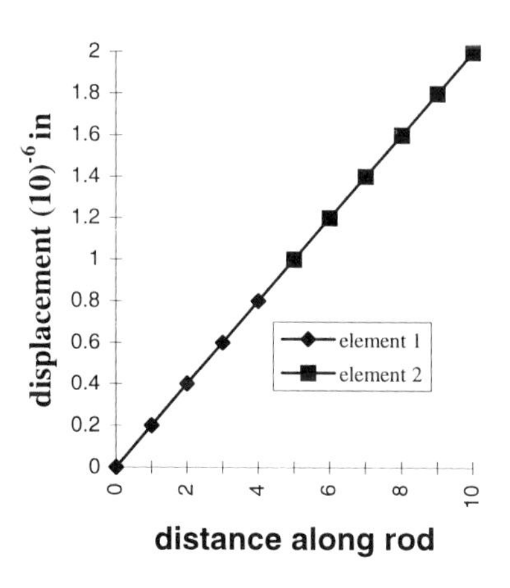

Figure 2.10 Displacement in Steering Link, Two Element Model

Comparing the results of Figure 2.10 to those in Figure 2.6, it should be obvious that one element is just as accurate as two. That is, when the exact displacement is linear, and a linear displacement assumption is employed, the closed-form solution can be obtained with a single element; using more elements does not improve the solution.

Using the definitions of strain given by (1.5.6) and (1.5.7), note that a linear function of displacement translates into constant strain, since normal strain is the first derivative of displacement. With Hooke's law for uniaxial stress, (1.5.9), constant strain translates into constant stress. Therefore, when displacement in a structure varies linearly, stress can be constant, and one linear element can produce the same result as many. It will be shown in Section 2.4 that additional linear elements are needed if the exact displacement is non-linear, as occurs in regions of a structure where stress is not constant.

Calculating Reaction Forces at Restrained Nodes

The top equation from (2.2.33) was removed, since the reaction force was unknown. This equation may now be solved, using the known value of nodal displacement. Recalling (2.2.33):

$$\frac{EA}{5}\begin{bmatrix} -1 & 0 \\ 2 & -1 \\ -1 & 1 \end{bmatrix}\begin{Bmatrix} U_2 \\ U_3 \end{Bmatrix} = \begin{Bmatrix} F_1 \\ 0 \\ P \end{Bmatrix}$$

Extracting the top row from the system above:

$$\frac{EA}{5}\left(-U_2 + 0U_3\right) = F_1$$

Or, rearranging the above and substituting in the known value of displacement:

$$F_1 = \frac{EA}{5}\left(-U_2\right) = \frac{30(10)^6(5)}{5}\left[-1.0(10)^{-6}\right] = -30lb_f$$

This same result can be determined (in this case) by simply noting that the summation of external forces acting upon the body must be zero:

$$\begin{aligned} &\sum F_X = P + F_1 = 0 \\ &30 + F_1 = 0 \\ &F_1 = -30lb_f \end{aligned}$$

2.3 The Global Stiffness Matrix

Key Concept: A stiffness matrix can be established for each element, independently. Subsequently, all of the *element stiffness matrices* can be combined to represent the stiffness of the entire structure. The matrix formed by assembling all of the element stiffness matrices is termed the *global stiffness matrix.*

As shown in Example 2.2, an expression for total potential energy can be established for an entire structure, and then minimized, yielding displacement values for every node in the model. The nodal values are then substituted back into each element's displacement interpolation function, yielding an expression for displacement that is valid within the domain of the respective element.

Establishing an expression for the total potential energy of the entire structure, and generating the associated global stiffness matrix, was not difficult in Example 2.2, since only two simple elements were used. However, in a finite element model with thousands of elements, it is impractical to develop a single expression that represents the total potential energy for an entire structural system.

Total potential energy for an entire structure can be expressed by summing energy terms from each element. For instance, in Example 2.2, two elements were used, and the total potential energy for the entire structure was given by (2.2.24):

$$\tilde{\Pi} = {}^1\tilde{\Pi} + {}^2\tilde{\Pi} = \left(\frac{1}{2}\right)\int_0^5 EA\left(\frac{d\,{}^1\tilde{U}}{dX}\right)^2 dX + \left(\frac{1}{2}\right)\int_5^{10} EA\left(\frac{d\,{}^2\tilde{U}}{dX}\right)^2 dX - F_1U_1 - F_2U_2 - F_3U_3$$

In general, energy terms from any number of elements can be summed. To establish the equilibrium equations, the expression for total potential energy must be minimized with respect to every nodal DOF, and set to zero; this can be justified by considering the sum rule of differentiation:

$$\frac{d\tilde{\Pi}}{d\underline{\underline{D}}} = \frac{d\,{}^1\tilde{\Pi}}{d\underline{\underline{D}}} + \frac{d\,{}^2\tilde{\Pi}}{d\underline{\underline{D}}} + \cdots + \frac{d\,{}^{nel}\tilde{\Pi}}{d\underline{\underline{D}}} = 0 \qquad \underline{\underline{D}} = \textit{vector of all nodal D.O.F.} \tag{2.3.1}$$

The derivative in (2.3.1) contains *nel* terms, where *nel* is the number of elements in the model. For the above to hold true in all cases, each term, representing the derivative of TPE for one particular element e, must be equated to zero:

$$\frac{d\,{}^e\tilde{\Pi}}{d\underline{\underline{D}}} = 0 \qquad or:\ d\,{}^e\tilde{\Pi} = 0 \tag{2.3.2}$$

The above suggests that an expression for the total potential energy of a single element can be established, then minimized, yielding the equilibrium equations for that particular element. Since the global displacement vector contains a total n nodal degrees of freedom, (2.3.2) actually consists of n partial derivatives:

$$d^e\tilde{\Pi} = \frac{\partial\, ^e\tilde{\Pi}}{\partial D_1} dD_1 + \frac{\partial\, ^e\tilde{\Pi}}{\partial D_2} dD_2 + \cdots + \frac{\partial\, ^e\tilde{\Pi}}{\partial D_n} dD_n = 0 \quad n = all\ DOF \tag{2.3.3}$$

The expression above states that for each element e, the partial derivative of total potential energy with respect to each term in the *global* displacement vector must be zero. If the above is to be true for all cases, each term must be zero:

$$\frac{\partial\, ^e\tilde{\Pi}}{\partial D_1} = 0 \qquad \frac{\partial\, ^e\tilde{\Pi}}{\partial D_2} = 0 \quad \cdots \quad \frac{\partial\, ^e\tilde{\Pi}}{\partial D_n} = 0 \tag{2.3.4}$$

However, the expression for total potential energy for element e only contains the nodal DOF's that belong to element e. Therefore, if there exist *ne* DOF's related to element e, only *ne* partial derivatives will yield non-trivial equations. For instance, if $^e\tilde{\Pi}$ is a function of only two nodal DOF's, say U_a and U_b, the equation formed by taking the partial of $^e\tilde{\Pi}$ with respect to a U_c is identically zero. So, for an element that has two nodal DOF's, two partial derivatives are sufficient to describe the element equilibrium equations:

$$\begin{aligned} \frac{\partial\, ^e\tilde{\Pi}}{\partial D_a} &= 0 \\ \frac{\partial\, ^e\tilde{\Pi}}{\partial D_b} &= 0 \end{aligned} \tag{2.3.5}$$

As suggested by (2.3.5), an element with *ne* nodal degrees of freedom requires *ne* equations to fully express equilibrium displacement. Notice that (2.3.5) actually represents a system of equations, of order *(ne) by (ne)*, with each partial derivative in (2.3.5) associated with one row in the system. Generating all *ne* partial derivatives in accordance with (2.3.5) renders the element equilibrium equations:

$$\underline{\underline{^eK}}\, \underline{^eD} = \underline{^eF}$$

Computing Global Stiffness Matrix

An effective way to establish a global stiffness matrix is to first generate individual stiffness matrices for each element, based upon (2.3.5), then combine all the element matrices to form the global matrix:

$$\underline{\underline{K}} = \sum_{e=1}^{e=nel} {}^{e}\underline{\underline{K}} \qquad nel \equiv \text{total number of elements}$$

Example 2.3—Developing Individual Stiffness Matrices

This example reexamines the two-element finite element analysis of the link presented in Example 2.2, and illustrates how individual element stiffness matrices can be formed and later combined to yield a global stiffness matrix.

Step 1—Idealize and Define Elements

Two elements will again be used for the steering link problem, as illustrated in Figure 2.11. Two expressions will be used for total potential energy, one for each element:

$$ {}^{1}\Pi = \frac{1}{2}\int_{{}^{1}X_a}^{{}^{1}X_b} EA\left(\frac{d\,{}^{1}\tilde{U}}{dX}\right)^2 dX - {}^{1}F_a\,{}^{1}U_a - {}^{1}F_b\,{}^{1}U_b \qquad (2.3.6)$$

$$ {}^{2}\Pi = \frac{1}{2}\int_{{}^{2}X_a}^{{}^{2}X_b} EA\left(\frac{d\,{}^{2}\tilde{U}}{dX}\right)^2 dX - {}^{2}F_a\,{}^{2}U_a - {}^{2}F_b\,{}^{2}U_b \qquad (2.3.7)$$

Notice that with two separate functionals, the variables must be precisely designated to avoid confusion. In (2.3.6) and (2.3.7), all of the variables that apply to Element 1 are preceded by a 1 superscript and those that apply to Element 2 have a 2 superscript. Since E and A are (in this problem) the same for each element, a superscript will not be used.

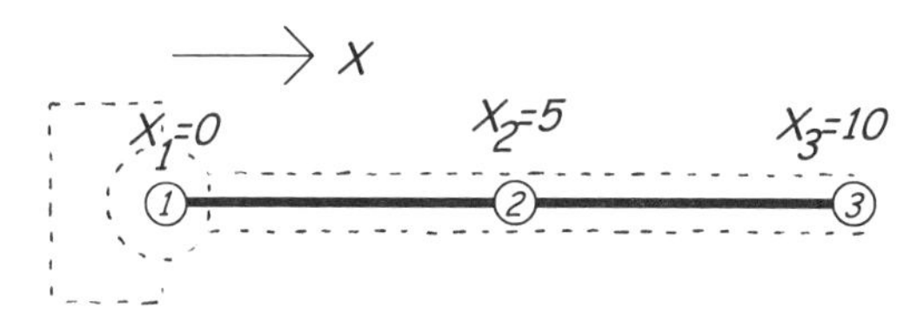

connectivity table

element	node a	node b
1	1	2
2	2	3

Figure 2.11 Link with Two Elements: Element Stiffness Matrix Example

Steps 2 and 3

As in Example 2.2.2, linear displacement assumptions are assumed for each element, and then transformed into interpolation functions:

$$ {}^1\tilde{U}(X) = \left(\frac{{}^1X_b - X}{{}^1H} \right) {}^1U_a + \left(\frac{X - {}^1X_a}{{}^1H} \right) {}^1U_b \tag{2.3.8} $$

$$ {}^2\tilde{U}(X) = \left(\frac{{}^2X_b - X}{{}^2H} \right) {}^2U_a + \left(\frac{X - {}^2X_a}{{}^2H} \right) {}^2U_b \tag{2.3.9} $$

Step 4—Substitute the Interpolation Function into Functional

Substituting the derivative of (2.3.8) into (2.3.6), and the derivative of (2.3.9) into (2.3.7):

$$ {}^1\Pi = \frac{1}{2} \int_{{}^1X_a}^{{}^1X_b} EA \left(\frac{{}^1U_b - {}^1U_a}{{}^1H} \right)^2 dX - {}^1F_a {}^1U_a - {}^1F_b {}^1U_b \tag{2.3.10} $$

$$ {}^2\Pi = \frac{1}{2} \int_{{}^2X_a}^{{}^2X_b} EA \left(\frac{{}^2U_b - {}^2U_a}{{}^2H} \right)^2 dX - {}^2F_a {}^2U_a - {}^2F_b {}^2U_b \tag{2.3.11} $$

Step 5—Integrate the Functional

Equations (2.3.10) and (2.3.11) are integrated, first removing all variables from the integrand:

$$ {}^1\Pi = \frac{1}{2} EA \left(\frac{{}^1U_b - {}^1U_a}{{}^1H} \right)^2 \left({}^1X_b - {}^1X_a \right) - {}^1F_a {}^1U_a - {}^1F_b {}^1U_b \tag{2.3.12} $$

$$ {}^2\Pi = \frac{1}{2} EA \left(\frac{{}^2U_b - {}^2U_a}{{}^2H} \right)^2 \left({}^2X_b - {}^2X_a \right) - {}^2F_a {}^2U_a - {}^2F_b {}^2U_b \tag{2.3.13} $$

Noting that the element length, H, is the difference between the two nodal coordinates, (2.3.12) and (2.3.13) can be reduced to:

$$ {}^1\Pi = \frac{1}{2} \frac{EA}{{}^1H} \left({}^1U_b - {}^1U_a \right)^2 - {}^1F_a {}^1U_a - {}^1F_b {}^1U_b \tag{2.3.14} $$

$$ {}^2\Pi = \frac{1}{2} \frac{EA}{{}^2H} \left({}^2U_b - {}^2U_a \right)^2 - {}^2F_a {}^2U_a - {}^2F_b {}^2U_b \tag{2.3.15} $$

Step 6—Minimize, Yielding Finite Element Equilibrium Equations
Each expression for total potential energy above is minimized with respect to each nodal displacement, and equated to zero, yielding the equilibrium equations for each element. Minimizing (2.3.14), then setting the result to zero:

$$\frac{\partial\, {}^1\Pi}{\partial\, {}^1U_a} = \frac{EA}{{}^1H}\left({}^1U_b - {}^1U_a\right)(-1) - {}^1F_a = 0 \tag{2.3.16}$$

$$\frac{\partial\, {}^1\Pi}{\partial\, {}^1U_b} = \frac{EA}{{}^1H}\left({}^1U_b - {}^1U_a\right)(\ 1) - {}^1F_b = 0 \tag{2.3.17}$$

Element equilibrium equations (2.3.16) and (2.3.17) can be cast into matrix form:

$$\frac{EA}{{}^1H}\begin{bmatrix} 1 & -1 \\ -1 & 1 \end{bmatrix}\begin{Bmatrix} {}^1U_a \\ {}^1U_b \end{Bmatrix} = \begin{Bmatrix} {}^1F_a \\ {}^1F_b \end{Bmatrix} \tag{2.3.18}$$

The above is equivalent to:

$$\underline{\underline{{}^1K}}\,\underline{{}^1D} = \underline{{}^1F}$$

In other words, performing the differentiation prescribed by (2.3.16) and (2.3.17) renders the element equilibrium equations above, as discussed earlier. Following the same process used for Element 1, the equilibrium equations for Element 2 are:

$$\frac{EA}{{}^2H}\begin{bmatrix} 1 & -1 \\ -1 & 1 \end{bmatrix}\begin{Bmatrix} {}^2U_a \\ {}^2U_b \end{Bmatrix} = \begin{Bmatrix} {}^2F_a \\ {}^2F_b \end{Bmatrix} \tag{2.3.19}$$

From the element equilibrium equations, the individual element stiffness matrices for Elements 1 and 2 are seen to be:

$${}^1K = \frac{EA}{{}^1H}\begin{bmatrix} 1 & -1 \\ -1 & 1 \end{bmatrix} \tag{2.3.20}$$

$${}^2K = \frac{EA}{{}^2H}\begin{bmatrix} 1 & -1 \\ -1 & 1 \end{bmatrix} \tag{2.3.21}$$

The element stiffness matrices need to be combined to form the global stiffness matrix. First, the connectivity table is consulted to obtain the equivalent global variables in (2.3.18) and (2.3.19):

$$\frac{EA}{{}^{1}H}\begin{bmatrix} 1 & -1 \\ -1 & 1 \end{bmatrix}\begin{Bmatrix} U_1 \\ U_2 \end{Bmatrix} = \begin{Bmatrix} F_1 \\ F_2 \end{Bmatrix} \tag{2.3.22}$$

$$\frac{EA}{{}^{2}H}\begin{bmatrix} 1 & -1 \\ -1 & 1 \end{bmatrix}\begin{Bmatrix} U_2 \\ U_3 \end{Bmatrix} = \begin{Bmatrix} F_2 \\ F_3 \end{Bmatrix} \tag{2.3.23}$$

To allow matrix addition of (2.3.20) and (2.3.21), corresponding components must be in the same location in each matrix. Hence, the individual element stiffness matrices above are expanded, using zeros as "place holders":

$$\frac{EA}{{}^{1}H}\begin{bmatrix} 1 & -1 & 0 \\ -1 & 1 & 0 \\ 0 & 0 & 0 \end{bmatrix}\begin{Bmatrix} U_1 \\ U_2 \\ U_3 \end{Bmatrix} = \begin{Bmatrix} F_1 \\ F_2 \\ F_3 \end{Bmatrix} \tag{2.3.24}$$

$$\frac{EA}{{}^{2}H}\begin{bmatrix} 0 & 0 & 0 \\ 0 & 1 & -1 \\ 0 & -1 & 1 \end{bmatrix}\begin{Bmatrix} U_1 \\ U_2 \\ U_3 \end{Bmatrix} = \begin{Bmatrix} F_1 \\ F_2 \\ F_3 \end{Bmatrix} \tag{2.3.25}$$

Notice that in (2.3.24), a null equation is added as the bottom row. This acts as a "placeholder" so that when combined with (2.3.25), the corresponding components of the matrices are added together. The zeros in the bottom row of (2.3.24) represent the fact that force F_3 has no effect upon the equilibrium equations for Element 1, since Node 3 does not belong to Element 1. Equation 2.3.23 is expanded in a similar fashion, noting that F_1 has no effect on Element 2; the result is a null equation in the top row of (2.3.25). Since both of the matrices above now have corresponding components, they can be added together, if the substitution ${}^{1}H = {}^{2}H$ is made:

$$\frac{EA}{{}^{1}H}\begin{bmatrix} 1 & -1 & 0 \\ -1 & 2 & -1 \\ 0 & -1 & 1 \end{bmatrix}\begin{Bmatrix} U_1 \\ U_2 \\ U_3 \end{Bmatrix} = \begin{Bmatrix} F_1 \\ F_2 \\ F_3 \end{Bmatrix} \tag{2.3.26}$$

The global stiffness matrix in (2.3.26) is generated using the relationship:

$$\underline{\underline{K}} = \sum_{e=1}^{e=nel} {}^{e}\underline{\underline{K}} \qquad nel \equiv \text{total number of elements}$$

In this particular case, the global stiffness is formed by combining (2.3.24) and (2.3.25), noting that the element lengths are equal:

$$\underline{\underline{K}} = \sum_{e=1}^{e=2} {}^{e}\underline{\underline{K}} = \frac{EA}{{}^{1}H}\begin{bmatrix} 1 & -1 & 0 \\ -1 & 1 & 0 \\ 0 & 0 & 0 \end{bmatrix} + \frac{EA}{{}^{1}H}\begin{bmatrix} 0 & 0 & 0 \\ 0 & 1 & -1 \\ 0 & -1 & 1 \end{bmatrix} = \frac{EA}{{}^{1}H}\begin{bmatrix} 1 & -1 & 0 \\ -1 & 2 & -1 \\ 0 & -1 & 1 \end{bmatrix}$$

Step 7—Solve Resulting Equilibrium Equations

The global system of equations (2.3.26) is equivalent to (2.2.30), and can be solved in the same manner that was shown in Example 2.2.

This concludes the example of generating and combining element stiffness matrices. In practice, finite element software generates individual element stiffness matrices and then forms the global stiffness matrix in a similar fashion as shown in the example above, with a few changes.

When programming finite element software, the stiffness matrix *for the rod element* is generally computed once, using closed-form integration, assuming that the cross sectional area is constant. Hence, the element equilibrium equations for a rod element, ${}^{e}\underline{\underline{K}}\,{}^{e}\underline{D} = {}^{e}\underline{F}$, are simply:

$$\frac{{}^{e}E\,{}^{e}A}{{}^{e}H}\begin{bmatrix} 1 & -1 \\ -1 & 1 \end{bmatrix}\begin{Bmatrix} {}^{e}U_a \\ {}^{e}U_b \end{Bmatrix} = \begin{Bmatrix} {}^{e}F_a \\ {}^{e}F_b \end{Bmatrix} \tag{2.3.27}$$

$e = \textit{element number}$

Preceding superscripts, indicating the element number, appear on the modulus and cross sectional area terms to account for the general case where rods may be composed of differing material properties and/or cross section.

Another fundamental difference between finite element software and the manually performed Example 2.3 is that, using FEA software, the element matrices are not expanded as shown in (2.3.24) and (2.3.25); doing so would result in wasteful use of computer memory, since many zero components would be stored. In practice, an algorithm is employed to simply add the respective components of all the element stiffness matrices together, rendering the global stiffness matrix.

2.4 *h*-Convergence

Key Concept: When the exact displacement in a structure is a linear function of the structure's spatial coordinates, a single linear element may be sufficient. However, when displacement is not a linear function, additional elements may be linked together to form an approximation. The term *h*-convergence is used to describe the process where the finite element approximation becomes an increasingly accurate representation of the true displacement as an increasing number of smaller elements is used.

As shown in Section 2.3, the accuracy of a finite element solution is not increased by using additional elements if a single element's displacement assumption matches the true displacement. The finite element method's general nature is better revealed when applied in a situation where the exact displacement is a not a linear function of the structure's spatial coordinates.

When Is Displacement Within a Structure Not a Linear Function?
In the last example, the steering link was idealized as a uniform rod, and displacement was found to be a linear function of the structure's spatial coordinates. When is displacement not a linear function of the structure's spatial coordinates? It is not possible to give a simple, all-encompassing answer to this question, however, if we restrict the question a specific type of structure and loading, some general conclusions can be drawn. Consider a rod under uniaxial tensile loading, with the governing differential equation, (1.2.1):

$$\frac{d}{dX}\left(EA\frac{dU}{dX}\right)=0 \tag{2.4.1}$$

Upon integrating and manipulating (2.4.1), it is found that:

$$U(X)=\int\left(\frac{c_1}{EA}\right)dX$$

Notice that if the elastic modulus and cross sectional area are constant, they can be removed from the integrand, and *U(X)* will simply be a linear function of *X*. Conversely, if either the modulus or the cross sectional area are not constant, then the opportunity exists for *U(X)* to be something other than a linear function.

An Analysis Where Displacement Is Not a Linear Function

When the exact displacement in a structure is a linear function, one element may be all that is required. However, if the exact displacement in a structure is not linear, many linear elements may be required to allow the finite element approximation to converge to the exact. The concept of convergence is illustrated by the next two examples, where a shaft of variable cross section is analyzed.

Example 2.4—Analysis of a Non-Uniform Shaft, One Element

The displacement in the non-uniform shaft shown in Figure 2.12 is evaluated using a model employing a single linear element, as illustrated in Figure 2.13.

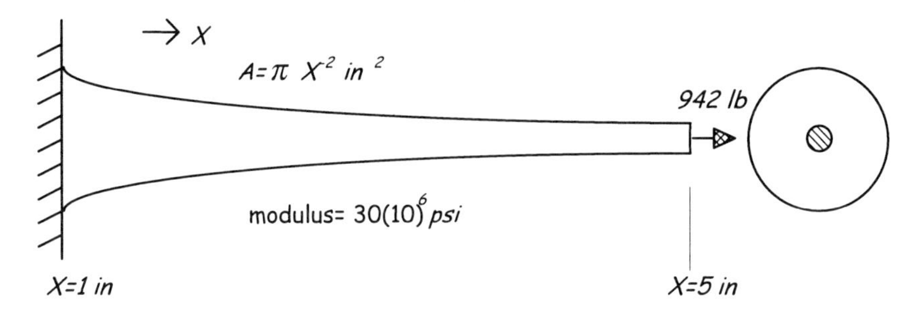

Figure 2.12 Non-Uniform Shaft

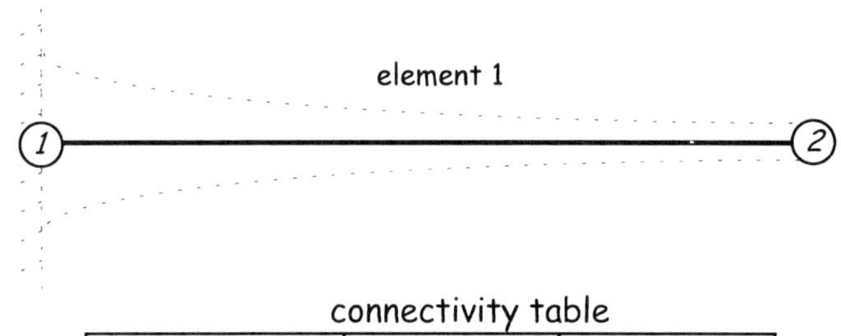

connectivity table

element	node a	node b
1	1	2

$$\tilde{U}(X) = \left(\frac{X_b - X}{X_b - X_a} \right) U_a + \left(\frac{X - X_a}{X_b - X_a} \right) U_b$$

Figure 2.13 Finite Element Model, Non-Uniform Shaft, One Element Model

It is often desirable to check a finite element model with a closed-form solution, when possible. In general, a closed-form solution will provide only a rough estimation of the exact value, for if an accurate closed-form solution were available, there would little need to construct a finite element model.

The cross sectional area of the non-uniform shaft in Figure 2.12 is described by a function of the axial coordinate, i.e., $A=\pi X^{-2}$. From the previous discussion, it is noted that a non-linear function of displacement is anticipated if the cross section of a shaft is not uniform, other things being equal. A closed-form solution for displacement in the non-uniform shaft of Figure 2.12 can be found by integrating Equation 1.2.1 and applying the boundary conditions:

$$U(0)=0 \qquad \left.\left(EA\frac{dU}{dX}\right)\right|_{X=L}=P$$

In this problem, $L=5$ and $P=942$. The expression for displacement resulting from the differential equation given by (1.2.1) is:

$$U(X)=\frac{X^3-1}{3}(10)^{-5}\ in \quad (1\le X\le 5) \tag{2.4.1}$$

As with the steering link, a uniaxial stress idealization will be employed for the non-uniform shaft, and one, 2-node rod finite element will be used (Figure 2.13). Referring to the connectivity table, and substituting the known variables into the given interpolation function:

$$\tilde{U}(X)=\left(\frac{5-X}{4}\right)(U_1)+\left(\frac{X-1}{4}\right)(U_2) \tag{2.4.2}$$

The interpolation function given in Figure 2.13 was originally defined by (2.2.7). With the interpolation function for this problem given by (2.4.2), we return to the rod element equilibrium equations defined in Section 2.3, Equation 2.3.27:

$$\frac{{}^eE\,{}^eA}{{}^eH}\begin{bmatrix}1 & -1\\ -1 & 1\end{bmatrix}\begin{Bmatrix}{}^eU_a\\ {}^eU_b\end{Bmatrix}=\begin{Bmatrix}{}^eF_a\\ {}^eF_b\end{Bmatrix} \tag{2.4.3}$$

Referring to the connectivity table in Figure 2.13, the appropriate global variables are used in Equation 2.4.3:

$$\frac{{}^1E\,{}^1A}{{}^1H}\begin{bmatrix} 1 & -1 \\ -1 & 1 \end{bmatrix}\begin{Bmatrix} U_1 \\ U_2 \end{Bmatrix} = \begin{Bmatrix} F_1 \\ F_2 \end{Bmatrix}$$

The values for the constants are substituted into the equation above:

$$\frac{30(10)^6\left({}^1A\right)}{4}\begin{bmatrix} 1 & -1 \\ -1 & 1 \end{bmatrix}\begin{Bmatrix} U_1 \\ U_2 \end{Bmatrix} = \begin{Bmatrix} F_1 \\ 942 \end{Bmatrix} \tag{2.4.4}$$

What variable should be used for the cross sectional area term in (2.4.4)? There exist several options; one is to use a mean value. The mean value of the function *f(x)*, on the domain $a \le x \le b$, is given as:

$$\overline{f(x)} = \frac{1}{b-a}\int_a^b f(x)\,dx \tag{2.4.5}$$

Since the average cross sectional area is required, the expression πX^{-2} is substituted for *f(x)* in (2.4.5):

$$\overline{A} = \frac{1}{4}\int_1^5 \pi\, X^{-2}\, dX = \left(\frac{-\pi}{4}\right)\frac{1}{X}\Bigg|_1^5 = 0.2\pi \tag{2.4.6}$$

Substituting (2.4.6) for the value of ${}^1A$ in (2.4.4), and simplifying:

$$4.71(10)^6\begin{bmatrix} 1 & -1 \\ -1 & 1 \end{bmatrix}\begin{Bmatrix} U_1 \\ U_2 \end{Bmatrix} = \begin{Bmatrix} F_1 \\ 942 \end{Bmatrix} \tag{2.4.7}$$

Using $U_1=0$ in the bottom equation of (2.4.7):

$$4.71(10)^6 U_2 = 942 \qquad \therefore U_2 = 2.0(10)^{-4} \tag{2.4.8}$$

Using $U_2 = 2.0(10)^{-4}$ and U_1=0 in (2.4.2):

$$\tilde{U}(X) = \left(\frac{5-X}{4}\right)\underset{\substack{\Uparrow \\ 0}}{(U_1)} + \left(\frac{X-1}{4}\right)\underset{\substack{\Uparrow \\ 2.0(10)^{-4}}}{(U_2)} \tag{2.4.9}$$

$$\therefore \tilde{U}(X) = 0.5(X-1)\ (10)^{-4} in$$

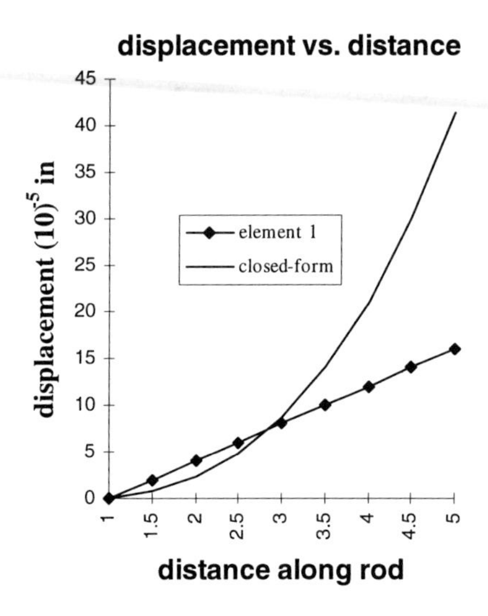

Figure 2.14 Finite Element Approximation for Non-Uniform Shaft, One Element

Equation 2.4.9 is plotted in Figure 2.14. Notice that there exists a substantial error between the finite element result and the closed-form solution, with the maximum error occurring at the end of the rod.

The reason for the error between the closed-form solution and the finite element model is that a linear displacement assumption and average cross sectional area were employed in the finite element model, while in reality, the cross section varies with distance along the axis of the shaft. Since the cross section varies, the exact displacement is a non-linear function of *X*. Error is introduced because the finite element model can only render a linear approximation for the exact (3rd order polynomial) equilibrium displacement.

To better understand how a combination of linear solutions can approximate a non-linear function, another finite element analysis, this time using two linear elements, will be performed.

Example 2.5—Analysis of a Non-Uniform Shaft, Two Elements

This example is essentially the same as Example 2.4, except that two linear finite elements are used instead of one (Figure 2.15). As before, a uniaxial stress idealization will be employed.

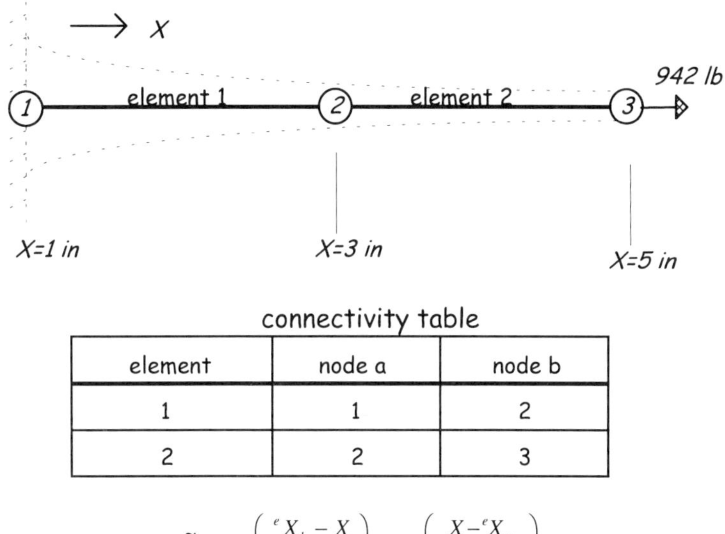

element	node a	node b
1	1	2
2	2	3

$$ {}^{e}\tilde{U}(X) = \left(\frac{{}^{e}X_b - X}{{}^{e}X_b - {}^{e}X_a} \right) {}^{e}U_a + \left(\frac{X - {}^{e}X_a}{{}^{e}X_b - {}^{e}X_a} \right) {}^{e}U_b $$

Figure 2.15 Finite Element Model, Non-Uniform Shaft, Two Elements

Since two elements are used in this case, we must be careful in defining the variables that are to be used in the interpolation functions—notice that superscripts are used to distinguish variables associated with a given element's interpolation function. Using the connectivity table and constants given, the interpolation function shown in Figure 2.15 renders interpolation functions for both of the elements:

$$ {}^{1}\tilde{U}(X) = \left(\frac{3 - X}{2} \right)(U_1) + \left(\frac{X - 1}{2} \right)(U_2) $$

$$ {}^{2}\tilde{U}(X) = \left(\frac{5-X}{2}\right)(U_2) + \left(\frac{X-3}{2}\right)(U_3) \tag{2.4.10} $$

With the displacement interpolation functions defined, the element equilibrium equations are considered. The equilibrium equations for the rod element defined in (2.3.27) are again employed:

$$ \frac{{}^{e}E\ {}^{e}A}{{}^{e}H}\begin{bmatrix} 1 & -1 \\ -1 & 1 \end{bmatrix}\begin{Bmatrix} {}^{e}U_a \\ {}^{e}U_b \end{Bmatrix} = \begin{Bmatrix} {}^{e}F_a \\ {}^{e}F_b \end{Bmatrix} $$

Referring to the connectivity table in Figure 2.15, the appropriate global variables are substituted into the equation above:

$$ \frac{{}^{1}E\,{}^{1}A}{{}^{1}H}\begin{bmatrix} 1 & -1 \\ -1 & 1 \end{bmatrix}\begin{Bmatrix} U_1 \\ U_2 \end{Bmatrix} = \begin{Bmatrix} F_1 \\ F_2 \end{Bmatrix} \qquad \frac{{}^{2}E\,{}^{2}A}{{}^{2}H}\begin{bmatrix} 1 & -1 \\ -1 & 1 \end{bmatrix}\begin{Bmatrix} U_2 \\ U_3 \end{Bmatrix} = \begin{Bmatrix} F_2 \\ F_3 \end{Bmatrix} \tag{2.4.11} $$

The values for the constants are substituted into (2.4.11):

$$ \frac{30(10)^6({}^{1}A)}{2}\begin{bmatrix} 1 & -1 \\ -1 & 1 \end{bmatrix}\begin{Bmatrix} U_1 \\ U_2 \end{Bmatrix} = \begin{Bmatrix} F_1 \\ F_2 \end{Bmatrix} \qquad \frac{30(10)^6({}^{2}A)}{2}\begin{bmatrix} 1 & -1 \\ -1 & 1 \end{bmatrix}\begin{Bmatrix} U_2 \\ U_3 \end{Bmatrix} = \begin{Bmatrix} F_2 \\ F_3 \end{Bmatrix} \tag{2.4.12} $$

As in the last example, the average cross sectional area will be used for each element. Referring to (2.4.6):

$$ \overline{{}^{1}A} = \frac{1}{2}\int_1^3 \pi X^{-2}\,dx = \left(\frac{-\pi}{2}\right)\frac{1}{X}\bigg|_1^3 = 0.333\pi \qquad \overline{{}^{2}A} = \frac{1}{2}\int_3^5 \pi X^{-2}\,dx = 0.0667\pi \tag{2.4.13} $$

Substituting (2.4.13) into (2.4.12) and simplifying:

$$ 1.57(10)^7\begin{bmatrix} 1 & -1 \\ -1 & 1 \end{bmatrix}\begin{Bmatrix} U_1 \\ U_2 \end{Bmatrix} = \begin{Bmatrix} F_1 \\ F_2 \end{Bmatrix} \qquad 3.14(10)^6\begin{bmatrix} 1 & -1 \\ -1 & 1 \end{bmatrix}\begin{Bmatrix} U_2 \\ U_3 \end{Bmatrix} = \begin{Bmatrix} F_2 \\ F_3 \end{Bmatrix} \tag{2.4.14} $$

Following the process of expanding the matrices shown in Example 2.3, the equations in (2.4.14) are also expanded:

$$1.57(10)^7 \begin{bmatrix} 1 & -1 & 0 \\ -1 & 1 & 0 \\ 0 & 0 & 0 \end{bmatrix} \begin{Bmatrix} U_1 \\ U_2 \\ U_3 \end{Bmatrix} = \begin{Bmatrix} F_1 \\ F_2 \\ F_3 \end{Bmatrix} \tag{2.4.15}$$

$$3.14(10)^6 \begin{bmatrix} 0 & 0 & 0 \\ 0 & 1 & -1 \\ 0 & -1 & 1 \end{bmatrix} \begin{Bmatrix} U_1 \\ U_2 \\ U_3 \end{Bmatrix} = \begin{Bmatrix} F_1 \\ F_2 \\ F_3 \end{Bmatrix} \tag{2.4.16}$$

Element 1, having a stiffness matrix coefficient of $1.57(10)^7$, is more stiff than Element 2. The reason for this is that Element 1 represents the left half of the shaft, which has a larger average cross sectional area—the larger cross section results in greater stiffness.

Since the two stiffness matrices have different coefficients, it is convenient to manipulate the coefficient and components of one of the matrices so that both can be summed in a simple fashion. Operating on (2.4.15):

$$1.57(10)^7 \begin{bmatrix} 1 & -1 & 0 \\ -1 & 1 & 0 \\ 0 & 0 & 0 \end{bmatrix} \Rightarrow 3.14(10)^6 \begin{bmatrix} 5 & -5 & 0 \\ -5 & 5 & 0 \\ 0 & 0 & 0 \end{bmatrix} \tag{2.4.17}$$

Equations (2.4.16) and (2.4.17) can now be summed:

$$3.14(10)^6 \begin{bmatrix} 5 & -5 & 0 \\ -5 & 5 & 0 \\ 0 & 0 & 0 \end{bmatrix} + 3.14(10)^6 \begin{bmatrix} 0 & 0 & 0 \\ 0 & 1 & -1 \\ 0 & -1 & 1 \end{bmatrix} = 3.14(10)^6 \begin{bmatrix} 5 & -5 & 0 \\ -5 & 6 & -1 \\ 0 & -1 & 1 \end{bmatrix}$$

The resulting equilibrium equations are:

$$3.14(10)^6 \begin{bmatrix} 5 & -5 & 0 \\ -5 & 6 & -1 \\ 0 & -1 & 1 \end{bmatrix} \begin{Bmatrix} U_1 \\ U_2 \\ U_3 \end{Bmatrix} = \begin{Bmatrix} F_1 \\ F_2 \\ F_3 \end{Bmatrix} \tag{2.4.18}$$

Note that $F_2=0$ and $F_3=P$, which allows (2.4.18) to be expressed as:

$$3.14(10)^6\begin{bmatrix} 5 & -5 & 0 \\ -5 & 6 & -1 \\ 0 & -1 & 1 \end{bmatrix}\begin{Bmatrix} U_1 \\ U_2 \\ U_3 \end{Bmatrix}=\begin{Bmatrix} F_1 \\ 0 \\ P \end{Bmatrix} \qquad (2.4.19)$$

The top equation (2.4.19) is removed, since F_1 is unknown, and the first column is removed, because U_1 is zero. The result is a linear system of equations with two unknowns and two equations:

$$3.14(10)^6\begin{bmatrix} 6 & -1 \\ -1 & 1 \end{bmatrix}\begin{Bmatrix} U_2 \\ U_3 \end{Bmatrix}=\begin{Bmatrix} 0 \\ P \end{Bmatrix} \qquad (2.4.20)$$

The simultaneous solution to (2.4.20) can be shown to be:

$$U_2 = 6.0(10)^{-5} \qquad U_3 = 36.0(10)^{-5} \qquad (2.4.21)$$

Using (2.4.21) and $U_1=0$ in the interpolation functions from (2.4.10):

$${}^1\tilde{U}(X)=\left(\frac{3-X}{2}\right)\underset{\overset{\Uparrow}{0}}{(U_1)}+\left(\frac{X-1}{2}\right)\underset{\overset{\Uparrow}{6.0(10)^{-5}}}{(U_2)} \Rightarrow {}^1\tilde{U}(X)=(3X-3)\,(10)^{-5} in \qquad (2.4.22)$$

$${}^2\tilde{U}(X)=\left(\frac{5-X}{2}\right)\underset{\overset{\Uparrow}{6.0(10)^{-5}}}{(U_2)}+\left(\frac{X-3}{2}\right)\underset{\overset{\Uparrow}{36.0(10)^{-5}}}{(U_3)} \Rightarrow {}^2\tilde{U}(X)=(15X-39)\,(10)^{-5} in \qquad (2.4.23)$$

An approximate solution is formed by employing (2.4.22) and (2.4.23); the results are plotted in Figure 2.16. Compared to the single element model of Figure 2.14, a significant improvement is realized by simply adding one additional element. One might visualize that with the addition of several more linear elements, a fairly good approximation for displacement can be achieved.

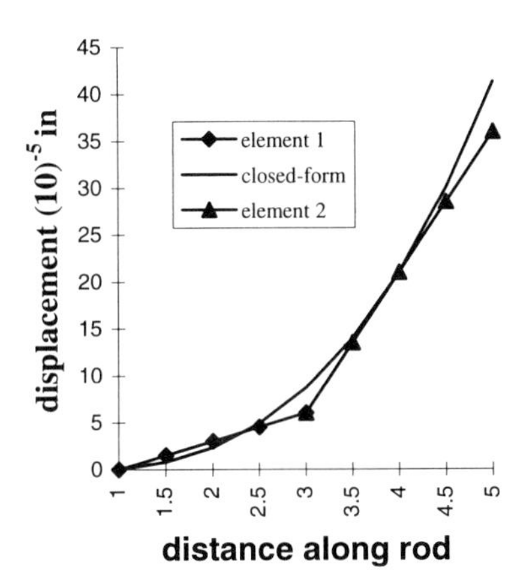

Figure 2.16 Non-Uniform Shaft, Two Elements

The Nature of Finite Element Displacement Interpolation

If the finite element approximation for displacement contains the same mathematical form as the exact displacement function, the finite element method can predict the true displacement. However, the displacement assumption typically *does not* contain all the terms necessary to allow the idealized structure to deform in exactly the same manner as the actual structure. In some cases, no analytical expression exists for the exact displacement. Therefore, it can be said that the finite element idealization is typically more constrained or *stiffer* than the actual structure (Fraeijs de Veubeke [4]). Although this is true, caution must be used when making generalizations based upon this fact.

Since the finite element idealization is stiffer, one might naturally (but incorrectly) assume that the finite element interpolation function must yield values of displacement that are everywhere less than the true displacement. However, observing the results in Figure 2.16 notice that Element 1 *overpredicts* displacement until the X-coordinate is approximately 2.5 inches, then *underpredicts* displacement for the rest of the element. Furthermore, as will be shown in Section 2.5, this finite element model both overpredicts and underpredicts stress, depending upon where (on the element) stress is observed. Care must be taken when making generalizations regarding finite element stress and the interpolated displacement results, particularly when observing results in a localized area. However, it may be fair to say that the finite element method tends to underestimate peak stresses when a structure is subjected to force (as opposed to displacement) boundary conditions.

h-Convergence

As the number of elements used in a finite element model approaches infinity, the error between the finite element solution for displacement and the exact displacement is expected to approach zero. (Likewise, the error between predicted and exact energy is also expected to approach zero.) This is the essence of *h*-convergence.

At any given point in a structural finite element model, a predicted value of displacement can be either too large or too small, making the evaluation of convergence difficult. In light of this, proofs of *h*-convergence are stated in terms of *norms*. Using norms, a formal proof of convergence shows that the total potential energy computed by the finite element method converges to the exact value as the size of the largest element dimension approaches zero. Convergence in displacement is anticipated with convergence in energy, although the two metrics need not converge at the same rate. An example of using an energy norm to predict error is given by Reddy [5, p. 134], and a similar example is considered below; norms are also discussed by Bathe [2, p. 58]. Other interesting reading related to the subject of finite element convergence can be found in Oliveira [6], Felippa [7], Verma [8], and Melosh [9].

Rate of Convergence

A properly refined mesh (Chapter 4) will exhibit convergence of displacement in accordance with the equation:

$$\textit{displacement convergence} \equiv \frac{e_2}{e_1} = O\left(\frac{h_2}{h_1}\right)^{p+1} \tag{2.4.24}$$

The equation above states that convergence is defined in terms of the error associated with two solutions, e_1 and e_2. The relative error is a function of the largest element size in each of the models, h_1 and h_2, and the order of the interpolation function, p; O denotes "of the order." The metrics e_1 and e_2 in (2.4.24) are computed using a particular type of norm called the L_2 ("*L-two*") norm. For a simple problem such as the non-uniform shaft of Example 2.4, the L_2

norm can be expressed as:

$$e = \sqrt{\int_a^b \left(u_{ex} - \tilde{u}\right)^2 dx} \tag{2.4.25}$$

$u_{ex} \equiv$ exact displacement $\qquad \tilde{u} \equiv$ predicted displacement

To illustrate displacement convergence using the L_2 norm, consider that in the non-uniform shaft of Examples 2.4 and 2.5, the maximum element length was changed from four inches to two. Since a linear interpolation function was employed, the following values are identified:

$$h_1 = 4.0 \quad h_2 = 2.0 \quad p = 1 \tag{2.4.26}$$

Using the values above in (2.4.24):

$$\frac{e_2}{e_1} = O\left(\frac{2}{4}\right)^2 = O(0.250) \tag{2.4.27}$$

The equation above suggests that the error using 2 inch long elements should be on the order of one-quarter the amount when a 4 inch long element is used. The finite element solution using one element (Example 2.4) was found to be:

$$\tilde{u} = 0.5(x-1)10^{-4} \tag{2.4.28}$$

It is convenient to simplify (2.4.28):

$$\tilde{u} = (5x-5)10^{-5} \qquad (2.4.28)$$

Recalling the exact solution:

$$u_{ex} = \left(\frac{x^3 - 1}{3}\right)10^{-5} \tag{2.4.29}$$

Substituting (2.4.29) and (2.4.28) into (2.4.25) and noting that the element was defined on the interval from x=1 to x=5:

$$e_1 = \sqrt{\int_1^5 \left(\left[\left(\frac{x^3 - 1}{3} \right) - (5x - 5) \right] 10^{-5} \right)^2 dx} = 14.3(10)^{-5} \tag{2.4.30}$$

Using two elements for the same problem, Example 2.5 showed that:

$$^1\tilde{u} = (3x - 3)10^{-5} \qquad ^2\tilde{u} = (15x - 39)10^{-5} \tag{2.4.31}$$

Using (2.4.31) in (2.4.25), and modifying the limits on the integral:

$$e_2 = \sqrt{\int_1^3 \left(\left[\left(\frac{x^3 - 1}{3} \right) - (3x - 3) \right] 10^{-5} \right)^2 dx + \int_3^5 \left(\left[\left(\frac{x^3 - 1}{3} \right) - (15x - 39) \right] 10^{-5} \right)^2 dx} = 3(10)^{-5} \tag{2.4.32}$$

Computing the ratio of the two norms:

$$\frac{e_2}{e_1} = \frac{3.00(10)^{-5}}{14.3(10)^{-5}} = 0.210$$

Thus, displacement convergence in the problem illustrated by Examples 2.4 and 2.5 is predicted quite well using (2.4.24). Stress convergence may be computed using a similar equation:

$$stress\ convergence = O\left(\frac{h_2}{h_1} \right)^{p+1-r} \tag{2.4.33}$$

$$r \equiv order\ of\ derivative$$

Convergence of stress is controlled by both the order of the interpolation polynomial and the highest order of the derivative(s) used to defined strain. Notice that stress is not expected to converge as quickly as displacement for a given degree of interpolation. Although this might seem to be an interesting but trivial fact, it does have a practical application: If one is analyzing a structure for maximum displacement, a somewhat coarse mesh might be sufficient whereas a more refined mesh would be required for accurate stress analysis.

The discussion of convergence above is applicable to solutions that are smooth, i.e., where discontinuities in displacement, stress, or strain are not present.

Summary of Finite Element *h*-Convergence

It was shown that with the correct choice of displacement assumption, a solution that is identical to the closed-form solution can be found using a single finite element. In other problems where the displacement assumption is of lower order than the exact displacement, the error between the exact displacement and the finite element solution is expected to decrease with an increasing number of smaller elements, other things being equal; this is the essence of *h*-convergence. The term exact displacement does not typically mean *closed-form solution* for displacement, since a closed-form solution simply may not exist.

What happens if an element uses a higher order displacement assumption when the exact displacement is only linear? For instance, what if the displacement assumption for a given element is a quadratic polynomial:

$$\tilde{U}(X) = a_1 + a_2 X + a_3 X^2 \tag{2.4.26}$$

If the exact displacement is known to be $U(X) = 5X$, and the displacement assumption shown in (2.4.26) is used, the process of finite element analysis would force the displacement assumption to assume the following values:

$$a_1 = 0 \qquad a_2 = 5 \qquad a_3 = 0$$

Therefore, if the form of the exact displacement and the form of displacement assumption are identical, or if the exact displacement *is a subset of the displacement assumption,* the finite element solution and the exact displacement can be identical. A common approach in solving analysis problems with the finite element method is to use lower order displacement assumptions (linear or quadratic) and increase the number of elements in areas where the displacement is expected to be a non-linear function. Recall that a non-linear displacement response is anticipated in areas of a structure where the stress is changing.

Other Types of Convergence (*p*-Convergence)

Convergence can also be achieved by other means. Depending on the type of problem, one could increment the polynomial order of the finite element displacement assumption until the correct solution is obtained. This is termed *p*-convergence, or the *p*-method of finite element analysis. Convergence is discussed in Strang [10], while Szabó [11] gives a detailed explanation of *h*-, *p*-, and *hp*-convergence; also see Babuska [12] for details of *p*- and *hp*-convergence.

2.5 Calculating Stress Using Displacement

Key Concept: The finite element method is an effective means of providing an approximate solution for displacement in complex structures. Once displacement is known, stress and strain may be computed.

In Section 2.2, it was mentioned that one requirement for the displacement assumption is that it must satisfy the essential boundary conditions of the element. This requirement has the effect of constraining the displacement assumption such that when finite elements are joined together, displacement is continuous at adjacent nodes. However, recall that no constraint was imposed upon *stress values* at adjacent nodes, and this allows stress to be discontinuous between elements. To illustrate this point, consider the two-element solution for the non-uniform shaft of Example 2.5, shown again in Figure 2.17. Note that the displacement at X=3, where the two elements are joined, is continuous.

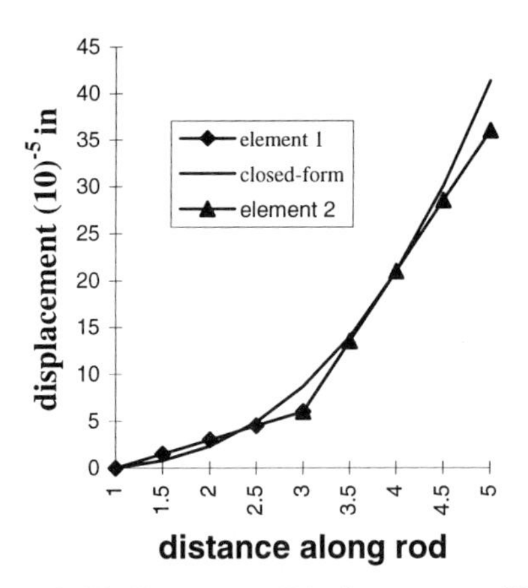

Figure 2.17 Continuous Displacement at Nodes

The solutions for displacement were given by (2.4.22) and (2.4.23):

$$^{1}\tilde{U}(X) = (3X - 3)\,(10)^{-5}$$
$$^{2}\tilde{U}(X) = (15X - 39)\,(10)^{-5}$$

Using the equation for uniaxial stress from Chapter 1, stress can be calculated using the equations above. From (1.5.12):

$$\tau_{uni} = E\frac{dU}{dX}$$

Using the finite element solutions, stress is calculated on each element separately. For stress in Element 1:

$$^{1}\tau_{uni} \approx E\frac{d\,^{1}\tilde{U}}{dX} = (30)10^{6}\frac{d}{dX}\left[(3X-3)\,(10)^{-5}\right] = 900\,psi \tag{2.5.1}$$

For Element 2:

$$^{2}\tau_{uni} \approx E\frac{d\,^{2}\tilde{U}}{dX} = (30)10^{6}\frac{d}{dX}\left[(15X-39)\,(10)^{-5}\right] = 4500\,psi \tag{2.5.2}$$

Using the closed-form solution for displacement, the exact stress is shown to be:

$$\tau_{uni} = E\frac{dU}{dX} = (30)10^{6}\frac{d}{dX}\left[\frac{X^{3}-1}{3}\right]10^{-5} = 300X^{2}\ psi \tag{2.5.3}$$

Equations (2.5.1) through (2.5.3) are plotted in Figure 2.18. Notice that stress computed by the linear finite elements is constant along the entire element but, more importantly, notice that stress is discontinuous at $X=3$, which is the location of the node that joins Elements 1 and 2.

The reason for the stress discontinuity is that the displacement assumption is only required to have $p-1$ derivative continuity across adjacent elements. Since derivatives in the functional for this problem are first order, $p-1$ is zero, therefore, no derivative continuity is required. In the problems considered, first order derivatives in the functional represent strain. Using the displacement based finite element method, the absence of $p-1$ derivative continuity typically corresponds to the absence of strain continuity in the solution of a solid mechanics problem. In this example the discontinuous strain results in a discontinuity of stress at the node that joins the two elements.

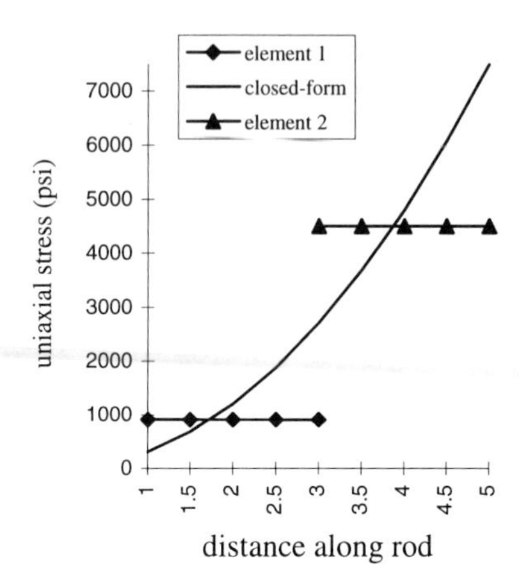

Figure 2.18 Stress in a Non-Uniform Shaft

Is Stress Discontinuous When Higher Order Elements Are Used?

The stress equation for each element, using a quadratic displacement assumption, would show a linear variation along the length of the element but would also be discontinuous at the nodes. This is because the use of higher order displacement assumptions does not necessarily invoke continuity of slope between elements. If the functional contains first order derivatives, then strain continuity (hence, stress continuity) is not enforced.[3] In this manner, the finite element method simplifies problems and provides an approximate solution by ignoring continuity requirements on some derivatives of the dependent variable.

Isn't Stress Continuity Required for Accurate Stress Analysis?

Using a stress-based approach to finite element analysis (instead of the displacement-based approach) elements can be designed to enforce stress continuity across element boundaries. However, it has been found this approach can result in elements that are overly-stiff, since in some cases stress may be (theoretically) discontinuous. For instance, in the case of the stepped shaft (steering link) shown in Figure 1.3, the stress at the step is theoretically infinite,

---

[3] If the displacement functions for two adjacent elements are identical, stress can be continuous cross the element boundary.

and imposing continuous stress across element boundaries amounts to an artificial constraint.

Why Do Stress Contour Plots Appear Continuous When Post-Processing?
When displaying finite element stress results using state-of-the-art finite element graphical software, stress contours may appear continuous even if a jump exists between elements. This occurs when the contours are rendered using graphical smoothing techniques which blend stress discontinuities between elements. Some finite element software allows the analyst to choose between element stress values or "smoothed" stress values, which are averaged across element boundaries. The analyst should consult suitable software documentation and be aware of the technique that is being employed for a given stress plotting routine. Related topics are discussed in Hinton [13], Ward [14], and Russell [15].

If Finite Elements Are Overly Stiff, Is Stress Underpredicted?
It was noted in Section 2.4 that since a finite element displacement assumption does not typically contain all the possible modes of deformation, the models are generally too stiff. However, it was also shown that this does not result in underprediction of interpolated displacement everywhere. In a finite element model with n degrees of freedom, the collection of all the nodal displacements can be considered a vector with n components. It is the *norm* of the displacement vector that will reflect the overly stiff nature of the finite element solution; displacement at certain points may be either overpredicted or underpredicted. Likewise, stress calculated by FEA values may overpredict or underpredict the exact stress values.

Figure 2.18 indicates that Element 1 overpredicts stress for X-coordinate values less than 1.73, then underpredicts at greater values of X. Similarly, Element 2 both underpredicts and overpredicts. Care must be taken when drawing general conclusions about the displacement and stress response of finite element models.

2.6 Finite Element Equilibrium

Key Concept: The finite element method simplifies solutions by allowing discontinuities in strain (and stress) at nodes and boundaries of adjacent elements. Since force is a function of stress, forces can also be discontinuous, such that computed nodal forces are typically not in equilibrium.

The fact that only $p-1$ derivatives of displacement need be continuous can cause discontinuities in metrics such as stress, strain, and force. To investigate this phenomenon, it is helpful to reexamine the continuity requirements for finite element displacement assumptions.

Essential and Natural Boundary Conditions

Consider again the two types of boundary conditions associated with the finite element formulation: *essential boundary conditions,* and *natural boundary conditions*. In solid mechanics problems, the essential boundary conditions are related to the kinematics of a structure, i.e., translation and rotation of the structure, while natural boundary conditions are related to forces acting upon the structure. The essential boundary conditions are identified by examining p, the highest order derivative in the functional. Displacement, and $p-1$ order derivatives of displacement, are termed the essential boundary conditions, and they are required to be continuous at the nodes and boundaries of each element. As a result of this constraint, the essential boundary conditions of the structure are satisfied, at least in an approximate sense.

Natural Boundary Conditions

Although satisfaction of the essential boundary conditions is typically required in finite element models, satisfaction of the natural boundary conditions is not. To illustrate this, consider Example 2.4, where a non-uniform shaft was analyzed using one element; the finite element solution for displacement was given as (2.4.9):

$$\tilde{U}(X) = 0.5(X-1)\,(10)^{-4}\,in \tag{2.6.1}$$

Also recall the natural boundary condition used for the differential equation from Example 2.4:

$$\left(EA\frac{dU}{dX}\right)\bigg|_{X=L} = P \tag{2.6.2}$$

The natural boundary condition above states that at the right end of the shaft, the product of the modulus, cross sectional area, and first derivative must be equal to the applied force, 942 lb_f. Substituting for the E, A, L, and P variables in (2.6.2):

$$\left(30(10)^6\pi\, X^{-2}\frac{dU}{dX}\right)\bigg|_{X=5} = 942 \tag{2.6.3}$$

Now, if the finite element solution for displacement, (2.6.1), is substituted into (2.6.3), does the equality prevail?

$$\left(30(10)^6 \pi X^{-2} \frac{d}{dX} 0.5(X-1)(10)^{-4} in\right)\Bigg|_{X=5} \overset{?}{=} 942$$

Performing the algebra above and substituting in $X=5$ into the equation:

$$30(10)^6 \pi (5)^{-2}\left[0.5(10)^{-4}\right] \overset{?}{=} 942$$

Simplifying the equation above:

$$189 \overset{?}{=} 942 \qquad (2.6.4)$$

In other words, using the approximate solution for displacement generated by the finite element method, the natural boundary condition is not satisfied as illustrated by the fact that the left side of (2.6.4) is not equal to the right.

Satisfying the Differential Equation of Equilibrium

The natural boundary condition is not satisfied in Example 2.4.1. Will the finite element solution for displacement satisfy the governing differential of equilibrium? Recalling (1.2.1), the governing differential equation for a uniaxially loaded rod:

$$\frac{d}{dX}\left(EA\frac{dU}{dX}\right) = 0 \qquad (2.6.5)$$

Substituting E and A from Example 2.4.1 into (2.6.5):

$$\frac{d}{dX}\left(30(10)^6 \pi X^{-2}\frac{dU}{dX}\right) = 0 \qquad (2.6.6)$$

Substituting the finite element approximation for displacement into (2.6.6):

$$\frac{d}{dX}\left(30(10)^6 \pi X^{-2}\frac{d}{dX}\left[0.5(X-1)(10)^{-4}\right]\right) \overset{?}{=} 0 \qquad (2.6.7)$$

Simplifying (2.6.7):

$$\frac{d}{dX}\left(4.71(10)^3 X^{-2}\right) \stackrel{?}{=} 0 \tag{2.6.8}$$

$$-9.42(10)^3 X^{-3} \stackrel{?}{=} 0 \tag{2.6.9}$$

Equation 2.6.9 indicates that the governing differential equation is not satisfied, since the left hand side does not equal the right. As a result, equilibrium is not satisfied. The same can be shown for the two-element model of the non-uniform shaft. The net result of this out-of-equilibrium condition is that nodal forces, calculated using the finite element solution, do not balance across element boundaries.

Calculating Nodal Forces

Consider the stress values for the two-element analysis of the non-uniform shaft in Example 2.5. In the linear elastic range, stress can be calculated using (1.5.1), shown again below:

$$S_{XX} = \frac{P}{A_0} \tag{2.6.10}$$

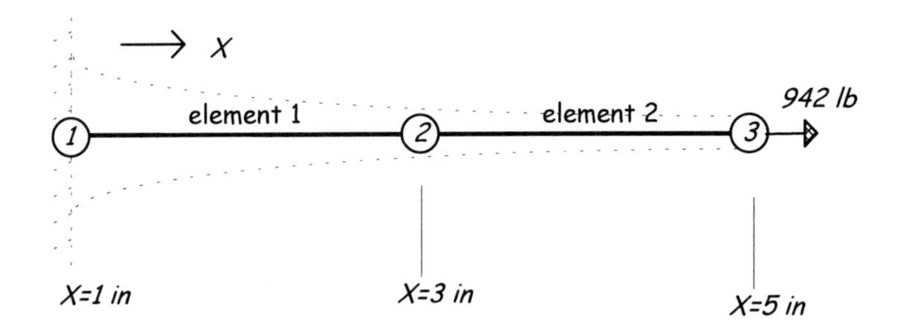

Figure 2.19 Shaft Finite Element Model, Two Elements

To evaluate the forces in the rod, (2.6.10) is rearranged as:

$$P = S_{XX} A_0 \tag{2.6.11}$$

Noting that $A_0 = \pi X^{-2}$, and the closed-form solution for normal stress in the shaft, the force is computed to be:

$$P = S_{XX} A_0 = 300 X^2 \left(\pi X^{-2}\right) = 942\ lb_f \tag{2.6.12}$$

The force in the shaft is constant, since the bar is not accelerating and no other forces are present. Now, consider the forces calculated using the finite element predictions for stress.

For Elements 1 and 2, the finite element model predicts, using (2.5.1) and (2.5.2), stress values of:

$$^{1}\tau_{uni} = 900\,psi \tag{2.6.13}$$

$$^{2}\tau_{uni} = 4500\,psi \tag{2.6.14}$$

Using the finite element stress calculated for Element 1 in place of S_{XX} in (2.6.11), and the X -coordinate value of 3.0 to compute A_0, FEA predicts the force at Node 2 to be:

$$P = 900\left(\pi\ 3^{-2}\right) = 314\ lbf \tag{2.6.14}$$

Doing the same as above but using the finite element stress from Element 2:

$$P = 4500\left(\pi\ 3^{-2}\right) = 1571\ lbf \tag{2.6.15}$$

It is seen that neither calculation matches the closed-form solution, (2.6.12). Furthermore, the forces calculated at the interface of the elements are not equal, which is to say, *inter-element force equilibrium* is not satisfied when forces are calculated using the finite element stress values. *External* equilibrium is satisfied since the *global equilibrium equations,* using the externally applied loads, were solved:

$$\underline{\underline{K}}\,\underline{D} = \underline{F}$$

The ratio of inter-element forces (or stresses) between adjacent elements is one metric that can be used to determine if a given finite element mesh is fine enough for suitable accuracy. If a smooth stress transition within the structure characterizes the exact solution to a given solid mechanics problem, the stress jumps between elements can be used as a measure of the suitability of a given mesh. In such cases, a large stress jump indicates a need for further mesh refinement while small stress jumps suggest a converged solution.

2.7 Matrix Operations

Key Concept: Because the finite element method leads to a system composed of many linear algebraic equations, it is convenient to express the equations in matrix form, then perform matrix operations to obtain a solution. A brief review of matrix notation related to the types of matrices encountered using finite element methods is given.

Finite element equilibrium equations are expressed as a set of simultaneous, linear, algebraic equations. For instance, the global equilibrium equations from Example 2.2, Equations 2.2.27–2.2.29, are given as:

$$\frac{EA}{5}\left(U_1 - U_2 + 0U_3\right) = F_1$$
$$\frac{EA}{5}\left(-U_1 + 2U_2 - U_3\right) = F_2$$
$$\frac{EA}{5}\left(0U_1 - U_2 + U_3\right) = F_3$$

Substituting into the above the known values:

$$3.0(10)^7\left(U_1 - U_2 + 0U_3\right) = F_1$$
$$3.0(10)^7\left(-U_1 + 2U_2 - U_3\right) = 0$$
$$3.0(10)^7\left(0U_1 - U_2 + U_3\right) = P$$

For clarity, the constant k will be used to denote the coefficient in the equations above:

$$\begin{aligned} k\left(U_1 - U_2 + 0U_3\right) &= F_1 \\ k\left(-U_1 + 2U_2 - U_3\right) &= 0 \\ k\left(0U_1 - U_2 + U_3\right) &= P \end{aligned} \qquad (2.7.1)$$

Where:

$$k = 3.0(10)^7$$

The equations in (2.7.1) are considered *simultaneous*, since the unknown displacement variables must satisfy all three equations at the same time, and *linear*, since the unknown variables are of the power unity and square matrix contains only constants. The term *algebraic* indicates that the equations can be solved using algebra, as opposed to requiring techniques of differential or integral calculus, for instance. Equations 2.7.1 can be cast into "*KD=F*" matrix form:

$$k\begin{bmatrix} 1 & -1 & 0 \\ -1 & 2 & -1 \\ 0 & -1 & 1 \end{bmatrix}\begin{Bmatrix} U_1 \\ U_2 \\ U_3 \end{Bmatrix} = \begin{Bmatrix} F_1 \\ 0 \\ P \end{Bmatrix} \tag{2.7.2}$$

Such that:

$$\underline{\underline{K}} = k\begin{bmatrix} 1 & -1 & 0 \\ -1 & 2 & -1 \\ 0 & -1 & 1 \end{bmatrix}$$

Using finite element software, equilibrium equations such as (2.7.2) are solved for nodal displacements using a digital computer. Matrix methods provide a convenient way to cast the equilibrium equations into a form so that they can be solved using computerized techniques. Before discussing the solution to simultaneous linear equations, a few definitions are given. Consider the set of simultaneous, linear equations below:

$$\begin{aligned} 2x - 4y &= 5 \\ 10x - 15y &= 20 \end{aligned} \tag{2.7.3}$$

Equations 2.7.3 can be expressed in matrix form as:

$$\begin{bmatrix} 2 & -4 \\ 10 & -15 \end{bmatrix}\begin{Bmatrix} x \\ y \end{Bmatrix} = \begin{Bmatrix} 5 \\ 20 \end{Bmatrix} \tag{2.7.4}$$

Using matrix notation, (2.7.4) can also be expressed as:

$$\underline{\underline{A}}\,\underline{\underline{B}} = \underline{\underline{C}} \tag{2.7.5}$$

Where:

$$\underline{\underline{A}} = \begin{bmatrix} 2 & -4 \\ 10 & -15 \end{bmatrix} \quad \underline{\underline{B}} = \begin{Bmatrix} x \\ y \end{Bmatrix} \quad \underline{\underline{C}} = \begin{Bmatrix} 5 \\ 20 \end{Bmatrix} \tag{2.7.6}$$

The double underscore is used in this text to represent matrices, although the use of boldface characters is quite common. A *matrix* is a collection of *elements* (no relation to finite elements, generally speaking), arranged in rows and columns. The elements can be numerical values, equations, or variables. For instance, the elements of the *A*-matrix above are the numerical values 2, –4, 10, and –15. The *A*-matrix shown is a square matrix, where the number of rows, *m*, is equal to the number of columns, *n*. In general, the elements of a matrix are defined using subscripts to identify their location. The first subscript of the element indicates the row, the second indicates the column; for instance:

$$\underline{\underline{D}} = \begin{bmatrix} d_{11} & d_{12} & \cdots & d_{1n} \\ d_{21} & d_{22} & \cdots & d_{2n} \\ d_{31} & d_{32} & \cdots & d_{3n} \\ d_{m1} & d_{m2} & \cdots & d_{mn} \end{bmatrix} \tag{2.7.7}$$

In general, the number of rows of a particular matrix need not equal the number of columns (*m* need not equal *n)* but the finite element equilibrium equations always contain a *square matrix,* where *m* does equal *n*. The size of a matrix is described by its *order*; the order of the *D*-matrix in (2.7.7) is *m by n* while the order of the *A*-matrix in (2.7.6) is *2 by* 2. If a matrix is square, it is common to simply identify it as a square matrix of order *m*. Hence, the *A*-matrix is a square matrix of order 2. The *main diagonal* of a square matrix consists of the elements that have equal values of *m* and *n,* hence, the main diagonal of (2.7.7) consists of the elements d_{11}, d_{22}, ...d_{mn}.

Matrix Operations

Matrices can be *added* only if they are of the same order. Hence, the *A* and *B* matrices shown in (2.7.6) cannot be added although matrices *B* and *C*, being of the same order, can be summed:

$$\underline{\underline{B}} + \underline{\underline{C}} = \begin{Bmatrix} x+5 \\ y+20 \end{Bmatrix} \tag{2.7.8}$$

Now consider the *multiplication* of matrices A and B shown in (2.7.6). The *A*-matrix can be post multiplied by the *B*-matrix if the number of columns in one is equal to the number of rows in the other. Thus, the matrix product $\underline{\underline{AB}}$ is defined as:

$$\underline{\underline{AB}} = \begin{bmatrix} 2 & -4 \\ 10 & -15 \end{bmatrix} \begin{Bmatrix} x \\ y \end{Bmatrix} = \begin{Bmatrix} 2x - 4y \\ 10x - 15y \end{Bmatrix} \tag{2.7.9}$$

Notice that m rows in the A-matrix yields *m rows* in the resultant; n columns in the B-matrix results in *n columns* in the resultant. The *transpose* operation specifies that the rows of a matrix are changed into columns, or vice versa.

Consider the A-matrix in (2.7.6) and its transpose, denoted as A^{T}:

$$\underline{\underline{A}} = \begin{bmatrix} 2 & -4 \\ 10 & -15 \end{bmatrix} \qquad \underline{\underline{A}}^T = \begin{bmatrix} 2 & 10 \\ -4 & -15 \end{bmatrix} \tag{2.7.10}$$

It can also be shown that, given two matrices A and B:

$$\left(\underline{\underline{AB}}\right)^T = \underline{\underline{B}}^T \underline{\underline{A}}^T \tag{2.7.11}$$

A symmetrical matrix is equal to its transpose. While the A-matrix of (2.7.10) is not symmetrical, the square matrix in (2.7.2) is:

$$\underline{\underline{K}} = 3.14(10)^6 \begin{bmatrix} 1 & -1 & 0 \\ -1 & 2 & -1 \\ 0 & -1 & 1 \end{bmatrix} \qquad \underline{\underline{K}}^T = 3.14(10)^6 \begin{bmatrix} 1 & -1 & 0 \\ -1 & 2 & -1 \\ 0 & -1 & 1 \end{bmatrix}$$

The concept of the *identity matrix* is useful when performing matrix algebra. The identity matrix contains all zeros except for the value of unity on all diagonal terms:

$$\underline{\underline{I}} \equiv \begin{bmatrix} 1 & 0 & 0 & \cdots & 0 \\ 0 & 1 & 0 & \cdots & 0 \\ 0 & 0 & 1 & \cdots & 0 \\ 0 & 0 & 0 & \cdots & 0 \\ 0 & 0 & 0 & \cdots & 1 \end{bmatrix}$$

The product of any given matrix and its inverse is equal to the identity matrix:

$$\underline{\underline{A}}^{-1}\underline{\underline{A}} \equiv \underline{\underline{I}} \qquad \textit{also: } \underline{\underline{A}}\,\underline{\underline{A}}^{-1} \equiv \underline{\underline{I}}$$

The relationships above assume that the inverse of the *A*-matrix exists. If it does not exist, the *A*-matrix is termed *singular,* and cannot be inverted. For an example of computing the identity matrix, consider the *A*-matrix defined in (2.7.10), and its inverse:

$$\underline{\underline{A}} = \begin{bmatrix} 2 & -4 \\ 10 & -15 \end{bmatrix} \qquad \underline{\underline{A}}^{-1} = \frac{1}{10}\begin{bmatrix} -15 & 4 \\ -10 & 2 \end{bmatrix}$$

The equation for computing the inverse of a square matrix is given in Selby [16, p. 125]. Now, pre-multiplying the *A*-matrix by its inverse yields an identity matrix of order two:

$$\underline{\underline{A}}^{-1}\underline{\underline{A}} = \frac{1}{10}\begin{bmatrix} -15 & 4 \\ -10 & 2 \end{bmatrix}\begin{bmatrix} 2 & -4 \\ 10 & -15 \end{bmatrix} = \begin{bmatrix} 1 & 0 \\ 0 & 1 \end{bmatrix}$$

It can be shown that post-multiplying the *A*-matrix by its inverse has the same effect. It is noted that finding the inverse of a large matrix is often a difficult task.

As mentioned, the identity matrix is useful in matrix algebra. For instance, given the system of equations below, assume that we wish to solve for the *D*-matrix:

$$\underline{\underline{K}}\,\underline{\underline{D}} = \underline{\underline{F}}$$

Using matrix algebra, each side of the equation above is multiplied by the *K*-matrix inverse:

$$\underline{\underline{K}}^{-1}\underline{\underline{K}}\,\underline{\underline{D}} = \underline{\underline{K}}^{-1}\underline{\underline{F}}$$

Following the identity matrix rule, the product of the *K*-matrix multiplied by its inverse is equal to the identity matrix. Using this relationship in the above:

$$\underline{\underline{I}}\,\underline{\underline{D}} = \underline{\underline{K}}^{-1}\underline{\underline{F}}$$

It can be shown that any matrix operated on by the identity matrix is equal to itself, hence the above is equivalent to:

$$\underline{\underline{D}} = \underline{\underline{K}}^{-1}\underline{\underline{F}}$$

The operations above would suggest that to compute the nodal displacements for a given finite element model, global equilibrium equations are manipulated as shown, using the inverse of the *K*-matrix. However, attempting to do so for any problem of practical interest would involve inverting a rather large *K*-matrix. In practice, matrix inversion is not performed, and the equilibrium equations are typically solved with some type of Gaussian elimination scheme. The reader unfamiliar with fundamental matrix operations is advised to consult suitable reference; selected topics in matrix algebra are reviewed by Bathe [2].

Vectors

A vector can be considered a matrix of order *m by 1* or *1 by n*. If a vector is of the order *m by 1,* it is called a *column vector*. For example, the *B*-matrix in (2.7.6) can be termed a column vector:

$$\underset{2\ by\ 1}{\underline{\underline{B}}} = \begin{Bmatrix} x \\ y \end{Bmatrix} \tag{2.7.12}$$

A *1 by n* vector is called a row vector; for example:

$$\underset{1\ by\ 2}{\underline{\underline{B}}^T} = \{x \quad y\} \tag{2.7.13}$$

Considering the definition of the transpose operation given above, it should be clear that the operation of transposing a column vector results in a row vector, and vice versa:

$$\underline{\underline{F}} = \begin{Bmatrix} f_1 \\ f_2 \\ \cdot \\ \cdot \\ \cdot \\ f_n \end{Bmatrix} \qquad \underline{\underline{F}}^T = \{f_1 \quad f_2 \quad \dots \quad f_n\} \tag{2.7.14}$$

Solving a System of Simultaneous, Linear, Algebraic Equations
The finite element equilibrium equations form a system of *simultaneous, linear, algebraic equations*. A straightforward way of solving such a system is to use simple substitution. Consider the system of equations below:

$$2x-4y=5 \tag{2.7.15}$$

$$10x-15y=20 \tag{2.7.16}$$

Solving (2.7.15) for x:

$$x=\frac{5+4y}{2} \tag{2.7.17}$$

Substituting (2.7.17) into (2.7.16):

$$10\left(\frac{5+4y}{2}\right)-15y=20 \qquad \therefore y=-1$$

Substituting y= −1 into (2.7.15) yields:

$$2x-4(-1)=5 \qquad \therefore x=\frac{1}{2} \tag{2.7.18}$$

While the substitution method works well for this small system of equations, it becomes increasingly impractical as the order of the system increases. In addition, this method is not readily adaptable to computer programming.

Gaussian Elimination
There exist two broad classes of procedures for solving a system of linear equations, namely, *direct* and *iterative* methods. Each has its advantages and disadvantages, but since the direct method is currently more common in linear, static finite element problems, it will be considered here. For example, assume that one wishes to solve the system of equations below for the unknown B-vector:

$$\underline{\underline{A}}\,\underline{B}=\underline{C}$$

Assume that B and C are vectors while A is a square, symmetric, nonsingular matrix. If the order of the square matrix is small, it is elementary to express the system of equations above as:

$$\underline{\underline{A}}^{-1}\underline{\underline{A}}\underline{\underline{B}} = \underline{\underline{A}}^{-1}\underline{\underline{C}}$$

Or, using the identity matrix, the above reduces to:

$$\underline{\underline{B}} = \underline{\underline{A}}^{-1}\underline{\underline{C}}$$

However, as the order of the square matrix increases, it becomes increasingly difficult to perform the inversion process. Hence, another method, called Gaussian elimination, is employed. Gaussian elimination uses matrix manipulation and is readily adapted to computer programming, unlike the substitution and inversion methods above. Consider again the system of simultaneous linear equations from the previous example:

$$\begin{aligned} 2x - 4y &= 5 \\ 10x - 15y &= 20 \end{aligned}$$

In matrix form, the equations above can be expressed as:

$$\begin{bmatrix} 2 & -4 \\ 10 & -15 \end{bmatrix} \begin{Bmatrix} x \\ y \end{Bmatrix} = \begin{Bmatrix} 5 \\ 20 \end{Bmatrix} \tag{2.7.19}$$

Using Gaussian elimination to solve (2.7.19), the rows are manipulated so that the last row of the square matrix results in only one nonzero entry, and that entry occupies the position a_{22}. In general, given a square matrix of order n, only one non-zero entry is allowed in the bottom row, and it must occupy the position a_{nn}. The second row from the bottom is then forced to have only two nonzero entries, $a_{n-1,n-1}$ and $a_{n-1,n}$; the third row from the bottom is required to have only three non-zero entries, and so on. In short, during the Gaussian elimination process the square matrix and the vector on the right hand side are manipulated to render the square matrix in the form of an upper triangle of non-zero elements. For instance, solving the system of equations $\underline{\underline{A}}\underline{\underline{B}} = \underline{\underline{C}}$, the values of a_{ii} and c_i are manipulated to transform the square matrix into upper triangular form:

$$\begin{bmatrix} a_{11} & a_{12} & a_{13} & a_{14} \\ a_{21} & a_{22} & a_{33} & a_{44} \\ a_{31} & a_{32} & a_{33} & a_{34} \\ a_{41} & a_{42} & a_{43} & a_{44} \end{bmatrix} \begin{Bmatrix} b_1 \\ b_2 \\ b_3 \\ b_4 \end{Bmatrix} = \begin{Bmatrix} c_1 \\ c_2 \\ c_3 \\ c_4 \end{Bmatrix} \Rightarrow \begin{bmatrix} a'_{11} & a'_{12} & a'_{13} & a'_{14} \\ 0 & a'_{22} & a'_{23} & a'_{24} \\ 0 & 0 & a'_{33} & a'_{34} \\ 0 & 0 & 0 & a'_{44} \end{bmatrix} \begin{Bmatrix} b_1 \\ b_2 \\ b_3 \\ b_4 \end{Bmatrix} = \begin{Bmatrix} c'_1 \\ c'_2 \\ c'_3 \\ c'_4 \end{Bmatrix}$$

(2.7.20)

Note that the B-vector is not modified, since this is the entity that we wish to solve for. The process to generate the upper triangular form begins by examining the first (leftmost) column of the square matrix. The top element in the leftmost column of the square matrix, a_{11}, is to be non-zero; every other element in column one is then forced to zero. Elements in the square matrix are forced to zero by:

- Operating on the left most column (Column 1) first
- Adding to one row a copy of another row. The copied row can be multiplied by a scalar to force certain elements in the matrix to take on the value of zero
- Leaving the original row unaltered
- Through several steps, all entries in Column 1 are forced to zero, except a_{11}, before moving to Column 2
- All entries in Column 2 are then forced to zero, except for a_{12} and a_{22}
- The process is repeated for each remaining column

Another way of stating the objective of Gaussian elimination is that for upper triangular form, each column n of a square matrix is allowed to have at most n non-zero entries, and these non-zero entries must occupy the top n rows. In this regard, observe the manipulated portion of (2.7.20), on the right hand side. Notice that Column 1 has only one entry and it is located at the top of the column, while Column 2 has two entries, and they occupy the top two positions, and so on. While we consider the upper triangular form, a lower triangular form could be utilized as well.

To illustrate Gaussian elimination, the solution to (2.7.19) is performed by examining the first column of the square matrix and attempting to force the element underneath the first entry (i.e., a_{21}) to zero. Recalling (2.7.19):

$$\begin{bmatrix} 2 & -4 \\ 10 & -15 \end{bmatrix} \begin{Bmatrix} x \\ y \end{Bmatrix} = \begin{Bmatrix} 5 \\ 20 \end{Bmatrix} \qquad (2.7.21)$$

The first row is:

$$\text{Row 1: } (2 \quad -4 \quad 5)$$

A copy is made of the coefficients in the first row and this copy is multiplied by negative five:

$$-5(2 \quad -4 \quad 5) = (-10 \quad 20 \quad -25)$$

The equation above is added to the bottom row of (2.7.21), resulting in:

$$\begin{bmatrix} 2 & -4 \\ 0 & 5 \end{bmatrix} \begin{Bmatrix} x \\ y \end{Bmatrix} = \begin{Bmatrix} 5 \\ -5 \end{Bmatrix} \tag{2.7.22}$$

It should be evident that the vector containing the unknown variables (x and y) is not manipulated. Now the bottom row equation in (2.7.22) can easily be solved, yielding $y=-1$. This value can be substituted into the top equation of (2.7.22), to yield $x=1/2$.

Matrix Methods and FEA

In certain cases the system of equations cannot be solved, since some systems have no unique solution. This will be illustrated in the next example in which Gaussian elimination is applied to the finite element equilibrium equations from Example 2.2, which were also used in the discussion of simultaneous equations in this section, in the form of (2.7.2):

$$k \begin{bmatrix} 1 & -1 & 0 \\ -1 & 2 & -1 \\ 0 & -1 & 1 \end{bmatrix} \begin{Bmatrix} U_1 \\ U_2 \\ U_3 \end{Bmatrix} = \begin{Bmatrix} F_1 \\ 0 \\ P \end{Bmatrix} \tag{2.7.23}$$

For convenience, both sides of (2.7.23) are divided by k:

$$\begin{bmatrix} 1 & -1 & 0 \\ -1 & 2 & -1 \\ 0 & -1 & 1 \end{bmatrix} \begin{Bmatrix} U_1 \\ U_2 \\ U_3 \end{Bmatrix} = \begin{Bmatrix} F^* \\ 0 \\ P^* \end{Bmatrix} \qquad F^* \equiv \frac{F_1}{k} \quad P^* \equiv \frac{P}{k} \tag{2.7.24}$$

Again, starting with Column 1, only one element is allowed in that column (only n elements are allowed in column n, and the column must be filled from the top down). A copy of the first row of (2.7.24) is added to the second row to force the element k_{21} to zero:

$$\text{Row 1: } \{1 \quad -1 \quad 0 \quad F^*\} \tag{2.7.25}$$

Adding a copy of Row 1 to Row 2 in (2.7.24):

$$\begin{bmatrix} 1 & -1 & 0 \\ 0 & 1 & -1 \\ 0 & -1 & 1 \end{bmatrix} \begin{Bmatrix} U_1 \\ U_2 \\ U_3 \end{Bmatrix} = \begin{Bmatrix} F^* \\ F^* \\ P^* \end{Bmatrix} \tag{2.7.26}$$

Column 1 is now in the correct form, so Column 2 is examined. Again, column n is allowed to have n non-zero entries, and the non-zero entries must reside in the top n rows. Since Column 2 has three non-zero entries but only two are allowed, k_{32} needs to be eliminated. A copy of Row 2 in (2.7.26) is added to Row 3:

$$\text{Row 2: } (0 \quad 1 \quad -1 \quad F^*)$$

Adding the copy of Row 2 to Row 3 in (2.7.26):

$$\begin{bmatrix} 1 & -1 & 0 \\ 0 & 1 & -1 \\ 0 & 0 & 0 \end{bmatrix} \begin{Bmatrix} U_1 \\ U_2 \\ U_3 \end{Bmatrix} = \begin{Bmatrix} F^* \\ F^* \\ P^* + F^* \end{Bmatrix} \tag{2.7.27}$$

$$\Uparrow$$
$$k_{33} = 0$$

The first example of Gaussian elimination produced the correct form, (2.7.22), with a_{22}=5; in the current example, however, we find that the analogous element in the square matrix, k_{33}, is zero. If k_{33} in (2.7.27) were non-zero, we would proceed as before:

$$U_3 = \frac{P^* + F^*}{k_{33}} \tag{2.7.28}$$

However, in this example k_{33} *is* zero, and (2.7.28) cannot be solved for a unique value of U_3, since division by zero is undefined. Whenever the square matrix has a row that is all zeros, the system of equations cannot be solved because the matrix is singular. A singular matrix occurs when:

- Any term on the main diagonal is zero, or becomes zero during the Gaussian elimination procedure (this implies that any row with all zeros makes the matrix singular)
- Two rows are identical

Diagonal terms cannot be zero since they are used in the denominator when calculating the nodal displacement, as shown in (2.7.28). The terms used in the denominator during Gaussian elimination are termed *pivots*.

One can determine whether or not a square matrix is singular by computing its *matrix determinant*—a zero value for the determinant indicates a singular matrix. As far as finite element analysis is concerned, a singular stiffness matrix cannot produce a *unique* solution for the unknown vector of nodal displacements regardless of whether one uses matrix inversion, Gaussian elimination, or any other scheme.

Singular Matrices and Boundary Conditions

Why is the system of equations given by (2.7.27) singular? The singular nature of (2.7.27) has both a mathematical and physical explanation; the physical explanation will be considered first.

Singular Matrix—Physical Explanation

Consider an *unrestrained* structure loaded with a point force, as shown in Figure 2.20, below.

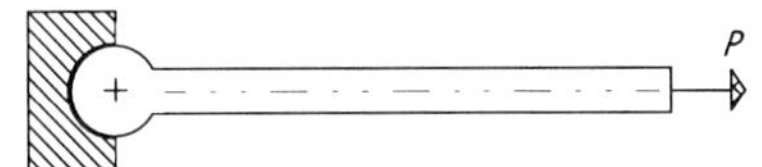

Figure 2.20 An Unrestrained Steering Link

In the figure above, only one external force exists, so the summation of forces is simply P. Since the summation of external forces is not equal to zero, the body will accelerate at a rate inversely proportional to its mass. However, if an external restraint is specified, a reaction force develops. For instance, if a restraint is imposed on the left end of the steering link structure above, a reaction force equal and opposite to the applied load, P, will be generated. The summation of external forces is then zero.

The impending motion of the structure, due to a lack of displacement restraints, can be considered a *rigid body motion*, i.e., the motion of a body without any associated strain energy. In other words, the body moves (displaces) but does not deform. Since the finite element equations (in linear static analysis) are based upon strain energy alone, the equilibrium equations become undefined when strain energy is zero. When using commercial finite element software to solve a simple static analysis problem, a singularity warning is usually provided to the user when rigid body motion is imminent.

Can the finite element method be used to solve problems associated with accelerating bodies? The answer is yes. In addition, the finite element method is often used for modal and forced vibration analysis. Fundamental to solving problems of dynamics using finite element analysis is the creation of a *mass matrix*, in addition to the finite element stiffness matrix. Finite element solutions to dynamics problems are beyond the scope of this text but are considered in Bathe [2] and Cook [17]; a very informative paper covering various aspects of dynamics problems is provided by Clough [18].

Singular Matrices—Mathematical Explanation

To investigate the mathematical aspect of matrix singularities, consider the model from Example 2.2, which was idealized by two rod elements. Given that U_1=0, a unique solution for displacement at Nodes 2 and 3 can be found:

$$U_1 = 0 \qquad U_2 = 1.0(10)^{-6} \qquad U_3 = 2.0(10)^{-6} \tag{2.7.29}$$

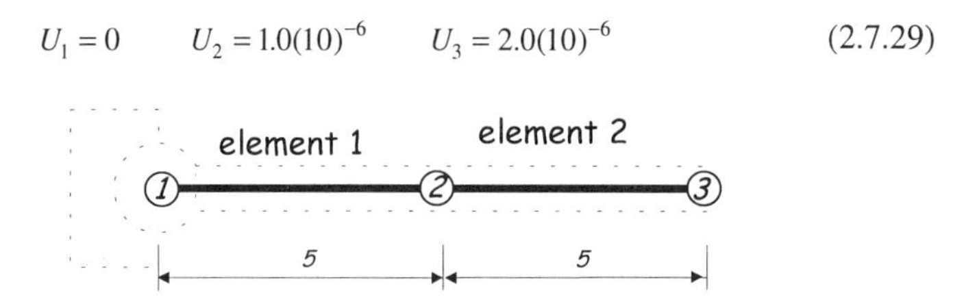

Figure 2.21 Finite Element Model from Example 2.2

Using the nodal displacement values above, the stretch in each element is calculated. For Element 1:

$$\Delta U = U_2 - U_1 = 1.0(10)^{-6} \tag{2.7.30}$$

For Element 2:

$$\Delta U = U_3 - U_2 = 1.0(10)^{-6} \tag{2.7.30}$$

The engineering strain for both elements is:

$$strain \approx \frac{\Delta L}{L_0} = \frac{\Delta U}{\Delta X} = \frac{1.0(10)^{-6}}{5} = 0.2(10)^{-6} \tag{2.7.31}$$

It is interesting to note what happens when U_1 is not specified. Without a unique value U_1, there exist infinite solutions for U_2 and U_3, since a stretch of $1.0(10)^{-6}$ inch can be realized for each element with infinite combinations of U_2 and U_3. For example, one such solution occurs if $U_1 = 1000(10)^{-6}$:

$$\begin{aligned} &U_1 = 1000(10)^{-6} \qquad U_2 = 1001(10)^{-6} \qquad \Delta U_{ele1} = U_2 - U_1 = 1.0(10)^{-6} \\ &U_2 = 1001(10)^{-6} \qquad U_3 = 1002(10)^{-6} \qquad \Delta U_{ele2} = U_3 - U_2 = 1.0(10)^{-6} \end{aligned} \tag{2.7.32}$$

Another valid solution can be found using $U_1 = 1.0(10)^{-6}$:

$$\begin{aligned} &U_1 = 1.0(10)^{-6} \qquad U_2 = 2.0(10)^{-6} \qquad \Delta U_{ele1} = U_2 - U_1 = 1.0(10)^{-6} \\ &U_2 = 2.0(10)^{-6} \qquad U_3 = 3.0(10)^{-6} \qquad \Delta U_{ele2} = U_3 - U_2 = 1.0(10)^{-6} \end{aligned} \tag{2.7.33}$$

In either case above, the strain is, as before:

$$strain \approx \frac{\Delta U}{\Delta X} = \frac{1.0(10)^{-6}}{5} = 0.2(10)^{-6} \tag{2.7.34}$$

Whenever the stiffness matrix is singular, there exist an infinite number of nodal displacements that could provide the same stretch, all rendering the same value for strain. When this occurs, we say that there exists no unique solution to the problem.

The discussion above indicates that when using the finite element method, boundary conditions are required, as they were required to obtain a unique solution using a differential equation and the classical Ritz approach. The imposition of boundary conditions and the associated matrix manipulations will be considered in more depth in Chapter 6.

Positive Definite Stiffness Matrix and the Associated Strain Energy

It will be shown (Chapter 3, Equation 3.5.13) that the strain energy for a given finite element model is expressed in terms of the stiffness matrix and the displacement vector:

$$Strain\ Energy = \frac{1}{2} \underline{D}^T \underline{\underline{K}} \underline{D} \tag{2.7.35}$$

If strain energy (a scalar value) is greater than zero for a given displacement vector, the K-matrix is said to be *positive definite.* When the K-matrix of a structure is positive definite, the structure is considered to be properly *restrained* and *stable,* meaning that neither rigid body motion nor structural collapse are imminent.

To better understand the concept of positive definiteness, consider the scalar equation for strain energy for a uniaxially loaded rod element:

$$Strain\ Energy = \frac{1}{2} ku^2$$

This is the same form of equation as that which describes the strain energy in a linear spring. For a uniaxially loaded rod element, the "spring constant" was derived in Section 1.6:

$$k = \frac{EA}{L}$$

It should be apparent that as the magnitude of displacement u increases, the strain energy increases, regardless of whether or not the displacement becomes more negative or more positive. Only if k becomes non-positive (i.e., less than or equal to zero) can strain energy not increase from its equilibrium value.

It can be shown that the components of the matrix equation given by (2.7.35) contain the same type of quadratic terms as the scalar strain energy equation. Hence, by analogy, an increase in the magnitude of displacement should be accompanied by an increase in strain energy when the K-matrix is "greater than zero." While simple observation reveals if a scalar (such as stiffness) is greater than zero, the same cannot be said about a matrix. However, it can be shown that when the strain energy computed by (2.7.35) is greater than zero, increasing the magnitude of the displacement vector will bring about an associated increase in strain energy. Again, when the strain energy is computed and found to be greater than zero, the K-matrix is termed positive-definite, which is analogous to having a scalar stiffness that is greater than zero.

When a structure is not properly restrained, or is unstable as in the case of structural collapse, the strain energy is no longer greater than zero. Assuming that the displacement vector does not contain all zeros (which would be a trivial problem), a zero value of strain energy suggests the existence of rigid body motion while a negative value suggests that the structure is collapsing. A requirement for the solution of finite element equilibrium equations for a linear, static, stable structure is that the stiffness matrix must be positive definite, which renders the strain energy greater than zero for any non-zero displacement.

Rigid Body Modes of Deformation

When a structure is not properly restrained, rigid body modes of displacement are possible. Rigid body modes occur when, in an improperly restrained finite element model, the nodal displacements are such that the body displaces without straining. In such a case, strain energy remains zero, while displacements are non-zero.

Rigid body modes of displacement in a finite element model may be identified by solving the system of equations:

$$\frac{1}{2}\underline{\underline{D}}^T \underline{\underline{K}}\,\underline{\underline{D}} = 0 \tag{2.7.36}$$

Operating on Equation 2.7.36 determines what combinations of nodal displacements will render zero strain energy. In other words, using Eigenvector analysis to solve the system of equations given by (2.7.36), one can determine the displacement vector (or vectors) that allow motion of the structure without causing strain.

An Example of a Rigid Body Mode

For a simple example, recall the finite element analysis of the steering link given by Example 2.4, shown again in Figure 2.22. The equilibrium equations for this finite element model, (2.2.18) were shown to be:

$$\frac{EA}{H}\begin{bmatrix} 1 & -1 \\ -1 & 1 \end{bmatrix}\begin{Bmatrix} U_a \\ U_b \end{Bmatrix} = \begin{Bmatrix} F_a \\ F_b \end{Bmatrix} \qquad \underline{\underline{K}} = \frac{EA}{H}\begin{bmatrix} 1 & -1 \\ -1 & 1 \end{bmatrix}$$

$\rightarrow X$

X_a (a) U_a F_a — X_b (b) U_b F_b

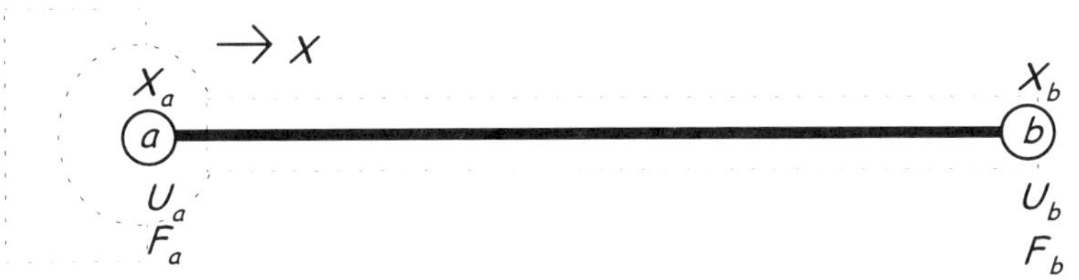

Figure 2.22 Rigid Body Mode of Link

When the nodal displacements associated with a rigid body mode are substituted into (2.7.36), the strain energy is zero. An Eigenvector analysis[4] of the element above will show that two modes of displacement, one flexible and the other rigid, are possible for this element.

The rigid body mode of displacement associated with the element may be expressed as:

$$ {}^{e}\underline{\underline{D}} = \begin{Bmatrix} U_a \\ U_b \end{Bmatrix} = \begin{Bmatrix} 1 \\ 1 \end{Bmatrix} \tag{2.7.37} $$

The magnitude of the rigid body displacement vector above is arbitrary; only the relative magnitude of the nodal displacements is significant. It is convenient to simply set the magnitude of the largest nodal displacement to unity, and then proportion all the other nodal displacement values relative to this value. Notice that (2.7.37) implies that the rod element is not stretched, since Node 1 has displaced the same amount as Node 2. If (2.7.37) truly represents a displacement vector that characterizes a rigid body mode, substituting it into (2.7.35) should result in zero. Hence, substituting (2.7.37) and the K-matrix shown in Figure 2.22 into (2.7.35):

$$ \textit{Strain Energy} = \frac{1}{2}\underline{\underline{D}}^T \underline{\underline{K}}\,\underline{\underline{D}} = \frac{1}{2}\{1 \quad 1\}\frac{EA}{H}\begin{bmatrix} 1 & -1 \\ -1 & 1 \end{bmatrix}\begin{Bmatrix} 1 \\ 1 \end{Bmatrix} \overset{?}{=} 0 $$

Performing the matrix algebra in the equation above:

$$ \frac{1}{2}\{1 \quad 1\}\frac{EA}{H}\begin{bmatrix} 1 & -1 \\ -1 & 1 \end{bmatrix}\begin{Bmatrix} 1 \\ 1 \end{Bmatrix} = \frac{1}{2}\{1 \quad 1\}\frac{EA}{H}\begin{Bmatrix} 1-1 \\ -1+1 \end{Bmatrix} = \frac{1}{2}\{1 \quad 1\}\frac{EA}{H}\begin{Bmatrix} 0 \\ 0 \end{Bmatrix} $$

Simplifying:

$$ \frac{1}{2}\{1 \quad 1\}\frac{EA}{H}\begin{Bmatrix} 0 \\ 0 \end{Bmatrix} = \frac{EA}{2H}\{1 \quad 1\}\begin{Bmatrix} 0 \\ 0 \end{Bmatrix} = \frac{EA}{2H}\{(0)(1) + (0)(1)\} = 0 $$

---

[4] The equation $\det\left({}^{e}\underline{\underline{K}} - \lambda\, \underline{\underline{I}}\right) = 0$ is solved for n values of λ; n is the order of the K-matrix. One Eigenvector, $\underline{\underline{D}}_i$, is computed for each λ_i using the equation $\underline{\underline{K}}\,\underline{\underline{D}}_i = -\lambda\, \underline{\underline{D}}_i$. (A value of unity is chosen for the first component of each $\underline{\underline{D}}_i$.)

The equations above suggest that when the displacement vector that represents rigid body displacement is substituted into the strain energy equation, the strain energy is zero (as expected). The number of rigid body modes that an element can display is a function of how many nodal degrees of freedom the element has. Eigenvector analysis can be performed on either an entire structure or on individual elements, to determine both the rigid and flexible modes of deformation that an element (or an entire structure) can display; see Bathe [2, p. 168]. When the finite element formulation of a structure includes a matrix that describes its mass distribution, Eigenvector analysis can provide a means of investigating the free vibration characteristics of a structure; see Cook [17].

Zero-Energy Modes of Deformation

The strain energy of a finite element model can be zero even if the model is properly restrained. This occurs when a *zero-energy mode* is present. A zero-energy mode is associated with the manner in which the strain energy term of the total potential energy expression is integrated. In short, elements are sometimes under-integrated to make them less stiff in some modes of deformation but, unfortunately, a zero-energy mode of deformation can occur as a result. See Cook [17, p. 190] for an introduction to zero-energy modes, and MacNeal [19] for an in-depth discussion on reduced integration and spurious modes of deformation.

Structural Collapse

Another phenomenon associated with strain energy occurs when the loading and displaced configuration of a structure are such that strain energy is less than zero. Hence, during structural collapse:

$$\text{Strain Energy} = \frac{1}{2} \underline{\underline{D}}^T \underline{\underline{K}}\,\underline{\underline{D}} < 0$$

When a loaded structure is properly restrained and stable, strain energy is greater than zero. However, during structural collapse, the structure assumes a bending mode of deformation where the strain energy is less than zero. Collapse, a dynamic phenomenon, is characterized by the presence of compressive membrane forces at some point during the loading process.

An example of collapse occurs when a slender column is subjected to a compressive, uniaxial force. If the compressive force is increased to sufficient magnitude, the stress state in the column will eventually change from uniaxial compression to bending, given some imperfection in either the geometry of the column or the direction of the load. If at some point the associated bending stiffness goes to zero, the column will collapse. If an *infinitesimally small* imperfection causes the stress in the column to switch from membrane to bending, *bifurcation* has occurred. If bifurcation occurs and at the same time the bending stiffness goes to zero, the resulting collapse of the column is considered an example of bifurcation buckling. The terms *membrane* and *bending stress* are reviewed in Chapter 4.

Bifurcation buckling implies that the pre-buckling deformation of the column is infinitesimally small, while the phenomenon of collapse does not include this restriction. Some structures may bifurcate without immediate collapse, in which case the structure will need to be analyzed using so-called postbuckling analysis. (Perhaps "post-bifurcation" analysis would be a better term.) Finite element procedures to analyze buckling and collapse were briefly considered in Section 1.6.

The Nature of Finite Element Global Matrices—Square, Symmetric, Sparse

Finite element global stiffness matrices are square, typically symmetric and sparse. Why are the matrices square? The expression for total potential energy contains one term for each degree of freedom. This results in one matrix *column* for each DOF. As shown in Example 2.2, one derivative of total potential energy is generated for each DOF; this results in one equation, that is, one matrix *row* for each nodal DOF. Hence, for each column there exists one row, which is a square matrix by definition.

Matrix Symmetry

Why are the matrices encountered in finite element models symmetric? Consider a tensile load, *F,* imposed upon a properly restrained, elastic structure, resulting in displacement of magnitude D. If the load is replaced with load *negative F* (sense reversed) it is assumed that displacement of equal magnitude but opposite sense to D will result. This phenomenon is described by the Maxwell-Betti reciprocal theory; see Timoshenko [20]. The reason for symmetric matrices in linear analysis of stable, elastic structures is related to the fact that the properties of this type of structure are such that reversing the loading sense will result in displacement of equal magnitude but opposite sense. When would the matrix not be symmetric? In a collapse problem, for instance a slender column with a compressive load, the stiffness matrix is not symmetric because compressive loads promote bending deformation while tensile loads result in membrane deformation only.

Exploiting Stiffness Matrix Symmetry

The symmetry of the global finite element stiffness matrix can be exploited using a matrix decomposition scheme. Using decomposition, only the entries in the upper (or lower) triangular area of a symmetric matrix need be computed and manipulated, since the other entries are known by symmetry.

$$\underline{\underline{K}} = \begin{bmatrix} a_{11} & a_{12} & a_{13} & a_{14} \\ & a_{22} & a_{33} & a_{44} \\ & & a_{33} & a_{34} \\ & \text{symmetric} & & a_{44} \end{bmatrix}$$

A general discussion of decomposition methods is provided by Bronson [21]; Bathe [2, p. 440] discusses decomposition methods related to finite procedures.

Matrix Sparseness

Sparseness refers to the fact that the global stiffness matrix typically contains many zeros. Recall the finite element model used in Example 2.2, as illustrated again in Figure 2.23. The reason for the zeros in the stiffness matrix of this example, (2.7.2), is that only two nodal displacements affect the nodal forces at Nodes 1 and 3. That is to say, only two displacement variables are needed to calculate either F_1 or P.

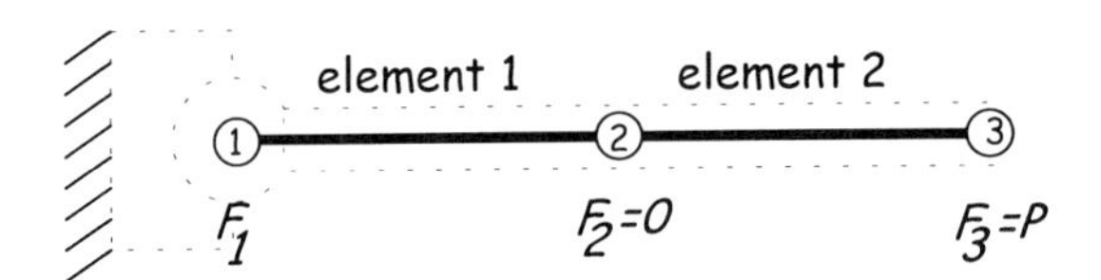

Figure 2.23 Finite Element Model from Example 2.2.2

To further understand the existence of the zeros in (2.7.2), reconsider the finite element model in Figure 2.23 as a combination of linear springs, illustrated in Figure 2.24.

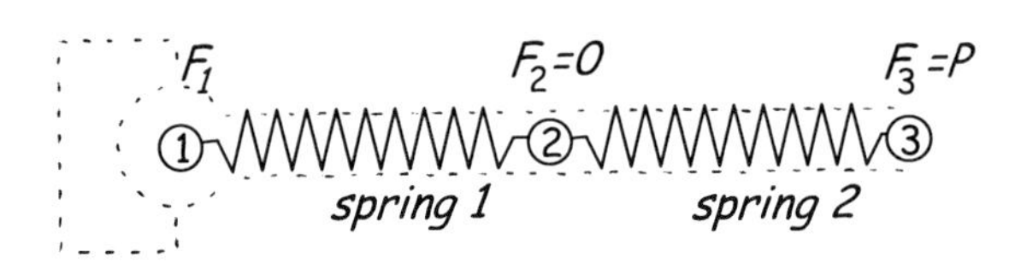

Figure 2.24 Finite Element Model as Linear Springs

The global stiffness matrix for this problem, (2.7.2), has two zero entries, k_{13} and k_{31}:

$$3.0(10)^7 \begin{bmatrix} 1 & -1 & 0 \\ -1 & 2 & -1 \\ 0 & -1 & 1 \end{bmatrix} \begin{Bmatrix} U_1 \\ U_2 \\ U_3 \end{Bmatrix} = \begin{Bmatrix} F_1 \\ 0 \\ P \end{Bmatrix}$$

Equation 2.7.2 describes the equilibrium equations of the springs in Figure 2.24. The first row of (2.7.2) equates to the nodal force at Node 1. Displacements at Nodes 1 and 2 are responsible for the force in Spring 1, and its associated force at Node 1. A zero appears in the third column of Row 1, indicating that displacement at Node 3 has no effect on the force in Spring 1, hence, no effect on the force at Node 1. However, displacement at *all the nodes* does affect the force at Node 2, since both springs are attached to Node 2. Due to the fact that the movement of all nodes affects the spring force at Node 2, no zeros appear in Row 2 of the stiffness matrix. Finally, Row 3 is equated to the force at Node 3, which is attached to Spring 2, only. Since the force in Spring 2 is influenced only by displacement at Nodes 2 and 3, a zero appears in the position k_{31} of the stiffness matrix. This indicates that displacement at Node 1 has no effect upon the force in Spring 2, and its associated force at Node 3.

In general, since the force in any one element (or "spring") is influenced by the displacement at only a relatively few nodes, the global stiffness matrix typically contains zeros. In finite element models with many elements, there typically exists a very large number of zeros. Consider the stiffness matrix for ten rod elements, shown below.

$$\begin{bmatrix} 1 & -1 & 0 & 0 & 0 & 0 & 0 & 0 & 0 & 0 & 0 \\ -1 & 2 & -1 & 0 & 0 & 0 & 0 & 0 & 0 & 0 & 0 \\ 0 & -1 & 2 & -1 & 0 & 0 & 0 & 0 & 0 & 0 & 0 \\ 0 & 0 & -1 & 2 & -1 & 0 & 0 & 0 & 0 & 0 & 0 \\ 0 & 0 & 0 & -1 & 2 & -1 & 0 & 0 & 0 & 0 & 0 \\ 0 & 0 & 0 & 0 & -1 & 2 & -1 & 0 & 0 & 0 & 0 \\ 0 & 0 & 0 & 0 & 0 & -1 & 2 & -1 & 0 & 0 & 0 \\ 0 & 0 & 0 & 0 & 0 & 0 & -1 & 2 & -1 & 0 & 0 \\ 0 & 0 & 0 & 0 & 0 & 0 & 0 & -1 & 2 & -1 & 0 \\ 0 & 0 & 0 & 0 & 0 & 0 & 0 & 0 & -1 & 2 & -1 \\ 0 & 0 & 0 & 0 & 0 & 0 & 0 & 0 & 0 & -1 & 1 \end{bmatrix}$$

Exploiting Matrix Sparseness

To compute equilibrium displacement, it is generally necessary to solve the finite element equilibrium equations in the form of a linear system of algebraic equations:

$$\underline{\underline{K}}\,\underline{D} = \underline{F} \qquad (2.7.38)$$

The K-matrix is a square, sparse, symmetric matrix representing the global stiffness of the structure being analyzed. The external forces are usually specified, such that the components of the F-vector are known; the objective is to compute the unknown values of nodal displacement.

Finite element software developers make use of the fact that in addition to using matrix symmetry, further computational efficiency in solving the equilibrium equations can typically be gained by exploiting the sparseness of the global stiffness matrix. That is, since the stiffness matrix typically contains many zero values, schemes have been designed to reduce computation time by first manipulating the location of the zeros, then implementing a solution scheme that takes advantage of a smaller system of equations, with fewer zeros.

Schemes to manipulate and solve a system of equations with a sparse matrix fall under two major categories, direct and iterative, as illustrated in Figure 2.25. Each of the methods for sparse matrices will be considered briefly.

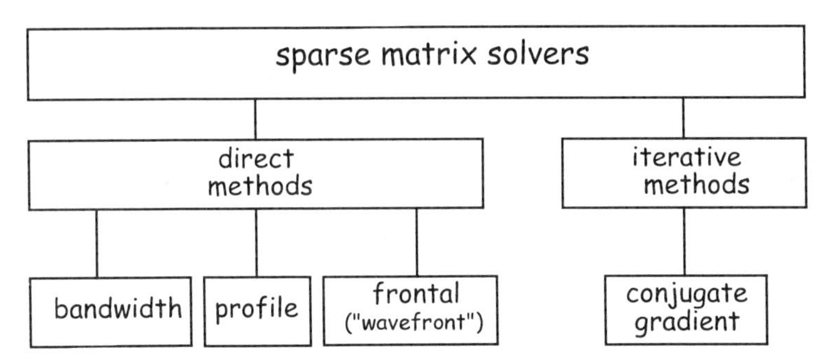

Figure 2.25 Solving Sparse Matrices

Direct Methods

Direct methods are some form of Gaussian elimination. To take advantage of the sparseness of the stiffness matrix, an optimization scheme is typically invoked to reduce the number of zeros that are included in the computations. In general, all three direct methods still process some zeros in the computation process, even after the optimization scheme is invoked. In contrast, the advantage of iterative methods is that they can produce a solution without processing any zeros.

Bandwidth and profile optimization schemes manipulate the node numbers in a finite element model to reduce the number of zeros that are processed, while the frontal ("wavefront") method manipulates the element numbers to achieve the same end.

$$\begin{array}{l} \quad \vert \quad\quad \leftrightarrow \quad\quad \vert \text{semi-bandwidth} = 4 \\ \underline{\underline{K}} = \begin{bmatrix} 1 & 0 & 0 & -1 & 0 \\ & 2 & -1 & 0 & 0 \\ & & 2 & -1 & 0 \\ & & & 2 & -1 \\ & & & & 1 \end{bmatrix} \end{array} \tag{2.7.39}$$

A. Direct Method—Bandwidth Optimization: A bandwidth optimizer attempts to reduce the maximum *semi-bandwidth* in any row of the global stiffness matrix. For example, consider the global stiffness matrix for a small finite element model given by (2.7.39). The *semi-bandwidth* is defined as the distance from the main diagonal to the farthest off-diagonal, non-zero entry, measured in terms of the number of columns (including the diagonal term). The semi-bandwidth of the stiffness matrix shown is equal to four.

When the non-zero terms of the stiffness matrix are clustered near the main diagonal, the semi-bandwidth is small, and the stiffness matrix is termed *banded*. Banded matrices can ignore any zeros that appear in columns that are outside of the semi-bandwidth, thus requiring less computer storage. Using bandwidth techniques, the stiffness matrix from (2.7.39) could be stored as in (2.7.40).

$$\begin{bmatrix} 1 & 0 & 0 & -1 \\ 2 & -1 & 0 & 0 \\ 2 & -1 & 0 & 0 \\ 2 & -1 & 0 & 0 \\ 1 & 0 & 0 & 0 \end{bmatrix} \tag{2.7.40}$$

The zeros in the shaded region of the matrix above are added as place-holders; they are not present in the original stiffness matrix. Bandwidth optimization is discussed in Cuthill [22].

B. Direct Method—Profile Optimization: Although the total storage requirements are reduced using the bandwidth technique, a significant number of zeros are still stored. Another method, using *profile reduction techniques*, endeavors to store only the entries in each *column* that are non-zero. For instance, referring to the original stiffness matrix, (2.7.39), only the entries shown below in the shaded portions of the columns would be stored:

$$\begin{bmatrix} 1 & 0 & 0 & -1 & 0 \\ & 2 & -1 & 0 & 0 \\ & & 2 & -1 & 0 \\ & & & 2 & -1 \\ & & & & 1 \end{bmatrix}$$

Although the profile method *attempts* to store only non-zero entries, zeros that are trapped between two non-zero entries in a column are still stored. In this example, the profile optimization scheme stores fewer zeros than the bandwidth method. The implementation of profile schemes is outlined in Jennings [23], Gibbs [24], and Lewis [25].

C. Direct Method—Frontal Optimization: Frontal optimization schemes have two distinguishing characteristics when compared to bandwidth and profile optimization. Firstly, the frontal (or "wavefront") method never assembles the entire stiffness matrix but instead eliminates equations from the solution process as soon as they are solved. While this requires less computer memory (RAM) than the other schemes, it also increases computer I/O operations. Hence, while the frontal method provides a convenient means of solving large finite element problems, the efficiency may be constrained by the computer's ability to handle intensive I/O operations. The frontal method also differs from the other solution methods by the fact that it manipulates the element numbers in a finite element model instead of the node numbers. The implementation of frontal schemes is discussed by Irons [26], Levy [27], Sloan [28], and Sloan [29].

Iterative Methods

While direct methods attempt to remove as many zeros as possible from the solution process, some (often many) zeros still remain. Using an iterative method, no zeros need be processed. This can lead to a significant decrease in the amount of time needed to solve certain finite element equilibrium equations. The most dramatic reduction in computation time is realized when solving problems with volume elements. When solving large problems using surface elements, direct methods may be faster and use substantially less RAM. The only advantage of an iterative method in such a case is some reduction in hard disk storage requirements; see Ramsay [30] for details.

A disadvantage of iterative methods is that they do not lend themselves to the analysis of multiple load cases, for instance, where the force vector in (2.7.38) would be assigned different values. This is in contrast to the direct method, where once the K-matrix is factorized it can be used to determine the displacement for any number of load cases. The advantages of iterative methods are illustrated by Bathe [31]. The Conjugate Gradient iterative method is discussed in Fried [32] and Reid [33], while the Gauss-Seidel iterative method is discussed in Bathe [2]. A comparison of theoretical computation times for direct and iterative methods is found in Parsons [34]. It is perhaps interesting to note that in the early days of FEA, iterative solvers were apparently used until direct solvers came into fashion; Felippa [7].

The discussion of solver technology above is provided to help the analyst evaluate claims made by FEA software vendors about the "superior" performance of a particular type of solver routine.

2.8 Common Finite Element Errors

Key Concept: There exist several sources of possible error using the finite element method—some of the more common errors are outlined in this section.

Errors associated with engineering analysis may be considered under either of two categories, namely, *idealization error* and *discretization error*.

1.0 Idealization error
  1.1 Posing the problem
  1.2 Establishing boundary conditions
  1.3 Stress-strain assumption
  1.4 Geometric simplification
  1.5 Specifying material behavior
  1.6 Loading assumptions

2.0 Discretization error
  2.1 Imposing boundary conditions
  2.2 Displacement assumption
  2.3 Poor strain approximation due to element distortion
  2.3 Feature representation
  2.5 Numerical integration

2.6 Matrix ill-conditioning
2.7 Degradation of accuracy during Gaussian elimination
2.8 Lack of inter-element displacement compatibility
2.9 Slope discontinuity between elements

Many of the errors above could be considered "operator error," where incorrect data (boundary conditions, forces, material properties) are entered into the model. The interpretation of results may also be considered a potential source of error—incorrect conclusions are easily drawn from an analysis if care is not exercised when interpreting results. A general discussion of some of the errors above will be considered.

Idealization Error

As discussed in Section 1.6, one must be able to *understand the physical nature* of an analysis problem well enough to conceive a proper idealization. From the idealization, a mathematical model may be derived. If the mathematical model is simple enough, a closed-form mathematical solution can be performed. However, in many cases the mathematical nature is such that only an approximate solution is applicable.

One might view mechanical idealization as a process through which a complex, physical problem is translated into a simplified, mathematical model. *Engineering assumptions are always required* in the process of idealization, and it is the quality of these assumptions that control the value of the idealization and the analysis results. The importance of this statement cannot be over-emphasized. Engineering judgment determines whether a given idealization reflects reality or simply results in an expensive exercise in futility. Among other errors, inaccuracies often introduced into the idealization via:

- Establishing boundary conditions
- Specifying Material behavior

Idealization Error—Establishing Boundary Conditions

There are innumerable examples of how the specification of improper boundary conditions can lead to either no results or poor results, as in the case that will be illustrated below.

It is impractical to review every possible manner in which one can error when establishing boundary conditions. The finite element analyst must gain a sufficient theoretical understanding of mechanical idealization principles so that he can understand what boundary conditions are applicable in any particular case. Consider a bracket loaded by the weight of a servo motor, as illustrated in Figure 2.26.

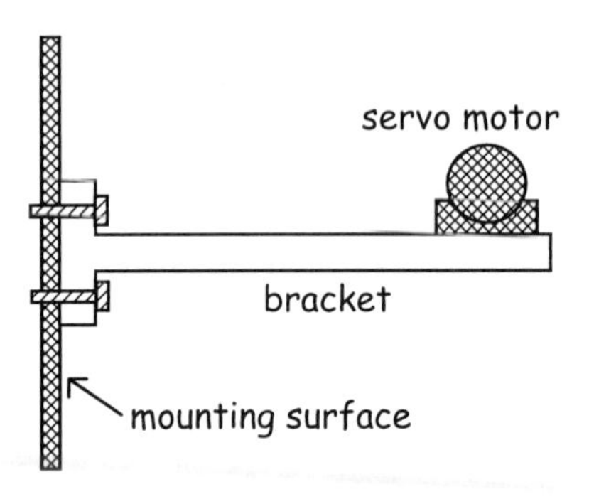

Figure 2.26 Servo Motor Bracket

What displacement boundary conditions should be imposed in this case? The most simple assumption would be to assume that the bracket is rigidly attached to the mounting surface, as shown in Figure 2.27, below.

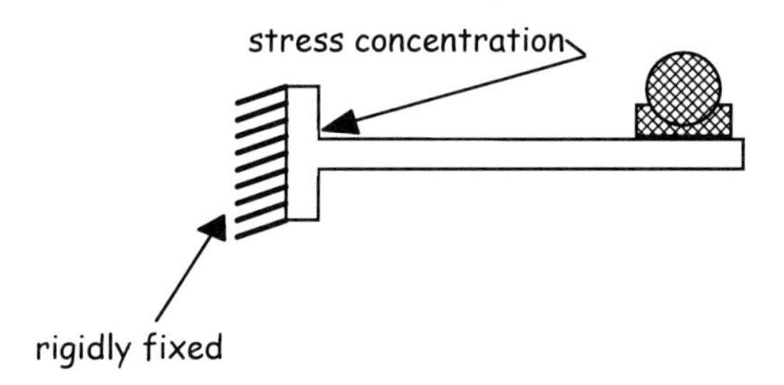

Figure 2.27 Restrained Bracket

Can you see any issues with such a boundary condition assumption? Using the displacement restraint illustrated in Figure 2.27, several things are not accounted for: deformation of the mounting surface, and deformation of the mounting bolts, for instance. One might imagine that a very flexible mounting surface would result in most of the displacement taking place in the surface instead of in the bracket. In addition, significant stretching of the mounting bolts could occur, depending upon their strength relative to the bracket and mounting surface. Finally, the restraints shown in Figure 2.27 do not account for the compressive stress that the bolt heads impart to the bracket. This compressive stress, in the vicinity of the stress concentration might have a significant impact upon the resulting stress values.

Idealization Error—Specifying Material Behavior
Some error may be introduced due to the presumption of certain material properties. For instance, if Poisson's ratio is assigned a value 0.5 (or nearly so) problems arise because the generalized Hooke's law equations for three-dimensional stress, (1.5.14), become undefined. An elastomeric material, a natural rubber for instance, is characterized by a Poisson's ratio value of 0.5, which indicates that the material is incompressible. Many metals, steel for instance, are also considered incompressible when they are deformed past the point of elastic behavior and enter the plastic (or elastic-plastic) material behavior region.

Discretization Error
Using finite element techniques, a mechanical idealization is first conceived and then *discretized*. Discretization is the process where the idealization, having an infinite number of DOF's, is replaced with a model having finite number of DOF's. In the following, errors associated with discretization will be briefly discussed.

Element Errors Associated with Discretization
- Imposing essential boundary conditions
- Displacement assumption
- Poor strain approximation due to element distortion
- Feature representation

Global Errors Associated with Discretization
- Numerical integration
- Matrix ill-conditioning
- Degradation of accuracy during Gaussian elimination
- Lack of inter-element displacement compatibility
- Slope discontinuity between elements

Element Error—Imposing Essential Boundary Conditions
One common discretization problem deals with the proper imposition of essential boundary conditions. The simple rod element, used in one-dimensional space, has only one DOF per node; as such, simply restraining displacement at one node is sufficient to inhibit rigid body motion, thereby avoiding a singularity in the stiffness matrix. (Recall that rigid body motion is displacement of a body in such a manner that no strain energy is induced.) However, when one considers finite elements in three-dimensional space, with nodes having three translational DOF's,

restraining rigid body motion can be more complicated. For example, consider the 8-node brick element in three-dimensional space depicted in Figure 2.28. What are the minimum nodal displacement restraints required to prevent rigid body motion of this element?

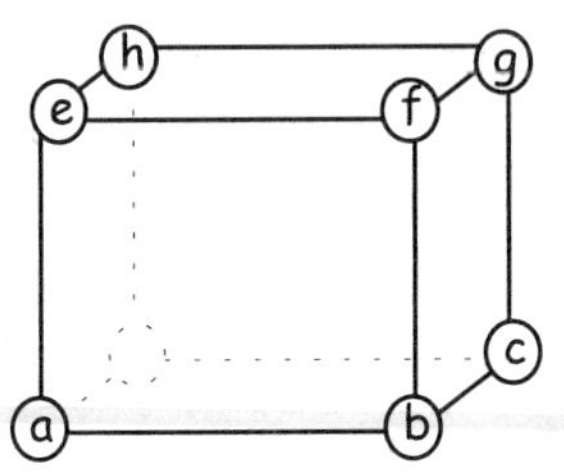

Figure 2.28 Brick Element in 3-D Space

In three dimensional space, there exists the potential for six rigid body modes. The 8-node brick element has three nodal degrees of freedom U, V, and W, representing displacement in the X, Y, and Z coordinate directions, respectively. To remove the rigid body modes of displacement, a first guess might be to restrain all displacement at any one of the nodes. Physically, this seems logical, since a rigid restraint at any point would seem to adequately restrain the structure. Unfortunately, physical logic does not always translate into good finite element discretization.

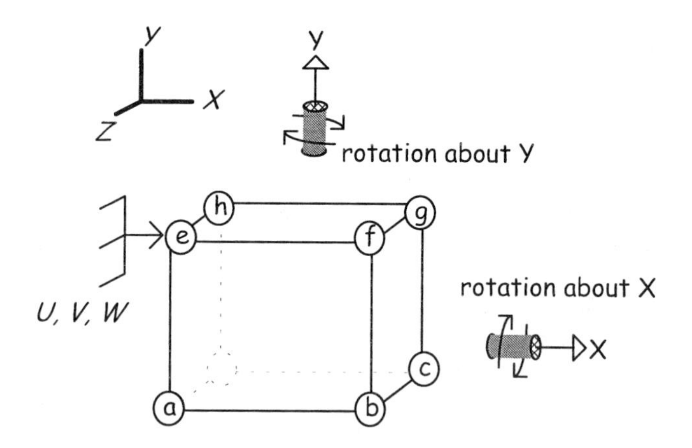

Figure 2.29 Restraining One Node

Totally restraining any one node of the finite element above will allow the body to rigidly rotate, mathematically speaking.

For example, the displacement at *Node e* in Figure 2.29 is totally restrained in all three coordinate directions, however, rigid body rotation will still result. The resulting rigid body rotations about the *X* and *Y* axes are illustrated in Figure 2.29; although not illustrated, rotation also occurs about the *Z* axis.

It may be helpful (in this example) to consider a displacement restraint at a single node as a ball and socket joint that attempts to constrain a body that is subjected to gravitational loading. Although a ball and socket joint at *Node e* (Figure 2.30) would restrain translational motion at the node, the body would tend to rotate.

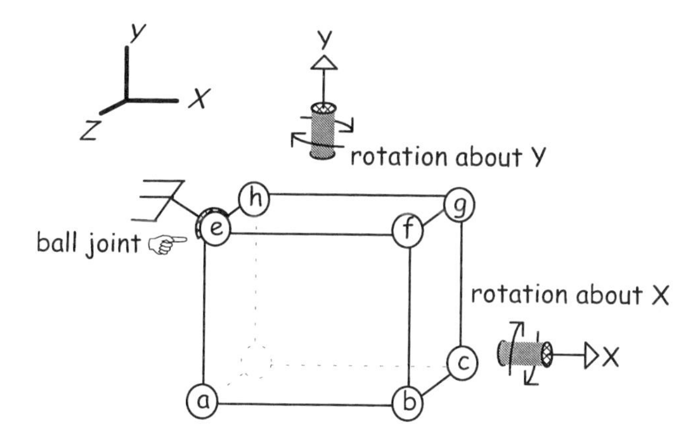

Figure 2.30 Nodal Restraint as a Ball and Socket Joint

Now, imagine that another support, restraining motion in only the *Y* direction, is introduced at *Node b*; see Figure 2.31. This additional restraint, in combination with the existing *Y* restraint at *Node e*, would provide a "resisting moment" to inhibit rotation about the *Z* axis. (An additional restraint at any one of Nodes *b*, *c*, *f*, or *g* will also provide the necessary "resisting moment.") Similarly, restraining *X* and *Y* translations at *Node h* provides "resisting moments" to inhibit rotation about the *Y* and *X* axes, respectively. These restraints are also illustrated in Figure 2.31. There are other combinations of translational restraints that would prevent rigid body motion of the element but in any case, more than three translational restraints are needed to prevent the three translations and three rotations that characterize rigid body motion in three-dimensional space.

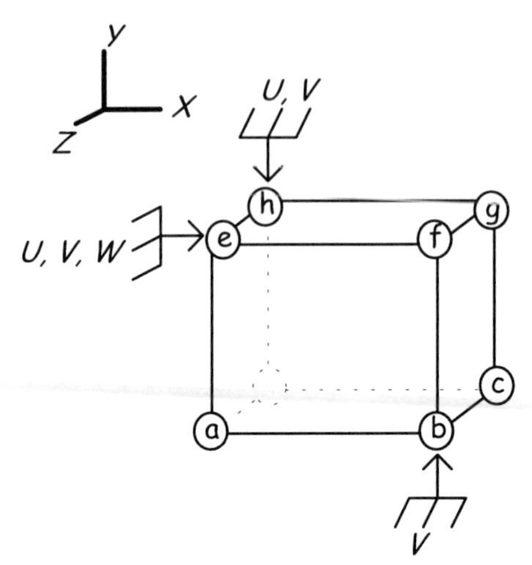

Figure 2.31 Inhibiting Rigid Body Motion

The discussion above illustrates a few basics concerning the importance of imposing the correct boundary conditions. Insufficient boundary conditions render the stiffness matrix singular, meaning that a unique solution for displacement cannot be found. Most finite element software programs will issue a warning, "*singular system encountered*" or perhaps, "*non-positive definite system encountered.*" The latter warning refers to the fact that the strain energy is not greater than zero using the given boundary conditions, suggesting that the structure is either unstable, as in the case of structural collapse, or not properly restrained, such that rigid body modes of displacement are possible. The singular system warning typically indicates rigid body motion.

Element Error—Displacement Assumption

As discussed, if the exact form of displacement is not contained in the finite element displacement assumption, an error occurs. The *h*-method of finite element analysis endeavors to minimize this error by using lower order displacement assumptions (typically linear or quadratic) then refining the model using more, smaller elements. The process of using more, smaller elements to gain increased accuracy is known as *h*-convergence. In the limit, as the largest dimension of the element approaches zero, the finite element approximation for displacement is assumed to approach the true displacement, further assuming that no other errors are introduced, either by poor initial idealization or by cumulative mathematical errors. Convergence is discussed in more depth in Chapter 4.

Element Error—Poor Strain Approximation Due to Element Distortion
Distorted elements influence the accuracy of the finite element approximation for strain. For instance, some bending elements (beams, plates, shells) compute transverse shear strain. This type of element may have difficulty computing shear strain when the element becomes very thin. Many software developers have been able to correct this particular element distortion problem. Element distortion will be considered further in Chapters 5 and 7.

Element Error—Feature Representation
Feature representation error results when the element boundary is unable to replicate the exact boundary of the structure being modeled. For example, consider the finite element model using four-node quadrilateral surface elements shown in Figure 2.32.

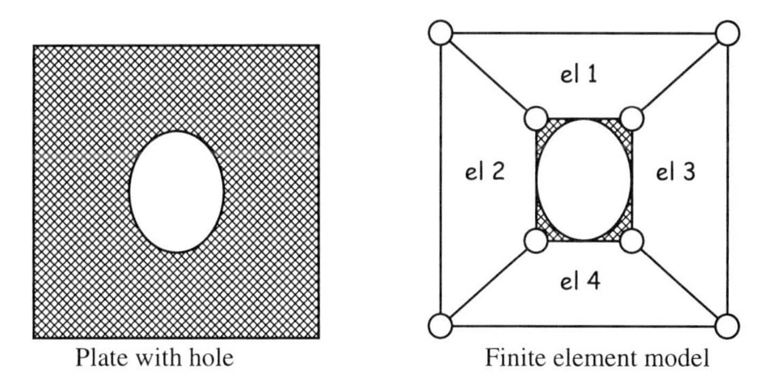

Figure 2.32 Discretization Error

Notice that using just four elements, a significant error in feature representation exists, as illustrated in the cross-hatched area that shows through on the figure on the right. Using more, smaller elements, it should be obvious that this type of error can be reduced.

Global Errors
Global errors are those associated with the assembled finite element model. Even if each element exactly represents the displacement within the boundary of a particular element, the assembled model may not represent the displacement within the entire structure, due to global errors. Consider first the global error associated with how the strain energy portion of the TPE functional is integrated.

Global Error—Numerical Integration
In Example 2.1, step 5 of the finite element process specifies that the integral portion of the total potential energy expression (the strain energy portion) requires integration. Since all of the variables within the integral are constant, integration is easily accomplished in this example using closed-form techniques. However, when higher order displacement assumptions are used, closed-form integration can be quite tedious, even in one-dimensional problems. In many situations, the integrand in the strain energy expression is so complicated that closed-from integration is either impractical or impossible. Therefore, except for the most simple finite elements, numerical integration is employed. The use of numerical integration instead of closed-form integration introduces another possible source of error. Numerical integration will be considered in more depth in Chapter 5.

Global Error—Matrix Ill-Conditioning
When very flexible elements are connected to very rigid elements, the finite element solution can render poor solutions for displacement due to round off error. Ill-conditioning errors typically manifest themselves during the solution phase of an analysis. An illustration of ill-conditioning is considered in the finite element model shown in Figure 2.33.

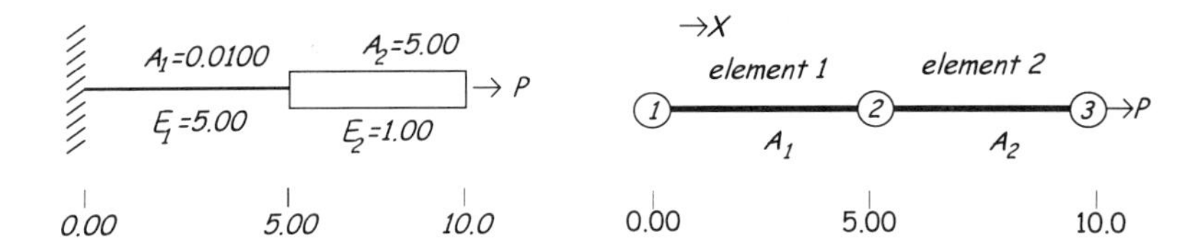

Figure 2.33 Structure with an Ill-Conditioned Stiffness Matrix

A stepped shaft, characterized by two different cross sections, is depicted in Figure 2.33. Two rod elements will be used to model the structure, with the expanded equilibrium equations for Element 1 given as:

$$\frac{{}^{1}E\,{}^{1}A}{{}^{1}H}\begin{bmatrix} 1 & -1 & 0 \\ -1 & 1 & 0 \\ 0 & 0 & 0 \end{bmatrix}\begin{Bmatrix} U_1 \\ U_2 \\ U_3 \end{Bmatrix}=\begin{Bmatrix} F_1 \\ F_2 \\ F_3 \end{Bmatrix}$$

The equilibrium equations for Element 2 are:

$$\frac{{}^2E\,{}^2A}{{}^2H}\begin{bmatrix} 0 & 0 & 0 \\ 0 & 1 & -1 \\ 0 & -1 & 1 \end{bmatrix}\begin{Bmatrix} U_1 \\ U_2 \\ U_3 \end{Bmatrix} = \begin{Bmatrix} F_1 \\ F_2 \\ F_3 \end{Bmatrix}$$

Substituting the known variables into the equations above, the stiffness matrices are:

$${}^1\underline{\underline{K}} = \frac{5.00(0.0100)}{5.00}\begin{bmatrix} 1 & -1 & 0 \\ -1 & 1 & 0 \\ 0 & 0 & 0 \end{bmatrix} = \begin{bmatrix} .0100 & -.0100 & 0 \\ -.0100 & .0100 & 0 \\ 0 & 0 & 0 \end{bmatrix}$$

$${}^2\underline{\underline{K}} = \frac{1.00(5.00)}{5.00}\begin{bmatrix} 0 & 0 & 0 \\ 0 & 1 & -1 \\ 0 & -1 & 1 \end{bmatrix} = \begin{bmatrix} 0 & 0 & 0 \\ 0 & 1.00 & -1.00 \\ 0 & -1.00 & 1.00 \end{bmatrix}$$

The element stiffness matrices are added together to represent the global stiffness:

$$\begin{bmatrix} .0100 & -.0100 & 0.00 \\ -.0100 & 1.01 & -1.00 \\ 0.00 & -1.00 & 1.00 \end{bmatrix}\begin{Bmatrix} U_1 \\ U_2 \\ U_3 \end{Bmatrix} = \begin{Bmatrix} F_1 \\ 0.00 \\ P \end{Bmatrix} \tag{2.8.1}$$

Boundary conditions are now imposed upon the global system of equations, (2.8.1). Since U_1=0 and F_1 is unknown, the *3 by 3* system in (2.8.1) is replaced by a *2 by 2* system:

$$\begin{bmatrix} 1.01 & -1.00 \\ -1.00 & 1.00 \end{bmatrix}\begin{Bmatrix} U_2 \\ U_3 \end{Bmatrix} = \begin{Bmatrix} 0.00 \\ P \end{Bmatrix} \tag{2.8.2}$$

Using Gaussian elimination, (2.8.2) is manipulated to yield:

$$\begin{bmatrix} 1.01 & -1.00 \\ 0.00 & .00990 \end{bmatrix}\begin{Bmatrix} U_2 \\ U_3 \end{Bmatrix} = \begin{Bmatrix} 0.00 \\ P \end{Bmatrix} \qquad \therefore U_3 = 101\,P \tag{2.8.3}$$

While performing the numerical operations associated with Gaussian elimination, round-off errors are often introduced in some (or all) of the numerical values. Consider the effects of a small error in the k_{11} component of the stiffness matrix given by (2.8.2):

$$\begin{bmatrix} 1.02 & -1.00 \\ -1.00 & 1.00 \end{bmatrix} \begin{Bmatrix} U_2 \\ U_3 \end{Bmatrix} = \begin{Bmatrix} 0.00 \\ P \end{Bmatrix} \tag{2.8.4}$$

Using Gaussian elimination to solve the system containing the small error:

$$\begin{bmatrix} 1.02 & -1.00 \\ 0.00 & .0196 \end{bmatrix} \begin{Bmatrix} U_2 \\ U_3 \end{Bmatrix} = \begin{Bmatrix} 0.00 \\ P \end{Bmatrix} \qquad \therefore U_3 = 51.0\ P \tag{2.8.5}$$

It may be surprising to note that a 1% error in one of the entries of the stiffness matrix is responsible for a 50% error in displacement. The large change in nodal displacement due to a small numerical error is attributed to *matrix ill-conditioning*. If all the numbers in the stiffness matrix, and associated numerical operations, were exact, there would be no error. However, using a digital computer, digit accuracy can be a concern, and the severity of the error depends upon the ratio of largest to smallest numbers used to solve the stiffness matrix, along with the word length of the computer.

Whenever a flexible member is joined to a relatively stiff member, ill-conditioning may be present. Many finite element software programs calculate a matrix conditioning number (when solving the equilibrium equations) to guide the analyst in this area of concern. Another result of matrix ill-conditioning is that reaction forces are incorrectly computed, as discussed in Chapter 6. See Cook [17] and Bathe [2, p. 481] for more on matrix ill-conditioning.

Global Error—Degradation of Accuracy During Gaussian Elimination

As shown, Gaussian elimination is a process where mathematical operations are performed in a stepwise fashion and, at each step, a quotient is formed with the *pivot* (previously discussed) in the denominator. With each step, some accuracy is lost, due to rounding; Cook [17, p. 551] illustrates this principle.

Global Error—Lack Of Inter-Element Displacement Compatibility

The issue of element compatibility deals with the fact that differing classes of elements represent displacement differently. For instance, a 4-node surface element for plane stress applications uses a linear polynomial displacement assumption while an 8-node element for the same application uses a quadratic displacement assumption. Without any modifications, joining two of these elements together would typically result in a discontinuity of displacement across the inter-element boundary. The same problem exists when joining other elements that have differing order displacement assumptions. The analyst should consult suitable software documentation, specific to the individual software, to correctly join dissimilar elements. Compatibility is considered again in Chapter 4.

Global Error—Slope Discontinuity Between Elements

The finite element method can be considered a method that provides an approximate solution for displacement by using a combination of lower-order solutions. For example, if a cubic polynomial describes the exact displacement within a structure, the finite element method may use several linear pieces as an approximation as shown in Example 2.5. Whereas linear approximations may well represent the displacement *within* a particular element, slope (derivative) continuity is typically not enforced between elements. Of what consequence is this?

Recall that normal strains, in the examples shown, are a function of first order derivatives of displacement.[5] Since slope continuity is not enforced, strain continuity is not enforced. Furthermore, since stress is a function of strain, the discontinuity of strain can result in a discontinuity of stress at element interfaces, as illustrated in Figure 2.18. The bottom line is this: the finite element method sacrifices slope (hence, perhaps stress) continuity that a closed-form solution would require, in return for an approximate solution for displacement. In a region of a finite element model where a stress discontinuity is present, forces computed using the stress values at the nodes of two adjacent elements are typically not in equilibrium. However, forces computed using the *average* stress will be in equilibrium. Thus, force equilibrium is maintained in an average sense.

A paper by Citipitioglu [35] considers accuracy, error indicators, and convergence; the same topics with application to the *p*-method of finite elements, are discussed in a more recent publication, Sazbó [11].

References

1. Ugural, A.C., Fenster, S.K., *Advanced Strength And Applied Elasticity*, Elsevier, N.Y., 1987
2. Bathe, K., *Finite Element Procedures In Engineering Analysis*, Prentice-Hall, Inc., Englewood Cliffs, N.J., 1982
3. Courant, R., "Variational Methods for the Solution of Problems of Equilibrium and Vibrations," The Collegiate Press, George Banta Publishing Co., January, 1943

[5] Strains are typically expressed in terms of either 1st or 2nd order derivatives using the finite element approach. Strain in terms of 2nd order derivatives will be considered in Chapter 4.

4. Fraeijs de Veubeke, B.M., "Upper and Lower Bounds in Matrix Strucutral Analysis," AGARDograph No. 72, p. 165–201, Pergamon Press, Oxford, 1964
5. Reddy, J.N., *An Introduction To The Finite Element Method*, McGraw-Hill, N.Y., 1984
6. Oliveira, E.R. Arantes, "Theoretical Foundations Of Thc Finite Element Method," Int. J. Solids Structures, Vol. 4, pp. 929–952
7. Felippa, C.A., Clough, R.W., "The Finite Element Method In Solid Mechanics," in SIAM-AMS, Proceedings, Vol. 2, American Mathematics Society, 1970
8. Verma, A., Melosh, R.J., "Numerical Tests For Assessing Finite Element Model Convergence," Int. J. Numer. Methods In Engineering, pp. 843–857, June, 1987
9. Melosh, R.J., "Principles For The Design Of Finite Element Meshes," in "State-Of-The-Art Surveys on Finite Element Analysis Technloogy," ASME, Chapter 3, 1983
10. Strang, G., Fix, G.J., *An Analysis of the Finite Element Method*, Prentice-Hall, Englewood Cliffs, NJ, 1973
11. Szabó, B., Babuska, I., *Finite Element Analysis*, John Wiley & Sons, Inc.,N.Y., 1991
12. Babuska, I., "The *p*- and *hp*- Versions Of The Finite Element Method The State Of The Art," in Finite Elements: Theory And Application, D.L. Doyer et al. (eds), Springer-Verlag, Berlin, 1988
13. Hinton, E., Campbell, J.S., "Local And Global Smoothing Of Discontinuous Finite Element Functions Using A Least Squares Method," Int. J. Num. Meth. in Eng., 8, p. 461–480, 1974
14. Ward, P., "Adaptive Methods And Error Analysis," presented at the Third International Conference On Quality Assurance and Standards In Finite Element Analysis, Stratford-upon-Avon, England, 10 September 1991
15. Russell, R., "Don't Trust The Pretty Pictures," in Machine Design, p. 68–84, 23 May 1996
16. Selby, S.M., ed., *CRC Standard Mathematical Tables*, Twentieth Edition, The Chemical Rubber Co., Cleveland, OH, 1972
17. Cook, R.D., *Concepts And Applications Of Finite Element Analysis*, John Wiley & Sons, N.Y., 1989
18. Clough, R.W., "Analysis of Structural Vibrations and Dynamic Response," Recent Advances in Matrix Methods of Structural Analysis & Design, University of Alabama Press, Huntsville, 1971
19. MacNeal, R.H., *Finite Elements: Their Design And Performance*, Marcel Dekker, Inc., N.Y., 1994
20. Timoshenko, S.P., *Theory Of Elasticity*, McGraw-Hill, N.Y., 1987

21. Bronson, R., *Matrix Algebra: An Introduction*, 2nd ed., Academic Press, N.Y., 1991
22. Cuthill, E., McKee, J., "Reducing The Bandwidth Of Sparse, Symmetric Matrices," Proc. 24th Nat. Conf. Assoc. Comput. Mach., Acm Pub P69, N.Y., pp. 157–172, 1969
23. Jennings, A., *Matrix Computation for Engineers and Scientists,* John Wiley & Sons, N.Y., 1977
24. Gibbs, N.E., "A Hybrid Profile Reduction Algorithm," ACM Trans. Math. Software 2, pp. 378–387, 1976
25. Lewis, J.G., "Implementation of the Gibbs-Poole-Stockmeyer and Gibbs-King Algorithms," ACM Tans. Math. Software, 8, pp. 180–189, 1982
26. Irons, B.M., "A Frontal Solution Program," Int. J. Num. Meth. in Eng., Vol. 2, p. 5–32, 1970
27. Levy, R., "Resequencing of the Structural Stiffness Matrix to Improve Computational Efficiency," Jet Propulsion Lab. Quart. Tech. Rev., Vol. 1, pp. 61–70, 1971
28. Sloan, S.W., "An Algorithm For Profile And Wavefront Reduction Of Sparse Matrices," Int. J. Num. Meth. Engng., Vol 23, pp. 239–251, 1986
29. Sloan, S.W., Ng, W.S., "A Direct Comparison of Three Algorithms for Reducing Profile and Wavefront," Computers & Structures, Vol. 33, No.2, pp. 411–419, 1989
30. Ramsay, A., "Iterative FEA Solvers," BenchMark, Alex Ramsay, Editor, NAFEMS, Glasgow, p. 14–15, March, 1994
31. Bathe, K.J., Walczak, J., Zhang, H., "Some Recent Advances for Practical Finite Element Analysis," Computers & Structures, Vol. 47, No. 45, pp. 511–521, 1993
32. Fried, I., "A Gradient Computational Procedure for the Solution of Large Problems Arising from the Finite Element Discretization Method," International Journal for Numerical Methods in Engineering, Vol. 2, pp. 477–494, 1970
33. Reid, J.K., "On the Method of Conjugate Gradients for the Solution of Large Sparse Systems of Linear Equations," Conference on Large Sparse Sets of Linear Equations, St. Catherine's College, Oxford, pp. 231–254, 1970
34. Parsons, D., "Iterative Methods and Finite Elements: When Will Convergence Occur?" USACM Bulletin, Vol 7, No. 3, September, 1994
35. Citipitioglu, E., "Accuracy Assessment for Finite Element Stress Analysis," in Proceeding of the Second Conference on Computing in Civil Engineering, ASCE, New York, p. 246–257, 1980

Chapter 3

Generalized Finite Element Equilibrium Equations

"You must be the change you want to see in the world."

—Gandhi

3.1 Review

Key Concept: While the finite element equilibrium equations for a one-dimensional rod element (Chapter 2) provided a means of illustrating some basic finite element concepts, they have limited practical use. In contrast, a *generalized expression* for finite element equilibrium equations provides a basis for establishing many different types of elements that can be used to solve a broad range of solid mechanics problems, in both two- and three-dimensional space.

To minimize complexity, only finite element concepts in one-dimensional space were considered in Chapters 1 and 2. However, most finite element software is designed to analyze problems posed in two- and three-dimensional space. Chapter 3 establishes generalized equilibrium equations that allow the finite element method to be applied in a wide range of solid mechanics problems, from simple one-dimensional problems, to complex three-dimensional stress analysis.

Before discussing how generalized equilibrium equations are developed, a brief review of highlights from Chapters 1 and 2 is given.

Functional for a Uniaxial Stress State

A functional applicable to problems involving a point loaded rod in a state of uniaxial stress was established in Chapter 1. In Chapter 2, this same functional was used to generate the finite element equilibrium equations for a 2-node rod element. Recalling (2.2.3), the functional representing total potential energy for the 2-node rod element:

$$\tilde{\Pi} = \frac{1}{2}\int_{X_a}^{X_b} EA\left(\frac{d\tilde{U}}{dX}\right)^2 dX - F_a U_a - F_b U_b \qquad (3.1.1)$$

Substituting the displacement interpolation function for the 2-node rod element into (3.1.1), integrating, and then minimizing, the equilibrium equations for this simple element are established:

$$\frac{EA}{H}\begin{bmatrix} 1 & -1 \\ -1 & 1 \end{bmatrix}\begin{Bmatrix} U_a \\ U_b \end{Bmatrix} = \begin{Bmatrix} F_a \\ F_b \end{Bmatrix} \quad (3.1.2)$$

The equilibrium equations in matrix form, above, are analogous to the *scalar* equilibrium equation of a linear spring:

$$KD = F \quad (3.1.3)$$

Displacement of the spring is denoted as D, the spring constant denoted as K, and the external force, F. By analogy, the *element* equilibrium equations from (3.1.2) can be expressed in matrix form as:

$$\underline{\underline{{}^eK}}\,\underline{\underline{{}^eD}} = \underline{\underline{{}^eF}} \quad (3.1.4)$$

Comparing (3.1.4) and (3.1.3), the element stiffness *matrix*, $\underline{\underline{{}^eK}}$, is analogous to the spring constant, K. For the 2-node rod element, comparing (3.1.2) with (3.1.4) suggests that the element stiffness matrix is:

$$\underline{\underline{{}^eK}} = \frac{EA}{H}\begin{bmatrix} 1 & -1 \\ -1 & 1 \end{bmatrix} \quad (3.1.5)$$

As Example 2.5 illustrated, individual element stiffness matrices can be combined to form a stiffness matrix that represents an entire structure. The combination of individual matrices is termed the *global* stiffness matrix:

$$\underline{\underline{K}} = \underline{\underline{{}^1K}} + \underline{\underline{{}^2K}} + \ldots + \underline{\underline{{}^{nel}K}}$$

Using the global stiffness matrix, along with the global displacement and global force vector, the equilibrium equations for the entire structure are given as:

$$\underline{\underline{K}}\,\underline{\underline{D}} = \underline{\underline{F}}$$

Applying boundary conditions, the global system of equations is solved for the unknown nodal displacements using matrix methods and a digital computer. When all of the nodal displacements are known, the interpolation functions from all the elements are fully defined. Used together, the assemblage of interpolation functions provides an approximate solution for displacement within an entire structure. If desired, one can compute stress and strain for any particular element, using the respective displacement interpolation function.

A Generic Functional is Desirable

It should be apparent that generating equilibrium equations for each element, as given by (3.1.4), is a fundamental aspect of the finite element method. It should also be clear that these elemental equilibrium equations are established from a functional that represents the total potential energy that is defined on the domain ("within the boundary") of a given element.

While the functional given by (3.1.1) is directly applicable to the problem of uniaxial stress in a point loaded rod, a generic functional, one that applies to many types of stress-strain idealizations, would be more useful. A generic functional provides a means of generating equilibrium equations for many different types of finite elements.

3.2 Generic Functional and Strain Energy

Key Concept: With a generic functional, equilibrium equations for many different types of elements can be established. Section 3.2 reveals which components comprise the strain energy portion of a generic functional.

Many different types of stress-strain idealizations can be considered if a generic functional can be established. Rigorous proofs of existence and derivations associated with functionals will not be considered in this text. Instead, the functional that was used for the simple one-dimensional rod will be examined, and a "generic functional" established. The advantage of the generic functional is that it can be applied to a broad range of solid mechanics problems, not just uniaxial stress. More rigorous (and elegant) mathematical approaches to establishing functionals are found in texts on variational methods, such as Reddy [1], and also in Huebner [2].

A generic functional is considered in Section 3.2, while Sections 3.3 and 3.4 reveal how this functional is cast into finite element terms, then minimized, yielding a generalized form of equilibrium equations that can be applied to many types of finite elements. A significant amount of matrix manipulation is required to arrive at a generic functional.

Establishing a Generic Functional

The strain energy term in a generic functional must account for a general state of stress. It was noted in Section 1.5 that to express a general state of stress, six stress components (three normal and three shear) are required. To establish a functional that can account for a general state of stress, it may be helpful to first review the functional that was used for uniaxial stress.

Recall (1.4.1), the functional used in conjunction with the idealization of a rod under axial load, as shown in Section 1.4:

$$\Pi = \frac{1}{2}\int_0^L EA\left(\frac{dU}{dX}\right)^2 dX - PU(L) \tag{3.2.1}$$

The functional above describes total potential energy for the entire rod. The integral term in (3.2.1) represents strain energy while the second term represents work potential (i.e., potential energy) of the external loads:

$$\Pi = (Strain\ Energy) + (Work\ Potential) \tag{3.2.2}$$

To obtain a generic functional, (3.2.1) will be modified to account for a general state of stress, using six stress components—modification of the strain energy term will be considered first. Observe that the strain energy term in (3.2.1) can be alternately expressed as:

$$Strain\ Energy = \frac{1}{2}\int_0^L \left(\frac{dU}{dX}\right) E \left(\frac{dU}{dX}\right) A dX \tag{3.2.3}$$

Equation 3.2.3 actually represents the integral of strain multiplied by stress, integrated over the entire volume of the body. This fact is revealed by considering the following definitions:

$$\tau_{uni} = \tau_{XX} = E\frac{dU}{dX} \qquad \varepsilon_{XX} = \frac{dU}{dX} \qquad dV = AdX$$

Substituting the definitions above into (3.2.3), an integral is established, with the integrand consisting of strain multiplied by stress:

$$Strain\ Energy = \frac{1}{2}\int_V \varepsilon_{XX}\tau_{XX}\, dV \tag{3.2.4}$$

Strain energy is equal to one-half the integral of strain multiplied by stress:

$$Strain\ Energy = \frac{1}{2}\int_V (Strain)(Stress)\, dV \tag{3.2.5}$$

The expression for strain energy in (3.2.4) contains only two factors: normal strain in the *X*-coordinate direction and normal stress; this is sufficient to describe the *strain energy density* in a uniaxially loaded body. An elementary discussion of strain energy is given by Higdon [3], while a discussion of strain energy density, as applied to variational problems, is given in Reddy [1].

To account for a *general* state of stress, six stress components and six strain components are required. A general expression for strain energy density should therefore include the following terms:

$$\varepsilon_{XX}\tau_{XX}+\varepsilon_{YY}\tau_{YY}+\varepsilon_{ZZ}\tau_{ZZ}+\gamma_{XY}\tau_{XY}+\gamma_{YZ}\tau_{YZ}+\gamma_{ZX}\tau_{ZX}$$

The terms above are conveniently expressed using matrix notation:

$$\varepsilon_{XX}\tau_{XX}+\varepsilon_{YY}\tau_{YY}+\varepsilon_{ZZ}\tau_{ZZ}+\gamma_{XY}\tau_{XY}+\gamma_{YZ}\tau_{YZ}+\gamma_{ZX}\tau_{ZX}\equiv\underline{\underline{\varepsilon}}^{T}\underline{\underline{\tau}} \tag{3.2.6}$$

Where:

$$\underline{\underline{\varepsilon}}^{T}=\begin{bmatrix}\varepsilon_{XX} & \varepsilon_{YY} & \varepsilon_{ZZ} & \gamma_{XY} & \gamma_{YZ} & \gamma_{ZX}\end{bmatrix} \qquad \underline{\underline{\tau}}=\begin{Bmatrix}\tau_{XX}\\ \tau_{YY}\\ \tau_{ZZ}\\ \tau_{XY}\\ \tau_{YZ}\\ \tau_{ZX}\end{Bmatrix} \tag{3.2.7}$$

Using the matrix notation shown in (3.2.6), the strain energy given by (3.2.5) can be expressed as:

$$Strain\ Energy=\frac{1}{2}\int_{V}\underline{\underline{\varepsilon}}^{T}\,\underline{\underline{\tau}}\,dV \tag{3.2.8}$$

Strain energy is expressed as an integral, with the integrand containing a matrix product of strain and stress. Note that using (3.2.8), one is not limited to problems of uniaxial stress, since a general state of stress can be represented using the six terms that the matrix product represents. In its present form (3.2.8) is not associated with the finite element method. To cast the expression for strain energy into finite element terms, a few more topics, namely, constitutive relationships and their associated material matrices, need to be covered.

3.3 The Material Matrix

Key Concept: A *material matrix* transforms strain into stress, with each type of stress-strain assumption having a unique material matrix. To establish a generic functional in finite element terms, it is convenient to employ a material matrix.

Constitutive Equations

The mathematical relationship between stress and strain is termed a *constitutive equation*. In solid mechanics problems, the constitutive equations describe how strain is transformed into stress, via the material matrix:

$$\{Stress\} = \begin{bmatrix} Material \\ Matrix \end{bmatrix} \{Strain\}$$

In general, the constitutive equations are expressed in terms of stress and strain vectors, along with a matrix of constants that is dependent upon the particular type of material and the state of stress. This matrix is termed the *material matrix*. In the case of uniaxial stress, the constitutive equation is simply stated as a scalar equation:

$$\tau_{uni} = E\varepsilon_{XX} \tag{3.3.1}$$

The uniaxial stress state can be expressed as a scalar equation since the stress, strain, and material matrices contain only one component each. Other constitutive equations are not so simple. For example, a state of plane stress in a linear, elastic, homogeneous, isotropic material is described by three stress and strain values, an elastic modulus, and Poisson's ratio:

$$\begin{aligned} \tau_{XX} &= \frac{E}{1-v^2}\left(\varepsilon_{XX} + v\varepsilon_{YY}\right) \\ \tau_{YY} &= \frac{E}{1-v^2}\left(v\varepsilon_{XX} + \varepsilon_{YY}\right) \\ \tau_{XY} &= \frac{E}{1-v^2}\left(\frac{1-v}{2}\gamma_{XY}\right) \end{aligned} \tag{3.3.2}$$

The three equations above, representing the constitutive equations for plane stress, are conveniently expressed using matrix notation:

$$\begin{Bmatrix} \tau_{XX} \\ \tau_{YY} \\ \tau_{XY} \end{Bmatrix} = \frac{E}{1-\nu^2}\begin{bmatrix} 1 & \nu & 0 \\ \nu & 1 & 0 \\ 0 & 0 & \frac{1-\nu}{2} \end{bmatrix}\begin{Bmatrix} \varepsilon_{XX} \\ \varepsilon_{YY} \\ \gamma_{XY} \end{Bmatrix} \tag{3.3.3}$$

Generally speaking, constitutive equations in solid mechanics problems take the form:

$$\underline{\underline{\tau}} = \underline{\underline{E}}\,\underline{\underline{\varepsilon}} \tag{3.3.4}$$

The material matrix can be considered a means by which strain is transformed into stress. When the stress state is uniaxial, the "material matrix" is simply a scalar, equal to the modulus of elasticity. Some common constitutive relationships used in solid mechanics problems are given in Table 3.1. The relationships given in the table assume that the material is linear, elastic, homogeneous, and isotropic. The number of components in the material matrix is determined by the type of stress-strain assumption, while the values for the variables (the elastic modulus and Poisson's ratio) are determined by the type of material used in the structure.

Returning to the issue of strain energy in the generic functional, recall the expression for strain energy given by (3.2.8):

$$\textit{Strain Energy} = \frac{1}{2}\int_V \underline{\underline{\varepsilon}}^T \underline{\underline{\tau}}\, dV \tag{3.3.5}$$

Substituting (3.3.4), the expression for the constitutive relationship, into (3.3.5):

$$\textit{Strain Energy} = \frac{1}{2}\int_V \underline{\underline{\varepsilon}}^T \underline{\underline{E}}\,\underline{\underline{\varepsilon}}\, dV \tag{3.3.6}$$

Upon substituting a *strain-displacement* matrix into (3.3.6), the resulting expression for strain energy will be cast into a form that is applicable to displacement based finite element methods.

Table 3.1 Common Constitutive Relationships

constitutive relationships $\underline{\underline{\tau}} = \underline{\underline{E}}\,\underline{\underline{\varepsilon}}$

uniaxial stress

$$\underline{\underline{\tau}} = \tau_{XX} \qquad \underline{\underline{E}} = E \qquad \underline{\underline{\varepsilon}} = \varepsilon_{XX}$$

Euler-Bernoulli beam stress

$$\underline{\underline{\tau}} = \tau_{XX} \qquad \underline{\underline{E}} = E \qquad \underline{\underline{\varepsilon}} = \varepsilon_{XX}$$

plane stress

$$\begin{Bmatrix} \tau_{XX} \\ \tau_{YY} \\ \tau_{XY} \end{Bmatrix} \frac{E}{1-\nu^2} \begin{bmatrix} 1 & \nu & 0 \\ \nu & 1 & 0 \\ 0 & 0 & \dfrac{1-\nu}{2} \end{bmatrix} \begin{Bmatrix} \varepsilon_{XX} \\ \varepsilon_{YY} \\ \gamma_{XY} \end{Bmatrix}$$

plane strain

$$\begin{Bmatrix} \tau_{XX} \\ \tau_{YY} \\ \tau_{XY} \end{Bmatrix} \frac{E}{(1+\nu)(1-2\nu)} \begin{bmatrix} 1 & \dfrac{\nu}{1-\nu} & 0 \\ \dfrac{\nu}{1-\nu} & 1 & 0 \\ 0 & 0 & \dfrac{1-2\nu}{2(1-\nu)} \end{bmatrix} \begin{Bmatrix} \varepsilon_{XX} \\ \varepsilon_{YY} \\ \gamma_{XY} \end{Bmatrix}$$

axisymmetric stress

$$\begin{Bmatrix} \tau_{RR} \\ \tau_{\theta\theta} \\ \tau_{ZZ} \\ \tau_{ZR} \end{Bmatrix} \frac{(1-\nu)E}{(1+\nu)(1-2\nu)} \begin{bmatrix} 1 & \dfrac{\nu}{1-\nu} & 0 \\ \dfrac{\nu}{1-\nu} & 1 & 0 \\ 0 & 0 & \dfrac{1-2\nu}{2(1-\nu)} \end{bmatrix} \begin{Bmatrix} \varepsilon_{RR} \\ \varepsilon_{\theta\theta} \\ \varepsilon_{ZZ} \\ \gamma_{ZR} \end{Bmatrix}$$

plate bending

$$\begin{Bmatrix} \tau_{XX} \\ \tau_{YY} \\ \tau_{XY} \end{Bmatrix} \frac{E}{1-\nu^2} \begin{bmatrix} 1 & \nu & 0 \\ \nu & 1 & 0 \\ 0 & 0 & \dfrac{1-\nu}{2} \end{bmatrix} \begin{Bmatrix} \varepsilon_{XX} \\ \varepsilon_{YY} \\ \gamma_{XY} \end{Bmatrix}$$

three-dimensional stress

$$\begin{Bmatrix} \tau_{XX} \\ \tau_{YY} \\ \tau_{YY} \\ \tau_{XY} \\ \tau_{YZ} \\ \tau_{ZX} \end{Bmatrix} \frac{E}{(1+\nu)(1-2\nu)} \begin{bmatrix} 1-\nu & \nu & \iota & 0 & 0 & 0 & \\ \nu & 1-\nu & \nu & 0 & 0 & 0 & \\ \nu & \nu & 1-\nu & 0 & 0 & 0 & \\ 0 & 0 & 0 & \dfrac{1}{2(1+\nu)} & 0 & 0 & 0 \\ 0 & 0 & 0 & 0 & \dfrac{1}{2(1+\nu)} & 0 & \\ 0 & 0 & 0 & 0 & 0 & \dfrac{1}{2(1+\nu)} & \end{bmatrix} \begin{Bmatrix} \varepsilon_{XX} \\ \varepsilon_{YY} \\ \varepsilon_{YY} \\ \gamma_{XY} \\ \gamma_{YZ} \\ \gamma_{ZX} \end{Bmatrix}$$

3.4 The Strain-Displacement Matrix

Key Concept: Using the finite element method, strain is approximated by a function of each element's nodal displacement variables. The *strain-displacement matrix* transforms the finite element nodal displacement variables into the finite element approximation for strain.

The strain energy term given by (3.3.6) is not associated with finite elements. To introduce finite element variables into the functional, a *B-matrix*, otherwise known as the *strain-displacement matrix*, is employed.

The *B*-matrix transforms finite element nodal variables into an approximate expression for strain.

$$\begin{Bmatrix} Strain \\ Vector \end{Bmatrix} = \begin{bmatrix} B \\ Matrix \end{bmatrix} \begin{Bmatrix} Nodal \\ Displacements \end{Bmatrix}$$

In general:

$$\underline{\underline{\varepsilon}} \approx \underline{\underline{B}}\, {}^{e}\underline{\underline{D}} \tag{3.4.1}$$

For an example of a particular *B*-matrix, consider strain in a 2-node rod element, where the nodal displacement vector contains just two components:

$$\underline{\underline{\varepsilon}} = \varepsilon_{XX} \approx \underline{\underline{B}} \begin{Bmatrix} U_a \\ U_b \end{Bmatrix}$$

How is the *B*-matrix determined for the 2-node rod element? Recalling the expression for normal strain, in terms of the finite element approximation for axial displacement:

$$\varepsilon_{XX} = \frac{\partial U}{\partial X} \approx \frac{\partial \tilde{U}}{\partial X} \tag{3.4.2}$$

Noting that the length of a rod element is defined as *H*, the interpolation function for the 2-node rod element from Chapter 2 is given as:

$$\tilde{U}(X) = \left(\frac{X_b - X}{H}\right) U_a + \left(\frac{X - X_a}{H}\right) U_b \tag{3.4.3}$$

Substituting (3.4.3) into (3.4.2):

$$\varepsilon_{XX} \approx \frac{\partial}{\partial X}\left(\left(\frac{X_b - X}{H}\right)U_a + \left(\frac{X - X_a}{H}\right)U_b\right) = \left(\frac{-1}{H}\right)U_a + \left(\frac{1}{H}\right)U_b \tag{3.4.4}$$

In matrix form, (3.4.4) may be expressed as:

$$\varepsilon_{XX} \approx \begin{bmatrix} \frac{-1}{H} & \frac{1}{H} \end{bmatrix} \begin{Bmatrix} U_a \\ U_b \end{Bmatrix} \tag{3.4.5}$$

The matrix that transforms the rod's nodal displacements into strain is given as:

$$\underline{\underline{B}} = \begin{bmatrix} \frac{-1}{H} & \frac{1}{H} \end{bmatrix} \tag{3.4.6}$$

Note that the number of rows in the *B*-matrix is determined by the component(s) of strain computed by a particular type of element; the number of columns is determined by the number of nodal DOF's that element has. The *B*-matrix is not influenced by the material properties of the structure, as is the material matrix.

Regardless of the type of element used, strain can be expressed in terms of a strain-displacement matrix and a vector of nodal displacement variables, as shown in (3.4.1). Substituting (3.4.1) into the expression for strain energy, (3.3.6):

$$Strain\ Energy = \frac{1}{2}\int_V \underline{\varepsilon}^T \underline{\underline{E}}\, \underline{\varepsilon}\, dV \approx \frac{1}{2}\int_V \left(\underline{\underline{B}}\, {}^e\underline{\underline{D}}\right)^T \underline{\underline{E}} \left(\underline{\underline{B}}\, {}^e\underline{\underline{D}}\right) dV \tag{3.4.7}$$

With the introduction of the *B*-matrix into the strain energy term, (3.4.7) now refers to the strain energy in *one particular finite element*. As such, the integral is evaluated over the volume of a single element. Rearranging the transposed[1] term in (3.4.7):

$$Strain\ Energy \approx \frac{1}{2}\int_V {}^e\underline{\underline{D}}^T\, \underline{\underline{B}}^T\, \underline{\underline{E}} \left(\underline{\underline{B}}\, {}^e\underline{\underline{D}}\right) dV \tag{3.4.8}$$

---

[1] Starting with the product of vectors *A* and *B*, the transpose of their product is given as: $\left(\underline{\underline{A}}\underline{\underline{B}}\right)^T = \underline{\underline{B}}^T \underline{\underline{A}}^T$

Equation (3.4.8) expresses strain energy in terms of finite element nodal displacement variables, the strain-displacement matrix, and the material matrix. Recalling the expression for total potential energy, (3.2.2):

$$\Pi = (Strain\ Energy) + (Work\ Potential)$$

Using the approximation for strain energy from (3.4.8) in the above:

$$^{e}\Pi = \frac{1}{2}\int_V \underline{^{e}D}^T\ \underline{\underline{B}}^T \underline{\underline{E}}\ \left(\underline{\underline{B}}\ \underline{^{e}D}\right) dV + (Work\ Potential) \tag{3.4.9}$$

As discussed in Appendix B, the work potential terms consist of nodal forces multiplied by nodal displacements:

$$Work\ Potential = (-1)\left({}^{e}D_a\,{}^{e}F_a + {}^{e}D_b\,{}^{e}F_b + \ldots\right) \tag{3.4.10}$$

The nodal displacement variable ${}^{e}D_a$ represents the displacement at Node a in any coordinate direction *(X, Y,* or *Z)* for element *e.* The nodal force term, ${}^{e}F_a$, represents a force at Node a, acting in the same direction as the displacement variable, ${}^{e}D_a$. It is also noted that in some elements, displacement variables can be of the rotational or translational type, while the "force" variable can represent a force or a force couple (moment). Using (3.4.10) in (3.4.9):

$$^{e}\tilde{\Pi} = \frac{1}{2}\int_V \underline{^{e}D}^T\ \underline{\underline{B}}^T \underline{\underline{E}}\ \left(\underline{\underline{B}}\ \underline{^{e}D}\right) dV - \left({}^{e}D_a\,{}^{e}F_a + {}^{e}D_b\,{}^{e}F_b + \ldots\right) \tag{3.4.11}$$

Equation 3.4.11 represents the total potential energy (approximately, in general) for many different types of elements; this is what we will call the "generic functional." Section 3.5 discusses how (3.4.11) is minimized to yield the generalized equilibrium equations, which can be used for many different types of elements.

3.5 Generalized Finite Element Equilibrium Equations

Key Concept: Given a generic expression for total potential energy (a "generic functional"), a general form for the finite element equilibrium equations can be established through minimization. This general form can be applied to many different types of solid mechanics problems by choosing appropriate material and strain-displacement matrices.

All that remains to obtain a generalized expression for finite element equilibrium equations is to minimize the generic total potential energy functional given by (3.4.11):

$$^{e}\tilde{\Pi} = \frac{1}{2}\int_V \underline{\underline{^{e}D}}^T \; \underline{\underline{B}}^T \underline{\underline{E}} \; \left(\underline{\underline{B}}\,\underline{\underline{^{e}D}}\right) dV - \left(^{e}D_a \, ^{e}F_a + {}^{e}D_b \, ^{e}F_b + \ldots\right)$$

It is convenient to arrange the above in the following form:

$$^{e}\tilde{\Pi} = \frac{1}{2}\underline{\underline{^{e}D}}^T \left[\int_V \underline{\underline{B}}^T \underline{\underline{E}}\,\underline{\underline{B}} \; dV\right] \underline{\underline{^{e}D}} - \left(\underline{\underline{^{e}D}}^T \; \underline{\underline{^{e}F}}\right) \tag{3.5.1}$$

It is possible to remove the displacement vector and its transpose from the integral of (3.5.1) because the nodal displacements are considered to be independent variables, as opposed to being functions of *X*, *Y*, and *Z*. The strain displacement matrix and the material matrix *can* be functions of the spatial coordinates, and therefore must be kept inside the integral, if the equations are to represent the general case. Given that an element has *ne* nodal degrees of freedom, it can be shown that for a broad class of solid mechanics problems, the bracketed term is always a symmetric, square matrix of order *ne*. This matrix, shown in (3.5.2), will be called the *Z*-matrix, for now.

$$\underline{\underline{^{e}Z}} \equiv \left[\int_V \underline{\underline{B}}^T \underline{\underline{E}}\,\underline{\underline{B}} \; dV\right] \tag{3.5.2}$$

It will be shown that the bracketed term in (3.5.2) is equivalent to the element stiffness matrix. Using (3.5.2) in (3.5.1) and rearranging:

$$^{e}\tilde{\Pi} = \frac{1}{2}\left(\underline{\underline{^{e}D}}^T \; \underline{\underline{^{e}Z}}\,\underline{\underline{^{e}D}}\right) - \left(\underline{\underline{^{e}D}}^T \; \underline{\underline{^{e}F}}\right) \tag{3.5.3}$$

To obtain the finite element equilibrium equations, (3.5.3) must be differentiated with respect to each nodal displacement variable, then set to zero. As discussed in

Chapter 2, and illustrated by (2.3.5), an element having *ne* nodal displacement variables generates *ne* equilibrium equations in the minimization process:

$$\frac{\partial^e\tilde{\Pi}}{\partial D_a} = \frac{\partial}{\partial D_a}\left(\frac{1}{2}\left({}^e\underline{\underline{D}}^T\ {}^e\underline{\underline{Z}}\ {}^e\underline{\underline{D}}\right) - \left({}^e\underline{\underline{D}}^T\ {}^e\underline{\underline{F}}\right)\right)$$

$$\frac{\partial^e\tilde{\Pi}}{\partial D_b} = \frac{\partial}{\partial D_b}\left(\frac{1}{2}\left({}^e\underline{\underline{D}}^T\ {}^e\underline{\underline{Z}}\ {}^e\underline{\underline{D}}\right) - \left({}^e\underline{\underline{D}}^T\ {}^e\underline{\underline{F}}\right)\right)$$

$$\vdots$$

Total of *ne* equations

The above again illustrates why the finite element equilibrium equations produce a square matrix: the number of columns is dictated by the number of nodal DOF's, *ne*, while *ne* is also the number of rows, since one row is generated each time a derivative is taken.

A more concise way of stating that (3.5.3) is to be differentiated *ne* times is:

$$\frac{\partial\ {}^e\tilde{\Pi}}{\partial\ {}^e\underline{\underline{D}}} = \frac{\partial}{\partial\ {}^e\underline{\underline{D}}}\left(\frac{1}{2}\ {}^e\underline{\underline{D}}^T\ {}^e\underline{\underline{Z}}\ {}^e\underline{\underline{D}} - {}^e\underline{\underline{D}}^T\ {}^e\underline{\underline{F}}\right)$$

Notice that the equation above states that differentiation is performed with respect to the entire *vector* of nodal displacements for a given element, not just a single component in the vector. The equation above can be expressed using two terms, the first term representing the derivative of the strain energy, while the second term represents differentiation of the work potential terms:

$$\frac{\partial\ {}^e\tilde{\Pi}}{\partial\ {}^e\underline{\underline{D}}} = \frac{\partial}{\partial\ {}^e\underline{\underline{D}}}\left(\frac{1}{2}\ {}^e\underline{\underline{D}}^T\ {}^e\underline{\underline{Z}}\ {}^e\underline{\underline{D}}\right) + \frac{\partial}{\partial\ {}^e\underline{\underline{D}}}\left(-\ {}^e\underline{\underline{D}}^T\ {}^e\underline{\underline{F}}\right) \quad (3.5.4)$$

The first term on the right hand side of (3.5.4) represents a *quadratic form*, and it can be shown[2] that differentiating this type of matrix equation with respect to a vector renders the following:

$$\frac{\partial}{\partial\, {}^e\underline{\underline{D}}}\left(\frac{1}{2}\,{}^e\underline{\underline{D}}^T\,{}^e\underline{\underline{Z}}\,{}^e\underline{\underline{D}}\right) = {}^e\underline{\underline{Z}}\,{}^e\underline{\underline{D}} \tag{3.5.5}$$

Operating on the second term in (3.5.4), minimization of the work potential terms with respect to all of the nodal variables (for the element) renders:

$$\frac{\partial}{\partial\, {}^e\underline{\underline{D}}}\left(-\,{}^e\underline{\underline{D}}^T\,{}^e\underline{\underline{F}}\right) = -\,{}^e\underline{\underline{F}} \tag{3.5.6}$$

If the operation in (3.5.6) above is not clear, consider an example with two vectors, $\underline{\underline{x}}^T = \{x_1 \;\; x_2\}$ and $\underline{\underline{y}} = \begin{Bmatrix} y_1 \\ y_2 \end{Bmatrix}$. Performing the matrix multiplication:

$$\underline{\underline{x}}^T\,\underline{\underline{y}} = x_1 y_1 + x_2 y_2 \tag{3.5.7}$$

Differentiating (3.5.7) with respect to x_1, and then with respect to x_2, yields the two components:

$$\frac{\partial\left(\underline{\underline{x}}^T\,\underline{\underline{y}}\right)}{\partial x_1} = y_1$$

$$\frac{\partial\left(\underline{\underline{x}}^T\,\underline{\underline{y}}\right)}{\partial x_2} = y_2$$

Hence, the differentiation of the matrix product in (3.5.7) with respect to the x-vector renders two components which together comprise the y-vector:

$$\frac{\partial\left(\underline{\underline{x}}^T\,\underline{\underline{y}}\right)}{\partial\,\underline{\underline{x}}} = \begin{Bmatrix} y_1 \\ y_2 \end{Bmatrix} = \underline{\underline{y}}$$

---

[2] Consider the scalar expression (1/2 *xkx)*. Using the product rule for differentiation, *d/dx (1/2 xkx)=1/2 dx/dx (kx)* + *1/2 (xk) dx/dx*, which is simply *kx*. Analogously, it can be shown that the derivative of $1/2\left(\underline{\underline{x}}^T\,\underline{\underline{k}}\,\underline{\underline{x}}\right)$ with respect to $\underline{\underline{x}}$ is simply $\underline{\underline{k}}\,\underline{\underline{x}}$, if $\underline{\underline{k}}$ is symmetric; Cook [4, p. 590].

Substituting (3.5.5) and (3.5.6) into (3.5.4):

$$\frac{\partial\, {}^{e}\tilde{\Pi}}{\partial\, {}^{e}\underline{\underline{D}}} = {}^{e}\underline{\underline{Z}}\, {}^{e}\underline{\underline{D}} - {}^{e}\underline{\underline{F}} \tag{3.5.8}$$

The equation above represents the differentiation of the total potential energy attributed to Element *e,* with respect to all the nodal displacement variables associated with that element. Recall that to minimize, the derivative of total potential energy is to be set to zero. Hence, setting (3.5.8) equal to a null vector ("zero") and rearranging:

$${}^{e}\underline{\underline{Z}}\, {}^{e}\underline{\underline{D}} = {}^{e}\underline{\underline{F}} \tag{3.5.9}$$

The equilibrium equations for Element *e,* expressed in matrix form, are thus defined by (3.5.9). Replacing the *Z*-matrix with its equivalent expression, (3.5.2):

$$\left[\int_V \underline{\underline{B}}^T\, \underline{\underline{E}}\, \underline{\underline{B}}\, dV\right] {}^{e}\underline{\underline{D}} = {}^{e}\underline{\underline{F}} \tag{3.5.10}$$

Equation (3.5.10) represents the *generalized finite element equilibrium equations*, and can be used with many types of elements to solve linear, static, solid mechanics problems. One substitutes into (3.5.10) the material and the strain-displacement matrices associated with the type of idealization that the element is to characterize. Chapter 4 considers how strain-displacement matrices are established for several simple elements.

Although only point loads have been considered so far, the force vector in (3.5.10) can also account for other types of loading on the element. This is discussed in Chapter 6 and in Appendix B.

Note that (3.5.10) is the same form as the scalar spring equation, $KD=F$:

$${}^{e}\underline{\underline{K}}\, {}^{e}\underline{\underline{D}} = {}^{e}\underline{\underline{F}} \tag{3.5.11}$$

The bracketed term represents the stiffness matrix of the element. Hence, comparing (3.5.11) with (3.5.10), the general expression for the stiffness of an individual finite element is:

$${}^{e}\underline{\underline{K}} \equiv \left[\int_V \underline{\underline{B}}^T\, \underline{\underline{E}}\, \underline{\underline{B}}\, dV\right] \tag{3.5.12}$$

Strain Energy in Matrix Form

Strain energy, and the positive definite nature of the K-matrix, was discussed in Chapter 2, Equation 2.7.35. With the matrix expressions derived in this chapter, we can now derive the *matrix* expression for strain energy. Substituting (3.5.12) into (3.5.1), total potential energy is expressed in terms of finite element entities:

$$^{e}\tilde{\Pi} = \frac{1}{2}\,{}^{e}\underline{\underline{D}}^{T}\;{}^{e}\underline{\underline{K}}\;{}^{e}\underline{\underline{D}} - \left({}^{e}\underline{\underline{D}}^{T}\;{}^{e}\underline{\underline{F}}\right)$$

Recalling that the first term in the expression for total potential energy represents strain energy, we have:

$$\textit{Strain Energy} = \frac{1}{2}\,{}^{e}\underline{\underline{D}}^{T}\;{}^{e}\underline{\underline{K}}\;{}^{e}\underline{\underline{D}} \tag{3.5.13}$$

The expression for strain energy above can be used to check if the K-matrix is positive definite, given a non-zero displacement vector, as discussed in Chapter 2.

To provide an illustration of how (3.5.12) is employed, the stiffness matrix for the 2-node rod element will be generated. However, one additional topic, *shape function notation*, will be covered first.

Shape Functions

Interpolation functions can become quite complex, and it is convenient to use alternate notation to express them. For example, consider the familiar 2-node line element used for uniaxial stress applications, shown in Figure 3.1.

Figure 3.1 A 2-Node Rod Element

The interpolation function for the element above is given as:

$$\tilde{U}(X) = \left(\frac{X_b - X}{H}\right)U_a + \left(\frac{X - X_a}{H}\right)U_b$$

It is convenient to express the bracketed functions in each term above as *shape functions*:

$$\tilde{U}(X) = N_1 U_a + N_2 U_b \qquad N_1 = \left(\frac{X_b - X}{H} \right) \quad N_2 = \left(\frac{X - X_a}{H} \right) \tag{3.5.14}$$

The shape functions of the type shown in (3.5.14) are *Lagrange polynomials.* Although linear shape functions are shown, Lagrange polynomials are not limited to first-order polynomials. Characteristic of this type of shape function is that at one and only one node in the model, the value of the shape function takes on the value of unity. At all other nodes, the value of the shape function is zero; elsewhere on the element, the shape function takes on intermediate values. This behavior is illustrated in Figure 3.2.

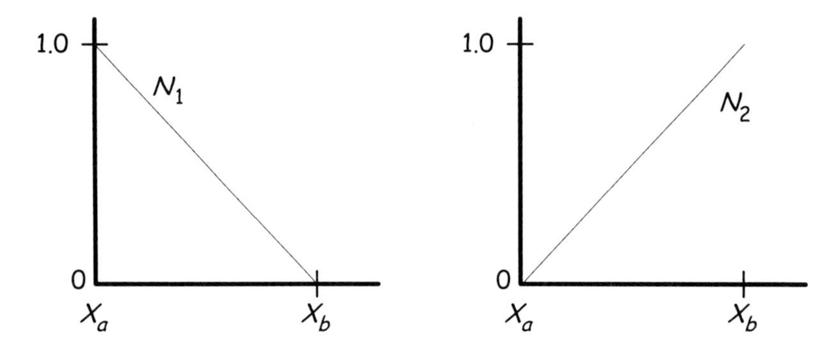

Figure 3.2 Shape Functions for a 2-Node Rod Element

It will be shown in Chapter 5 that for *isoparametric* elements, the shape functions (N_i) define both the *shape* of the element and the displacement within the element, in terms of nodal variables; hence the term "shape function." The interpolation function shown in (3.5.14) is expressed using shape function notation because the strain-displacement matrices are more easily manipulated as a result. Consider how the strain-displacement matrix is expressed for the two-node element above using shape functions. Recalling the expression for normal strain, and using

(3.5.14):

$$\varepsilon_{XX} \approx \frac{\partial \tilde{U}}{\partial X} = \frac{\partial}{\partial X}\left(N_1 U_a + N_2 U_b\right)$$

or:

$$\varepsilon_{XX} \approx \frac{\partial N_1}{\partial X} U_a + \frac{\partial N_2}{\partial X} U_b = \begin{bmatrix} \frac{\partial N_1}{\partial X} & \frac{\partial N_2}{\partial X} \end{bmatrix} \begin{Bmatrix} U_a \\ U_b \end{Bmatrix}$$

Since $\underline{\underline{\varepsilon}} \approx \underline{\underline{B}}\, {}^e\underline{\underline{D}}$, the equation above suggests that the strain-displacement matrix for the 2-node rod element, in terms of shape functions, is expressed as:

$$\underline{\underline{B}} = \begin{bmatrix} \frac{\partial N_1}{\partial X} & \frac{\partial N_2}{\partial X} \end{bmatrix} \tag{3.5.15}$$

Equation 3.5.15 illustrates that the finite element strain-displacement matrix contains derivatives of shape functions. The *B*-matrix can be thought of as the medium through which the finite element nodal displacement variables are transformed into strain, via differential operators. Differentiating the shape functions for the 2-node rod element and substituting into (3.5.15), the result is of course equivalent to (3.4.6).

Element Equilibrium Equations Using the Generalized Form

The generalized form of the element equilibrium equations will now be employed to generate the equilibrium equations for a 2-node rod element. Recalling the general form, (3.5.10):

$$\left[\int_V \underline{\underline{B}}^T\ \underline{\underline{E}}\ \underline{\underline{B}}\, dV\right] {}^e\underline{\underline{D}} = {}^e\underline{\underline{F}}$$

Since the 2-node rod element calculates uniaxial stress only, the material matrix is simply equal to the modulus of elasticity, *E*, as shown in Table 3.1. Using the modulus, and substituting (3.5.15) into the general form above:

$$\left[\int_V \begin{Bmatrix} \frac{\partial N_1}{\partial X} \\ \frac{\partial N_2}{\partial X} \end{Bmatrix} E \begin{bmatrix} \frac{\partial N_1}{\partial X} & \frac{\partial N_2}{\partial X} \end{bmatrix} dV\right] {}^e\underline{\underline{D}} = {}^e\underline{\underline{F}} \tag{3.5.16}$$

As suggested by Figure 3.1, the displacement and force vectors for this element have two components, such that (3.5.16) becomes:

$$\left[\int_V \begin{Bmatrix} \dfrac{\partial N_1}{\partial X} \\ \dfrac{\partial N_2}{\partial X} \end{Bmatrix} E \begin{bmatrix} \dfrac{\partial N_1}{\partial X} & \dfrac{\partial N_2}{\partial X} \end{bmatrix} dV \right] \begin{Bmatrix} U_a \\ U_b \end{Bmatrix} = \begin{Bmatrix} F_a \\ F_b \end{Bmatrix} \tag{3.5.17}$$

Referring to (3.5.14), the derivatives for the shape functions are calculated and substituted into (3.5.17):

$$\left[\int_V \begin{Bmatrix} \dfrac{-1}{H} \\ \dfrac{1}{H} \end{Bmatrix} E \begin{bmatrix} \dfrac{-1}{H} & \dfrac{1}{H} \end{bmatrix} dV \right] \begin{Bmatrix} U_a \\ U_b \end{Bmatrix} = \begin{Bmatrix} F_a \\ F_b \end{Bmatrix} \tag{3.5.18}$$

The integral in (3.5.18) is evaluated over the volume of a single element. For the case of an idealized rod element with uniform cross section A, the substitution $dV = A\,dX$ is introduced into (3.5.18), rendering:

$$\left[\int_{X_a}^{X_b} \begin{Bmatrix} \dfrac{-1}{H} \\ \dfrac{1}{H} \end{Bmatrix} E \begin{bmatrix} \dfrac{-1}{H} & \dfrac{1}{H} \end{bmatrix} A dX \right] \begin{Bmatrix} U_a \\ U_b \end{Bmatrix} = \begin{Bmatrix} F_a \\ F_b \end{Bmatrix} \tag{3.5.19}$$

Performing the integration and matrix multiplication as indicated in (3.5.19), and recalling that the length of the element (H) is X_b–X_a, the resulting equilibrium equations are:

$$\frac{EA}{H} \begin{bmatrix} 1 & -1 \\ -1 & 1 \end{bmatrix} \begin{Bmatrix} U_a \\ U_b \end{Bmatrix} = \begin{Bmatrix} F_a \\ F_b \end{Bmatrix}$$

The expression above is the same as given by (3.1.2), which was generated by minimizing the functional given by (3.1.1). This completes the example of how the generalized expression for finite element equilibrium equations can be employed to generate equilibrium equations for a single finite element.

Chapter 3 Summary

In Chapter 3, a generic functional was established, with strain energy expressed using a material matrix and a strain-displacement matrix. The generic functional was minimized, rendering a generalized form of equilibrium equations, applicable to many different types of stress-strain assumptions.

Strain-Displacement and Material Matrices
The strain-displacement matrix (B-matrix) transforms finite element nodal displacement variables into strain, while the material matrix (E-matrix) transforms strain into stress. For the 2-node rod element:

$$\underline{\underline{B}} = \begin{bmatrix} \frac{-1}{H} & \frac{1}{H} \end{bmatrix} \qquad \underline{\underline{E}} = E$$

A generic functional can be expressed in terms of the B- and E-matrices, then minimized, rendering an expression for the element equilibrium equations. The resulting equilibrium equations are general, and can be applied to elements that are used to solve problems from an entire class of problems, viz., linear-elastic problems of solid mechanics.

Shape Functions
Finite element interpolation functions can become quite complicated, and it was shown that shape function notation is handy to express interpolation functions in compact form. For a 2-node rod element:

$$\tilde{U}(X) = N_1 U_a + N_2 U_b \qquad N_1 = \left(\frac{X_b - X}{H}\right) \quad N_2 = \left(\frac{X - X_a}{H}\right)$$

It is also convenient to define the interpolation functions in terms of a shape function matrix and the nodal displacement vector:

$$\tilde{U}(X) = N_1 U_a + N_2 U_b \qquad or: \quad \tilde{U}(X) = \underline{\underline{N}}\, {}^e\underline{\underline{D}}$$

For the 2-node rod element, the shape function matrix is simply:

$$\underline{\underline{N}} \equiv \begin{bmatrix} N_1 & N_2 \end{bmatrix} \tag{3.5.20}$$

Element Equilibrium Equations
A generalized form of equilibrium equations was given by (3.5.10):

$$\left[\int_V \underline{\underline{B}}^T\, \underline{\underline{E}}\, \underline{\underline{B}}\, dV\right] {}^e\underline{\underline{D}} = {}^e\underline{\underline{F}}$$

For a 2-node rod element, (3.5.10) in terms of shape functions yields:

$$\left[\int_V \begin{Bmatrix} \dfrac{\partial N_1}{\partial X} \\ \dfrac{\partial N_2}{\partial X} \end{Bmatrix} E \begin{bmatrix} \dfrac{\partial N_1}{\partial X} & \dfrac{\partial N_2}{\partial X} \end{bmatrix} dV \right] {}^e\underline{\underline{D}} = {}^e\underline{\underline{F}}$$

Using the derivatives of the shape functions, components of the force and displacement vectors, and the definition of element length, the above yields:

$$\frac{EA}{H}\begin{bmatrix} 1 & -1 \\ -1 & 1 \end{bmatrix}\begin{Bmatrix} U_a \\ U_b \end{Bmatrix} = \begin{Bmatrix} F_a \\ F_b \end{Bmatrix}$$

Perhaps most important is (3.5.12), which will be used extensively in the next chapter to compute the element stiffness matrix for several different types of elements:

$${}^e\underline{\underline{K}} \equiv \left[\int_V \underline{\underline{B}}^T \; \underline{\underline{E}} \; \underline{\underline{B}} \, dV\right]$$

References

1. Reddy, J.N., *Energy and Variational Methods in Applied Mechanics*, John Wiley & Sons, N.Y., 1984
2. Huebner, K.H., *The Finite Element Method For Engineers*, John Wiley & Sons, N.Y., 1982
3. Higdon, A., *Mechanics of Materials*, John Wiley & Sons, N.Y., 1985
4. Cook, R.D., *Concepts And Applications Of Finite Element Analysis*, John Wiley & Sons, N.Y., 1989

Chapter 4

Simple Elements

"Whatever is done from love always occurs beyond good and evil."
—Nietzsche

4.1 Review

Key Concept: A generalized expression for finite element equilibrium equations, and the associated stiffness matrix formula for individual finite elements (considered in Chapter 3), provides a basis to develop many different types of elements.

In Chapter 3, a general expression for the equilibrium equations related to a single finite element was established, and a formula for computing individual element stiffness matrices defined. Recall that to compute an element stiffness matrix, both a material matrix and a strain-displacement are required. Material matrices for several common types of mechanical idealizations were given in Chapter 3, Table 3.1. The purpose of Chapter 4 is to establish *strain-displacement* matrices for a few simple elements, then show how these matrices, along with the material matrices listed in Chapter 3, can be used to compute various element stiffness matrices.

Before beginning the discussion of strain-displacement matrices and other associated topics, a brief overview of some highlights from Chapter 3 is given.

Assembling Element Stiffness Matrices

In practice, it is convenient to generate a stiffness matrix for each element, separately, then combine all the individual matrices to form a *global stiffness matrix*:

$$\underline{\underline{K}} = {}^{1}\underline{\underline{K}} + {}^{2}\underline{\underline{K}} + \cdots {}^{nel}\underline{\underline{K}} \qquad nel = \text{total number of elements} \tag{4.1.1}$$

Global Stiffness

The global stiffness matrix, along with the global displacement and force vectors, constitutes the equilibrium equations for an entire structure:

$$\underline{\underline{K}}\,\underline{\underline{D}} = \underline{\underline{F}} \tag{4.1.2}$$

With boundary conditions applied, the global system can be solved for nodal displacements, using matrix methods. The resulting nodal displacement values are then used to establish an approximate, piecewise continuous expression that characterizes displacement in the entire structure.

The discussion in Chapter 3 focused upon the generalized equilibrium equations for a single element as shown in (3.5.11):

$$^{e}\underline{\underline{K}}\;^{e}\underline{\underline{D}} = {}^{e}\underline{\underline{F}}$$

From the generalized equilibrium equations, an expression for computing individual element stiffness was established:

$$^{e}\underline{\underline{K}} \equiv \left[\int_V \underline{\underline{B}}^T\; \underline{\underline{E}}\;\underline{\underline{B}}\, dV\right] \tag{4.1.3}$$

The equation above indicates that the stiffness for an individual element is a function of the strain displacement matrix (B-matrix) and the material matrix (E-matrix) integrated over the volume of the *element*. The equation can be applied to many types of elements by supplying the appropriate B- and E-matrices. When strain displacement matrices corresponding to the E-matrices given in Table 3.1 are established, one can generate the associated element stiffness matrices.

Stiffness Matrices for Various Types of Elements

The focus of Chapter 4 will be to show how the element stiffness matrix equation, (4.1.3), is applied to several different types of *simple* finite elements. Before developing the stiffness matrices, a brief overview of some common finite elements is provided. In Section 4.3, general requirements for finite elements are discussed.

4.2 Some Common, Simple Finite Elements

Key Concept: The least complex finite elements ("simple elements") have straight boundaries and fewer nodes, while more complex elements allow element boundaries to be curved, and have more nodes. Simple elements typically have lower order displacement assumptions than the more complicated versions.

Finite elements appear in a variety of shapes, employ a varying number of nodes, and have been designed to model many different types of idealizations. Observe some common finite elements as illustrated in Figure 4.1. A brief overview of these common elements, used for solid mechanics problems, follows.

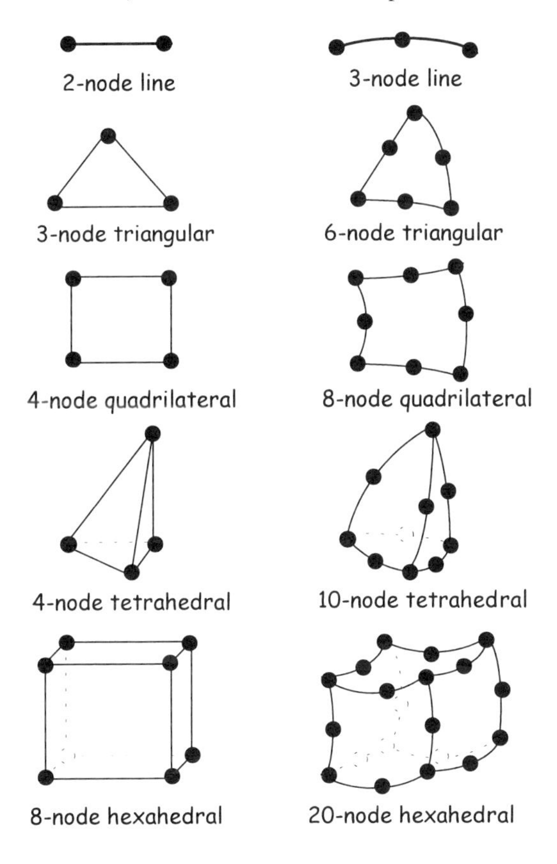

Figure 4.1 Some Common Finite Elements

All of the elements considered in this text are basic finite elements, without modifications to improve their performance. Many of them are considered only for illustrative purposes, and would not be recommended for general use due to their poor performance in terms of accuracy, computer resources required, and speed of solution. Typically, a finite element software developer modifies the basic element formulations to gain improved performance. Some of the modifications are proprietary, others are well documented. MacNeal [1] discusses the shortcomings of finite element design while illustrating techniques that are used to improve element performance.

Describing Finite Elements

The following is a list of attributes that can be used to define finite elements, with the last three attributes dealing specifically with displacement based finite elements used to solve solid mechanics problems:

- The number of nodes
- The shape: quadrilateral, triangular, hexahedral, etc.
- Geometric characteristic: line, surface, or volume
- Compatibility between elements
- The order of the element's displacement assumption
- The stress/strain idealization that the element characterizes
- The idealization type: continuum or structural

For example, consider an element used for plane stress applications, such as the 3-node element on the left hand side of Figure 4.1. This element could be described as a "three-node, triangular, surface element for plane stress applications." Implicit in this description is that the element:

- Has a linear displacement assumption
- Is of a class of elements called continuum elements

Analysts will often shorten the description of the 3-node, triangular surface element even further, and simply refer to it as a "3-node Tri." The attributes listed above will be considered in more depth in this chapter.

In this text, elements that have fewer nodes, lower order displacement assumptions, and straight boundaries will be called "simple elements." Figure 4.1 shows both simple and more complicated elements, with the most simple elements on the left hand side. Although elements having mid-side nodes are shown with curved boundaries, there are advantages to using this type of element with straight sides only; more on this topic later.

Displacement Assumption Order

The order of the displacement assumption in *h*-type finite elements is typically controlled by the number of nodes that the element has on its boundary. Elements that are developed to be used for a common type of idealization, plane stress for instance, often have both a lower order displacement assumption and a higher order version. For example, when used for plane stress, the 3-node, triangular surface element contains a linear displacement assumption while the 6-node, triangular surface element, used for the same application, has a quadratic displacement assumption.

Continuum Approach Versus Mechanics of Materials

Stress-strain idealizations can be based upon either a continuum approach (using elasticity theory) or a mechanics of materials approach. One significant difference between the two approaches is that elasticity theory makes use of four principles:

1. Hooke's law
2. Force equilibrium
3. Strain compatibility
4. Satisfaction of boundary conditions

While all four principles are considered when using the continuum approach, strain compatibility it is not always considered using the mechanics of material approach. As such, the mechanics of materials approach is less rigorous although under certain loading and geometric restrictions it can render the same results as elasticity theory. Strain compatibility relationships can be established by computing certain derivatives of normal and shear strain, as given in Chapter 1, Equations 1.5.6 and 1.5.7; see Ugural [2] for details.

Another distinguishing characteristic between continuum and structural idealizations is that structural idealizations generally apply to structures that are intended to perform a particular structural function; beams, plates, and shells for instance. When a structure is designed to perform a specific function, some components of stress are often neglected, since they are presumed insignificant when the structure is used for its intended purpose. Structural idealizations, having some stress components neglected, are not well suited for representing general, three-dimensional stress states; in such cases, the continuum approach is more suitable. However, for load-deflection (or similar) analysis, a mechanics of materials approach is often considered both appropriate and accurate.

Bending and Membrane Deformation

Structural idealizations (beams, plates, and shells) are typically used when a structure exhibits bending deformation in response to external loads. How does bending deformation differ from membrane (extensional) deformation? To answer this question, consider the slender member shown in Figure 4.2

When a bending load is applied to a slender structure, the structure responds with *bending deformation*, as illustrated on the left hand side of Figure 4.2. This type of deformation results in flexural (bending) stress, where the magnitude of the normal stress varies linearly with the distance from the neutral surface, i.e., in the depth direction. In contrast, if a uniform load (tensile or compressive) were applied to the right end of the structure, with the resultant passing through the centroid of all resisting cross sections, *membrane deformation* would occur. In such a case, the normal stress on any cross section does not vary through the depth.[1] This behavior is illustrated by the structure on the right in Figure 4.2.

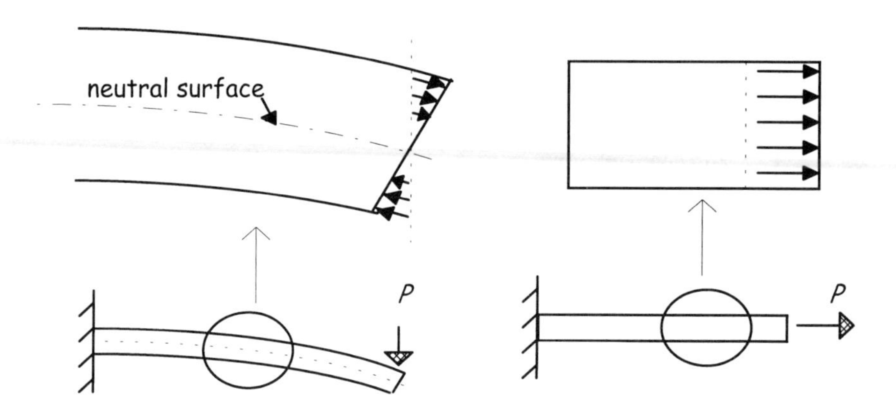

Figure 4.2 Bending and Membrane Deformation

The structure shown on the left hand side of Figure 4.2 is flexible due to its geometric properties, and responds in a bending mode of deformation if loaded in the manner shown. Notice that the material above the neutral surface is in a state of tension while the material below is in compression. This behavior is characteristic of *bending deformation*: to one side of the neutral surface, the material is in tension, while on the other side, the material is in compression. Material on the neutral surface is left un-deformed. Bending deformation can also occur in plate and shell structures.

The beam on the left hand side of Figure 4.2 deforms on planes that are normal to the direction of the load, and also on planes *parallel* to the load. However, it can be shown that for slender beams, normal stress in the depth direction, along with transverse shear stress, can typically be neglected. In general, *stress in the thickness direction in plate and shells, and in both the depth and width directions*

[1] Stress is uniform except near the restrained end, where Poisson's ratio affects the state of stress. Membrane stress is induced, assuming that bifurcation does not occur when a compressive load is applied.

in beams, is presumed zero. As a result, the stress state directly in the vicinity of the load is generally not characterized properly. These localized effects are discounted using a mechanics of materials approach.

Beams, plates, and shells can exhibit bending since they are relatively shallow (the term "thin" is used with plates and shells) and therefore flexible in response to loads which are applied normal to the neutral surface. Shell structures, in contrast to plates, can simultaneously exhibit both bending and membrane deformation.

The interested reader is referred to Ugural [2] for more insight on the topics of bending versus membrane deformation, mechanics of materials, and elasticity theory.

Comparing Continuum and Structural Elements

Continuum elements are finite elements designed to model idealizations based upon a continuum approach (using elasticity theory) while *structural elements* are typically designed using a mechanics of materials approach as a theoretical basis. Beam, plate, and shell elements are examples of structural elements. Lately, however, some shell elements have been developed using what is essentially an continuum approach; Dvorkin [3].

Using a mechanics of materials approach, one assumes that stress in certain orientations of the structure can be neglected. Hence, when finite elements are designed to characterize structural plates and shells, stress in the thickness direction is presumed zero; in simple beam elements, stress in the *depth and width directions* is presumed zero.

While common continuum elements have, at each node, translational DOF's only, structural elements have both translational and rotational nodal DOF's. The rotational DOF's allow a structural element to meet the essential boundary conditions associated with bending deformation. Recall from Chapter 1 that specification of the slope (rotation) at the end of a beam structure is essential to obtaining a correct solution to beam bending problems. This applies to plates and shells as well. Furthermore, the addition of rotational DOF's allows structural elements to maintain inter-element compatibility of *p–1* order derivatives, at least at adjacent nodes of the element.

Elements in which first derivative continuity is maintained at the nodes and across element boundaries are called C^1 ("see-one") elements, since **C**ontinuity of **1**st order derivatives is enforced. Commonly used continuum elements do not require derivative continuity, and are often termed C^0 ("see-zero") elements, since zero order derivative continuity (i.e., no derivative continuity) is required.

Infinitesimal Strains Presumed

It has been established that, with the presumption of infinitesimal strains, deformation in a structure is presumed to be so small that the deformed configuration of the structure is essentially identical to the un-deformed configuration. When deformation is infinitesimal, one can consider stress and strain in a deformed structure using the same coordinate system that was defined in the un-deformed structure. As increasing deformation causes a body to shift position relative to the original coordinate system, error is introduced. (This is also the case if large rigid body motion takes places). In addition, the small strain metrics defined in Chapter 1 also become increasingly errant with larger deformation. In large deformation problems, non-linear solution procedures are generally required.

Common Elements

The least complex structural element is the beam element, while the least complex continuum element is the rod element. Table 4.1 summarizes some of the more common elements used in solid mechanics problems, for both structural and continuum idealizations; other often-used elements exist.

Table 4.1 Common Element Types

continuum elements		structural elements	
type	characteristic	type	characteristic
rod/ truss	line	beam	line
plane stress	surface	plate	surface
plane strain	surface	shell	surface
axisymmetric	surface		
3-D stress	volume		

Why Not Use Volume Elements for All Applications?

While a three-dimensional element could theoretically model any of the mechanical idealizations that have been mentioned so far, there are several reasons why solid elements are not always the best choice. One reason for avoiding three-dimensional elements is that they are computationally expensive. That is to say, a significant amount of computational energy needs to be expended to generate the stiffness matrix for a single three-dimensional element.

Another reason for avoiding the use of three-dimensional elements is that the geometry of certain types of structures would require the use of many solid elements when far fewer surface elements could suffice. This is due to the fact that finite elements must be utilized with certain regard to their geometric proportions, meaning that *distorted* elements are undesirable. Distortion will be covered in some detail in Chapter 7 but for now, consider that one measure of distortion deals with element *aspect ratio*. An eight-node, hexagonal, volume element used to model a three-dimensional stress state is depicted in both undistorted and distorted configurations in Figure 4.3. The aspect ratio of an element can be considered to be the ratio of the largest dimension of any one side

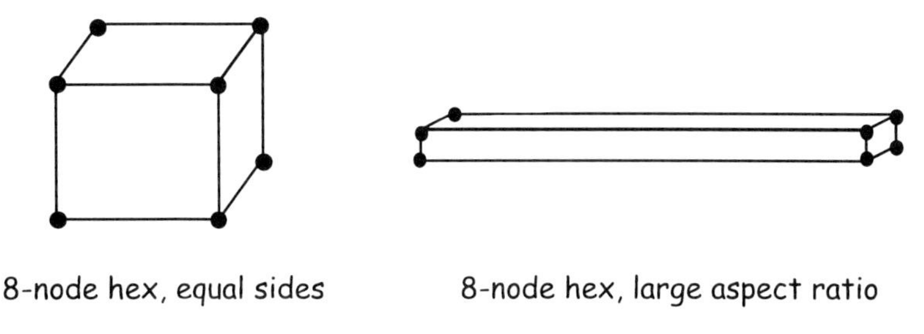

Figure 4.3 Distortion in an 8-Node Hexagonal Volume Element

to the smallest. Some analysts choose to keep the aspect ratio to a maximum of 4-to-1 to ensure suitable accuracy under various loading and distortion conditions.[2] Now, if 8-node, hexagonal volume elements were used to model a very thin structure, say the trunk lid on a typical automobile, a very large number of elements would be required if the 4-to-1 maximum aspect ratio is to be maintained. In other words, since the trunk structure is so thin, very many short-sided elements would be needed to maintain the desired four-to-one aspect ratio, since the largest dimension any side could be is four times the thickness.

Instead of presuming a three-dimensional stress state, and using the associated 8-node hexagonal volume elements, the trunk lid in Figure 4.4 might be idealized by assuming that shell theory characterizes the structural response. Accordingly, a 4-node quadrilateral surface element for shell applications may be utilized. Shell elements are often used in structures such as automotive body panels and aircraft skins when stress in the thickness direction can be neglected, and loads are carried by both bending and membrane forces.

[2] The 4-to-1 ratio is somewhat arbitrary. Aspect ratio is only one measure of distortion; others exist, and will be discussed in Chapter 7.

Figure 4.4 Surface Element Mesh on Trunk Lid

Shell elements, being of the surface type, have no geometrical thickness; the thickness of the structure is described by a mathematical constant. Therefore, regardless of how thin an actual structure is, when using shell elements the analyst need only maintain proper aspect ratio with respect to the length and width, not the thickness.[3] Thus, using the 4-to-1 maximum aspect ratio as before, far fewer surface elements would be needed, compared to the alternative of volume elements with the same aspect ratio criterion. Although 8-node volume and 4-node surface elements were used as examples in the above discussion, the same argument applies to other types of surface and volume elements as well. There are other reasons why volume elements may not be the best choice for a given type of idealization. The relative merits of individual element types will be summarized in Chapter 7.

4.3 Requirements for Finite Elements

Key Concept: To ensure *monotonic convergence* of the finite element solution, both the individual elements and the assemblage of elements ("the mesh") must meet certain requirements.

The term *h-convergence* is used in reference to the process where the predicted displacement in a finite element model better represents the exact displacement as a larger number of smaller elements is employed. Only *h*-type elements are

[3] A very small mathematical value of thickness (or depth in beams) initially caused a computational problem in elements that accounted for shear stress. Most finite element software developers have overcome this deficiency and have designed elements which accurately represent stress, even in the case where element thickness (or depth) is assigned a very small value.

considered in this text, thus, the term monotonic convergence herein implies convergence using the *h*-method of finite elements.

Displacement within a finite element is interpolated from the discrete values of displacement at the element's nodes. To generate a finite element interpolation function, one usually begins with a displacement assumption, typically a polynomial with adjustable terms (a_i's). Chapter 2 gave some rules for displacement assumptions to be used with one-dimensional finite elements. Recall that a displacement assumption must:

1. Have sufficiently differentiable terms
2. Be complete
3. Have linearly independent terms
4. Satisfy the essential boundary conditions of the element

If a displacement assumption used for simple one-dimensional problems satisfies the requirements above, the finite element model will exhibit desirable convergence characteristics, namely, monotonic convergence.

Since the finite element examples considered up to this point have been one-dimensional, the four requirements above have sufficed. However, when more complex finite elements are considered, the requirements above need to be replaced with a broader set of rules, for both the individual displacement assumptions and the mesh.

Monotonic Convergence Using the Displacement Based, *h*-Method

When the predicted displacement in a finite element model does not characterize the exact displacement, better accuracy is often obtained by using a greater number of smaller elements. This pre-supposes that the idealization and analysis procedures are properly invoked. The process of modifying the model to employ more, smaller elements is termed mesh refinement. With mesh refinement, the finite element solution is expected to converge, monotonically, to the exact solution.

What does the term "monotonic convergence" mean, and why is it important? To illustrate the concept of monotonic convergence, the graph in Figure 4.5 shows exact displacement, D, at some point in a structure. The dashed monotonic convergence line suggests that as the finite element mesh is refined, the finite element solution approaches (converges upon) the exact value. Note that for every increase in the number of elements, displacement on the monotonic line becomes a better approximation for the exact displacement. When convergence is monotonic, the analyst knows that each successive refinement yields a more accurate solution than the preceding one.

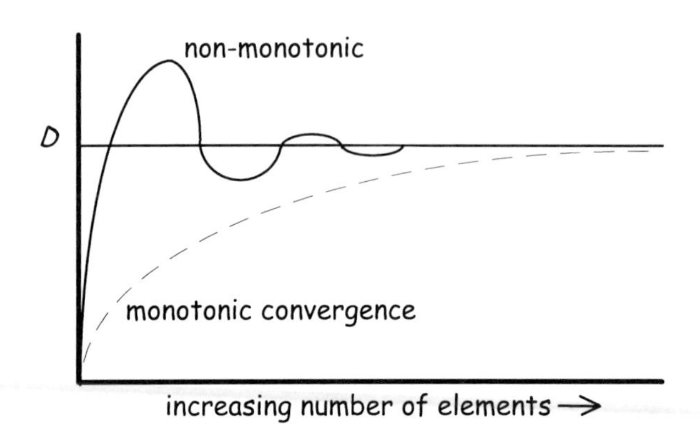

Figure 4.5 Convergence

The analyst can therefore intuit if further mesh refinement will yield significantly better results, or if the current level of refinement is sufficient. However, the same cannot be said when convergence is non-monotonic.

When convergence is non-monotonic, further mesh refinement may provide better *or worse* accuracy, making it difficult to determine the number of elements needed in a particular model. Monotonic convergence has an advantage in that the degree of mesh refinement needed for a particular level of accuracy can be inferred by observing how much the solution changes between each refinement. For example, if the change in displacement is small with each successive mesh refinement, the solution is assumed to have converged, and further refinement will not render a more accurate model. (This, of course, does not always mean that the correct answer has been obtained, only that the results will not be more accurate with further refinement.) In light of the advantages of monotonic convergence, finite elements are typically designed to ensure such behavior, although there are some exceptions which will be mentioned shortly. See Melosh [4] for a description of tests that are used for monotonic convergence.

Formal Proofs of Convergence

When using finite element methods based upon total potential energy, a formal proof of convergence in terms of energy exists, as discussed in Mikhilin [5], Oliveira [6], Felippa [7], Verma [8] and Melosh [9]. Hence, when finite elements meet certain requirements, it can be shown that the finite element approximation for minimum total potential energy approaches the exact value as the largest dimension of any element in a particular model approaches zero. In other words, if h is the largest element dimension of any element, the error between the minimum total potential energy computed using the finite element approximation for equilibrium displacement and the exact minimum total potential approaches zero as h approaches zero:

$$(\tilde{\Pi} - \Pi_{ex}) \to 0 \qquad \text{as } h \to 0 \tag{4.3.1}$$

$\tilde{\Pi}$ = finite element approximation for total potential energy at equilibrium
Π_{ex} = the exact minimum total potential energy, i.e., energy computed using exact equilibrium displacement
h = largest dimension of all the elements in the model

The minimum total potential energy computed using the exact equilibrium displacement is considered to be the exact minimum. Equation 4.3.1 suggests that the finite element approximation becomes a better estimator of minimum total potential energy as the element size diminishes, other things being equal. *A better approximation for equilibrium displacement is anticipated as the finite element approximation for minimum total potential energy converges to the exact*, since it is assumed that the exact displacement will yield the exact minimum. To quantify the error between the approximate and exact values of minimum total potential energy, energy norms are employed, as mentioned in Section 2.4.

As suggested by (4.3.1), monotonic convergence occurs when every reduction in the maximum value of h results in a smaller magnitude of the difference between the finite element approximation for total potential energy and the exact total potential energy. The study of convergence leads to the generalization that, when certain criteria are met:

- The finite element approximation for minimum total potential energy converges monotonically to the exact value of minimum total potential energy

- The finite element approximation for displacement monotonically converges to the exact equilibrium displacement

Although both energy and displacement predicted by the finite element method are expected to converge to the exact values, the *rates* of convergence for displacement and energy are not necessarily the same.

Requirements to Ensure Monotonic Convergence
To ensure monotonic convergence of displacement predicted by the finite element method, both the individual elements and the assemblage of elements ("the mesh") should meet the five requirements described in Table 4.2. Although the

property of geometric isotropy is not included in the table, it will be considered a general requirement for displacement assumptions.

Requirements 1 and 2 in Table 4.2 are equivalent to maintaining *completeness in energy,* while Requirement 3 ensures that inter-element compatibility is maintained. The topic of displacement assumption requirements is discussed by Felippa [7] while the concept of completeness in energy is considered by Key [10].

Table 4.2 Requirements for Monotonic Convergence

Requirements for Each Element's Displacement Assumption
1. Rigid Body Representation: The displacement assumption must be able to account for all rigid body displacement modes of the element.
2. Uniform Strain Representation: Constant strain states for all strain components specified in the constitutive equations of a particular idealization must be represented within the element as the largest dimension of the element approaches zero.
Requirements for the Mesh
3. Compatibility Between Elements: The dependent variable(s), and *p–1* derivatives of the dependent variable, must be continuous at the nodes and across the inter-element boundaries of adjacent elements.
4. Mesh Refinement: Each successive mesh refinement must contain all of the previous nodes and elements in their original location.
5. Uniform Strain Representation: The mesh must be able to represent uniform strain when boundary conditions that are consistent with a uniform strain condition are imposed.

If all of the requirements are met, the finite element solution will converge, monotonically, with mesh refinement. This presumes that other errors, such as computational round off error, are not introduced in the solution procedure.

If the individual elements and mesh meet all of the requirements *except* Requirement 3, compatibility between elements, the mesh can still converge. However, in such a case *monotonic* convergence is not assured.

Interestingly, a mesh of incompatible elements may converge with fewer elements than a mesh with compatible elements. The reason being that the displacement based finite element method typically produces a mesh that is too stiff, and, since incompatible elements have the effect of introducing gaps in the mesh, an incompatible mesh is made more flexible. The compatibility concept will be illustrated in Figure 4.9.

Understanding Convergence Requirements

Requirements for monotonic convergence are discussed in order to give the reader insight regarding fundamental requirements of finite elements used for solid mechanics applications.

1. Convergence and Rigid Body Representation

An element's displacement assumption must be able to represent rigid body motion. The rigid body requirement is important, since an element may need to characterize a portion of a structure that displaces without straining. For instance, consider the rod in Figure 4.6, modeled with two rod elements and three nodes; a force of magnitude P is applied at Node 2.

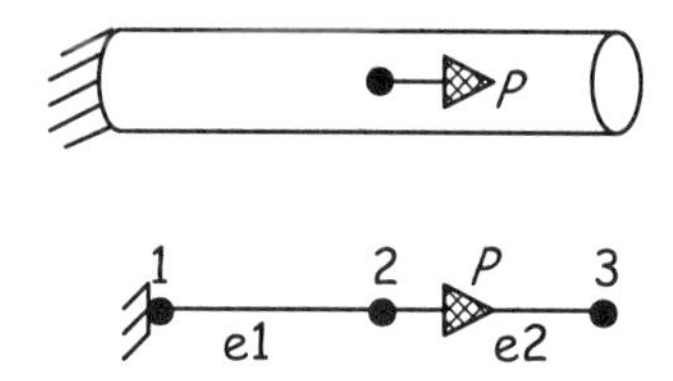

Figure 4.6 Rigid Body Motion of a Rod

Note that Element 2 must displace, although is it not strained—in other words, Element 2 experiences rigid body motion, where the displacement at Node 3 is equal to the displacement at Node 2.

In a one-dimensional element with rectangular coordinates, a polynomial displacement assumption must have a constant term to ensure that the type of rigid body motion described above is allowed. In such a case, the constant term a_1 in the polynomial $\tilde{U}(X) = a_1 + a_2 X$ allows the rigid body mode to occur, since, if a_2 assumes a value of zero, displacement is the same anywhere on the element, and is equal to the value a_1.

Difficulties characterizing rigid body motion are encountered when a local, non-rectangular coordinate system is used to define displacement assumptions, as typically occurs in the case of curved shells. Enforcing rigid body motion requirements often proves difficult in such cases. Rigid body motion in shell elements is discussed by Ashwell [11] and Haisler [12].

One can determine the nature of both rigid body and flexible modes of a finite element by performing an Eigenvector analysis of the element's stiffness matrix. Bathe [13, p. 168] illustrates both rigid body and flexible modes for a four-node surface element used for plane stress applications.

2. Convergence and Uniform Strain Representation Within an Element

An individual element must be able to represent a uniform (constant) strain state as the element size approaches zero. (This same requirement for the entire mesh will be considered shortly.) In Verma [8], the authors discuss the assertion that finite elements must not only represent constant strain states but also must exhibit the characteristic that these states are "preferred" over all other possible states.

Regardless of how much strain varies within in body, if small enough portions of a body are examined, strain will appear constant. This same characteristic is therefore desirable in a finite element as well. The terms that are required in a displacement assumption to allow it to characterize uniform strain can be determined by examining the strain components that appear in the constitutive equations.

Consider a 2-line element used for uniaxial stress with a local (x-coordinate) axis running along the length of the element. The constitutive equation for this element contains only one stress component, the normal stress in the x-axis direction:

$$\tau_{xx} \equiv E\varepsilon_{xx} = E\frac{du}{dx}$$

The constitutive equation above indicates that for this element, a single, first order derivative, du/dx, defines strain; other elements typically have additional strain components. If the displacement assumption for the 2-node line element for uniaxial stress contains at least a linear term, the constant (uniform) strain condition can be represented:

$$\tilde{u}(x) = a_1 + a_2 x \qquad \varepsilon_{xx} \approx \frac{d\tilde{u}}{dx} = a_2 \tag{4.3.2}$$

Alternately, using only a constant term for the displacement assumption in (4.3.2) would not allow constant strain to be represented. Likewise, choosing a displacement assumption that contained a constant and a quadratic term, without a linear term, would not allow constant strain to be represented. It is for these reasons that a complete polynomial (a polynomial with all terms present) is desirable for a displacement assumption. Leaving terms out of the displacement assumption can result in undesirable behavior of the element.

In the example of constant strain above, the first derivative is equal to a constant. Implicit in the uniform strain requirement is that displacement is continuous everywhere within the element. In other words, since displacement can be found by integrating strain terms, and strain in the element is required to be constant throughout the element, integrating the expression for strain (a constant) yields a linear function for displacement that is everywhere continuous within the element.

Some types of *structural* elements use an approach where normal strain terms in the constitutive equation are expressed in terms of second order derivatives (curvatures). In such a case, constant strain requires that second order derivatives must be continuous everywhere in the element. For example, consider an element where lateral displacement w is expressed as a quadratic function, and normal strain in bending is proportional to the second derivative:

$$\tilde{w}(x) = a_1 + a_2 x + a_3 x^2 \qquad \varepsilon_{xx} \propto \frac{d^2 \tilde{w}}{dx^2} = 2a_3 \tag{4.3.3}$$

If the quadratic expression for $\tilde{w}$ is not correctly represented within the element, strain will not be expressed as shown in (4.3.3). In other words, if the element fails to interpolate quadratic displacement properly, constant strain cannot be represented. Hence, implicit in the requirement of constant strain in these types of elements is that the displacement assumption must be able to properly represent quadratic displacement. However, *isoparametric elements* (discussed in Chapter 5) cannot properly represent quadratic displacement if the elements have curved boundaries, even if they are complete in quadratic terms; see MacNeal [1, p. 108]. In light of this, even though these elements could be used with curved sides, they cannot pass the constant strain requirement.

Other isoparametric elements cannot properly represent quadratic displacement within the element even when the boundaries are straight. These elements, the so-called serendipity elements, are missing a term (or terms) in the displacement assumption, such that, regardless of its shape, the element cannot properly represent a complete quadratic. Thus, these elements will not meet the uniform strain requirement if strain is expressed in terms of second order derivatives. While isoparametric elements can encounter difficulties characterizing higher order displacement, they have no problem characterizing linear displacement.

Equation 4.3.3 considers how *normal strain* in bending is affected when quadratic displacement is not properly represented. In other types of isoparametric elements, uniform *shear strain* may not be properly represented. For example, a fundamental flaw in *Mindlin* type plate and shell elements is that a state of uniform *transverse shear* strain cannot be properly represented. Mindlin elements will be mentioned in Section 4.6. There are other reasons why an element may have difficulty representing uniform strain, as will be shown in Section 4.6.

3. Convergence and Inter-Element Compatibility

The dependent variable, and $p-1$ order derivatives of the dependent variable, must be continuous at the nodes and across the boundaries of adjacent elements.

The value p represents the order of the highest order derivative found in the strain terms (shown in the constitutive equations) for a given type of idealization. Using line elements, inter-element compatibility is enforced simply by requiring displacement (and possibly derivative) continuity at all *nodes*. However, for two- and three-dimensional elements, compatibility becomes a somewhat more complicated issue, since continuity must be maintained *across element boundaries*, as well as at the nodes. In two- and three-dimensional elements, continuity at the nodes does not ensure continuity along the entire boundary.

When does inter-element compatibility become a problem? Incompatibility can be introduced in a finite element mesh when adjacent elements:

A. Are not properly connected at the nodes
B. Have differing order displacement assumptions
C. Have differing nodal DOF
D. Are of the incompatible type

A. Incompatibility due to Elements Not Properly Connected: Consider the finite element model for the non-uniform shaft as discussed in Section 2.4. If the mesh for this model were created using typical finite element mesh generation software, the initial mesh with two, 2-node line elements would actually contain four nodes. Figure 4.7 shows that Nodes 2 and 3 are *coincident nodes,* meaning that they have the same spatial coordinates but belong to separate elements. If left in this condition, the elements are not compatible since displacement continuity is not enforced between Elements 1 and 2. This particular example of an incompatible mesh would likely result in a singular matrix, since only Node 1 is restrained, leaving Element 2 completely unrestrained. However, it is possible to create a finite element mesh where a limited number of coincident nodes are present, and a singular matrix does not occur. In such cases the model may behave as if a crack is present between the elements.

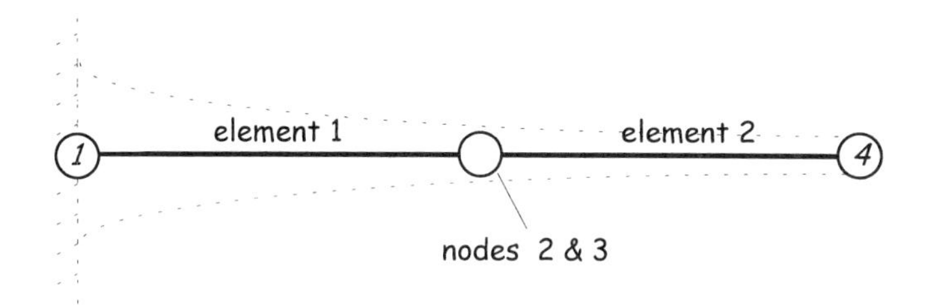

connectivity table before node merge

element	node a	node b
1	1	2
2	3	4

Figure 4.7 Model Before Node Merging and Renumbering

When creating finite element models, a *node merging procedure* is typically invoked, such that each pair of coincident nodes is replaced by a single node, and the connectivity table is updated to reflect the new connectivity.

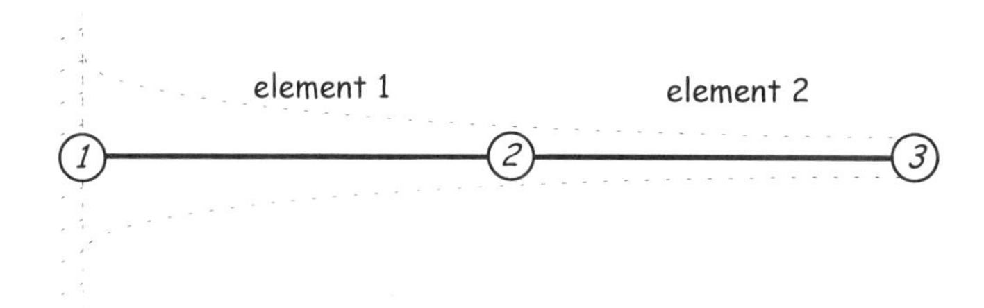

connectivity table after node merge

element	node a	node b
1	1	2
2	2	3

Figure 4.8 Merged and Renumbered Model

After the node merging procedure is performed, the nodes are typically renumbered, using sequential node numbers. In the present example, the merged, re-numbered finite element model would appear as in Figure 4.8. Some finite element meshing software programs will, by default, automatically perform the merging and renumbering.

B. Incompatibility due to Differing Order Displacement Assumptions: Another source of element incompatibility occurs when two adjacent elements have differing order displacement assumptions. Consider two elements in Figure 4.9, one having a linear displacement assumption and the other having a quadratic assumption. Notice that a gap between the elements occurs since the displacement for the linear element can only be represented by a straight line while displacement in the other element is a quadratic function. In this particular case, the incompatibility might also be considered to be of the first type: elements not properly connected. Adjacent elements that have differing order displacement assumptions can be used if a constraint equation (often called "multi" or "single-point" constraints) is invoked along the common edge.

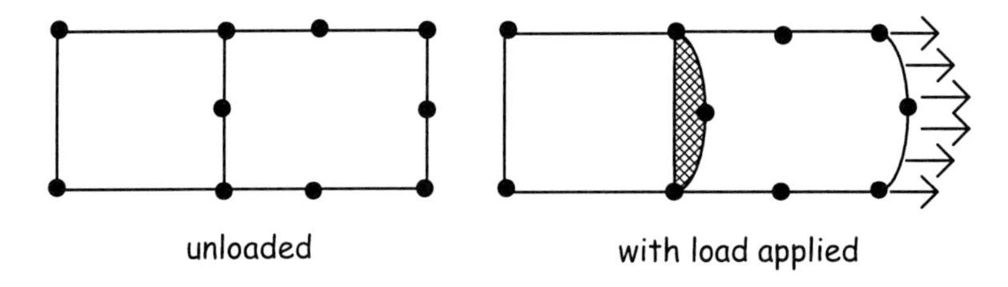

Figure 4.9 Incompatible Mesh Due to Displacement Assumption Mismatch

Displacement assumptions can also be mismatched when elements with higher order displacement assumptions are used. Notice in Figure 4.10 that the mid-side node of Element 1 is connected to corner nodes of Elements 2 and 3. Higher order elements must be matched such that the mid-side node of one element is connected to the mid-side node of the other.

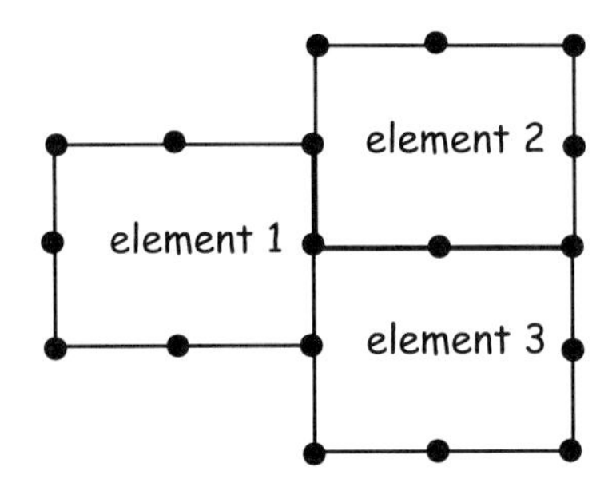

Figure 4.10 Incompatible Mesh Due to Mid-side Node Mismatch

C. Incompatibility due to Differing Nodal Variables: Incompatibility also occurs when elements having differing types of nodal DOF are joined. Figure 4.11 depicts a 2-node beam element attached to one node of a 8-node hexagonal volume element used for 3-D stress. Recall that continuum elements routinely have translational DOF only, while structural elements have both translational and rotational DOF's.

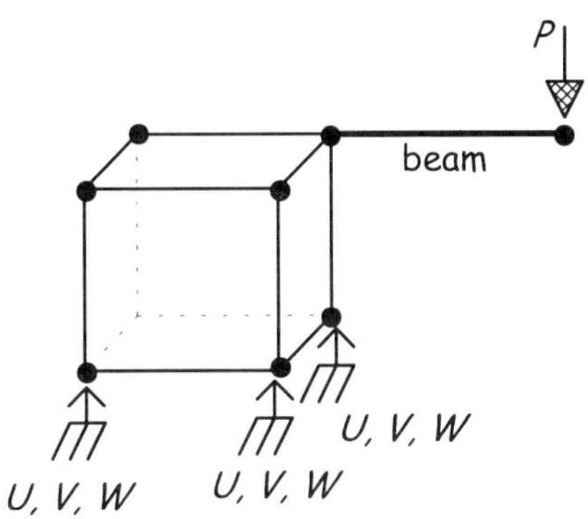

Figure 4.11 Incompatible Nodal DOF

When joining continuum and structural elements, a special constraint must be imposed upon the structural element's rotational DOF, least a singular stiffness matrix result. The 8-node brick element shown in Figure 4.11 is well restrained. However, its nodes do not have rotational DOF's, therefore, at the node where the beam is attached, there exists no DOF from the brick to couple with the rotational

DOF of the beam. As depicted, the beam element will experience rigid body rotation. Methods to join elements of different types are discussed in [15].

D. Incompatibility due to Incompatible Elements: The last type of incompatibility to be mentioned occurs when elements in a mesh are of the "incompatible type."

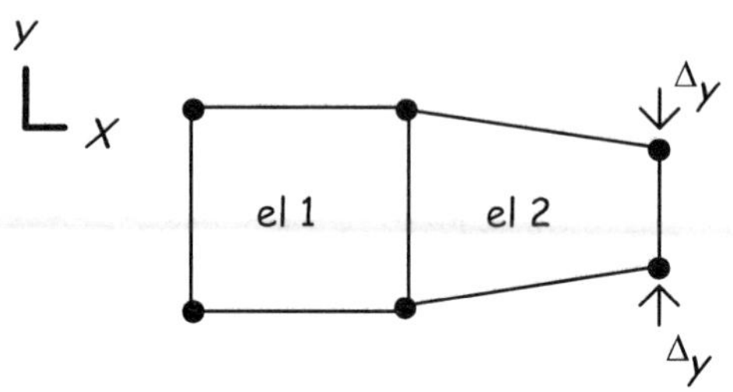

Figure 4.12 Incompatible Elements

Even when two identically formulated, incompatible elements are joined together, a gap in displacement can still exist. Incompatible elements are such that displacement along any given edge is a function of something other than just the nodal DOF's on the adjacent edge.

Consider the two elements in Figure 4.12. Both are identically formulated, 4-node quadrilateral surface elements designed with incompatible modes of displacement. Assume that two nodes of Element 2 are given a displacement of Δy as depicted in Figure 4.12. With no other nodes displaced, the elements would appear as shown. However, displacement for Element 2 would be *computed* as shown in Figure 4.13.

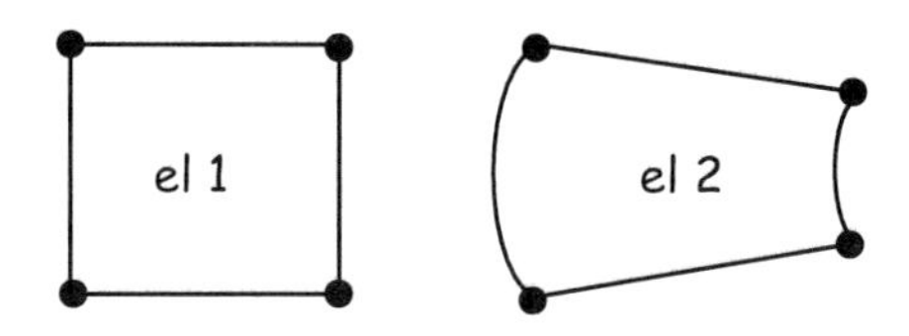

Figure 4.13 Computed Displacement

Notice that Element 1 has no change but Element 2 records quadratic displacement in the *X*-coordinate direction, along two edges. Also, note that displacement does not match along the edge shared by the two elements, even though the elements are of exactly the same type. (The elements in Figure 4.13 are shown separated for clarity.) Element 2 records a displacement along the shared edge even though the nodes on the shared edge have not displaced. Hence, the displacement along this edge is a function of something other than just the nodal DOF's on the edge. This behavior typifies the response of an incompatible element.

The elements shown in Figure 4.12 were intentionally designed to be incompatible. Adding the incompatible mode shown, sometimes called a bubble function, helps this element to perform better when used to model bending modes of deformation. Incompatible elements are discussed by Wilson [16], and Simo [17]. The performance of a 4-node quadrilateral element with a bubble function is considered in Section 7.3 of this text.

Some elements are incompatible without the intentional bubble functions, simply because they don't have enough nodal variables on a given edge to uniquely define displacement, or derivatives of displacement, on a particular edge. For example, Kirchhoff plate elements require continuity of displacement *and* continuity of first derivatives along any edge. Enforcing derivative continuity along an edge becomes difficult in these elements because, typically, there are not enough nodal variables on any one edge to adequately define the first derivative. For some Kirchhoff plate elements, if the expression for the derivative of transverse displacement along an edge is examined, it is seen that the expression is of higher order than the variables associated with the edge would allow. That is, if an edge has two *related* DOF's, a linear function along that edge can be uniquely defined; if three related DOF's are available, a quadratic is possible, and so on. Some elements (many Kirchhoff plates elements, for instance) may have only two related DOF's while a function of quadratic order is being used for interpolation. This issue will again be considered in Section 4.6. *Whenever displacement (or a derivative of displacement) along an edge is defined by something other than the nodal DOF's on that edge, an incompatibility is possible.*

Why is compatibility important? Using a compatible mesh, the finite element method produces a structural representation that is (on average) either too stiff, or correct. As a result, the analyst knows that the average displacement is either under predicted or correct. However, when incompatible elements are used, gaps appear between elements, as illustrated in Figure 4.9. Since the gaps have the effect of making the structure less stiff, average displacement can now be under-predicted, over-predicted, or exact. Hence, *monotonic* convergence is no longer assured although it is assumed that with suitable refinement, a mesh of incompatible elements will still converge.

4. Convergence and the Discretization Process

Each successive mesh refinement must contain all of the previous nodes and elements in their original location.

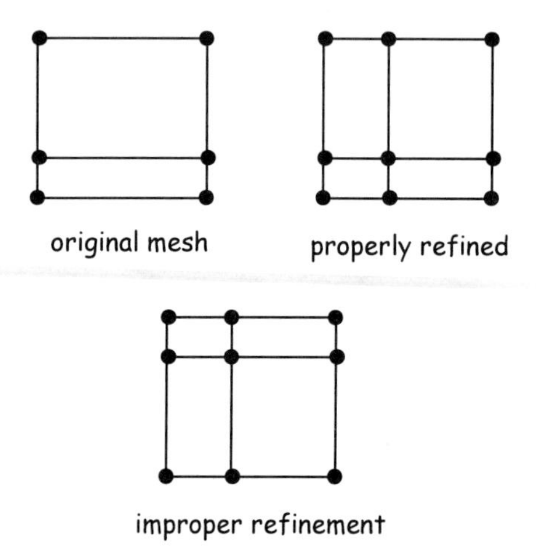

Figure 4.14 Mesh Refinement

The stress predicted by the finite element method can be very sensitive to the location of the nodes, especially in areas of high stress gradient. Care must be taken when refining a mesh so that the previous nodes are not moved from their original location. It is entirely possible that moving the original nodes during mesh refinement would produce less accurate results than the original mesh. To illustrate, consider the mesh arrangements shown in Figure 4.14. The properly refined mesh leaves all six nodes from the previous mesh in the same location, then adds three more nodes, defining a total of four elements. This refinement scheme allows the results of the new mesh to be *at least* as good as the original. Note that in the improperly refined model, two of the original nodes have been repositioned. Now, if most of the displacement occurs in the lower half of the mesh, moving nodes away from that area could result in a lower stress value than the original model predicted, even though the refined model has more elements.

5. *Convergence and Uniform Strain Representation Within Mesh*

The mesh must be able to represent a uniform strain state, when suitable boundary conditions are imposed. To test the uniform strain condition, a *Patch Test* has been devised. Using a patch test, a simple mesh is constructed, and boundary conditions imposed, such that a uniform state of strain should result. Either displacements or loads can be applied, depending upon what characteristics of the mesh are to be investigated.

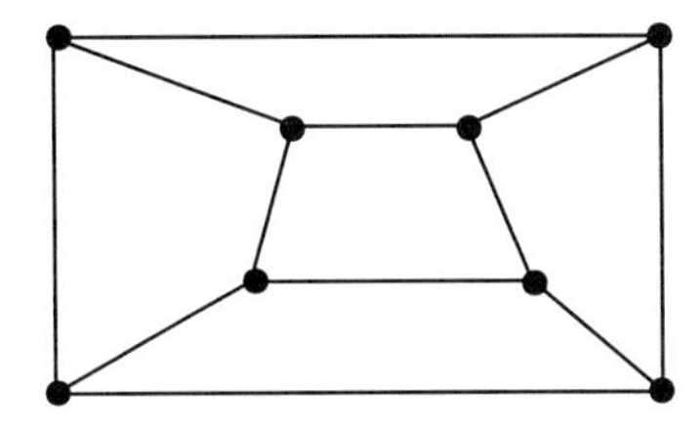

Figure 4.15 Patch Test for 2-D Elements

Figure 4.15 illustrates a mesh for a patch test of two-dimensional elements although, theoretically, substantial variation in the geometry of the elements used for a patch test is permitted.

The elements used in a patch test are constructed in a somewhat distorted manner to test an element's ability to perform under less than ideal conditions. The patch test can show if an element type is prone to undesirable performance characteristics when distorted—for instance, inability to represent uniform strain states.

Patch tests have been around for quite some time as discussed in Irons [18], Bazeley [19], and Irons [20]; a *de facto* standard patch test exists, as described in MacNeal [21]. In a detailed review of the patch test given by MacNeal [1], the author states that *all* well designed isoparametric elements with *linear* displacement assumptions can pass the patch test. However, only *some* isoparametric elements with *quadratic* displacement assumptions can pass the patch test, and these elements must have straight sides and complete polynomial displacement assumptions.

Polynomials for Displacement Assumptions

In an attempt to meet the requirements for monotonic convergence, complete polynomials of suitable order are typically used for finite element displacement assumptions. Note that the term "complete polynomial" implies that none of the polynomial terms, up to a certain order, are left out. This requirement will be relaxed somewhat for some elements. The term "complete polynomial" differs from the "complete in energy" terminology defined previously.

How is a polynomial displacement assumption of suitable order determined? For a particular type of idealization, the strain terms in the constitutive equations are examined to determine the order of the highest derivative, p. The minimum requirement for a complete polynomial displacement assumption is that all terms up to and including the *pth* order term need to be included.

What Terms Constitute a Complete Polynomial?

In problems where a single variable polynomial displacement assumption is sufficient, the terms needed for a complete polynomial are simply defined as:

$$\tilde{U}(X) = a_1 + a_2 X + a_3 X^2 + \ldots + a_n X^{n-1} \tag{4.3.4}$$

If the constitutive equations contain first order derivatives, $p=1$, and the minimum requirement for a complete polynomial to meet the rigid body and uniform strain requirements is:

$$\tilde{U}(X) = a_1 + a_2 X \tag{4.3.5}$$

Equation 4.3.4 is applicable to interpolation functions in one variable only. When a displacement assumption in two variables is required, the terms needed for a complete polynomial are described in terms of Pascal's triangle,[4] shown below.

$$\begin{array}{ccccccccc} & & & & 1 & & & & \\ & & & X & & Y & & & \\ & & X^2 & & XY & & Y^2 & & \\ & X^3 & & X^2Y & & XY^2 & & Y^3 & \\ X^4 & & X^3Y & & X^2Y^2 & & XY^3 & & Y^4 \end{array} \tag{4.3.6}$$

...more terms...

In a three-dimensional case, an analogous diagram may be defined in terms of *X*, *Y*, and *Z* variables as illustrated by Huebner [23, p. 134].

If a complete, second-order polynomial function of U in two variables is required, Pascal's triangle indicates that the polynomial should contain six terms:

$$\tilde{U}(X,Y) = a_1 + a_2 X + a_3 Y + a_4 X^2 + a_5 XY + a_6 Y^2 \tag{4.3.7}$$

[4] Blaise Pascal (1623–1662) was an inventor with achievements in the areas of mathematics, physics, and teaching methodology. His written work spans disciplines both scientific and theological, with significant and lasting impact upon French literature; Beardsley [22].

To eliminate all six a_i's in (4.3.7), an element with six nodal DOF's related to U would be required. In contrast, a complete *linear* polynomial in two variables consists of only three terms:

$$\tilde{U}(X,Y) = a_1 + a_2 X + a_3 Y \tag{4.3.8}$$

For a linear polynomial in two variables, a three node element, with one nodal displacement variable at each node, will allow the elimination of all three parameters, a_1, a_2, and a_3, in (4.3.8).

Geometric Isotropy

If a polynomial displacement assumption has geometric isotropy, it contains equal representation of variables in each coordinate direction. Hence, geometric isotropy concerns displacement assumptions that have either two or three variables. Consider the three term displacement assumption in two variables X and Y:

$$\tilde{U}(X,Y) = a_1 + a_2 X + a_3 Y \tag{4.3.9}$$

Notice that displacement in (4.3.9) is a linear function of the variables X and Y. Now, consider a six-term displacement assumption in two variables:

$$\tilde{U}(X,Y) = a_1 + a_2 X + a_3 Y + a_4 X^2 + a_5 XY + a_6 Y^2 \tag{4.3.10}$$

Suppose that we wish to design a 4-node compatible element, that interpolates $U(X,Y)$, and has one nodal displacement variable at each node. This dictates the use of a four-term displacement assumption. The displacement assumption in (4.3.9) contains too few terms, while the displacement assumption in (4.3.10) contains too many. As mentioned, in some elements we wish to deviate from the rule of complete polynomial displacement assumptions, as in the case of this 4-node element. The deviation from a complete polynomial will be allowed if the displacement assumption maintains geometric isotropy. The question is, which terms should be dropped from (4.3.10)? One candidate displacement assumption would be:

$$\tilde{U}(X,Y) = a_1 + a_2 X + a_3 Y + a_4 X^2$$

However, notice that displacement computed using this assumption would be a linear function of Y while having quadratic representation in X. This tends to make an element exhibit artificial directional characteristics, i.e., an-isotropic behavior. Examining (4.3.10), is it apparent that the only choice that would allow equal representation in both X and Y directions is:

$$\tilde{U}(X,Y) = a_1 + a_2 X + a_3 Y + a_4 XY \tag{4.3.11}$$

Using a displacement assumption with some terms missing, such as (4.3.11), will allow an element to display isotropic displacement behavior, at least in some restricted shapes. (This element does exhibit a compatibility problem, as will be discussed in Chapter 5.)

While geometric anisotropy is generally considered undesirable, some elements are purposely designed to exhibit this type of behavior—elements used in the field of fracture mechanics to model crack tip phenomena, for instance.

4.4 A 3-Node Surface Element for Plane Stress

Key Concept: The least complex surface element is the 3-node triangle. This element can be used for a variety of applications; the plane stress application will be considered in this section.

The 2-node line element for rod applications, considered in Chapters 2 and 3, has limited use. On the other hand, surface elements can be used in a relatively broad range of applications, such as plane stress, plane strain, axisymmetric solid, plate bending, and shells. In this and following sections, a 3-node, triangular surface element will be considered for several different applications.

Plane Stress Application

As indicated in the application notes at the end of Chapter 1, a plane stress idealization can be used for the case of a plate with a hole, when the loading

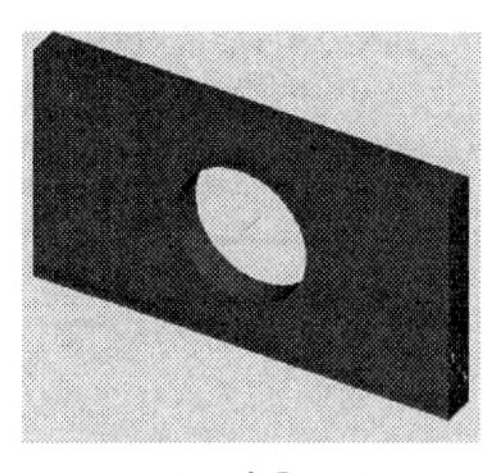

actual Part

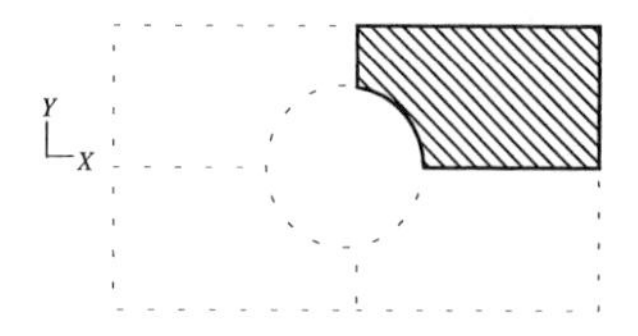

1/4 surface mesh boundary

Figure 4.16 Plate Structure, Plane Stress Assumption

induces in-plane stress only. Figure 4.16 again shows the plate with hole from Chapter 1; recall that a tensile load was applied to the plate.

Although 4-node quadrilateral surface elements were used in the plane stress application example at the end of Chapter 1, 3-node triangular surface elements can also be used for this problem, as illustrated below, where six elements are employed.

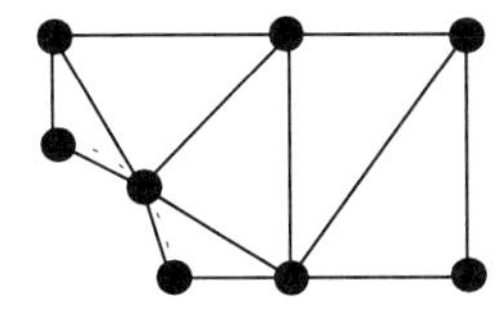

The 4-node quadrilateral surface element is generally preferred over the 3-node triangular element, in terms of accuracy; however, the 3-node triangular element has the advantage of representing curved geometry more easily. A 3-node element will be considered here, since it is slightly less complex. More complicated elements, such as the 4-node quadrilateral surface element, will be considered in Chapter 5.

The Element Stiffness Matrix Equation

Recall that to generate the stiffness matrix for an element, a material matrix (*E*-matrix) and a strain-displacement matrix (*B*-matrix) are required, as indicated by Equation 3.5.12, shown again below:

$$ {}^{e}\underline{\underline{K}} \equiv \left[\int_V \underline{\underline{B}}^T \, \underline{\underline{E}} \, \underline{\underline{B}} \, dV \right] \tag{4.4.1} $$

Since the material matrices for several types of applications are already known (Table 3.3.1), all that is required to generate stiffness matrices for different applications are the respective strain-displacement matrices. The strain-displacement matrix for a 3-node, triangular, surface element for plane stress is now considered.

The Constitutive Equations for Plane Stress

From Table 3.1, the constitutive equation ($\underline{\tau} = \underline{\underline{E}}\underline{\varepsilon}$) for plane stress appears as:

$$ \begin{Bmatrix} \tau_{XX} \\ \tau_{YY} \\ \tau_{XY} \end{Bmatrix} = \frac{E}{1-\nu^2} \begin{bmatrix} 1 & \nu & 0 \\ \nu & 1 & 0 \\ 0 & 0 & \dfrac{1-\nu}{2} \end{bmatrix} \begin{Bmatrix} \varepsilon_{XX} \\ \varepsilon_{YY} \\ \gamma_{XY} \end{Bmatrix} \tag{4.4.2} $$

To determine the strain-displacement matrix, the interpolation functions need to be established for the element. To determine what variables need to be interpolated, the strain vector in the constitutive equations is examined. Notice that the strain vector given in the constitutive equations for plane stress, (4.4.2), contains three components: normal strain in *X*, normal strain in *Y*, and shear strain *X*-*Y*. The strain vector can alternately be expressed in terms of first order partial derivatives:

$$\underline{\underline{\varepsilon}} = \begin{Bmatrix} \varepsilon_{XX} \\ \varepsilon_{YY} \\ \gamma_{XY} \end{Bmatrix} = \begin{Bmatrix} \dfrac{\partial U}{\partial X} \\ \dfrac{\partial V}{\partial Y} \\ \dfrac{\partial U}{\partial Y} + \dfrac{\partial V}{\partial X} \end{Bmatrix} \tag{4.4.3}$$

It is apparent that two displacement functions need to be defined for the case of plane stress: displacement in the *X*-coordinate direction (*U*) and displacement in the *Y*-coordinate direction (*V*).

Interpolation Functions for a 3-Node Surface Element, Plane Stress

Since the strain terms in (4.4.3) require two functions of displacement, *U(X,Y)* and *V(X,Y)*, a 3-node triangular element will be designed to interpolate both. A three-node element might take the shape of a right triangle, as illustrated in Figure 4.17.

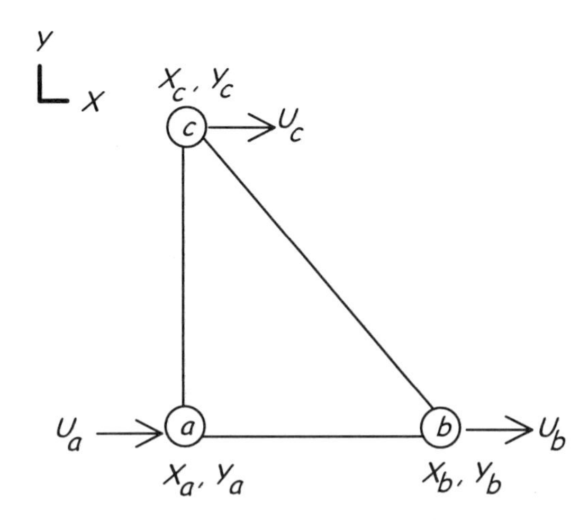

Figure 4.17 3-Node Triangular Element, X-Displacement

For other triangular shapes, the development is the same. Nodal displacements U and V are present at each node. However, since the interpolation function in the X-coordinate direction is presently being considered, only the displacement variables associated with $U(X,Y)$ are shown in Figure 4.17; note that the X and Y variables represent the spatial coordinates of each node. Referring to Pascal's triangle, (4.3.6), it is seen that the general form for a polynomial in X and Y is:

$$\tilde{U}(X,Y) = a_1 + a_2 X + a_3 Y + a_4 X^2 + a_5 XY + a_6 Y^2 + \cdots \tag{4.4.4}$$

As before, using the mathematical approach of developing interpolation functions, the parameters a_i are to be eliminated by constraining the displacement assumption to each nodal value of displacement. This is done by inserting into the displacement assumption the spatial coordinates at a particular node, and then equating the result to the respective nodal displacement variable. This constraint procedure is followed at each node. Since three nodal values of displacement associated with $U(X,Y)$ are defined for this element, three parameters can be eliminated. This dictates the use of a linear displacement assumption, since a linear displacement assumption in two variables employs three a_i's:

$$\tilde{U}(X,Y) = a_1 + a_2 X + a_3 Y \tag{4.4.5}$$

Constraining the displacement assumption to take on the values of nodal displacement:

$$\begin{aligned} \tilde{U}(X_a,Y_a) &= a_1 + a_2 X_a + a_3 Y_a = U_a \\ \tilde{U}(X_b,Y_b) &= a_1 + a_2 X_b + a_3 Y_b = U_b \\ \tilde{U}(X_c,Y_c) &= a_1 + a_2 X_c + a_3 Y_c = U_c \end{aligned} \tag{4.4.6}$$

The first of the three equations above states that when the spatial coordinates of Node a are introduced into the displacement assumption, the displacement assumption must be equal to the nodal value of displacement, U_a. Likewise, the other two equations are constrained to the nodal values of displacement at Nodes $b$and c, respectively. In matrix form, the equations given by (4.4.6) are stated as:

$$\begin{bmatrix} 1 & X_a & Y_a \\ 1 & X_b & Y_b \\ 1 & X_c & Y_c \end{bmatrix} \begin{Bmatrix} a_1 \\ a_2 \\ a_3 \end{Bmatrix} = \begin{Bmatrix} U_a \\ U_b \\ U_c \end{Bmatrix} \tag{4.4.7}$$

A more compact way of expressing (4.4.7) is:

$$\underline{\underline{A}}\underline{a} = \begin{Bmatrix} U_a \\ U_b \\ U_c \end{Bmatrix}$$

The A-matrix above is defined as:

$$\underline{\underline{A}} = \begin{bmatrix} 1 & X_a & Y_a \\ 1 & X_b & Y_b \\ 1 & X_c & Y_c \end{bmatrix} \tag{4.4.8}$$

The meaningless parameters (a_i's) can be solved for, in terms of both displacement and spatial variables at the nodes, by solving the system of equations above. The system is solved by inverting the square matrix in (4.4.7):

$$\begin{Bmatrix} a_1 \\ a_2 \\ a_3 \end{Bmatrix} = \begin{bmatrix} 1 & X_a & Y_a \\ 1 & X_b & Y_b \\ 1 & X_c & Y_c \end{bmatrix}^{-1} \begin{Bmatrix} U_a \\ U_b \\ U_c \end{Bmatrix} \tag{4.4.9}$$

Small square matrices (*3 by 3* or less) can be inverted easily using elementary matrix inversion methods. However, the algebra necessary to solve (4.4.9) is somewhat tedious and will be dispatched. Consider that after solving (4.4.9) for the parameters a_i, and substituting into the original displacement assumption, (4.4.5), the interpolation function for displacement in the X-coordinate direction can be expressed as:

$$\tilde{U}(X,Y) = N_1 U_a + N_2 U_b + N_3 U_c$$

The shape functions for this element are:

$$\begin{aligned} N_1 &= \frac{1}{2A}\left[X_b Y_c - X_c Y_b + (Y_b - Y_c)X + (X_c - X_b)Y\right] \\ N_2 &= \frac{1}{2A}\left[X_c Y_a - X_a Y_c + (Y_c - Y_a)X + (X_a - X_c)Y\right] \\ N_3 &= \frac{1}{2A}\left[X_a Y_b - X_b Y_a + (Y_a - Y_b)X + (X_b - X_a)Y\right] \end{aligned} \tag{4.4.10}$$

The A variable in the equations above is the surface area of the triangular element, which can be computed using the cross product of two vectors representing the sides of the triangle, as shown in Figure 4.18.

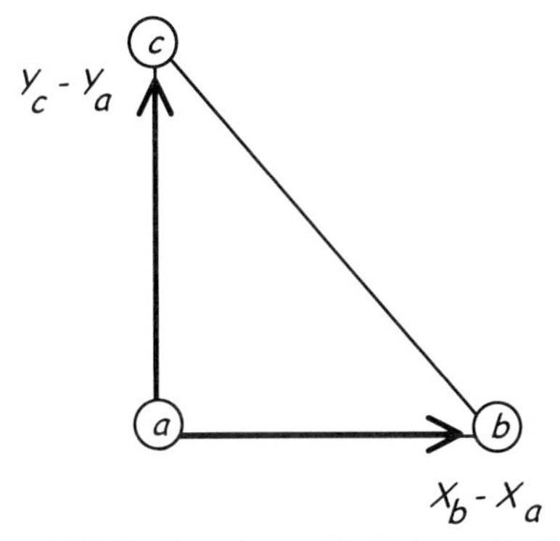

Figure 4.18 Surface Area of a Triangular Element

It so happens that the area of the triangle is equivalent to one-half the determinant of the square matrix given by (4.4.8). The determinant of the A-matrix and the cross product are related because the A-matrix contains the nodal coordinates that define the vectors on the two sides of the triangle. In the process of computing the determinant, the cross product of the vectors is computed, such that area of the triangle is expressed as:

$$Area = \frac{1}{2}\left(\det \underline{\underline{A}}\right) \tag{4.4.11}$$

The equation given by (4.4.11) is applicable to triangular surfaces of arbitrary orientation and shape, and can also be used when computing the stiffness matrix for 3-node triangular elements. A detailed explanation of computing area using the cross product of two vectors in given in Fisher [24, p. 595].

The same procedure that was used to generate the interpolation function for $U(X,Y)$ can be followed to generate the interpolation function $V(X,Y)$. The element is shown again in Figure 4.19, this time with nodal displacement variables associated with displacement in the Y-coordinate direction. Starting as before with Pascal's triangle, a linear displacement assumption in two variables is again expressed in terms of three unknown parameters:

$$\tilde{V}(X,Y) = b_1 + b_2 X + b_3 Y \tag{4.4.12}$$

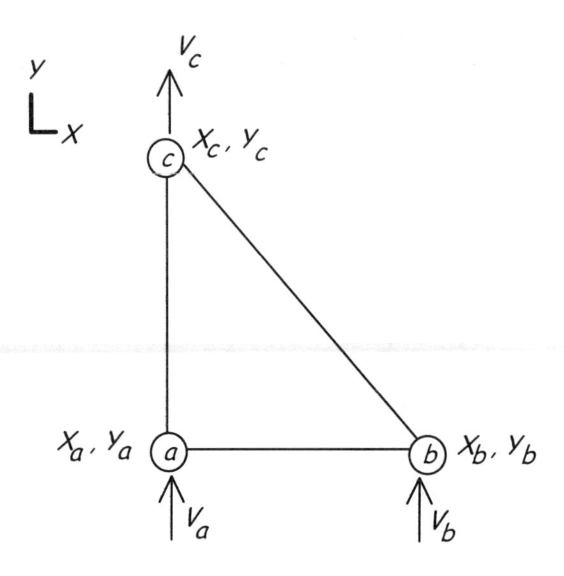

Figure 4.19 A 3-Node Triangular Element, Y-Displacement

Constraining the displacement assumption to take on the values of the nodal variables:

$$\begin{aligned}\tilde{V}(X_a, Y_a) &= b_1 + b_2 X_a + b_3 Y_a = V_a \\ \tilde{V}(X_b, Y_b) &= b_1 + b_2 X_b + b_3 Y_b = V_b \\ \tilde{V}(X_c, Y_c) &= b_1 + b_2 X_c + b_3 Y_c = V_c\end{aligned} \tag{4.4.13}$$

Hence, for the displacement interpolation in the *Y*-coordinate direction:

$$\tilde{V}(X,Y) = N_1 V_a + N_2 V_b + N_3 V_c$$

It might be apparent that, after manipulating (4.4.13), the same shape functions given by (4.4.10) are obtained. With the interpolation functions now defined for this element, the strain-displacement matrix can be developed.

Strain-Displacement Matrix for a 3-Node Surface Element, Plane Stress

An approximation for the strain components associated with the plane stress idealization may now be established. Interpolation functions for *U(X,Y)* and

$V(X,Y)$ are used to generate approximate expressions for normal strain in the X- and Y-coordinate directions, respectively:

$$\varepsilon_{XX} \approx \frac{\partial \tilde{U}}{\partial X} = \frac{\partial}{\partial X}\left(N_1 U_a + N_2 U_b + N_3 U_c\right) = \left(\frac{\partial N_1}{\partial X} U_a + \frac{\partial N_2}{\partial X} U_b + \frac{\partial N_3}{\partial X} U_c\right)$$

$$\varepsilon_{YY} \approx \frac{\partial \tilde{V}}{\partial Y} = \frac{\partial}{\partial Y}\left(N_1 V_a + N_2 V_b + N_3 V_c\right) = \left(\frac{\partial N_1}{\partial Y} V_a + \frac{\partial N_2}{\partial Y} V_b + \frac{\partial N_3}{\partial Y} V_c\right)$$

(4.4.14)

Note that the finite element approximation for normal strain is expressed in terms of partial derivatives of the shape functions. In a similar manner, shear strain may be expressed as:

$$\gamma_{XY} \approx \frac{\partial \tilde{U}}{\partial Y} + \frac{\partial \tilde{V}}{\partial X} = \frac{\partial N_1}{\partial Y} U_a + \frac{\partial N_1}{\partial X} V_a + \frac{\partial N_2}{\partial Y} U_b + \frac{\partial N_2}{\partial X} V_b + \cdots \tag{4.4.15}$$

Using (4.4.14) and (4.4.15), strain may be expressed in matrix form as:

$$\begin{Bmatrix} \varepsilon_{XX} \\ \varepsilon_{YY} \\ \gamma_{XY} \end{Bmatrix} \approx \begin{Bmatrix} \dfrac{\partial \tilde{U}}{\partial X} \\ \dfrac{\partial \tilde{V}}{\partial Y} \\ \dfrac{\partial \tilde{U}}{\partial Y} + \dfrac{\partial \tilde{V}}{\partial X} \end{Bmatrix} = \begin{bmatrix} \dfrac{\partial N_1}{\partial X} & 0 & \dfrac{\partial N_2}{\partial X} & 0 & \dfrac{\partial N_3}{\partial X} & 0 \\ 0 & \dfrac{\partial N_1}{\partial Y} & 0 & \dfrac{\partial N_2}{\partial Y} & 0 & \dfrac{\partial N_3}{\partial Y} \\ \dfrac{\partial N_1}{\partial Y} & \dfrac{\partial N_1}{\partial X} & \dfrac{\partial N_2}{\partial Y} & \dfrac{\partial N_2}{\partial X} & \dfrac{\partial N_3}{\partial Y} & \dfrac{\partial N_3}{\partial X} \end{bmatrix} \begin{Bmatrix} U_a \\ V_a \\ U_b \\ V_b \\ U_c \\ V_c \end{Bmatrix} \tag{4.4.16}$$

In other words, strain is expressed in terms of a strain-displacement matrix along with a vector of nodal displacements:

$$\underline{\underline{\varepsilon}} = \underline{\underline{B}}^{e} \underline{\underline{D}} \tag{4.4.17}$$

Comparing (4.4.16) and (4.4.17), the strain-displacement for the 3-node triangular surface element for plane stress takes the form of a *3 by 6* matrix:

$$\underline{\underline{B}} = \begin{bmatrix} \frac{\partial N_1}{\partial X} & 0 & \frac{\partial N_2}{\partial X} & 0 & \frac{\partial N_3}{\partial X} & 0 \\ 0 & \frac{\partial N_1}{\partial Y} & 0 & \frac{\partial N_2}{\partial Y} & 0 & \frac{\partial N_3}{\partial Y} \\ \frac{\partial N_1}{\partial Y} & \frac{\partial N_1}{\partial X} & \frac{\partial N_2}{\partial Y} & \frac{\partial N_2}{\partial X} & \frac{\partial N_3}{\partial Y} & \frac{\partial N_3}{\partial X} \end{bmatrix} \tag{4.4.18}$$

Notice that the strain-displacement matrix contains one row for every strain term in the strain vector, and one column for every nodal displacement variable. Since the strain vector for plane stress has three strain terms, three rows are present in (4.4.18); with a total of six nodal displacement variables, the B-matrix has six columns. Recalling the shape functions from (4.4.10), the derivatives shown in may be calculated:

$$\underline{\underline{B}} = \frac{1}{2A} \begin{bmatrix} Y_b - Y_c & 0 & Y_c - Y_a & 0 & Y_a - Y_b & 0 \\ 0 & X_c - X_b & 0 & X_a - X_c & 0 & X_b - X_a \\ X_c - X_b & Y_b - Y_c & X_a - X_c & Y_c - Y_a & X_b - X_a & Y_a - Y_b \end{bmatrix} \tag{4.4.19}$$

For this element, all the terms in the strain-displacement are constant. This is a result of two factors. First, the strain displacement matrix, (4.4.18), contains first order derivatives, and second, the shape functions contain only first order free variables, X and Y. For elements with higher order terms in the displacement assumption, first order derivative strain terms render a B-matrix that contains free variables, as well as constants.

Generating the Stiffness Matrix for a 3-Node Triangular Surface Element

The items required for computing the stiffness matrix are now available for the 3-node surface element for plane stress applications. Recalling the equation for the finite element stiffness matrix:

$$\underline{\underline{{}^eK}} \equiv \left[\int_V \underline{\underline{B}}^T \, \underline{\underline{E}} \, \underline{\underline{B}} \, dV \right] \tag{4.4.20}$$

Substituting both the material matrix from (4.4.2), and the strain-displacement matrix, (4.4.19), into (4.4.20), the element stiffness is expressed as:

$$
{}^{e}\underline{\underline{K}} = \int_V \left(\frac{1}{2A}\right)^2 \begin{bmatrix} Y_b - Y & 0 & X_c - X_b \\ 0 & X_c - X_b & Y_b - Y_c \\ Y_c - Y & 0 & X_a - X_c \\ 0 & X_a - X_c & Y_c - Y_a \\ Y_a - Y_b & 0 & X_b - X_a \\ 0 & X_b - X_a & Y_a - Y_b \end{bmatrix} \underline{\underline{E}} \begin{bmatrix} Y_b - Y_c & 0 & \cdots \\ 0 & X_c - X_b & \cdots \\ X_c - X_b & Y_b - Y_c & \cdots \end{bmatrix} dV
$$

$$
\textit{where:} \qquad E \equiv \frac{E}{1-\nu^2} \begin{bmatrix} 1 & \nu & 0 \\ \nu & 1 & 0 \\ 0 & 0 & \frac{1-\nu}{2} \end{bmatrix}
$$

(4.4.21)

Since the terms in the integrand are constants, the stiffness matrix for this element is simply:

$$
{}^{e}\underline{\underline{K}} = \left(\frac{1}{2A}\right)^2 \begin{bmatrix} Y_b - Y & 0 & X_c - X_b \\ 0 & X_c - X_b & Y_b - Y_c \\ Y_c - Y & 0 & X_a - X_c \\ 0 & X_a - X_c & Y_c - Y_a \\ Y_a - Y_b & 0 & X_b - X_a \\ 0 & X_b - X_a & Y_a - Y_b \end{bmatrix} \underline{\underline{E}} \begin{bmatrix} Y_b - Y_c & 0 & \cdots \\ 0 & X_c - X_b & \cdots \\ X_c - X_b & Y_b - Y_c & \cdots \end{bmatrix} \int_V dV
$$

Surface elements are often used with a constant value of thickness, t, such that the differential volume is expressed as $t\ dA$. Using this substitution in the equation above, and performing the integration, the stiffness matrix can be computed using the expression:

$$
{}^{e}\underline{\underline{K}} = \left(\frac{1}{2A}\right)^2 \begin{bmatrix} Y_b - Y & 0 & X_c - X_b \\ 0 & X_c - X_b & Y_b - Y_c \\ Y_c - Y & 0 & X_a - X_c \\ 0 & X_a - X_c & Y_c - Y_a \\ Y_a - Y_b & 0 & X_b - X_a \\ 0 & X_b - X_a & Y_a - Y_b \end{bmatrix} \underline{\underline{E}} \begin{bmatrix} Y_b - Y_c & 0 & \cdots \\ 0 & X_c - X_b & \cdots \\ X_c - X_b & Y_b - Y_c & \cdots \end{bmatrix} t\ A
$$

The area of the triangular surface element, A, shown in the equation above, can be replaced with one-half the determinant of the A-matrix as given by (4.4.11):

$$
{}^{e}\underline{\underline{K}} = \left(\frac{1}{2A}\right)^2 \begin{bmatrix} Y_b - Y & 0 & X_c - X_b \\ 0 & X_c - X_b & Y_b - Y_c \\ Y_c - Y & 0 & X_a - X_c \\ 0 & X_a - X_c & Y_c - Y_a \\ Y_a - Y_b & 0 & X_b - X_a \\ 0 & X_b - X_a & Y_a - Y_b \end{bmatrix} \underline{\underline{E}} \begin{bmatrix} Y_b - Y_c & 0 & \cdots \\ 0 & X_c - X_b & \cdots \\ X_c - X_b & Y_b - Y_c & \cdots \end{bmatrix} \frac{t \det \underline{\underline{A}}}{2}
$$

(4.4.22)

While it would be very tedious to perform the matrix algebra in (4.4.22) by hand, it is fairly simple to write a software routine to accomplish the same. Again, this element is typically not recommended due to its overly-stiff nature, which causes under prediction of displacement, stress, and strain. The force and displacement vectors for this element are defined as:

$$
{}^{e}\underline{\underline{D}} \equiv \begin{Bmatrix} U_a \\ V_a \\ U_b \\ V_b \\ U_c \\ V_c \end{Bmatrix} \qquad {}^{e}\underline{\underline{F}} \equiv \begin{Bmatrix} F_{Xa} \\ F_{Ya} \\ F_{Xb} \\ F_{Yb} \\ F_{Xc} \\ F_{Yc} \end{Bmatrix} \tag{4.4.23}
$$

Of What Consequence Are Constant Terms in the *B*-Matrix?

Since only constant terms appear in the B-matrix for the 3-node element above, (4.4.19), strain is constant over the entire element. In other words, a single element can characterize only one strain state: constant strain. If there were X and Y free variables in (4.4.19), in addition to the constants shown, then a single element could represent two types of strain: constant strain, and at least one other mode of linearly varying strain. As more terms are added to a given displacement assumption, a single element can represent more modes of displacement; this translates into the element also being able to represent more modes of strain, other things being equal.

Ideally, a single element would represent all possible modes of displacement, or at least all modes that a polynomial can represent. That is, displacement would be represented as a constant, linear, quadratic, or any higher order polynomial function. In such a case, a single element, or least far fewer elements, would be needed in many finite element models. Conversely, as an element becomes more limited in the number of displacement modes it can represent, it has the potential to be more stiff, other things being equal. The stiffness is of course artificial, since the actual structure can deform in an infinite number of modes.

One extreme example of artificial stiffness occurs when an element can exhibit only constant *displacement*; such elements are termed *rigid elements*, since they cannot deform at all:

$$\text{if } \; U(X) = C_1 \qquad \text{then } \varepsilon_{XX} = \frac{dU}{dX} = 0$$

Rigid elements are designed to perform in the manner described above—it is not a deficiency.

Other elements may deform sufficiently well in several modes of deformation but lack one (or a few) modes. Recall from the requirements for finite element displacement assumptions that all elements should, at the very least, be able to represent all constant strain modes. To do this, the displacement assumption must have at least one term such that when the derivatives needed to represent strain are computed, a constant term remains. As suggested by (4.3.3), when strains are represented with second order derivatives, then second order terms are needed in the displacement assumption to represent constant strain.

How Does Lack of Displacement Modes Translate into Increased Stiffness?
The non-uniform shaft from Example 2.4 provides an example of how a missing mode of deformation increases the stiffness of an element. In the example, it was shown that the exact displacement is characterized by a cubic polynomial, while the element can displace only in a linear fashion. Observing the results in Figure 2.14, notice that using one element, the displacement (on average) is significantly under predicted, with maximum values differing by 100%! In effect, the element is too stiff to allow it to deform as much as the actual shaft would because the element cannot represent a third order polynomial. If a third order term had been included in the element's interpolation function, it could deform exactly as required. When a single element is unable to characterize the exact displacement, the element will behave in an overly stiff manner.

For another example of overly stiff behavior, consider the bending of the cantilever beam in Figure 4.20.

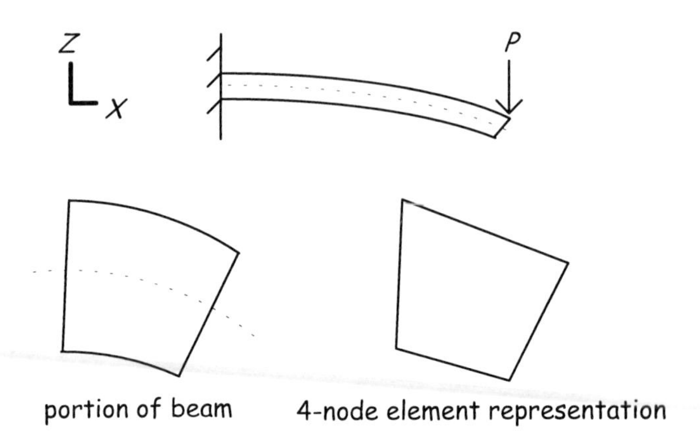

Figure 4.20 Approximating Bending with a Straight-sided Element

It should be clear from the first illustration in Figure 4.20 that the displacement in the *Z*-coordinate direction is not a linear function of *X*. The shape of the top surface has curvature, requiring at least a quadratic function to provide proper representation. Now, if a straight sided element with only a linear displacement assumption is used to model a structure in bending, the best it can do is to exhibit shear deformation, as shown. However, in very shallow beams, shear deformation is nearly zero. Hence, using the straight-sided element (without any modifications) would yield a very stiff element with regards to shallow beam bending, since the only mode of deformation available to the element is a mode that is essentially non-existent in the actual structure. This type of behavior is considered in Chapter 7, Example 7.1.

In general, when elements that can only represent a few modes of displacement are used to model a portion of a structure that has a high strain gradient, many elements will be necessary to represent the exact strain (and stress) within the structure. Since the 3-node triangular surface element for plane stress can represent only one mode of strain, i.e., uniform strain, the element is considered very stiff. An element exhibiting this type of behavior should be used with extreme caution since it can significantly underpredict displacement, and grossly underpredict stress and strain.

4.5 Other Simple Continuum Elements

Key Concept: Plane stress, plane strain, axisymmetric, and three-dimensional stress applications are typically analyzed using continuum elements.

As shown in Section 4.4, the 3-node surface element can be applied to a plane stress idealization. This type of element can be applied to a variety of mechanical idealizations: plane stress, plane strain, and axisymmetric bodies of revolution, for instance. A 4-node tetrahedral volume element is a continuum element that can be used for three-dimensional stress idealizations. These elements, used to model idealizations based upon some form of the generalized Hooke's law equations, are considered in this section.

A Plane Strain Application

With techniques similar to those used in the preceding section, the 3-node surface element can be easily expanded from the plane stress application to both plane strain and axisymmetric applications. Consider the finite element application from Chapter 1 where a water retention dam, subjected to hydraulic pressure along one surface, is approximated by a plane strain idealization.

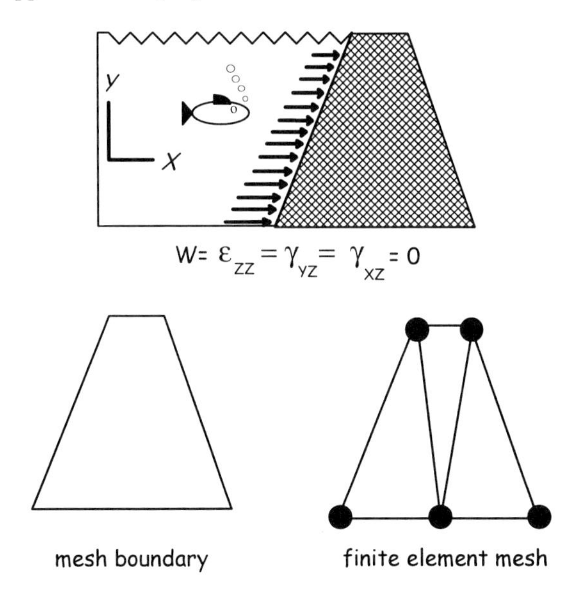

Figure 4.21 Finite Element Idealization of a Dam

Observing Figure 4.21, note that displacement through the thickness of the dam is presumed zero, rendering strains in that direction zero as well. A finite element model for this plane strain idealization could be constructed using three, 3-node triangular elements within the mesh boundary, as illustrated. Although 3-node, triangular elements for plane strain are used here, as mentioned, the 4-node quadrilateral surface element is a better element. However, the 3-node element will be used, since it is not quite as complex as the 4-node element.

It is again noted that the number of elements employed in the models considered in this text is not based upon accuracy; typically, the examples use the minimum number of elements possible while still conveying the rough shape of the structure under consideration.

The Element Stiffness Matrix Equation

To develop the stiffness matrix for the 3-node, plane strain element, (3.5.12) is again employed:

$$^{e}\underline{\underline{K}} \equiv \left[\int_V \underline{\underline{B}}^T \, \underline{\underline{E}} \, \underline{\underline{B}} \, dV\right] \tag{4.5.1}$$

Since the material matrix for plane strain was given in Table 3.1, all that is required to compute the element stiffness matrix for a plane strain element is the strain-displacement matrix. As before, to develop the strain-displacement matrix, interpolation functions need to be established. To determine what variables need to be interpolated, the constitutive equations for plane strain are examined.

The Constitutive Equations for Plane Strain

Recalling the plane strain constitutive equations from Table 3.1:

$$\underline{\underline{\tau}} = \begin{Bmatrix} \tau_{XX} \\ \tau_{YY} \\ \tau_{XY} \end{Bmatrix} \qquad \underline{\underline{E}} = \frac{E}{(1+\nu)(1-2\nu)} \begin{bmatrix} 1 & \frac{\nu}{1-\nu} & 0 \\ \frac{\nu}{1-\upsilon} & 1 & 0 \\ 0 & 0 & \frac{1-2\nu}{2(1-\nu)} \end{bmatrix} \qquad \underline{\underline{\varepsilon}} = \begin{Bmatrix} \varepsilon_{XX} \\ \varepsilon_{YY} \\ \gamma_{XY} \end{Bmatrix} \tag{4.5.2}$$

Normal stress in the thickness direction may be calculated separately after the in-plane strain components are determined, using the relationship:

$$\tau_{ZZ} = \frac{\nu E}{(1+\nu)(1-2\nu)} \left(\varepsilon_{XX} + \varepsilon_{YY}\right)$$

Comparing (4.5.2) with constitutive equations for plane stress, notice that the stress and strain vectors are identical. Recalling the definition for the strain terms in the plane stress constitutive equations:

$$\underline{\underline{\varepsilon}} = \begin{Bmatrix} \varepsilon_{XX} \\ \varepsilon_{YY} \\ \gamma_{XY} \end{Bmatrix} = \begin{Bmatrix} \dfrac{\partial U}{\partial X} \\ \dfrac{\partial V}{\partial Y} \\ \dfrac{\partial U}{\partial Y} + \dfrac{\partial V}{\partial X} \end{Bmatrix} \tag{4.5.3}$$

As in the case of plane stress, displacements U and V need to be interpolated. It should be apparent that the triangular element for plane strain, shown in Figure 4.22, has exactly the same requirements as the plane stress element, because the same nodal variables are used in both. Therefore, the process of generating displacement interpolation functions for the plane strain element is exactly the same as in Section 4.4, beginning with Equation 4.4.5.

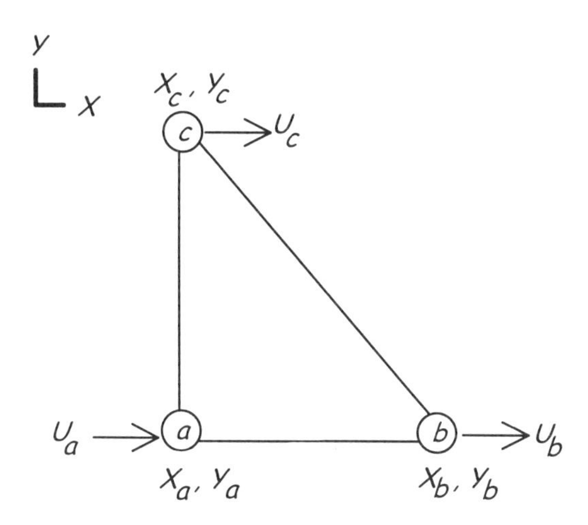

Figure 4.22 A 3-Node Triangular Element

The process of generating displacement interpolation functions for the plane strain element will not be shown because the steps are the same as before. Since the interpolation functions and strain vectors are the same for both plane stress and plane strain elements, the strain-displacement matrix (4.4.19) can be used for either. The only difference between these two elements is the constants that are used in the material matrix.

The Stiffness Matrix for a 3-Node, Triangular, Plane Strain Element

Substituting the strain displacement matrix (4.4.19), along with the material matrix from (4.5.2), into the element stiffness matrix equation (4.5.1):

$$
{}^{e}\underline{\underline{K}} = \int_V \left(\frac{1}{2A}\right)^2 \begin{bmatrix} Y_b - Y & 0 & X_c - X_b \\ 0 & X_c - X_b & Y_b - Y_c \\ Y_c - Y & 0 & X_a - X_c \\ 0 & X_a - X_c & Y_c - Y_a \\ Y_a - Y_b & 0 & X_b - X_a \\ 0 & X_b - X_a & Y_a - Y_b \end{bmatrix} \underline{\underline{\mathrm{E}}} \begin{bmatrix} Y_b - Y_c & 0 & \cdots \\ 0 & X_c - X_b & \cdots \\ X_c - X_b & Y_b - Y_c & \cdots \end{bmatrix} dV \tag{4.5.4}
$$

Where:

$$
\underline{\underline{E}} = \frac{E}{(1+\nu)(1-2\nu)} \begin{bmatrix} 1 & \frac{\nu}{1-\nu} & 0 \\ \frac{\nu}{1-\upsilon} & 1 & 0 \\ 0 & 0 & \frac{1-2\nu}{2(1-\nu)} \end{bmatrix}
$$

Comparing the stiffness matrix formula for plane stress, (4.4.21), with the stiffness formula for plane strain, (4.5.4), it is evident that once the element stiffness formula is developed for a plane stress, the stiffness for plane strain can be obtained by simply changing a few constants in the material matrix. Hence, for plane strain, we use (4.4.22) but change the terms in the material matrix. The force and displacement vectors for the 3-node plane strain element are the same as those used for plane stress:

$$
{}^{e}\underline{\underline{D}} \equiv \begin{Bmatrix} U_a \\ V_a \\ U_b \\ V_b \\ U_c \\ V_c \end{Bmatrix} \qquad {}^{e}\underline{\underline{F}} \equiv \begin{Bmatrix} F_{Xa} \\ F_{Ya} \\ F_{Xb} \\ F_{Yb} \\ F_{Xc} \\ F_{Yc} \end{Bmatrix}
$$

Axisymmetric Solid Application

Consider the solid ring subjected to internal pressure, as shown in Chapter 1.

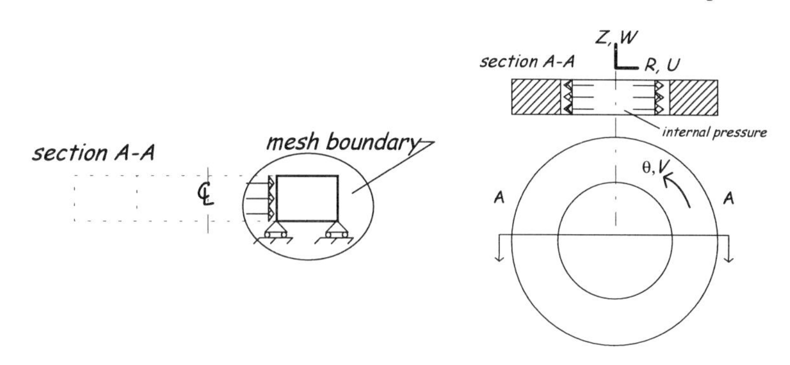

Figure 4.23 An Axisymmetric Ring

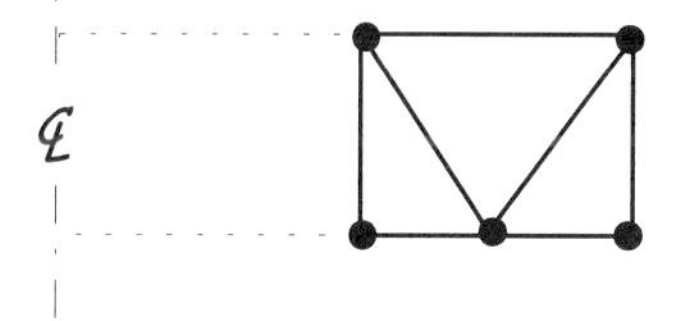

Figure 4.24 Ring Mesh

An axisymmetric idealization might also be used in the analysis of a section of a long pipe which is subjected to internal fluid pressure. Again, 3-node surface elements can be employed for modeling the axisymmetric solid application as shown in Figures 4.23 and 4.24. Using (4.5.1) to calculate element stiffness, a material matrix and displacement interpolation matrix need to be defined.

The Constitutive Equations for an Axisymmetric Solid of Revolution

Consider the constitutive equations for an axisymmetric solid:

$$\underline{\underline{\tau}} = \begin{Bmatrix} \tau_{RR} \\ \tau_{\theta\theta} \\ \tau_{ZZ} \\ \tau_{ZR} \end{Bmatrix} \quad \underline{\underline{E}} = \frac{(1-\nu)E}{(1+\nu)(1-2\nu)} \begin{bmatrix} 1 & \frac{\nu}{1-\nu} & \frac{\nu}{1-\nu} & 0 \\ \frac{\nu}{1-\nu} & 1 & \frac{\nu}{1-\nu} & 0 \\ \frac{\nu}{1-\nu} & \frac{\nu}{1-\nu} & 1 & 0 \\ 0 & 0 & 0 & \frac{(1-2\nu)}{2(1-\nu)} \end{bmatrix} \quad \underline{\underline{\varepsilon}} = \begin{Bmatrix} \varepsilon_{RR} \\ \varepsilon_{\theta\theta} \\ \varepsilon_{ZZ} \\ \gamma_{ZR} \end{Bmatrix} \tag{4.5.5}$$

The strain terms for the axisymmetric solid idealization are expressed as:

$$\underline{\underline{\varepsilon}} = \begin{Bmatrix} \varepsilon_{RR} \\ \varepsilon_{\theta\theta} \\ \varepsilon_{ZZ} \\ \gamma_{ZR} \end{Bmatrix} = \begin{Bmatrix} \dfrac{\partial U}{\partial R} \\ \dfrac{U}{R} \\ \dfrac{\partial W}{\partial Z} \\ \dfrac{\partial U}{\partial Z} + \dfrac{\partial W}{\partial R} \end{Bmatrix} \tag{4.5.6}$$

Notice that the hoop strain, $\varepsilon_{\theta\theta}$, is equal to the radial displacement, U, divided by the radial distance, R. It is evident from (4.5.6) that the only variables which need to be interpolated for this element are the radial displacement, U, and the axial displacement, W.

Figure 4.25 illustrates a 3-node triangular element with a coordinate system related to the axisymmetric problem. The hoop direction is orthogonal to the Z and R axes, acting out of the page. To generate the element stiffness matrix for this element, (4.5.1) again is employed:

$$^{e}\underline{\underline{K}} \equiv \left[\int_V \underline{\underline{B}}^T \, \underline{\underline{E}} \, \underline{\underline{B}} \, dV\right]$$

The strain-displacement can be developed once the interpolation functions are defined.

Interpolation Functions for a 3-Node Element, Axisymmetric Applications

Beginning with the interpolation function for the displacement in the radial direction, $\tilde{U}(R,Z)$, a three term displacement assumption can be employed since there exist three nodal variables related to displacement in the radial direction:

$$\tilde{U}(R,Z) = a_1 + a_2 R + a_3 Z \tag{4.5.7}$$

Constraining the displacement assumption to take on the nodal variables of displacement:

$$\begin{aligned} \tilde{U}(R_a, Z_a) &= a_1 + a_2 R_a + a_3 Z_a = U_a \\ \tilde{U}(R_b, Z_b) &= a_1 + a_2 R_b + a_3 Z_b = U_b \\ \tilde{U}(R_c, Z_c) &= a_1 + a_2 R_c + a_3 Z_c = U_c \end{aligned} \tag{4.5.8}$$

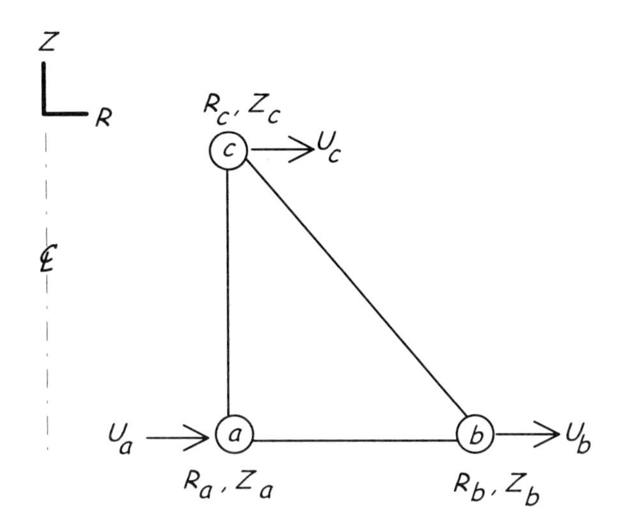

Figure 4.25 A 3-Node Triangular Element

It should be apparent that, except for the fact that the *X* and *Y* variables are replaced with *R* and *Z*, these are the same equations used in the 3-node element for plane strain and plane stress, (4.4.6). Hence, the radial displacement can be expressed as:

$$\tilde{U}(R,Z) = N_1 U_a + N_2 U_b + N_3 U_c \tag{4.5.9}$$

The resulting shape functions are the same as (4.4.10), with a change of variables:

$$\begin{aligned} N_1 &= \frac{1}{2A}\left[R_b Z_c - R_c Z_b + (Z_b - Z_c)R + (R_c - R_b)Z\right] \\ N_2 &= \frac{1}{2A}\left[R_c Z_a - R_a Z_c + (Z_c - Z_a)R + (R_a - R_c)Z\right] \\ N_3 &= \frac{1}{2A}\left[R_a Z_b - R_b Z_a + (Z_a - Z_b)R + (R_b - R_a)Z\right] \end{aligned} \tag{4.5.10}$$

Displacement in the *Z*-coordinate direction is likewise interpolated using the shape functions from (4.5.10):

$$\tilde{W}(R,Z) = N_1 W_a + N_2 W_b + N_3 W_c$$

Strain-Displacement Matrix, 3-Node Element, Axisymmetric Element
The three normal strains for the axisymmetric element are expressed as:

$$\varepsilon_{RR} = \frac{\partial U}{\partial R} \approx \frac{\partial}{\partial R}(N_1 U_a + N_2 U_b + N_3 U_c) = \frac{\partial N_1}{\partial R} U_a + \frac{\partial N_2}{\partial R} U_b + \frac{\partial N_3}{\partial R} U_c$$

$$\varepsilon_{\theta\theta} = \frac{U}{R} \approx \frac{1}{R}(N_1 U_a + N_2 U_b + N_3 U_c)$$

$$\varepsilon_{ZZ} = \frac{\partial W}{\partial Z} \approx \frac{\partial N_1}{\partial Z} W_a + \frac{\partial N_2}{\partial Z} W_b + \frac{\partial N_3}{\partial Z} W_c$$

The shear strain terms expressed as:

$$\gamma_{ZR} = \left(\frac{\partial U}{\partial Z}\right) + \left(\frac{\partial W}{\partial R}\right) \approx \left(\frac{\partial N_1}{\partial Z} U_a + \frac{\partial N_2}{\partial Z} U_b + \frac{\partial N_3}{\partial Z} U_c\right) + \left(\frac{\partial N_1}{\partial R} W_a + \frac{\partial N_2}{\partial R} W_b + \frac{\partial N_3}{\partial R} W_c\right)$$

The strain vector may be expressed in matrix form as:

$$\begin{Bmatrix} \varepsilon_{RR} \\ \varepsilon_{\theta\theta} \\ \varepsilon_{ZZ} \\ \gamma_{ZR} \end{Bmatrix} \approx \begin{bmatrix} \frac{\partial N_1}{\partial R} & 0 & \frac{\partial N_2}{\partial R} & 0 & \frac{\partial N_3}{\partial R} & 0 \\ \frac{N_1}{R} & 0 & \frac{N_2}{R} & 0 & \frac{N_3}{R} & 0 \\ 0 & \frac{\partial N_1}{\partial Z} & 0 & \frac{\partial N_2}{\partial Z} & 0 & \frac{\partial N_3}{\partial Z} \\ \frac{\partial N_1}{\partial Z} & \frac{\partial N_1}{\partial R} & \frac{\partial N_2}{\partial Z} & \frac{\partial N_2}{\partial R} & \frac{\partial N_3}{\partial Z} & \frac{\partial N_3}{\partial R} \end{bmatrix} \begin{Bmatrix} U_a \\ W_a \\ U_b \\ W_b \\ U_c \\ W_c \end{Bmatrix} \tag{4.5.11}$$

The shape functions in (4.5.10), and the associated derivatives, can substituted into (4.5.11). This exercise will be left for the interested reader. From (4.5.11), it is evident that the strain-displacement matrix for this element is:

$$\underline{\underline{B}} = \begin{bmatrix} \frac{\partial N_1}{\partial R} & 0 & \frac{\partial N_2}{\partial R} & 0 & \frac{\partial N_3}{\partial R} & 0 \\ \frac{N_1}{R} & 0 & \frac{N_2}{R} & 0 & \frac{N_3}{R} & 0 \\ 0 & \frac{\partial N_1}{\partial Z} & 0 & \frac{\partial N_2}{\partial Z} & 0 & \frac{\partial N_3}{\partial Z} \\ \frac{\partial N_1}{\partial Z} & \frac{\partial N_1}{\partial R} & \frac{\partial N_2}{\partial Z} & \frac{\partial N_2}{\partial R} & \frac{\partial N_3}{\partial Z} & \frac{\partial N_3}{\partial R} \end{bmatrix} \tag{4.5.12}$$

The strain-displacement matrix for the axisymmetric element contains only constants, just like the 3-node elements that were developed for plane stress and plane strain. The element stiffness matrix for the axisymmetric element is computed as:

$$^{e}\underline{\underline{K}} \equiv \int \begin{bmatrix} \frac{\partial N_1}{\partial R} & \frac{N_1}{R} & 0 & \frac{\partial N_1}{\partial Z} \\ 0 & 0 & \frac{\partial N_1}{\partial Z} & \frac{\partial N_1}{\partial R} \\ \frac{\partial N_2}{\partial R} & \frac{N_2}{R} & 0 & \frac{\partial N_2}{\partial Z} \\ 0 & 0 & \frac{\partial N_2}{\partial Z} & \frac{\partial N_2}{\partial R} \\ \frac{\partial N_3}{\partial R} & \frac{N_3}{R} & 0 & \frac{\partial N_3}{\partial Z} \\ 0 & 0 & \frac{\partial N_3}{\partial Z} & \frac{\partial N_3}{\partial R} \end{bmatrix} \underline{\underline{E}} \begin{bmatrix} \frac{\partial N_1}{\partial R} & 0 & \frac{\partial N_2}{\partial R} & 0 & \frac{\partial N_3}{\partial R} & 0 \\ \frac{N_1}{R} & 0 & \frac{N_2}{R} & 0 & \frac{N_3}{R} & 0 \\ 0 & \frac{\partial N_1}{\partial Z} & 0 & \frac{\partial N_2}{\partial Z} & 0 & \frac{\partial N_3}{\partial Z} \\ \frac{\partial N_1}{\partial Z} & \frac{\partial N_1}{\partial R} & \frac{\partial N_2}{\partial Z} & \frac{\partial N_2}{\partial R} & \frac{\partial N_3}{\partial Z} & \frac{\partial N_3}{\partial R} \end{bmatrix} dV$$

$$\textit{where:} \quad \underline{\underline{E}} \equiv \begin{bmatrix} 1 & \frac{v}{1-v} & \frac{v}{1-v} & 0 \\ \frac{v}{1-v} & 1 & \frac{v}{1-v} & 0 \\ \frac{v}{1-v} & \frac{v}{1-v} & 1 & 0 \\ 0 & 0 & 0 & \frac{(1-2v)}{2(1-v)} \end{bmatrix}$$

`The force and displacement vectors for this element are:

$$\underline{\underline{{}^eD}} \equiv \begin{Bmatrix} U_a \\ W_a \\ U_b \\ W_b \\ U_c \\ W_c \end{Bmatrix} \qquad \underline{\underline{{}^eF}} \equiv \begin{Bmatrix} F_{Ra} \\ F_{Za} \\ F_{Rb} \\ F_{Zb} \\ F_{Rc} \\ F_{Zc} \end{Bmatrix}$$

The axisymmetric element developed here actually represents the entire ring, with the surface revolved 360° around the Z axis. The associated limits of integration with respect to the circumferential angle are from zero to 2π. However, since it is presumed that one radial slice is like any other, the element is often considered to represent a one *radian* slice of the structure. As such, the integration of the element is more easily computed as:

$$ {}^e\underline{\underline{K}} \equiv \int_A \int_0^1 \underline{\underline{B}}^T \underline{\underline{E}}\,\underline{\underline{B}}\, d\theta\, dA \qquad (4.5.13)$$

Computational Issues Associated with the Axisymmetric Element

Note that terms with *1/R* appear in the second row of the strain-displacement matrix, (4.5.12). This term may require special consideration, depending upon the geometry of the structure, and how the integral is evaluated. If a prismatic ("solid") cylinder is being considered, *1/R* is undefined on the centerline of the cylinder, where $R=0$. Cook [15, p. 297] provides a discussion of some of the details regarding how elements of this type are designed with regard to the issue of the *1/R* term.

The next continuum element to be considered is the 4-node, tetrahedral volume element which can be used for three-dimensional stress applications.

Three-Dimensional Stress

Analysts will generally attempt to utilize idealizations that reduce the computer resources needed to build and analyze a structural model. When used appropriately, line and surface elements can substantially reduce the complexity of a model while providing suitable accuracy. However, in some cases, the geometry and loading of a structure are such that only a three-dimensional stress idealization is applicable. Elements that utilize fully three-dimensional constitutive equations are required in such a case. Figure 4.26 shows the mesh boundary for an alternator housing used for automotive applications.

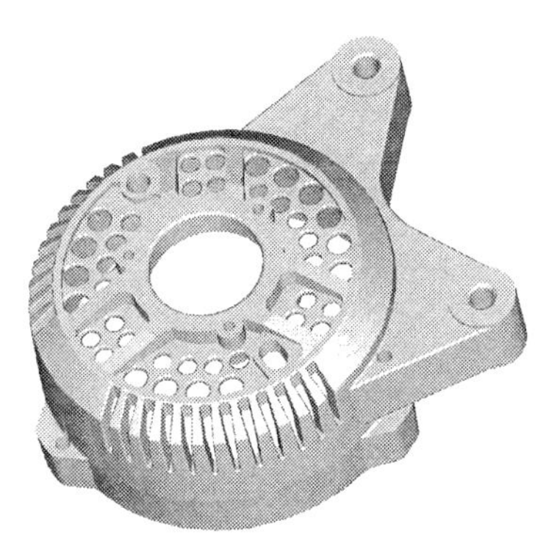

Figure 4.26 Three-Dimensional Mesh Boundary (Alternator Housing)

In service, the alternator housing is subjected to loads in three mutually perpendicular planes. The geometric complexity of the structure, along with the nature of the loading, suggests the use of a three-dimensional stress idealization.

The Constitutive Equations for Three-Dimensional Stress

The constitutive equations for three-dimensional stress, from Table 3.1, are expressed again, below.

$$\underline{\underline{E}} = \frac{E}{(1+\nu)(1-2\nu)}\begin{bmatrix} 1-\nu & \nu & \nu & 0 & 0 & 0 \\ \nu & 1-\nu & \nu & 0 & 0 & 0 \\ \nu & \nu & 1-\nu & 0 & 0 & 0 \\ 0 & 0 & 0 & \frac{1}{2(1+\nu)} & 0 & 0 \\ 0 & 0 & 0 & 0 & \frac{1}{2(1+\nu)} & 0 \\ 0 & 0 & 0 & 0 & 0 & \frac{1}{2(1+\nu)} \end{bmatrix}$$

$$\underline{\underline{\tau}} = \begin{Bmatrix} \tau_{XX} \\ \tau_{YY} \\ \tau_{YY} \\ \tau_{XY} \\ \tau_{YZ} \\ \tau_{ZX} \end{Bmatrix} \qquad \underline{\underline{\varepsilon}} = \begin{Bmatrix} \varepsilon_{XX} \\ \varepsilon_{YY} \\ \varepsilon_{YY} \\ \gamma_{XY} \\ \gamma_{YZ} \\ \gamma_{ZX} \end{Bmatrix}$$

When expressed in terms of derivatives, the strain terms indicate which displacement functions need to be interpolated:

$$\underline{\underline{\varepsilon}} = \begin{Bmatrix} \varepsilon_{XX} \\ \varepsilon_{YY} \\ \varepsilon_{YY} \\ \gamma_{XY} \\ \gamma_{YZ} \\ \gamma_{ZX} \end{Bmatrix} = \begin{Bmatrix} \dfrac{\partial U}{\partial X} \\ \dfrac{\partial V}{\partial Y} \\ \dfrac{\partial W}{\partial Z} \\ \dfrac{\partial U}{\partial Y} + \dfrac{\partial V}{\partial X} \\ \dfrac{\partial V}{\partial Z} + \dfrac{\partial W}{\partial Y} \\ \dfrac{\partial W}{\partial X} + \dfrac{\partial U}{\partial Z} \end{Bmatrix} \tag{4.5.14}$$

The strain terms above reveal that three displacement interpolation functions are required, one interpolation function for each coordinate direction. At each node of the three-dimensional element, three nodal displacement variables will be defined: U, V, and W, for displacement in the X, Y, and Z coordinate directions.

The simplest volume element that can be used to model a three-dimensional state of stress is a 4-node, tetrahedral volume element. Like the 3-node, triangular surface element, the "4-node tet" element performs poorly and is not recommended for general use. However, it is the most simple three-dimensional element, and is used here for illustrative purposes.

Figure 4.27 depicts a 4-node tetrahedral element. As before, the objective is to develop an expression for stiffness, using Equation 3.5.12, which requires the definition of a strain-displacement matrix. The strain-displacement matrix can be developed once the interpolation functions are known.

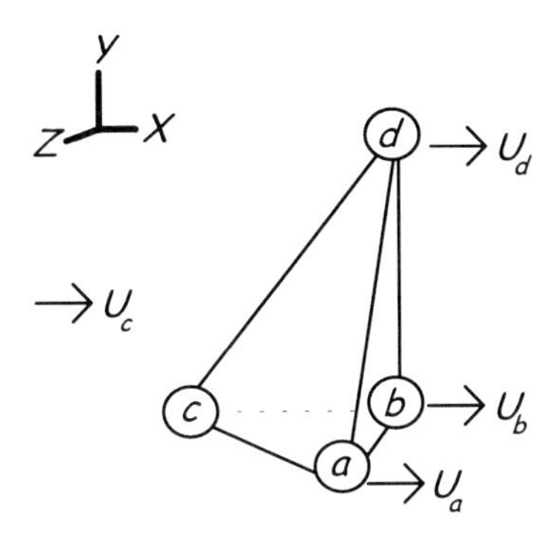

Figure 4.27 A 4-Node Tetrahedral Element

A 4-Node Volume Element—Interpolation Functions

From the illustration of the tetrahedral element in Figure 4.27, it is evident that four nodal variables that relate to displacement in the X-coordinate direction are available for interpolation. A four term polynomial can therefore be employed for this element:

$$\tilde{U}(X,Y,Z) = a_1 + a_2 X + a_3 Y + a_4 Z \qquad (4.5.15)$$

The terms in (4.5.15) are determined by examining the three-dimensional analogy to Pascal's triangle, previously mentioned. Following the usual procedure, the displacement assumption is constrained to the nodal variables of displacement:

$$\begin{aligned}
\tilde{U}(X_a, Y_a, Z_a) &= a_1 + a_2 X_a + a_3 Y_a + a_4 Z_a = U_a \\
\tilde{U}(X_b, Y_b, Z_b) &= a_1 + a_2 X_b + a_3 Y_b + a_4 Z_b = U_b \\
\tilde{U}(X_c, Y_c, Z_c) &= a_1 + a_2 X_c + a_3 Y_c + a_4 Z_c = U_c \\
\tilde{U}(X_d, Y_d, Z_d) &= a_1 + a_2 X_d + a_3 Y_d + a_4 Z_d = U_d
\end{aligned}$$

Hence, the system of equations that needs to be solved is:

$$\begin{aligned}
a_1 + a_2 X_a + a_3 Y_a + a_4 Z_a &= U_a \\
a_1 + a_2 X_b + a_3 Y_b + a_4 Z_b &= U_b \\
a_1 + a_2 X_c + a_3 Y_c + a_4 Z_c &= U_c \\
a_1 + a_2 X_d + a_3 Y_d + a_4 Z_d &= U_d
\end{aligned} \qquad (4.5.16)$$

In matrix form, (4.5.16) is expressed as:

$$\underline{\underline{A}}\underline{a} = \begin{Bmatrix} U_a \\ U_b \\ U_c \end{Bmatrix} \qquad \underline{\underline{A}} = \begin{bmatrix} 1 & X_a & Y_a & Z_a \\ 1 & X_b & Y_b & Z_b \\ 1 & X_c & Y_c & Z_c \\ 1 & X_d & Y_d & Z_d \end{bmatrix} \tag{4.5.17}$$

Solving the system of equations in (4.5.16) for the four unknown parameters (a_i's), then substituting back into the original displacement assumptions, a displacement interpolation function is expressed in terms of four shape functions:

$$\tilde{U}(X,Y,Z) = N_1 U_a + N_2 U_b + N_3 U_c + N_4 U_d$$

The process outlined above can be repeated to render interpolation functions for *V(X,Y,Z)* and *W(X,Y,Z)*. The shape functions include many constants but are of the form:

$$N_i = \frac{1}{6V}\left[e_i + f_i X + g_i Y + h_i Z\right] \qquad V = volume \tag{4.5.18}$$

Each of the constants (*e, f, g, h*) in (4.5.18) evolves from solving the system of equations shown in (4.5.16); in the general case, the constants actually consist of many terms. To obtain these terms, the inversion of the *4 by 4* matrix shown in (4.5.17) is required. While a somewhat simple computation can be performed to invert a *3 by 3* or smaller matrix, larger matrices require a more laborious technique, in general; see Selby [25, p. 125]. However, a 4-node tet element with coordinates conveniently defined can allow equations (4.5.16) to be easily solved. The constants in the shape functions can, in such a case, be expressed with just a few terms.

For example, consider the 4-node tet element shown in Figure 4.28. The volume of this particular element can be shown to be 0.333 units. Substituting the spatial coordinates shown in Figure 4.28 into (4.5.16):

$$\begin{aligned} a_1 + a_2(0) + a_3(0) + a_4(1) &= U_a \\ a_1 + a_2(1) + a_3(0) + a_4(0) &= U_b \\ a_1 + a_2(-1) + a_3(0) + a_4(0) &= U_c \\ a_1 + a_2(0) + a_3(1) + a_4(0) &= U_d \end{aligned} \tag{4.5.19}$$

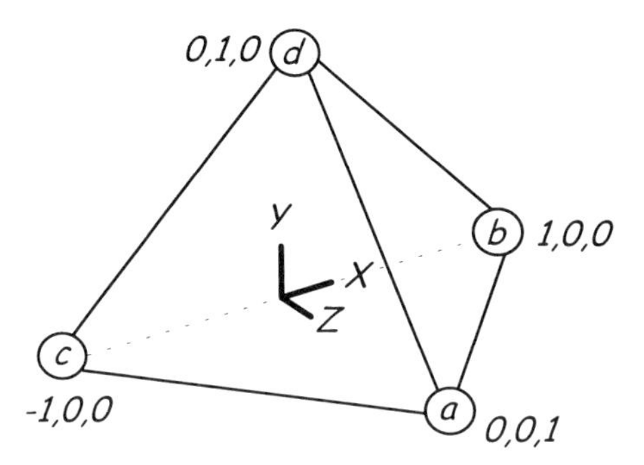

Figure 4.28 A 4-Node Tetrahedron, Simple Coordinates

Solving (4.5.19) for the a_i's then substituting back into (4.5.15) and rearranging, the shape functions for this particular 4-node tetrahedral element are:

$$N_1 = \frac{1}{2}(2Z) \qquad N_2 = \frac{1}{2}(1 + X - Y - Z)$$
$$N_3 = \frac{1}{2}(1 - X - Y - Z) \qquad N_4 = \frac{1}{2}(2Y) \tag{4.5.20}$$

Using the shape functions above, the interpolation function can be expressed as:

$$\tilde{U}(X,Y,Z) = \frac{1}{2}[(2Z)U_a + (1 + X - Y - Z)U_b + (1 - X - Y - Z)U_c + (2Y)U_d] \tag{4.5.21}$$

Note again that the interpolation function above is for the element shown in Figure 4.28 only—the shape functions for the general case are of the form (4.5.18) and are more difficult to express because of the many terms they contain. The validity of the interpolation function above can be verified by substituting in the coordinates of a particular node and then observing that the interpolation function assumes the value of the respective nodal displacement variable. The interpolation functions for *V(X,Y,Z)* and *W(X,Y,Z)* use the same shape functions.

Strain-Displacement Matrix for a 4-Node, Tetrahedral Element, 3-D Stress
Since the interpolation functions for the general, 4-node tet are quite complicated, the strain terms are also complicated. The strain-displacement matrix for this

element will not be given here; the interested reader is referred to Hoole [26, p. 399] for more details on the strain-displacement matrix for this type of element.

Element Stiffness for the 4-Node, Tetrahedral Element
The element stiffness matrix may be calculated in the usual fashion:

$$^{e}\underline{\underline{K}} = \int_V \underline{\underline{B}}^T \underline{\underline{E}}\,\underline{\underline{B}}\, dV$$

Since the strain vector in (4.5.14) contains six strain terms, the strain-displacement matrix has six rows. With a total of 12 nodal displacement variables, the strain-displacement matrix requires 12 columns. Hence, to compute the element stiffness matrix, the integral would appear as:

$$^{e}\underline{\underline{K}} = \int_V \begin{bmatrix} B_{1,1} & B_{2,1} & \cdots & B_{6,1} \\ B_{1,2} & \cdots & & B_{6,2} \\ \vdots & & & \vdots \\ B_{1,12} & \cdots & & B_{6,12} \end{bmatrix} \underline{\underline{E}} \begin{bmatrix} B_{1,1} & B_{1,2} & \cdots & B_{1,12} \\ B_{2,1} & \cdots & & B_{2,12} \\ \vdots & & & \vdots \\ B_{6,1} & \cdots & & B_{6,12} \end{bmatrix} dV$$

$$\underline{\underline{E}} \equiv \frac{E}{(1+\nu)(1-2\nu)} \begin{bmatrix} 1-\nu & \nu & \nu & 0 & 0 & 0 \\ \nu & 1-\nu & \nu & 0 & 0 & 0 \\ \nu & \nu & 1-\nu & 0 & 0 & 0 \\ 0 & 0 & 0 & \frac{1}{2(1+\nu)} & 0 & 0 \\ 0 & 0 & 0 & 0 & \frac{1}{2(1+\nu)} & 0 \\ 0 & 0 & 0 & 0 & 0 & \frac{1}{2(1+\nu)} \end{bmatrix}$$

Even the most simple three-dimensional element can become quite expensive in terms of the computational resources needed to produce the element stiffness matrix. Using the more common 8-node brick element, the strain-displacement matrix contains six rows and *24 columns*, further indicating that three-dimensional elements are very expensive. This again illustrates one reason why analysts typically avoid three-dimensional elements.

With three displacement variables at each node, the displacement vector for the

4-node tet contains a total of 12 components, as does the force vector.

$$
{}^{e}\underline{\underline{D}} \equiv \begin{Bmatrix} U_a \\ V_a \\ W_a \\ U_b \\ \vdots \\ W_d \end{Bmatrix} \qquad {}^{e}\underline{\underline{F}} \equiv \begin{Bmatrix} F_{Xa} \\ F_{Ya} \\ F_{Za} \\ F_{Xb} \\ \vdots \\ F_{Zd} \end{Bmatrix}
$$

4.6 Elementary Beam, Plate, and Shell Elements

Key Concept: Beam, plate, and shell structural elements are designed to characterize structures having specific proportions, subjected to particular types of loading.

Structural elements are designed to characterize structures that have specific geometric proportions and loading. For instance, long, slender structures subjected to loads that induce bending can often be modeled with beam elements, while thin, flat, "surface type" structures subjected to bending loads may be modeled with plate elements. Although only briefly mentioned in this text, shell structures are thin, *curved* surface type structures that may carry loads through both bending and membrane deformation, simultaneously. The relative proportion of bending to membrane deformation depends upon the loading, restraints, and the ratio of bending to membrane stiffness.

Even though plates and shells are relatively thin, structures modeled with these types of idealizations can take on a vast array of shapes and sizes. In addition, beams, plates, shells, and solid elements may be combined within a given model to characterize a wide range of structures.[5] The first structural element to be considered in this section is a simple 2-node line element for beam bending applications.

Euler-Bernoulli Beams

Beams are *slender* structural members, with one dimension significantly greater than the other two. Figure 4.29 shows a *cantilever* beam with a concentrated tip load p; other types of bending loads and end restraints are common. For instance, the top (or bottom) surface of the beam may be loaded with a distributed load, or a

---

[5] Caution must be exercised when combining elements of different types. Reference 15 considers details related to combining shell and solid elements.

concentrated load may be placed somewhere other than the tip. In another situation, both ends may be totally constrained to model a clamped beam.

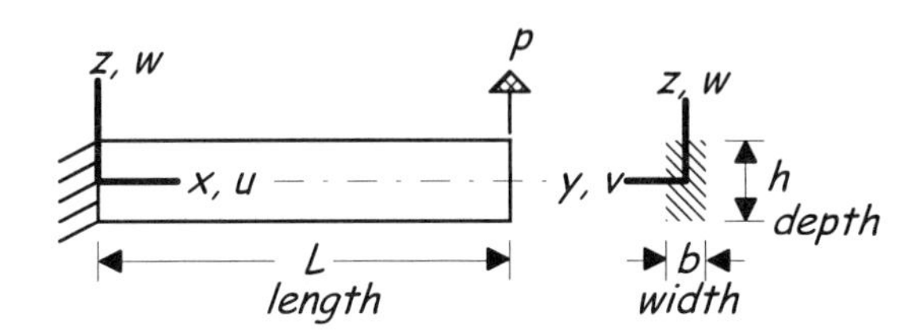

Figure 4.29 Cantilever Beam with Concentrated Load

For an *Euler-Bernoulli* beam, the loading must result in a force-couple (moment) that induces bending about the y-axis only. The effects of axial and/or torsional loads are not considered in the Euler-Bernoulli beam formulation. A review of the major assumptions associated with Euler-Bernoulli beams will be considered shortly.

A coordinate system that is attached to a structure (or finite element) may be called a *local coordinate system.* In this text, local coordinate systems will be denoted with lowercase variables. Note that the beam shown in Figure 4.29 uses a local coordinate system, attached to the structure at the left end, positive z pointing upwards. The displacements in the x and z coordinate directions are denoted as u and w, respectively.

Figure 4.30 illustrates a portion of a loaded beam with resulting bending moments, M; the deformed shape is greatly exaggerated.

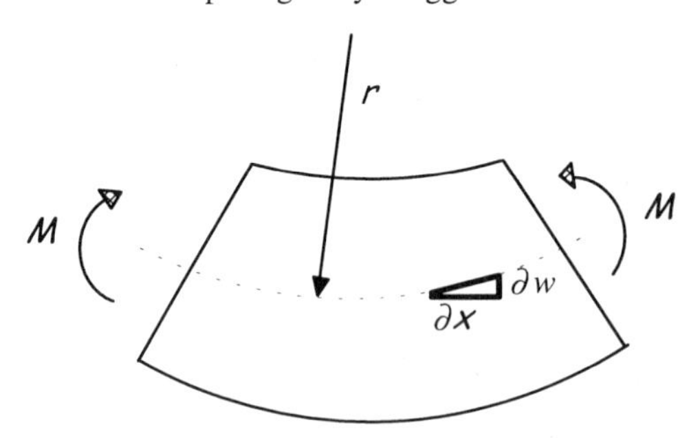

Figure 4.30 Portion of a Beam with Bending Moments

The beam is compressed on the top surface while the bottom surface is in tension. In this text, moments that tend to deform a beam in such a manner will be defined as positive. Notice that the slope of the neutral surface can be expressed as a derivative, $\partial w/\partial x$. The radius of curvature, r, is associated with the curvature of the neutral surface, which is depicted in Figure 4.30 as a dashed line. *Curvature* is defined as the inverse of the radius:

$$curvature \equiv \frac{1}{r}$$

When the radius of curvature (r) approaches infinity the *curvature* of the beam approaches zero. It is a relatively simple matter to show that the curvature of the neutral surface in an Euler-Bernoulli beam can be expressed (Higdon [27, p. 356]) as:

$$\frac{1}{r} = \frac{M}{EI} \tag{4.6.1}$$

Equation 4.6.1 states that curvature of the neutral surface is proportional to the bending moment, M, inversely proportional to the elastic modulus, E, and also inversely proportional to the second moment of cross sectional area, I. (Recall that for a rectangular cross section, $I=1/12\ bh^3$.)

From elementary calculus, curvature can also be related to the slope and the second derivative of displacement:

$$\frac{1}{r} = \frac{\frac{\partial}{\partial x}\left(\frac{\partial w}{\partial x}\right)}{\left(1+\left(\frac{\partial w}{\partial x}\right)^2\right)^{\frac{3}{2}}} \tag{4.6.2}$$

For small deformations, $\partial w/\partial x \ll 1$, and the denominator of (4.6.2) is essentially equal to unity. As such, the curvature of the beam, $1/r$, is approximately equal to the second derivative of transverse displacement:

$$\frac{1}{r} \approx \frac{\partial}{\partial x}\left(\frac{\partial w}{\partial x}\right) = \frac{\partial^2 w}{\partial x^2} \tag{4.6.3}$$

Incidentally, if slope $\partial w/\partial x$ is constant, curvature is zero, since curvature is equivalent to the change in slope. Substituting the right hand side of (4.6.3) into (4.6.1):

$$\frac{\partial^2 w}{\partial x^2} = \frac{M}{EI} \tag{4.6.4}$$

Equation 4.6.4 is known as the Euler-Bernoulli beam bending equation; it relates the second derivative of transverse displacement to three factors: the bending moment, elastic modulus, and the second moment of cross sectional area. Again, transverse displacement, w, refers to the displacement of the neutral surface; using the Euler-Bernoulli assumption, bending of the beam is characterized *entirely* by the deformed shape of the neutral surface. Equation 4.6.4 can be integrated twice, and, with the proper boundary conditions applied, provides (in some cases) a closed-form solution for the transverse displacement of an Euler-Bernoulli beam.

The Element Stiffness Matrix Equation for an Euler-Bernoulli Beam

As in the other examples, a material matrix (E-matrix) and a strain-displacement matrix (B-matrix) are required to compute the element stiffness matrix using Equation 3.5.12, shown again below:

$$^e\underline{\underline{K}} \equiv \left[\int_V \underline{\underline{B}}^T \, \underline{\underline{E}} \, \underline{\underline{B}} \, dV\right]$$

The constitutive equations will be examined to arrive at the material and strain-displacement matrices for the Euler-Bernoulli beam, but first, the assumptions associated with Euler-Bernoulli beams are considered.

Euler-Bernoulli Beam Assumptions

If Euler-Bernoulli ("elementary") beam theory is to be used, some restrictions must be imposed to ensure suitable accuracy. Six items related to the Euler-Bernoulli beam bending assumptions will be considered in brief:

1. Slenderness
2. Straight, narrow, and uniform beams
3. Small deformations
4. Bending loads only
5. Linear, isotropic, homogeneous material response
6. Only normal stress in the x-coordinate direction (τ_{xx}) significant

1. *Slender Beams*

Both shear and bending stress affect the transverse displacement of a beam; shear stress also causes plane sections of a beam to warp. The presumption a of slender beam allows one to discount the affects of shear stress upon transverse displacement and warping; Ugural [2].

Transverse Shear Stress Is Presumed Insignificant: If a beam is loaded with end moments only, no transverse shear exists.[6] However, if a beam is loaded with a transverse load, some level of transverse shear stress is present, and the magnitude of this stress is a function of how slender the beam is, other things being equal. Slenderness in beams may be characterized by the ratio of length to depth. If the ratio of length to depth is great enough, for example, if $L/h \geq 10$, transverse displacement caused by shear stress is typically neglected. For a tip loaded cantilever beam with $L/h=10$, the transverse displacement due to shear stress is about 1% of the transverse displacement due to bending.

Orthogonal Planes Remain Plane and Orthogonal: Without any shearing forces, cross sections originally plane and orthogonal to the neutral surface before loading remain so after loads are applied. In contrast, when transverse shear is present, rotation of orthogonal planes relative to the neutral surface, along with warping, occurs. A good depiction of warping in beam sections is provided by Higdon [27, p. 320].

Vibrating Beams: Although shear deformation is neglected in statically loaded, slender, narrow beams, the effects of shear deformation cannot be neglected in vibrating beams, even if the beam is slender and narrow. The Euler-Bernoulli assumption typically will not provide suitable accuracy over a wide range of frequency; Timoshenko beams may provide an increased measure of accuracy in such cases. Timoshenko beams for vibration problems are discussed by Thomas [28]; related discussion is provided by Huang [29].

2. *Narrow, Straight, and Uniform Beams*

If a beam is narrow, such that the width is not larger than the depth, deformation on any particular longitudinal slice of the beam is essentially the same as on any other slice. However, wide beams can undergo significant *anticlastic deformation*, where each longitudinal slice of the beam will experience a different amount of transverse displacement. Anticlastic action, a function of Poisson's ratio, is caused by the fact that material on one side of the neutral surface of the beam is in tension, while on the other side, the material is in compression. Bending a deep, wide rubber eraser will typically result in visible anticlastic action. Timoshenko beam theory is often used when beams are deep and/or wide. Anticlastic action is

---

[6] Transverse shear is equal to the first derivative of the bending moment. If end moments are applied to a beam, the bending moment is *constant* along the entire beam. Since the first derivative of the constant bending moment is zero, the shear is zero; Higdon [27, p. 266].

discussed by Ugural [2, p. 139]; deep and wide beams are discussed in Timoshenko [30, p. 354].

If a beam is curved instead of straight, a different approach is required, since stress is no longer linearly proportional to the distance from the midsurface. For an extreme case, consider a very curved "beam," such that an arch is formed, as depicted in Figure 4.31. In such a case, most of the load is carried by membrane action rather than bending; curved beams are discussed in Higdon [27]. In addition to being straight, an Euler-Bernoulli beam is assumed to be uniform, such that EI is constant along the beam.

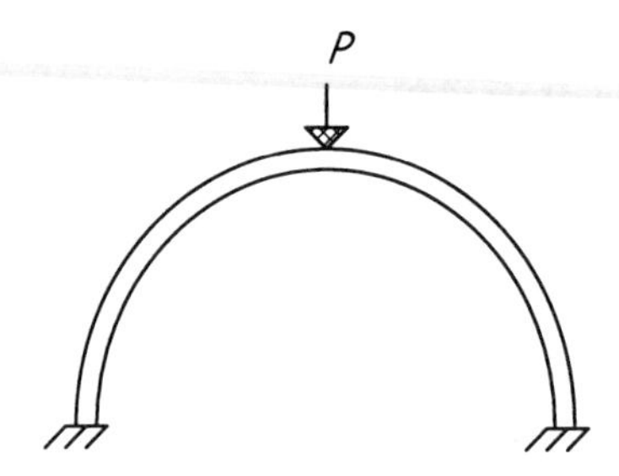

Figure 4.31 An Arch Structure

3. *Small Deformations*

Curvature can be approximated by the second derivative of displacement via (4.6.3) if the square of the slope is much less than unity. In addition, the small strain metrics are actually approximations that become less accurate if deformation is large. Recall from Chapter 1 that derivative squared terms are dropped from the Green-Lagrange strain metric when deformation is small. The absence of these terms is acceptable for small (infinitesimal) strains. The infinitesimal deformation assumption also implies that an originally straight and narrow beam is essentially straight and narrow when loaded.

4. *Bending Loads Only*

Equation 4.6.1 accounts for deformation due to bending only. That is, the beam is subjected either to loads normal to the neutral surface, or to moments about the y-axis, or perhaps a combination of the two. In any case, loads that cause the beam to twist, or cause any x- or y-axis displacement of the neutral surface, are not accounted for.

5. *Linear, Isotropic, Homogeneous Material Response*
It is assumed that the beam displaces as a linear function of the applied load, i.e., Hooke's law applies. In addition, the material is presumed isotropic and homogeneous, such that material properties (e.g., elastic modulus) do not vary with direction or location within the beam.

6. *Normal Stress (τ_{xx}) Is the Only Stress Component of Significance*
It is assumed that only normal stress along the x-axis occurs, and the magnitude of this stress varies linearly in the z-coordinate direction, from a value of zero on the neutral surface, to a maximum at the top and bottom surfaces. Since the normal stress is zero on the neutral surface, points on the neutral surface move only in the z-coordinate direction. Notice that stress in the depth direction (τ_{zz}) is zero even though the loads are applied in the depth direction. In other words, a slender beam *bends* in response to normal loads, while compression through the thickness is nearly zero. Ugural considers the relative magnitudes of various stress components in beams; Ugural [2, p. 156].

A 2-Node Line Element for Beam Applications
Assume that an element to model a slender, narrow beam, consistent with the Euler-Bernoulli assumptions above, is desired. Although a cantilever beam is used in the following development, the element can be used for other types of end restraints as well. Examine the illustration of the cantilever beam shown in Figure 4.32, and the mesh boundary in Figure 4.33.

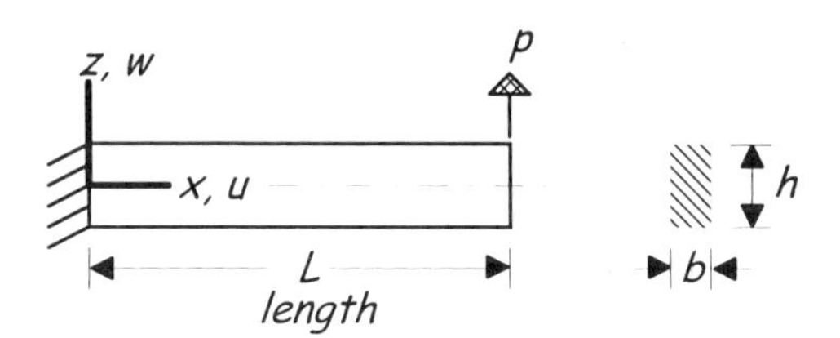

Figure 4.32 Cantilever Beam with Point Load

Since the beam has both uniform cross section and material properties, the geometry of the beam can be represented by a line mesh boundary, and beam elements created within the boundary. As with the 2-node line element for uniaxial stress (the rod element), the cross sectional properties for the beam element are specified by mathematical constants. For a beam, both the depth and width of the cross section must be specified, since the second moment of cross sectional area, I in (4.6.1), is calculated using both of these factors.

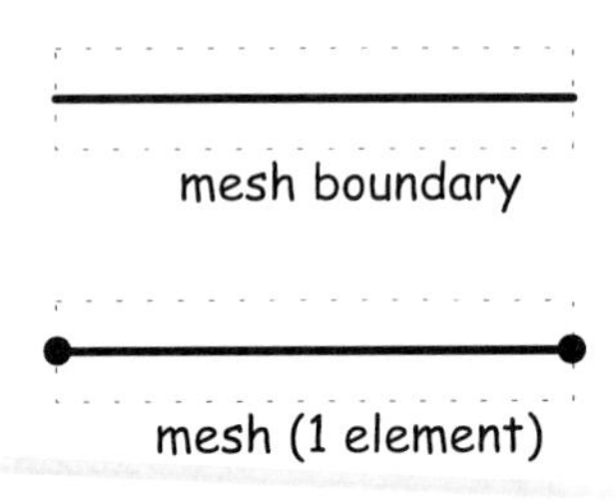

Figure 4.33 Mesh Boundary and Mesh for a Cantilever Beam

It will be shown that a two node line element for beam bending can exactly compute (within the limits of elementary beam theory) the displacement in a slender, narrow, point loaded cantilever beam.

The Constitutive Equations for Beam Bending

It is assumed that normal stress in the x-coordinate direction is the only stress component of significance in Euler-Bernoulli beams. From Table 3.1, the constitutive equations are:

$$\underline{\underline{\tau}} = \tau_{xx} \qquad \underline{\underline{E}} = E \qquad \underline{\underline{\varepsilon}} = \varepsilon_{xx} = \frac{\partial u}{\partial x} \tag{4.6.5}$$

However, using the Euler-Bernoulli assumption, strain in the beam is characterized by the curvature of the neutral surface, and normal strain in (4.6.5) can alternately be expressed as:

$$\varepsilon_{xx} = -z\frac{\partial^2 w}{\partial x^2} \tag{4.6.6}$$

The manner in which (4.6.6) is arrived at will be illustrated in the following. To begin, observe Figure 4.34 below.

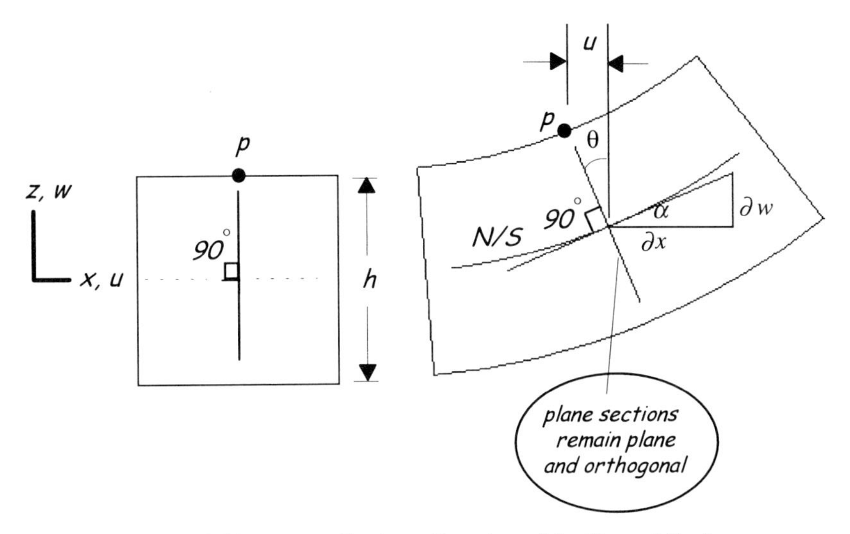

Figure 4.34 A Beam Section—Rotation of the Neutral Surface

Keep in mind that the deformed portion of the beam (illustrated on the right) is grossly exaggerated. The small deformation assumption specifies that the slope of the beam is so small that the configuration in the deformed state is essentially the same as the un-deformed shape

Consider rotation of the neutral surface, which will be defined by α. Note that:

$$tan\,\alpha = \frac{\partial w}{\partial x} \tag{4.6.7}$$

For small angles the cosine is nearly equal to unity, hence the tangent function is essentially equal to the sine function:

$$tan\,\alpha = \frac{sin\,\alpha}{cos\,\alpha} \approx \sin\alpha \qquad (\alpha << 1)$$

Recall that the sine of a small angle is approximately equal to the angle in radians:

$$\sin\alpha \approx \alpha \qquad (\alpha << 1)$$

Therefore, for small angles, the tangent function on the left hand side of (4.6.7) can be replaced by α:

$$\alpha \approx \frac{\partial w}{\partial x} \tag{4.6.8}$$

Equation 4.6.8 states that rotation (slope) of the neutral surface is approximately equal to the derivative of the transverse displacement with respect to x.

Now, consider the displacement of Point p in the x-coordinate direction as illustrated in Figure 4.34. The magnitude of this displacement is approximately equal to an arc of radius $h/2$, rotated through an angle θ:

$$u = -\left(\frac{h}{2}\right)\theta$$

In general, an arbitrary point at a distance z from the neutral axis will displace in the following manner:

$$u = -z\theta \tag{4.6.9}$$

The negative sign in (4.6.9) appears due to the orientation of the coordinate system, and because counter-clockwise rotation is defined as positive. Using the Euler-Bernoulli assumption, it is presumed that shear deformation is neglected, which means that a plane through a cross section of the beam, originally orthogonal to the neutral surface, is assumed to remain plane and orthogonal after the beam is loaded. Therefore, rotation of Point p through angle θ can be completely defined by the rotation of the neutral surface. In the absence of shear effects:

$$\theta = \alpha \tag{4.6.10}$$

Substituting (4.6.10) into (4.6.9):

$$u = -z\alpha \tag{4.6.11}$$

Using the definition of alpha given by (4.6.8) in (4.6.11):

$$u = -z\frac{\partial w}{\partial x} \tag{4.6.12}$$

Equation 4.6.12 suggests that axial displacement in Euler-Bernoulli beams is approximated by the slope of the neutral surface, and is a linear function of the distance from the neutral surface, z. Observe that at any point on the neutral surface (where $z = 0$) the axial displacement is zero, consistent with Euler-Bernoulli Assumption 6.

Referring to (4.6.6), it was stated that normal strain can be expressed in terms of a second order derivative. Using (4.6.12) and the definition of normal strain:

$$\varepsilon_{xx} = \frac{\partial u}{\partial x} = \frac{\partial}{\partial x}\left(-z\frac{\partial w}{\partial x}\right) = -z\frac{\partial^2 w}{\partial x^2} \tag{4.6.13}$$

So, for an Euler-Bernoulli beam element, strain can be defined in terms of a second order derivative such that p, the highest order derivative in the strain term, is equal to 2. In Table 4.2, it was stated that displacement and $p-1$ order derivatives need to be continuous at the nodes of adjacent elements. Note that this is the first element considered in this text that requires continuity of displacement *and* first order derivatives. Interpolation functions for an Euler-Bernoulli beam element are now developed.

Interpolation Functions for a 2-Node Line Element, Euler-Bernoulli Beam

According to (4.6.13), transverse displacement, w, is required to compute strain. Since uniform cross sectional properties are assumed, the geometry of the beam is easily represented using a 2-node line element as depicted in Figure 4.35, with mathematical constants to describe the cross section.

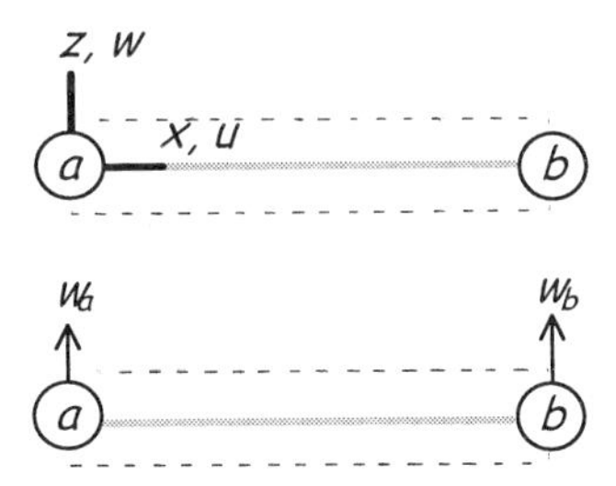

Figure 4.35 Hypothetical 2-Node Line Element, Translational DOF Only

Using the element above, a two-term displacement assumption could be employed:

$$\tilde{w}(x) = a_1 + a_2 x$$

Following the same procedures as before, the displacement assumption is constrained to the nodal variables, and the a_i's eliminated, yielding an interpolation function just like the rod element:

$$\tilde{w}(x) = N_1 w_a + N_2 w_b$$

However, when elements of the type above are joined together, there exists no constraint upon the first derivatives of displacement. In other words, using the interpolation function given, (without nodal variables of slope) continuity of *p–1 derivatives* is not enforced at the nodes of adjacent elements. In addition, there exists no easy way to invoke the slope boundary conditions associated with bending problems. Furthermore, using a linear function in (4.6.13), notice that strain is zero, hence the problem is trivial: Stress and strain are always zero in the beam. In short, a two-term displacement assumption cannot be used in this case.

To ensure continuity of translational displacement and *derivatives* at each node, an interpolation function that contains nodal variables for both translation and slope can be developed, starting with the element shown in Figure 4.36. The element on the left shows the nodal DOF's while the same element on the right is shown with the nodal forces and moments that exist at each node. In all of the elements considered so far, work potential terms have been expressed using only translation variables and forces. Intuitively, one might see that moments and rotations are the analogue for forces and translations.

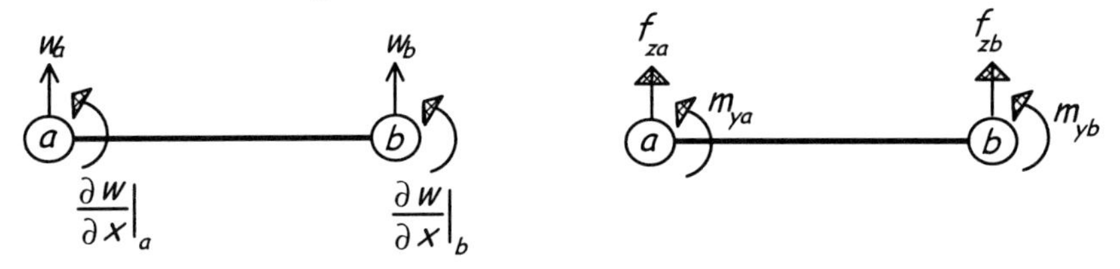

Figure 4.36 A 2-Node Line Element for Beam Applications

Derivatives are used in this element to represent rotation at the nodes. Since denoting the derivative at each node as shown on the left hand side of Figure 4.36

is slightly cumbersome, the derivatives will instead be denoted as:

$$\left.\frac{\partial w}{\partial x}\right|_a \equiv \left(w_{,x}\right)_a$$

Now, since four nodal DOF's exist for the element given in Figure 4.36, a four term displacement assumption can be used:

$$\tilde{w}(x) = a_1 + a_2 x + a_3 x^2 + a_4 x^3 \tag{4.6.14}$$

Notice that the first derivative of (4.6.14) is computed as:

$$\frac{\partial \tilde{w}}{\partial x} = w_{,x} = a_2 + 2a_3 x + 3a_4 x^2 \tag{4.6.15}$$

Using the same approach as always, the displacement assumption is equated to the nodal displacement variables to ensure displacement continuity, and in this case *p–1 derivative* continuity, at the nodes:

$$\begin{aligned}
\tilde{w}(x_a) &= a_1 + a_2 x_a + a_3 x_a^2 + a_4 x_a^3 = w_a \\
\left.\frac{\partial \tilde{w}}{\partial x}\right|_a &= a_2 + 2a_3 x_a + 3a_4 x_a^2 = \left(w_{,x}\right)_a \\
\tilde{w}(x_b) &= a_1 + a_2 x_b + a_3 x_b^2 + a_4 x_b^3 = w_b \\
\left.\frac{\partial \tilde{w}}{\partial x}\right|_b &= a_2 + 2a_3 x_b + 3a_4 x_b^2 = \left(w_{,x}\right)_b
\end{aligned} \tag{4.6.16}$$

Simplifying the system of equations above:

$$\begin{aligned}
a_1 + a_2 x_a + a_3 x_a^2 + a_4 x_a^3 &= w_a \\
a_2 + 2a_3 x_a + 3a_4 x_a^2 &= \left(w_{,x}\right)_a \\
a_1 + a_2 x_b + a_3 x_b^2 + a_4 x_b^3 &= w_b \\
a_2 + 2a_3 x_b + 3a_4 x_b^2 &= \left(w_{,x}\right)_b
\end{aligned} \tag{4.6.17}$$

With considerable work, the system of equations above may be solved for the a_i's, the a_i's substituted back into (4.6.14), and the result rearranged in terms of nodal

DOF's, yielding:

$$\tilde{w}(x) = N_1 w_a + N_2 \left(w_{,x}\right)_a + N_3 w_b + N_4 \left(w_{,x}\right)_b \tag{4.6.18}$$

Notice that displacement is now being interpolated from both displacement and slope variables. The shape functions in (4.6.18) can be shown to be:

$$\begin{aligned} N_1 &= 1 - 3\frac{\bar{x}^2}{l^2} + 2\frac{\bar{x}^3}{l^3} \\ N_2 &= \bar{x} - 2\frac{\bar{x}^2}{l^2} + \frac{\bar{x}^3}{l^3} \\ N_3 &= 3\frac{\bar{x}^2}{l^2} - 2\frac{\bar{x}^3}{l^3} \\ N_4 &= \frac{\bar{x}^2}{l} + \frac{\bar{x}^3}{l^2} \end{aligned} \tag{4.6.19}$$

The substitutions $\bar{x}$ and l in (4.6.19), defined below, are used to simplify the expressions for the shape functions:

$$\bar{x} \equiv x - x_a \qquad l \equiv x_b - x_a$$

Lowercase "L" indicates element length while uppercase "L" indicates the length of the entire structure. The interpolation functions in (4.6.19) contain x^3 terms, as they should, since the displacement assumption, (4.6.14), began with cubic terms.

The Shape of the Deformed Element

It is interesting to note that while the expression for displacement is a cubic function, this 2-node beam element can only assume the shape of a straight line when loaded. So, using a single beam element to model a cantilever beam, the slightly curved shape of the loaded beam will not be exhibited even though the deflection at the tip can be correctly computed. That is, the deformed shape will not appear to be correct using a single, 2-node line element. In addition, recalling (4.6.12):

$$u = -z\frac{\partial w}{\partial x}$$

Notice that at the neutral surface, $z=0$, therefore axial displacement on the neutral surface is zero. Since the element characterizes displacement at the neutral surface, the deformed element will not indicate any axial displacement, regardless of how much the beam deflects.

Strain-Displacement Matrix for a 2-Node Line Element, Beam Applications
With the displacement interpolation function now established, the strain-displacement matrix for the Euler-Bernoulli beam can be considered. Recall that in Euler-Bernoulli beams, normal strain is expressed as a second order derivative:

$$\varepsilon_{xx} = -z\frac{\partial^2 w}{\partial x^2} \tag{4.6.20}$$

Substituting (4.6.18) into the equation above:

$$\varepsilon_{xx} \approx -z\frac{\partial^2}{\partial x^2}\left\{N_1 w_a + N_2\left(w_{,x}\right)_a + N_3 w_b + N_4\left(w_{,x}\right)_b\right\} \tag{4.6.21}$$

Equation 4.6.21 can be expressed in the usual way, using the strain-displacement matrix:

$$\varepsilon_{xx} \approx \underline{\underline{B}}\,{}^e\underline{D} = -z\begin{bmatrix}\frac{\partial^2 N_1}{\partial x^2} & \frac{\partial^2 N_2}{\partial x^2} & \frac{\partial^2 N_3}{\partial x^2} & \frac{\partial^2 N_4}{\partial x^2}\end{bmatrix}\begin{Bmatrix} w_a \\ \left(w_{,x}\right)_a \\ w_b \\ \left(w_{,x}\right)_b \end{Bmatrix} \tag{4.6.22}$$

Hence, the strain-displacement matrix for this element is:

$$\underline{\underline{B}} = -z\begin{bmatrix}\frac{\partial^2 N_1}{\partial x^2} & \frac{\partial^2 N_2}{\partial x^2} & \frac{\partial^2 N_3}{\partial x^2} & \frac{\partial^2 N_4}{\partial x^2}\end{bmatrix} \tag{4.6.23}$$

From (4.6.19), the derivatives of the shape functions are computed as:

$$\begin{aligned} \frac{\partial^2 N_1}{\partial x^2} &= -\frac{6}{l^2} + 12\frac{\overline{x}}{l^3} \qquad & \frac{\partial^2 N_2}{\partial x^2} &= -\frac{4}{l} + \frac{6\overline{x}}{l^2} \\ \frac{\partial^2 N_3}{\partial x^2} &= \frac{6}{l^2} - \frac{12\overline{x}}{l^3} \qquad & \frac{\partial^2 N_4}{\partial x^2} &= -\frac{2}{l} + \frac{6\overline{x}}{l^2} \end{aligned} \tag{4.6.24}$$

Recalling the definition of $\bar{x}$, notice that the strain-displacement matrix for this element, (4.6.23), now contains the x variable, such that strain can vary along the length of the beam element. This is in contrast to all of the other elements considered so far, since they all had constants in their strain-displacement matrices. For convenience, the matrix of shape function derivatives is denoted as $\underline{\underline{B'}}$:

$$\begin{bmatrix} \frac{\partial^2 N_1}{\partial x^2} & \frac{\partial^2 N_2}{\partial x^2} & \frac{\partial^2 N_3}{\partial x^2} & \frac{\partial^2 N_4}{\partial x^2} \end{bmatrix} \equiv \begin{bmatrix} B_1' & B_2' & B_3' & B_4' \end{bmatrix} = \underline{\underline{B'}} \tag{4.6.25}$$

The strain-displacement matrix in (4.6.23) can therefore be expressed as:

$$\underline{\underline{B}} = -z\underline{\underline{B'}}$$

Using the expression above in the element stiffness matrix formula, (3.5.12):

$$\underline{\underline{{}^eK}} \equiv \int_V \underline{\underline{B}}^T \underline{\underline{E}}\ \underline{\underline{B}}\, dV = \int_V \left(-z\underline{\underline{B'}}\right)^T \underline{\underline{E}} \left(-z\underline{\underline{B'}}\right) dV$$

Recalling that the material matrix is simply equal to the elastic modulus for the Euler-Bernoulli beam, the above may be arranged as:

$$\underline{\underline{{}^eK}} = \int_V Ez^2 \left(\underline{\underline{B'}}\right)^T \underline{\underline{B'}}\, dV \tag{4.6.26}$$

Since the cross section of the beam is assumed constant, the differential volume in the equation above can be expressed as:

$$dV = A dx$$

Also, since the width of the beam is assumed to be equal to a constant, b:

$$dV = (b\, dz)\, dx$$

Using the expression for differential volume above in (4.6.26):

$$\underline{\underline{{}^eK}} \equiv \int_{x_a}^{x_b} \int_{\frac{-h}{2}}^{\frac{h}{2}} Ez^2 \left(\underline{\underline{B'}}\right)^T \underline{\underline{B'}}\, b\, dz\, dx$$

Since the limits on the inner integral are always from $-h/2$ to $h/2$, it is convenient to compute this integral separately:

$$^{e}\underline{\underline{K}} \equiv \int_{x_a}^{x_b} E\left(\underline{\underline{B'}}\right)^T \underline{\underline{B'}}\, dx \left(\int_{\frac{-h}{2}}^{\frac{h}{2}} b\, z^2 dz \right) \tag{4.6.27}$$

Notice that the bracketed integral in (4.6.27) is equivalent to the second moment of cross sectional area for a rectangular beam section:

$$I = \int_{\frac{-h}{2}}^{\frac{h}{2}} b\, z^2 dz \tag{4.6.28}$$

Recall that for a beam with a rectangular cross section, I is computed as:

$$I = \int_{\frac{-h}{2}}^{\frac{h}{2}} b\, z^2 dz = b \left.\frac{z^3}{3}\right|_{-\frac{h}{2}}^{\frac{h}{2}} = \frac{bh^3}{12}$$

Using (4.6.28) in (4.6.27):

$$^{e}\underline{\underline{K}} \equiv \int_{x_a}^{x_b} EI \left(\underline{\underline{B'}}\right)^T \underline{\underline{B'}}\, dx \tag{4.6.29}$$

Each term in the stiffness matrix can be calculated separately by performing the matrix multiplication indicated in (4.6.29). To do so, (4.6.29) is restated with the terms of the B'-matrix shown:

$$^{e}\underline{\underline{K}} \equiv \int_{x_a}^{x_b} EI \begin{Bmatrix} B_1' \\ B_2' \\ B_3' \\ B_4' \end{Bmatrix} \begin{bmatrix} B_1' & B_2' & B_3' & B_4' \end{bmatrix} dX$$

Performing the matrix algebra in the equation above, and noting that EI is presumed constant along the beam:

$$\underline{\underline{^eK}} \equiv EI \int_{x_a}^{x_b} \begin{bmatrix} B_1'B_1' & B_1'B_2' & \cdots & B_1'B_4' \\ B_2'B_1' & \cdots & \cdots & \cdots \\ \cdots & \cdots & \cdots & \cdots \\ B_4'B_1' & \cdots & \cdots & B_4'B_4' \end{bmatrix} dx \tag{4.6.30}$$

The dots are used in place of writing out every term. Each term in the element stiffness matrix above can be expressed as:

$$^eK_{mn} \equiv EI \int_{x_a}^{x_b} B_m' B_n' \, dx \tag{4.6.31}$$

The displacement and force vectors for the 2-node line element for beam bending are:

$$\underline{\underline{^eD}} = \begin{Bmatrix} w_a \\ (w_{,x})_a \\ w_b \\ (w_{,x})_b \end{Bmatrix} \qquad \underline{\underline{^eF}} = \begin{Bmatrix} f_{za} \\ m_{ya} \\ f_{zb} \\ m_{yb} \end{Bmatrix}$$

To gain an appreciation for the terms in the element stiffness matrix, the first term in the element stiffness matrix is computed using (4.6.31). Referring to (4.6.25) and (4.6.24), and making use of the definition of $\bar{x}$:

$$^eK_{11} = EI \int_{x_a}^{x_b} B_1'B_1' \, dx = EI \int_{x_a}^{x_b} \left(-\frac{6}{l^2} + 12\frac{\bar{x}}{l^3}\right)\left(-\frac{6}{l^2} + 12\frac{\bar{x}}{l^3}\right) dx = \frac{12EI}{l^2} \tag{4.6.32}$$

Performing all of the mathematics in (4.6.32) is quite tedious if done by hand, especially integrating all of the terms that result from the multiplication of the two terms in the integrand. Numerical integration schemes are often used in cases where closed-form integration is impractical. Numerical integration will be discussed in Chapter 5.

After computing the terms in (4.6.32), the element stiffness matrix for the 2-

node Euler-Bernoulli beam takes the form:

$$\underline{\underline{{}^{e}K}} \equiv \frac{EI}{l^3}\begin{bmatrix} 12l & 6l & -12 & 6l \\ 6l & 4l^2 & -6l & 2l^2 \\ -12 & -6l & 12 & -6l \\ 6l & 2l^2 & -6l & 4l^2 \end{bmatrix} \tag{4.6.33}$$

The next example shows that the element stiffness matrix given by (4.6.33) is exact (within elementary beam theory) for a point loaded cantilever beam.

Example 4.1—A Cantilever Beam Subjected to a Point Load at the Tip
Consider a finite element analysis of a cantilever beam, subjected to a concentrated tip load. The objective in this example is to find the displacement in the z-coordinate direction at the tip.

Since only one element is used in this example, a connectivity table is not needed. Using the stiffness matrix from (4.6.33), the equilibrium equations for this element are:

$$\frac{EI}{l^3}\begin{bmatrix} 12l & 6l & -12 & 6l \\ 6l & 4l^2 & -6l & 2l^2 \\ -12 & -6l & 12 & -6l \\ 6l & 2l^2 & -6l & 4l^2 \end{bmatrix}\begin{Bmatrix} w_a \\ (w_{,x})_a \\ w_b \\ (w_{,x})_b \end{Bmatrix} = \begin{Bmatrix} f_{za} \\ m_{ya} \\ f_{zb} \\ m_{yb} \end{Bmatrix} \tag{4.6.34}$$

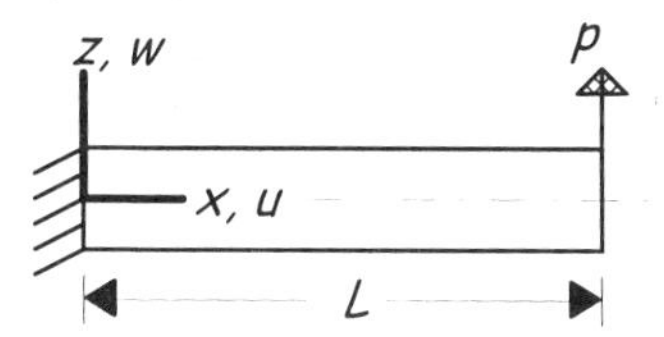

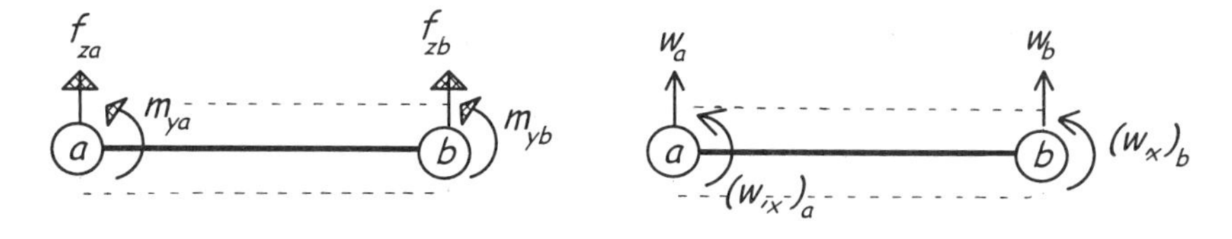

Figure 4.37 Cantilever Beam with Concentrated Tip Load

For the case of the cantilever beam in Figure 4.37, displacement and rotation at Node a are restrained, so these values are set to zero, as is the external moment at Node b. In addition, the force at Node b is set equal to P:

$$w_a = \left(w_{,x}\right)_a = m_{yb} = 0 \qquad f_{zb} = P \tag{4.6.35}$$

The conditions of (4.6.35) can be imposed upon (4.6.34):

$$\frac{EI}{l^3}\begin{bmatrix} 12l & 6l & -12 & 6l \\ 6l & 4l^2 & -6l & 2l^2 \\ -12 & -6l & 12 & -6l \\ 6l & 2l^2 & -6l & 4l^2 \end{bmatrix}\begin{Bmatrix} 0 \\ 0 \\ w_b \\ \left(w_{,x}\right)_b \end{Bmatrix} = \begin{Bmatrix} f_{za} \\ m_{ya} \\ P \\ 0 \end{Bmatrix} \tag{4.6.36}$$

One can see that the first two columns of the square matrix in (4.6.36) have no effect in this case, since the displacements at Node a are zero. Hence, an equivalent set of equations is expressed as:

$$\frac{EI}{l^3}\begin{bmatrix} -12 & 6l \\ -6l & 2l^2 \\ 12 & -6l \\ -6l & 4l^2 \end{bmatrix}\begin{Bmatrix} w_b \\ \left(w_{,x}\right)_b \end{Bmatrix} = \begin{Bmatrix} f_{za} \\ m_{ya} \\ P \\ 0 \end{Bmatrix} \tag{4.6.37}$$

The first two equations in (4.6.37) can be set aside, since they simply add two additional unknowns to the problem, being that the reaction force and moment at Node a are "unknown." Eliminating the first two rows of (4.6.37):

$$\frac{EI}{l^3}\begin{bmatrix} 12 & -6l \\ -6l & 4l^2 \end{bmatrix}\begin{Bmatrix} w_b \\ \left(w_{,x}\right)_b \end{Bmatrix} = \begin{Bmatrix} P \\ 0 \end{Bmatrix} \tag{4.6.38}$$

Solving the system of equations in (4.6.38) with l=L:

$$w_b = \frac{PL^3}{3EI} \qquad \left(w_{,x}\right)_b = \frac{PL^2}{2EI}$$

This is the same deflection that elementary beam theory predicts; Higdon [27, p. 708].

An Alternate Way to Express Constitutive Equations for a Beam Element
Recall that the standard element stiffness matrix formula, (3.5.12), is expressed as:

$$ {}^{e}\underline{\underline{K}} \equiv \int_{V} \underline{\underline{B}}^{T}\, \underline{\underline{E}}\, \underline{\underline{B}}\, dV $$

Using (4.6.29) to compute the stiffness of the Euler-Bernoulli beam element:

$$ {}^{e}\underline{\underline{K}} \equiv \int_{x_a}^{x_b} \left(\underline{\underline{B'}}\right)^{T} (EI)\, \underline{\underline{B'}}\, dx \qquad (4.6.39) $$

Comparing the standard formula for element stiffness with (4.6.39), the "material matrix" is defined as EI, and a B'-matrix is used instead of the B-matrix. Notice that the B'-matrix is not the same as the B-matrix: the former transforms nodal DOF into *curvature* while the latter translates nodal DOF directly into *strain.*

This concludes the discussion on Euler-Bernoulli beam elements. Although the beam element shown does not include torsional or shear affects, nor displacement in other coordinate directions, there are many robust beam elements available to the analyst today which do include these properties.

Kirchhoff Plates
There exists a large class of structures that are thin, i.e., structures where two dimensions are significantly larger than the third. Stamped metal brackets, sheet metal for automotive body components, aircraft skins, and many structural reinforcing members are often thin.

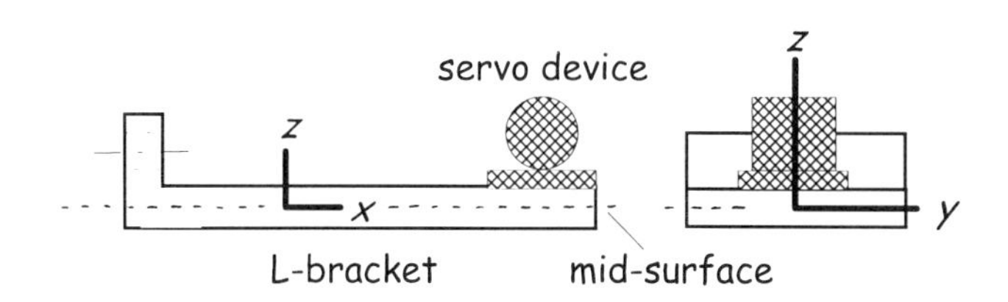

Figure 4.38 L-bracket with Servo Attached, Thin Plate Assumption

Consider the example of the servo mounting bracket from Chapter 1, shown again in Figure 4.38.

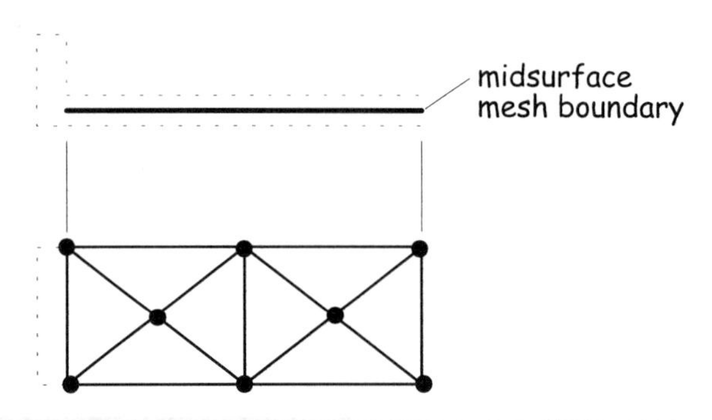

Figure 4.39 Finite Element Mesh of Servo Bracket

In some cases, only the horizontal portion of the bracket may need to be considered, and a finite element mesh, using 3-node plate elements, might be employed, as illustrated in Figure 4.39.

Although a finite element model using 4-node quadrilateral surface elements was depicted in Chapter 1, triangular surface elements can also be used. As illustrated in Figure 4.39, eight triangular elements are used for this crude mesh. A 4-node quadrilateral element is generally preferred over a 3-node triangular element, in terms of accuracy. However, since the 3-node element is less complex, it will be considered for illustrative purposes. The 3-node plate element considered here uses the *Kirchhoff approach*, which has largely been abandoned in favor of the more robust, and more complicated, *Mindlin approach.* Plates are flat structural members where two dimensions are significantly greater than the third. The analysis of plates can be simplified if certain assumptions are applied, the assumptions being somewhat analogous to the Euler-Bernoulli beam bending assumptions. Figure 4.40 shows some of the terminology associated with plate bending. In Kirchhoff plates, loading which causes bending about both the x- and y-coordinate axes is considered. The loading and geometry of Kirchhoff plates is such that the centroidal (*mid*) surface of the plate is the neutral surface of bending, and points on this surface move only in the z-coordinate direction.

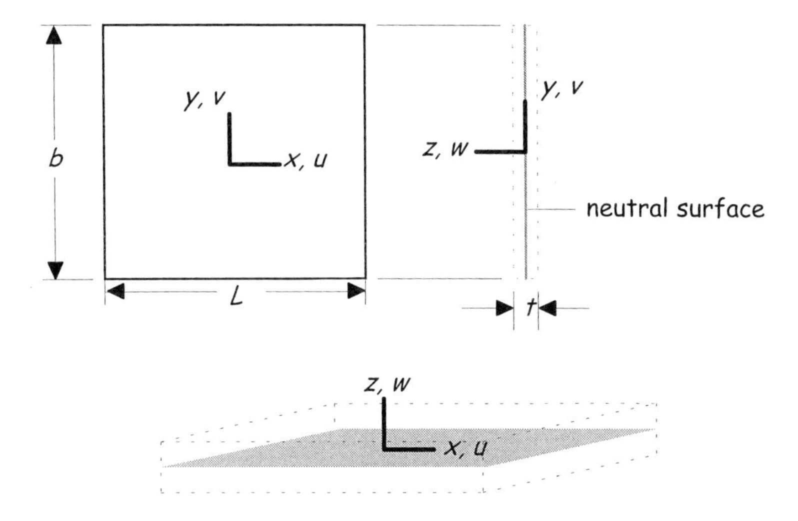

Figure 4.40 Thin Plate

Neither transverse shear deformation nor the effects of membrane deformation, i.e., in-plane stretching or compressing of the plate's midsurface, are considered. Analogous to the Timoshenko beam, there exist plate and shell elements which *can* account for shear stress. These elements, called *Mindlin* elements, will not be discussed in this text due to their complexity. Kirchhoff plates may be analyzed in much the same fashion as the previously considered Euler-Bernoulli beam.

Consider a portion of a plate viewed normal to the side, as shown in Figure 4.41, with the deformation greatly exaggerated. Using the same moment convention that was used for Euler-Bernoulli beams, moments that tend to deform a plate with compression on the top and tension on the bottom will be considered positive.

While beams exhibit bending in only one plane, plates can exhibit bending in two: the x–z plane and the y–z plane. As with Euler-Bernoulli beams, displacement in Kirchhoff plates is described by the deformation of the neutral surface. The slope of the neutral surface in the x–z plane is defined as $\partial w/\partial x$ while the slope in the y–z plane is $\partial w/\partial y$.

As with Euler-Bernoulli beams, curvature in plates can be expressed in terms of second order derivatives:

$$\frac{1}{r_{xz}} \approx \frac{\partial^2 w}{\partial x^2} \qquad \frac{1}{r_{yz}} \approx \frac{\partial^2 w}{\partial y^2} \tag{4.6.40}$$

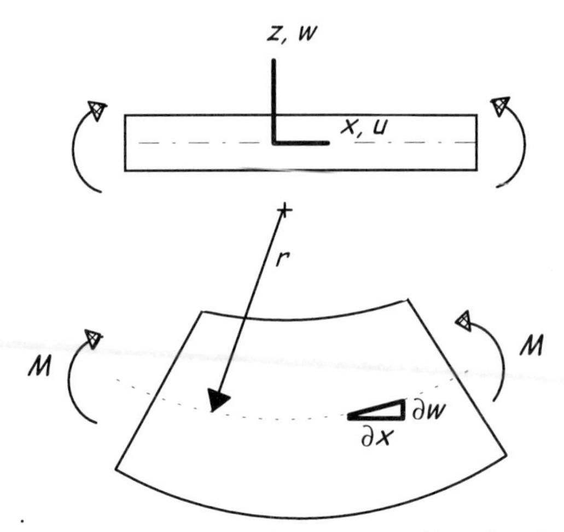

Figure 4.41 Portion of Plate with Exaggerated Bending Deformation

The Element Stiffness Matrix Equation

As in the other examples, a material matrix and a strain-displacement matrix are required to compute the element stiffness matrix using Equation 3.5.12:

$$^{e}\underline{\underline{K}} \equiv \left[\int_V \underline{\underline{B}}^T \; \underline{\underline{E}}\;\underline{\underline{B}}\, dV\right]$$

The constitutive equations will be examined to arrive at the material and strain-displacement matrices, but first, assumptions consistent with Kirchhoff plate elements are considered.

Kirchhoff Plate Bending Assumptions

Several restrictions must be imposed when using the Kirchhoff plate bending formulation if suitable accuracy is to be obtained; six related items will be considered briefly:

1. Thinness
2. Flatness and uniformity
3. Small deformation
4. No membrane deformation
5. Linear, isotropic, homogeneous materials
6. τ_{xx}, τ_{yy}, τ_{xy}: the only stress components of significance

1. *Thin Plates*
If the ratio of the in-plane dimensions to thickness is greater than ten, a plate may be considered thin; in other words, thin plates are such that $L/t \geq b/t \geq 10$.

Transverse Shear Stress Does Not Affect Transverse Displacement: Although transverse displacement of a plate loaded normal to its neutral surface is affected by both normal stress and transverse shear stress, it is assumed that the shear stress does not have a significant impact on the magnitude of transverse displacement. It can be shown that the magnitude of transverse displacement due to shear stress is small in thin plates.

Orthogonal Planes Remain Plane and Orthogonal: In a thin plate, any cross sectional plane, originally orthogonal to the neutral surface, is assumed to remain plane and orthogonal when the plate is loaded. As with deep beams, cross sections in thick plates, originally orthogonal and planar, will be neither under transverse load, due to shearing effects.

2. *Flat and Uniform*
Kirchhoff plates are presumed flat, and as such, loads are not carried by membrane deformation. If initial curvature were to exist, some of the load could be carried by membrane action, hence, curved plates (*shells*) require a different approach to account for the combination of simultaneous membrane and bending deformation.

3. *Small Deformation*
There are three reasons why simple plate analysis is restricted to small deformation. First, as with all the analyses mentioned in this text, it is assumed small (infinitesimal) strain measures are used. As mentioned in Chapter 1, the expression for strain actually contains second order terms, and these terms are truncated if the strains are presumed small.

A second reason for limiting plate analyses to small deformation is, like the Euler-Bernoulli beam, plate curvature can be expressed in terms of second order derivatives if the square of the slope is negligible. Thirdly, as a plate deflects, its transverse stiffness changes. If the deflection does not exceed a certain amount the transverse stiffness can be presumed constant.

Why does the transverse stiffness change with deflection? As a plate deforms into a curved (or a doubly curved) surface, transverse loads are resisted by both bending and membrane deformation. The addition of membrane deformation affects the transverse stiffness of the plate. For instance, consider Figure 4.42 where a ball is placed on a very thin "plate."

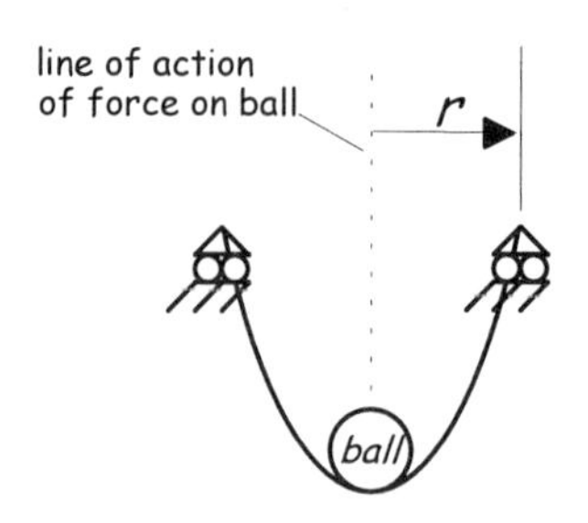

Figure 4.42 "Plate" Deflection

The plate is attached to rollers which allow the ends of the plate to move toward each other under load. Notice that as the amount of plate deflection increases, more load is carried by membrane tension and less by bending. As the supports move closer together, the moment about each mounting point decreases, since the lever arm distance, r, becomes shorter. As the moment decreases, the ball must be supported by an increasing amount of membrane deformation. One can see that if the plate were deformed such that the sides were totally vertical, the entire load would essentially be carried by membrane action alone. Indeed, a structure responding as depicted in Figure 4.42 may be termed a *membrane* instead of a plate, since its bending stiffness is very small, and the load is primarily carried by membrane forces.

The important thing to remember is that if the deflection of an initially flat plate does not exceed a certain amount, a bending load is mainly resisted by bending stiffness, while only an insignificant portion is resisted by membrane stiffness. In such a case, the transverse stiffness of the plate may be considered constant, and characterized by bending stiffness only. However, with increasing deflection, more load is resisted by membrane stiffness, and the presumption of constant stiffness with the associated neglect of membrane effects is no longer valid. How much transverse displacement is allowed before the stiffness of an initially flat plate deviates significantly from its initial, constant value? Roark's Handbook [31] suggests that transverse displacement less than one-half the plate thickness ($w < t/2$) should not impart excessive error in flat plate stiffness calculations.

The small deformation assumption implies that an initially flat, thin plate is essentially thin and flat when loaded. Shell theory is required to account for structures that carry loads through simultaneous bending and membrane deformation. Morely [32, p. 537] discusses the relative contribution of membrane versus bending stress in shells. A comparison of bending to membrane stress ratios in a spherical shell is given in Ugural [2, p. 428].

4. *No Membrane Deformation*
Kirchhoff plates do not account for membrane deformation. This means that at the neutral surface, there can be no displacement in the x- or y-coordinate directions, which would suggest the existence of membrane deformation.

5. *Linear, Isotropic, Homogeneous Material Response*
It is assumed that plate deformation is a linear function of the applied load, and that the material properties, uniform throughout the plate, are not directionally dependent.

6. *Significant Stress Components are τ_{xx}, τ_{yy}, and τ_{xy}*
Only three stress components can be non-zero, while all other stress components are presumed insignificant. In addition, on the neutral surface, *all* stress components are presumed zero. The magnitudes of the non-zero stress components, τ_{xx}, τ_{yy}, and τ_{xy}, increase linearly in the z-coordinate direction, from a value of zero at the neutral surface to a maximum at the top and bottom surfaces.

When plates are subjected to transverse loads, both transverse shear stress and normal stress in the z-coordinate direction are present. However, as in shallow beams, it is assumed that transverse shear stress in thin plates is relatively insignificant when compared to the normal stress in the longitudinal direction of the beam, hence:

$$\tau_{yz} = \tau_{zx} \approx 0$$

The normal stress in the thickness direction of the plate, τ_{zz}, is also presumed zero, for the same reason that stress in the depth direction of a shallow beam is zero. In other words, because a plate is relatively thin, loads transverse to the surface of the plate are resisted by bending of the plate, as opposed to compressing the plate in the thickness direction. In effect, the assumptions above suggest that plates can be presumed to be in a state of plane stress, with the appropriate constitutive equations given by Table 3.1. This assumption also applies to thin shells, as discussed by Koiter [33].

Shear Stress Affects the Edge Reaction Forces
It was mentioned that shear stress affects transverse displacement. In addition, shear stress affects reaction forces on restrained edges of plates and shells.

The restrained edges of shell and plate structures develop reaction forces. The magnitude of the reaction is dependent upon the behavior in the vicinity very near the restrained edge. This *boundary layer effect* cannot be captured without accounting for transverse shear stress. In addition, even in models that do account for shear effects, the boundary layer effect is difficult to model without special care. Schwab [34] discusses the use of *p*-type finite elements to model the boundary layer effect in plates.

The Constitutive Equations for Kirchhoff Plates

Since the only stress components of significance in Kirchhoff plates are τ_{xx}, τ_{yy}, and τ_{xy}, the constitutive equations for plate bending stress are the same as those for plane stress:

$$\begin{Bmatrix} \tau_{xx} \\ \tau_{yy} \\ \tau_{xy} \end{Bmatrix} = \frac{E}{1-\nu^2} \begin{bmatrix} 1 & \nu & 0 \\ \nu & 1 & 0 \\ 0 & 0 & \dfrac{1-\nu}{2} \end{bmatrix} \begin{Bmatrix} \varepsilon_{xx} \\ \varepsilon_{yy} \\ \gamma_{xy} \end{Bmatrix} \tag{4.6.41}$$

Notice that the strain vector contains three terms: normal stress in *x*, normal strain in *y*, and shear strain *x-y*. The strain vector can be expressed as:

$$\underline{\underline{\varepsilon}} = \begin{Bmatrix} \varepsilon_{xx} \\ \varepsilon_{yy} \\ \gamma_{xy} \end{Bmatrix} \tag{4.6.42}$$

Or, in terms of derivatives:

$$\underline{\underline{\varepsilon}} = \begin{Bmatrix} \dfrac{\partial u}{\partial x} \\ \dfrac{\partial v}{\partial y} \\ \dfrac{\partial u}{\partial y} + \dfrac{\partial v}{\partial x} \end{Bmatrix} \tag{4.6.43}$$

Notice that two displacements are required: *u(x,y)* and *v(x,y)*. However, as with the Euler-Bernoulli beam, displacement in the Kirchhoff plate will be expressed in terms of rotation of the neutral surface:

$$u = -z\frac{\partial w}{\partial x} \qquad v = -z\frac{\partial w}{\partial y} \tag{4.6.44}$$

Strain terms can then be expressed as:

$$\varepsilon_{xx} = -z\frac{\partial^2 w}{\partial x^2} \qquad \varepsilon_{yy} = -z\frac{\partial^2 w}{\partial y^2} \quad \gamma_{xy} = -2z\frac{\partial^2 w}{\partial x \partial y} \tag{4.6.45}$$

In this manner, only the transverse displacement of the neutral surface, w, need be interpolated to allow computation of the relevant strain terms.

A 3-Node Surface Element for Plate Bending

Assume that an element to model thin, flat plates, consistent with the Kirchhoff plate bending assumptions, is required, as shown in Figure 4.43; the dual arrows indicate that the right hand rule of rotation is employed. For instance, consider rotation about the y-axis at Node b:

$$(w,_x)_b = \frac{\partial w}{\partial x}$$

With the right thumb pointing in the direction of the dual arrow, i.e., along the negative y-axis, the rotational variable at Node b tends to curl the node upward, in the direction of the positive z-axis. This is defined as a positive rotation, and bending moments which bend a plate in this manner are considered positive.

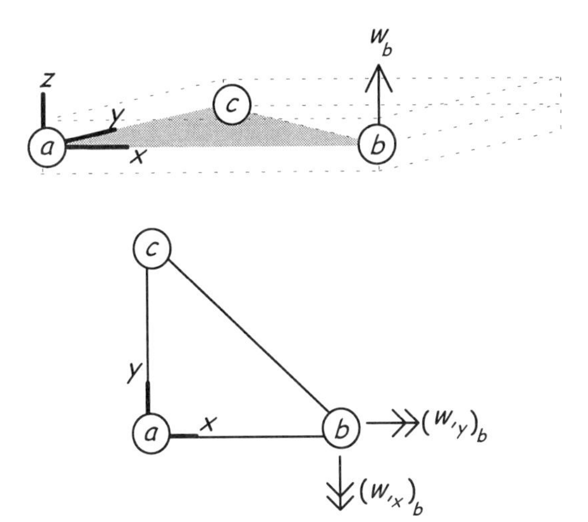

Figure 4.43 A 3-Node Kirchhoff Plate Bending Element

Interpolation Functions for a Kirchhoff Plate Bending Element

From (4.6.45), it is apparent that strains in a Kirchhoff plate can be computed solely on the basis of the transverse displacement. Hence, a 3-node triangular element will be designed to interpolate $w(x,y)$. As with the Euler-Bernoulli beam element, both translational displacement and slopes are required as nodal variables to ensure continuity of these values at the nodes. This is determined by the fact that second order derivatives in (4.6.45) are used to define strain, hence, $p-1=1$, such that continuity of first order derivatives is prescribed. Only the nodal DOF at Node *b* are shown in Figure 4.43, although there exist a total of nine nodal variables for this element. As such, a nine term displacement assumption can be employed. Referring to Pascal's triangle, (4.3.6), one candidate displacement assumption would take the form:

$$\tilde{w}(x,y) = a_1 + a_2 x + a_3 y + a_4 x^2 + a_5 xy + a_6 y^2 + a_7 x^3 + a_8 x^2 y + a_9 xy^2 \quad (4.6.46)$$

While nine terms are the most that can be used, a complete, cubic polynomial requires ten terms. Because (4.6.46) does not contain a y^3 term, this displacement assumption cannot maintain geometric isotropy, as discussed in Section 4.3. Another alternative would be to use:

$$\tilde{w}(x,y) = a_1 + a_2 x + a_3 y + a_4 x^2 + a_5 y^2 + a_6 x^3 + a_7 x^2 y + a_8 xy^2 + a_9 y^3 \quad (4.6.47)$$

This displacement assumption produces an element that, while geometrically isotropic, cannot pass the patch test since it cannot represent constant in-plane twisting strain. To represent constant twisting strain, an *xy* term is needed. To illustrate, observe that twisting strain is computed as γ_{xy}:

$$\gamma_{xy} \approx -2z\frac{\partial^2 \tilde{w}}{\partial x \partial y} = -2z\frac{\partial^2}{\partial x \partial y}\left(a_1 + a_2 x + a_3 y + a_4 x^2 + a_5 y^2 + a_6 x^3 + a_7 x^2 y + a_8 xy^2 + a_9 y^3\right)$$

Notice that only terms which have variables in both x and y survive the differentiation process above, yielding:

$$\gamma_{xy} \approx -2z\left(2a_7 x + 2a_8 y\right) \quad (4.6.48)$$

Equation 4.6.48 cannot represent constant strain because both terms are linear functions. If the term $a_5 xy$ were included in (4.6.47), a constant term would, after the differentiation process, appear in (4.6.48), and constant strain could then be properly represented.

For the purpose of illustration, we will continue using displacement assumption (4.6.47), taking note of the aforementioned deficiency. Constraining the displacement assumption to assume the values of the nodal displacement variables:

$$\tilde{w}(x_a, y_a) = a_1 + a_2 x_a + a_3 y_a + a_4 x_a^2 + a_5 y_a^2 + a_6 x_a^3 + a_7 x_a^2 y_a + a_8 x_a y_a^2 + a_9 y_a^3 = w_a$$

$$\left.\frac{\partial \tilde{w}}{\partial x}\right|_a = a_2 + 2a_4 x_a + 3a_6 x_a^2 + 2a_7 x_a y_a + a_8 y_a^2 = \left(w,_x\right)_a$$

$$\left.\frac{\partial \tilde{w}}{\partial y}\right|_a = a_3 + 2a_5 y_a + a_7 x_a^2 + 2a_8 x_a y_a + 3a_9 y_a^2 = \left(w,_y\right)_a$$

$$\tilde{w}(x_b, y_b) = a_1 + a_2 x_b + a_3 y_b + a_4 x_b^2 + a_5 y_b^2 + a_6 x_b^3 + a_7 x_b^2 y_b + a_8 x_b y_b^2 + a_9 y_b^3 = w_b$$

$$\vdots$$

$$\left.\frac{\partial \tilde{w}}{\partial y}\right|_c = a_3 + 2a_5 y_c + a_7 x_c^2 + 2a_8 x_c y_c + 3a_9 y_c^2 = \left(w,_y\right)_c$$

(4.6.49)

The equations given by (4.6.49) above represent a *9 by 9* system of equations that may be expressed in matrix form as:

$$\begin{bmatrix} 1 & x_a & y_a & x_a^2 & y_a^2 & x_a^3 & y_a^3 & x_a^2 y_a & x_a y_a^2 \\ 0 & 1 & 0 & 2x_a & 0 & 3x_a^2 & 0 & 2x_a y_a & y_a^2 \\ 0 & 0 & 1 & 0 & 2y_a & 0 & x_a^2 & 2x_a y_a & 3y_a^2 \\ \vdots & \vdots & \vdots & \vdots & \vdots & \vdots & \vdots & \vdots & \vdots \\ 0 & 0 & 1 & 0 & 2y_c & 0 & x_c^2 & 2x_c y_c & 3y_c^2 \end{bmatrix} \begin{Bmatrix} a_1 \\ a_2 \\ a_3 \\ \vdots \\ a_9 \end{Bmatrix} = \begin{Bmatrix} w_a \\ \left(w,_x\right)_a \\ \left(w,_y\right)_a \\ \vdots \\ \left(w,_y\right)_c \end{Bmatrix} \qquad (4.6.50)$$

Normally, (4.6.50) would be solved for the a_i's, the a_i's substituted back into the displacement assumption, and the result re-arranged in terms of shape functions:

$$\tilde{w}(x, y) = N_1 w_a + N_2 \left(w,_x\right)_a + N_3 \left(w,_y\right)_a + N_4 w_b + \cdots + N_9 \left(w,_y\right)_c$$

Using matrix notation, the above could then be expressed as:

$$\tilde{w}(x,y) = \underline{\underline{N}} \begin{Bmatrix} w_a \\ (w,_x)_a \\ (w,_y)_a \\ \vdots \\ (w,_y)_c \end{Bmatrix} \tag{4.6.51}$$

Where:

$$\underline{\underline{N}} \equiv \begin{bmatrix} N_1 & N_2 & N_3 & \cdots & N_9 \end{bmatrix}$$

However, the problem with proceeding as shown is that it is necessary to invert the *9 by 9* matrix in (4.6.50) to obtain a closed-form expression for the a_i's. The process of expressing the inverse of this matrix in terms of variables would be a formidable task, if done by hand.[7] Recall that it was difficult to invert the *4 by 4* matrix for the 4-node tetrahedral. For the Kirchhoff plate element, an alternate approach is employed. This same approach can be used with any element. The key to this approach is that instead of trying to invert a large matrix and manipulate the results to yield an interpolation function in terms of shape functions, the interpolation functions are never explicitly expressed.

To begin using the alternate approach, the displacement assumption is expressed in matrix form as:

$$\begin{aligned} \tilde{w}(x,y) &= a_1 + a_2x + a_3y + a_4x^2 + a_5y^2 + a_6x^3 \\ &+ a_7x^2y + a_8xy^2 + a_9y^3 \equiv \underline{\underline{Pa}} \end{aligned} \tag{4.6.52}$$

The components of the P-vector may be referred to as the *basis functions*, while the components of the a-vector are the *basis function coefficients*. The following definitions of the basis function vector and basis coefficient vector associated with

[7] A relatively new type of software product allows the solution of such problems in symbolic fashion. (Mathcad® by MathSoft Inc., Cambridge, MA, is one such product.) Alternately, large square matrices can be partitioned and inverses computed analytically; many terms would be involved for the general case; Selby [25, p. 125].

(4.6.52) are:

$$\underline{\underline{P}} \equiv \begin{bmatrix} 1 & x & y & x^2 & y^2 & x^3 & y^3 & x^2y & xy^2 \end{bmatrix} \qquad \underline{\underline{a}} \equiv \begin{Bmatrix} a_1 \\ a_2 \\ a_3 \\ \vdots \\ a_9 \end{Bmatrix} \tag{4.6.53}$$

It is helpful to define (4.6.50) as:

$$\underline{\underline{A}}\underline{a} = \begin{Bmatrix} w_a \\ (w,_x)_a \\ (w,_y)_a \\ w_b \\ \vdots \end{Bmatrix} \tag{4.6.54}$$

The A-matrix in (4.6.54) is given as:

$$\begin{bmatrix} 1 & x_a & y_a & x_a^2 & y_a^2 & x_a^3 & y_a^3 & x_a^2 y_a & x_a y_a^2 \\ 0 & 1 & 0 & 2x_a & 0 & 3x_a^2 & 0 & 2x_a y_a & y_a^2 \\ 0 & 0 & 1 & 0 & 2y_a & 0 & x_a^2 & 2x_a y_a & 3y_a^2 \\ \vdots & \vdots & \vdots & \vdots & \vdots & \vdots & \vdots & \vdots & \vdots \\ 0 & 0 & 1 & 0 & 2y_c & 0 & x_c^2 & 2x_c y_c & 3y_c^2 \end{bmatrix} \tag{4.6.55}$$

Re-arranging (4.6.54), the basis coefficient vector can be expressed as:

$$\underline{a} = \underline{\underline{A}}^{-1} \begin{Bmatrix} w_a \\ (w,_x)_a \\ (w,_y)_a \\ w_b \\ \vdots \end{Bmatrix}$$

Now, substituting the expression above into the right hand side of (4.6.52), an expression for displacement in the z-coordinate direction is obtained:

$$\tilde{w}(x,y) = \underline{\underline{P}}\,\underline{\underline{A}}^{-1}\begin{Bmatrix} w_a \\ (w_{,x})_a \\ (w_{,y})_a \\ w_b \\ \vdots \end{Bmatrix} \tag{4.6.56}$$

This is essentially the same process that has been followed in all of the previous examples, although the process was not stated in matrix form. The next step would be to find a closed-form expression for A^{-1}, substitute it into (4.6.56), perform the matrix algebra, then re-arrange in terms of shape functions and nodal displacement variables. Comparing (4.6.56) with (4.6.51), the shape function matrix is equivalent to:

$$\underline{\underline{N}} \equiv \underline{\underline{P}}\,\underline{\underline{A}}^{-1}$$

Due to the number of terms involved, the displacement interpolation function expressed by (4.6.56) will be left in its current form instead of trying to perform the algebra needed to express the shape functions explicitly.

Strain-Displacement Matrix for 3-Node Surface Element, Plate Applications
An approximation for the strain components associated with the Kirchhoff plate element may now be established. Recalling the strain metrics from (4.6.45):

$$\varepsilon_{xx} = -z\frac{\partial^2 w}{\partial x^2} \qquad \varepsilon_{yy} = -z\frac{\partial^2 w}{\partial y^2} \qquad \gamma_{xy} = -2z\frac{\partial^2 w}{\partial x \partial y} \tag{4.6.57}$$

Substituting (4.6.56) into (4.6.57):

$$\varepsilon_{xx} \approx -z\left(\frac{\partial^2 \underline{\underline{P}}}{\partial x^2}\right)\underline{\underline{A}}^{-1}\begin{Bmatrix} w_a \\ (w_{,x})_a \\ (w_{,y})_a \\ w_b \\ \vdots \end{Bmatrix} \qquad \varepsilon_{yy} \approx -z\left(\frac{\partial^2 \underline{\underline{P}}}{\partial y^2}\right)\underline{\underline{A}}^{-1}\begin{Bmatrix} w_a \\ (w_{,x})_a \\ (w_{,y})_a \\ w_b \\ \vdots \end{Bmatrix}$$

and:

$$\gamma_{xy} \approx -2z\left(\frac{\partial^2 \underline{\underline{P}}}{\partial x \partial y}\right)\underline{\underline{A}}^{-1}\begin{Bmatrix} w_a \\ \left(w,_x\right)_a \\ \left(w,_y\right)_a \\ w_b \\ \vdots \end{Bmatrix}$$

(4.6.58)

Comparing (4.6.57) and the three equations in (4.6.58), curvatures are expressed as:

$$\frac{\partial^2 w}{\partial x^2} \approx \left(\frac{\partial^2 \underline{\underline{P}}}{\partial x^2}\right)\underline{\underline{A}}^{-1}\begin{Bmatrix} w_a \\ \left(w,_x\right)_a \\ \left(w,_y\right)_a \\ w_b \\ \vdots \end{Bmatrix} \qquad \frac{\partial^2 w}{\partial y^2} \approx \left(\frac{\partial^2 \underline{\underline{P}}}{\partial y^2}\right)\underline{\underline{A}}^{-1}\begin{Bmatrix} w_a \\ \left(w,_x\right)_a \\ \left(w,_y\right)_a \\ w_b \\ \vdots \end{Bmatrix}$$

Why does the differentiation process in (4.6.58) apply only to the *P*-matrix? The reason is that only the *P*-matrix contains free variables that can be differentiated. The *A*-matrix contains nodal coordinates while the displacement vector contains only nodal displacement variables, neither of which contain spatial coordinate variables, x and y. Therefore, the only free variables in the expressions for strain above are contained in the *P*-matrix and the other terms are considered as constants. The strain vector for this element can be expressed in terms of derivatives of the *P*-vector, and arranged as shown in (4.6.59).

$$\underline{\underline{\varepsilon}} = \begin{Bmatrix} \varepsilon_{xx} \\ \varepsilon_{yy} \\ \gamma_{xy} \end{Bmatrix} \approx -z \begin{Bmatrix} \left(\dfrac{\partial^2 \underline{\underline{P}}}{\partial x^2} \right) \underline{\underline{A}}^{-1} \begin{Bmatrix} w_a \\ (w,_x)_a \\ (w,_y)_a \\ w_b \\ \vdots \end{Bmatrix} \\ \left(\dfrac{\partial^2 \underline{\underline{P}}}{\partial y^2} \right) \underline{\underline{A}}^{-1} \begin{Bmatrix} w_a \\ (w,_x)_a \\ (w,_y)_a \\ w_b \\ \vdots \end{Bmatrix} \\ \left(2 \dfrac{\partial^2 \underline{\underline{P}}}{\partial y \partial x} \right) \underline{\underline{A}}^{-1} \begin{Bmatrix} w_a \\ (w,_x)_a \\ (w,_y)_a \\ w_b \\ \vdots \end{Bmatrix} \end{Bmatrix} \tag{4.6.59}$$

Referring to (4.6.53), the derivatives of the P-matrix are computed as:

$$\frac{\partial^2 \underline{\underline{P}}}{\partial x^2} = \begin{bmatrix} 0 & 0 & 0 & 2 & 0 & 6x & 0 & 2y & 0 \end{bmatrix}$$
$$\frac{\partial^2 \underline{\underline{P}}}{\partial y^2} = \begin{bmatrix} 0 & 0 & 0 & 0 & 2 & 0 & 6y & 0 & 2x \end{bmatrix}$$
$$\frac{\partial^2 \underline{\underline{P}}}{\partial y \partial x} = \begin{bmatrix} 0 & 0 & 0 & 0 & 0 & 0 & 0 & 2x & 2y \end{bmatrix}$$

Using the above in (4.6.59) and re-arranging:

$$\underline{\underline{\varepsilon}} \approx -z \begin{bmatrix} 0 & 0 & 0 & 2 & 0 & 6x & 0 & 2y & 0 \\ 0 & 0 & 0 & 0 & 2 & 0 & 6y & 0 & 2x \\ 0 & 0 & 0 & 0 & 0 & 0 & 0 & 4x & 4y \end{bmatrix} \underline{\underline{A}}^{-1} \begin{Bmatrix} w_a \\ (w,_x)_a \\ (w,_y)_a \\ w_b \\ \vdots \end{Bmatrix} \tag{4.6.60}$$

The strain vector is expressed in terms of the strain-displacement matrix and the nodal DOF vector:

$$\underline{\underline{\varepsilon}} \approx \underline{\underline{B}}\,{}^{e}\underline{\underline{D}}$$

For this 3-node Kirchhoff plate element, strain is expressed as:

$$\underline{\underline{\varepsilon}} \approx \underline{\underline{B}} \begin{Bmatrix} w_a \\ (w,_x)_a \\ (w,_y)_a \\ w_b \\ \vdots \end{Bmatrix} \qquad (4.6.61)$$

Comparing (4.6.61) to (4.6.60), the strain-displacement matrix for this element is:

$$\underline{\underline{B}} = -z \begin{bmatrix} 0 & 0 & 0 & 2 & 0 & 6x & 0 & 2y & 0 \\ 0 & 0 & 0 & 0 & 2 & 0 & 6y & 0 & 2x \\ 0 & 0 & 0 & 0 & 0 & 0 & 0 & 4x & 4y \end{bmatrix} \underline{\underline{A}}^{-1} \qquad (4.6.62)$$

As with the strain-displacement matrix for the 2-node beam, the plate strain-displacement matrix above is expressed using a B'-matrix:

$$\underline{\underline{B}} = -z\underline{\underline{B'}} \qquad (4.6.63)$$

Where:

$$\underline{\underline{B'}} \equiv \begin{bmatrix} 0 & 0 & 0 & 2 & 0 & 6x & 0 & 2y & 0 \\ 0 & 0 & 0 & 0 & 2 & 0 & 6y & 0 & 2x \\ 0 & 0 & 0 & 0 & 0 & 0 & 0 & 4x & 4y \end{bmatrix} \underline{\underline{A}}^{-1} \qquad (4.6.64)$$

For this element, the terms in the strain-displacement are not constant and the A-matrix needs to be inverted. Small square matrices can be inverted using a closed-form solution. However, trying to invert the A-matrix in (4.6.64) would be very tedious, generally speaking. Numerical methods can be used to invert the A-matrix, if the variables in the A-matrix are replaced with their respective numerical values; Selby [25]. Alternately, symbolic mathematical software can be used. In any case, the terms in the B'-matrix are not shown explicitly.

Stiffness Matrix for a 3-Node, Triangular Kirchhoff Plate Element

The items required for computing the stiffness matrix are now available for the 3-node Kirchhoff element presently under consideration. Recalling the equation for the finite element stiffness matrix (3.5.12):

$$ {}^{e}\underline{\underline{K}} \equiv \left[\int_V \underline{\underline{B}}^T \; \underline{\underline{E}} \, \underline{\underline{B}} \, dV \right] $$

Using (4.6.63) in the above:

$$ {}^{e}\underline{\underline{K}} \equiv \int_V \left(-z\underline{\underline{B'}}\right)^T \; \underline{\underline{E}} \left(-z\underline{\underline{B'}}\right) dV \tag{4.6.65} $$

Substituting the material matrix shown in (4.6.39) into the above and re-arranging:

$$ {}^{e}\underline{\underline{K}} \equiv \int_V \left(\underline{\underline{B'}}\right)^T \frac{E}{1-\nu^2} \begin{bmatrix} 1 & \nu & 0 \\ \nu & 1 & 0 \\ 0 & 0 & \dfrac{1-\nu}{2} \end{bmatrix} \left(\underline{\underline{B'}}\right) z^2 \, dV \tag{4.6.66} $$

Substituting $dV = dx\, dy\, dz$ into (4.6.66):

$$ {}^{e}\underline{\underline{K}} \equiv \iiint \left(\underline{\underline{B'}}\right)^T \frac{E}{1-\nu^2} \begin{bmatrix} 1 & \nu & 0 \\ \nu & 1 & 0 \\ 0 & 0 & \dfrac{1-\nu}{2} \end{bmatrix} \left(\underline{\underline{B'}}\right) z^2 \, dx\, dy\, dz \tag{4.6.67} $$

As with the uniform beam, integration on z for a plate is always within the same limits. So, for a uniform thickness plate element, (4.6.67) can be expressed as:

$$ {}^{e}\underline{\underline{K}} \equiv \iint \left(\underline{\underline{B'}}\right)^T \frac{E}{1-\nu^2} \begin{bmatrix} 1 & \nu & 0 \\ \nu & 1 & 0 \\ 0 & 0 & \dfrac{1-\nu}{2} \end{bmatrix} \left(\underline{\underline{B'}}\right) dx\, dy \left(\int_{-t/2}^{t/2} z^2 dz \right) \tag{4.6.68} $$

Performing the integration on the z-domain, (4.6.68) reduces to:

$$\underline{\underline{{}^{e}K}} \equiv \iint \left(\underline{\underline{B'}}\right)^T \frac{Et^3}{12\left(1-\nu^2\right)} \begin{bmatrix} 1 & \nu & 0 \\ \nu & 1 & 0 \\ 0 & 0 & \frac{1-\nu}{2} \end{bmatrix} \left(\underline{\underline{B'}}\right) dx\, dy \tag{4.6.69}$$

Substituting the expression for the B'-matrix, (4.6.64), into (4.6.69):

$$\underline{\underline{{}^{e}K}} \equiv \iint \left(\underline{\underline{A}}^{-1}\right)^T \begin{bmatrix} 0 & 0 & 0 \\ 0 & 0 & 0 \\ 0 & 0 & 0 \\ 2 & 0 & 0 \\ 0 & 2 & 0 \\ 3x & 0 & 0 \\ 0 & 3y & 0 \\ 2y & 0 & 4x \\ 0 & 2x & 4y \end{bmatrix} \underline{\underline{E'}} \begin{bmatrix} 0 & 0 & 0 & 2 & 0 & 3x & 0 & 2y & 0 \\ 0 & 0 & 0 & 0 & 2 & 0 & 3y & 0 & 2x \\ 0 & 0 & 0 & 0 & 0 & 0 & 0 & 4x & 4y \end{bmatrix} \underline{\underline{A}}^{-1} dxdy \tag{4.6.70}$$

The E'-matrix in (4.6.70) is given as:

$$\underline{\underline{E'}} \equiv \frac{Et^3}{12\left(1-\nu^2\right)} \begin{bmatrix} 1 & \nu & 0 \\ \nu & 1 & 0 \\ 0 & 0 & \frac{1-\nu}{2} \end{bmatrix}$$

The displacement and force vectors for this element take the form:

$$\underline{\underline{{}^{e}D}} = \begin{Bmatrix} w_a \\ \left(w_{,x}\right)_a \\ \left(w_{,y}\right)_a \\ w_b \\ \vdots \\ \left(w_{,y}\right)_c \end{Bmatrix} \qquad \underline{\underline{{}^{e}F}} = \begin{Bmatrix} f_{za} \\ m_{ya} \\ m_{xa} \\ f_{zb} \\ \vdots \\ m_{yc} \end{Bmatrix}$$

The element above is essentially the same as the CKZ triangle that was developed in the early days of finite element analysis by Cheung, King, and Zienkiewicz; see Bazeley [19] and Cheung [35]. This element is not recommended for practical use due to the deficiencies previously mentioned.

Compatibility and the 3-Node CKZ Plate Element

The CKZ element is not compatible. Remember that for C^1 elements, displacement and *p–1* derivatives of displacement must be continuous at the nodes and across the boundaries of adjacent elements. Examining the strain terms for this plate element, (4.6.45), second order derivatives are noted. Hence, $p-1=1$, indicating that first order derivatives are required to be continuous at the nodes and across the element boundary.

To investigate the compatibility of the triangular plate element, consider the rotational DOF's along one edge, as shown in Figure 4.44. Recall the terms that comprise first order derivatives of w with respect to x:

$$\frac{\partial w}{\partial x} = a_2 + 2a_4 x + 3a_6 x^2 + 2a_7 xy + a_8 y^2 \qquad (4.6.72)$$

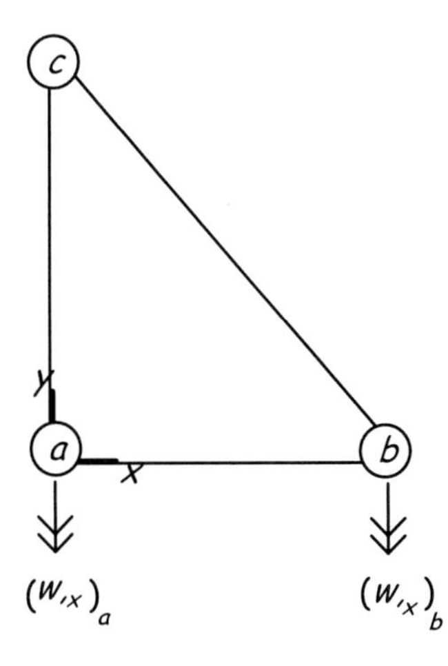

Figure 4.44 Rotational DOF Along Edge a-b

Along Edge *a-b*, the *y*-coordinate is zero:

$$\left.\frac{\partial w}{\partial x}\right|_{a-b} = a_2 + 2a_4 x + 3a_6 x^2 \tag{4.6.73}$$

Notice that the expression for the derivative is quadratic, however, only two DOF's related to the derivative exist along Edge *a-b*, hence, only a linear function for the derivative can be described in a compatible fashion. For this element the derivative is not compatible across the element boundary, since the derivative must be a function of something other than the local variables associated with that edge. Myriad schemes have been devised to produce compatible elements of the Kirchhoff type. As mentioned, this type of element has been largely abandoned in favor of Mindlin elements, which have the additional advantage of accounting for shear deformations.

The objective of introducing the 3-node Kirchhoff plate element was to allow the reader to gain some appreciation for the assumptions and equations used in conjunction with structural plates. Finite element software today typically employs elements that perform much better than a plate element using the CZK formulation. In practice, analysts often use shell elements instead of plates, since a shell can account for both bending and membrane deformation.

Structural Shells

Shells can be discussed only briefly in this volume. Developing finite elements for shell structures has proven to be a very challenging undertaking for finite element software developers, since these are perhaps the most complex elements used in solid mechanics problems.

What is a structural shell? A shell is a relatively thin, "surface type" structure that has initial curvature and can carry bending loads through a combination of bending and membrane deformation. Bending deformation in a shell is analogous to bending in a flat plate, while the membrane deformation is similar to that which occurs in a plane stress member. However, in an actual shell structure, the bending and membrane affects are coupled, such that the bending deformation in one portion of the structure affects the membrane in another.

Coupling in a shell is due to curvature. Thin arches or cylinders, automotive fuel tanks, and hollow spherical structures, such as hot air balloons, might be considered shell structures, as illustrated in Figure 4.45. What is a *thin* shell? If the thickness of a shell divided by the smallest radius of curvature is less than 1/20, the shell is considered thin, and the effects of shear stress can be neglected; Ugural [2, p. 428].

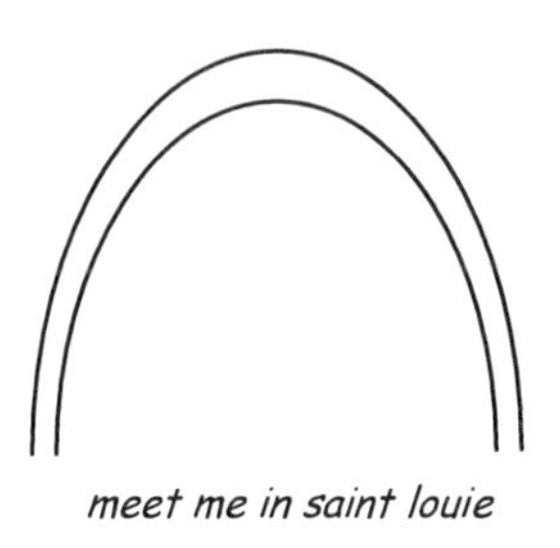

Figure 4.45 Shell Type Structures

As discussed by Morley [32], the error in neglecting shear stress in shell structures is of the order *t/R*, where *t* is the shell thickness and *R* is the smallest of the curvature radii.

Shells bare some similarity to plates, indeed, a plate structure might be considered a special case of a shell, where the geometry is flat and there is no membrane deformation. Hence, finite elements developed for thin shells can typically be used for thin plate applications although the reverse is not true: plate elements cannot be used for shells, since they have no mechanism to carry membrane loads.

How are shell elements developed? Three approaches have been taken to develop elements to model shell structures:

- Classical shell theory
- Modified 3-D elements with rotational DOF (General shell elements)
- A "shell element" formed by combining a flat plate and a plane stress element

The governing equations for classical shell theory are quite complex and contain approximations. Some of the earliest shell elements used classical shell theory as the underlying basis for developing elements of this type, however, this approach has largely been abandoned. An excellent discussion of general shell theory is given by Koiter [33].

A modified three-dimensional element can be used for shell applications. This type of element typically begins with a three-dimensional, hexagonal, isoparametric volume element with the associated three-dimensional constitutive equations based upon generalized Hooke's law. The element is then "degenerated" into a surface element by removing the through-the-thickness dimension. Since this element accounts for shear stress components, it is of the Mindlin type. This approach to shell elements employs none of the limiting assumptions of shell theory, because is it formulated using a continuum mechanics approach. (Normal stress in the thickness direction is, however, assigned a value of zero.) The degenerated three-dimensional elements appear to be the most effective. However, they are also the most complex and, for that reason, they will not be covered in this text.

The last approach to shell elements, combining a flat plate with a plane stress element, is the most simple approach and will be mentioned here. This combined element will be called the PPS element in this text (**P**late and **P**lane **S**tress), however, this element is commonly called a "flat shell" element. The term "flat shell" may be slightly misleading, however, because there exist 4-node (flat, i.e., non-curved) surface elements for shell applications that are formulated using the degenerated three-dimensional approach, as discussed by Dvorkin [3]. The PPS element, when compared to the degenerated three-dimensional element, is substantially different in both its design and performance.

How are flat elements used to model curved geometry? Shell structures can be described in a faceted arrangement, as shown in Figure 4.46. Here, the mid-surface of the hollow, thin-walled, semi-cone is meshed using 3-node PPS surface elements.

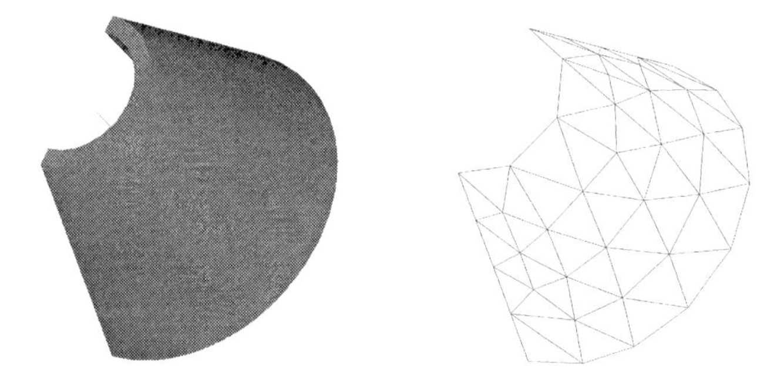

Figure 4.46 A Shell Structure Modeled Using a Faceted Arrangement

The PPS element used in the faceted arrangement above is a combination of a flat plate element and a plane stress element, generating an element with five DOF's per node, as shown in Figure 4.47.

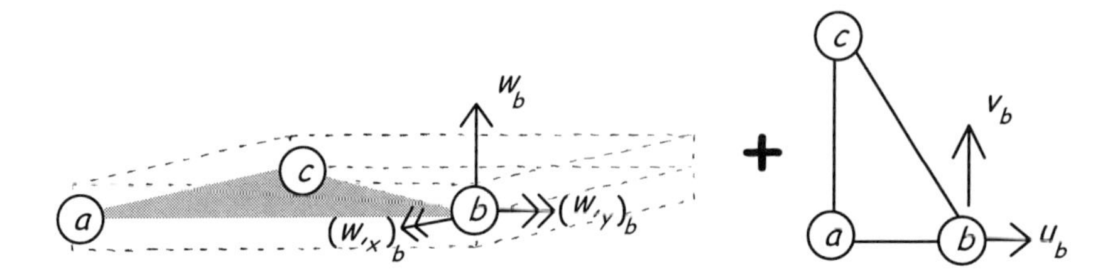

Figure 4.47 A PPS Element

The advantage of the approach above is its simplicity. Using the PPS approach (with a more robust plate element than the CKZ) allows this type of element to meet the requirements of rigid body motion and uniform strain. A disadvantage of the PPS element, and other flat elements used for shell applications, is that there exists no coupling (within a single element) between bending deformation and membrane deformation. The interpolation for the PPS element is described by:

$$u(x,y) = N_1 u_a + N_2 u_b + N_3 u_c$$
$$v(x,y) = N_1 v_a + N_2 v_b + N_3 v_c$$
$$w(x,y) = N_1 w_a + N_2 (w,_x)_a + N_3 (w,_y)_a + N_4 w_b + \cdots + N_9 (w,_y)_c$$

It is shown that, for this element, only transverse displacement is a function of nodal rotations—membrane displacement is not affected by rotation of the midsurface.

In more robust general shell elements, membrane displacement is tied to bending deformation within the element by virtue of the displacement assumption. For example, a 3-node general shell element with five DOF's at each node might interpolate displacement using a nine-term function:

$$\begin{aligned} u &= N_1 u_a + N_2 u_b + N_3 u_c + N_4 \alpha_a + N_5 \beta_a + N_6 \alpha_b + N_7 \beta_b + N_8 \alpha_c + N_9 \beta_c \\ v &= N_1 v_a + N_2 v_b + N_3 v_c + N_4 \alpha_a + N_5 \beta_a + N_6 \alpha_b + N_7 \beta_b + N_8 \alpha_c + N_9 \beta_c \\ w &= N_1 w_a + N_2 w_b + N_3 w_c + N_4 \alpha_a + N_5 \beta_a + N_6 \alpha_b + N_7 \beta_b + N_8 \alpha_c + N_9 \beta_c \end{aligned} \qquad (4.6.74)$$

A 4-node quadrilateral shell element using a similar type of displacement assumption is discussed in Dvorkin [3, p. 78]. In that reference, the nodal rotation variables, α and β, are independent variables, and as such are not based upon the slope of the neutral surface. This contrasts with Kirchhoff elements, which base nodal rotations upon the slope of the neutral surface, and use derivatives of transverse displacement to describe the slope.

An element using the interpolation functions given in (4.6.74) has the advantage of accounting for membrane deformation due to bending within the element. However, transverse displacement is still not coupled to membrane displacement. The only way for this type of coupling to occur within a single element is for the element to have initial curvature. This requires at least a six-node triangular element or an eight-node quadrilateral element. The initial curvature allows bending forces at one point in the element to be coupled with membrane forces in another part of the element.

Although there exists no coupling between rotation of the midsurface and membrane deformation *within* any single flat PPS element, there is a mechanism by which coupling can occur *between* elements in a mesh. For example, if flat elements having membrane and bending DOF's meet at an angle, some of the membrane deformation in one will be transformed into bending deformation of the other. Likewise with bending: bending deformation of one will be transferred into membrane deformation of the other. Figure 4.48 illustrates that the nodal force acting to bend Element 1, P, acts upon Element 2 through *both* bending and membrane forces.

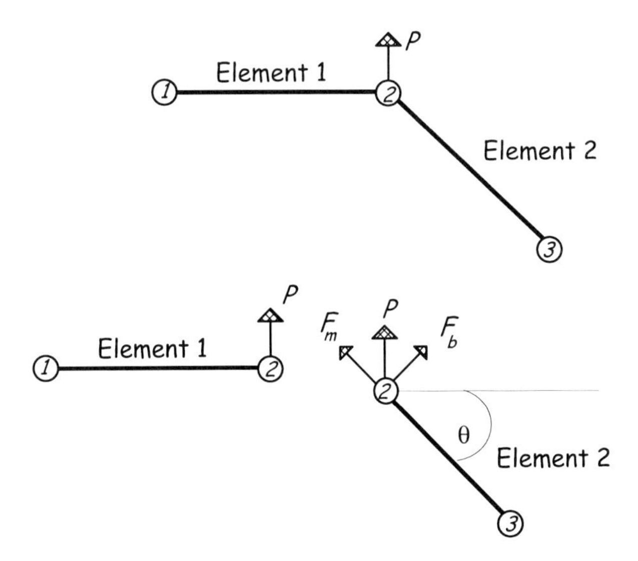

Figure 4.48 Coupling of Bending and Membrane Forces Between Flat Elements

The bending component acting on Element 2 is computed to be:

$$F_b = P\cos\theta$$

The membrane component is:

$$F_m = P\cos(90-\theta)$$

In this manner, coupling is introduced into the model at the mesh level, even though, within any single element, there is no coupling. The benefit of the coupling is that applied loads are resisted by the correct *type* of stiffness, such that the membrane component of the applied loads is resisted by membrane deformation while the bending component is resisted by bending deformation. This tends to result in a more accurate prediction of displacement within a curved structure.

Membrane Loads Affecting Transverse Displacement

Elements used to model shell structures can better approximate the exact stiffness of a structure by coupling bending and membrane deformation, either within or between elements. In other words, elements that allow coupling will have the characteristic of resisting bending loads with bending stiffness, and membrane loads with membrane stiffness. However, the change in bending stiffness due to membrane forces is still not accounted for, even if elements exhibit coupling. Is there a means to account for changes in transverse stiffness due to membrane loads? Yes. An *initial stress stiffness* matrix can be calculated to account for the changes in bending stiffness due to membrane loads.

The process associated with the initial stiffness matrix can be achieved in two steps, assuming that a linear system prevails. In the first step, a shell structure is loaded, and a stiffness matrix associated with the membrane effect is computed. The stiffness matrix associated with the membrane loads is called the initial stiffness matrix. In the second step of the two step process, an analysis procedure is again invoked, this time using both the initial stress stiffness matrix combined with the regular stiffness matrix. If the initial loads produce membrane compression, the initial stiffness matrix subtracts from the transverse stiffness terms. If membrane tension is produced, the initial stiffness matrix adds to the transverse stiffness matrix. One might correctly assume that if enough membrane compression were applied, the transverse stiffness will approach zero, and collapse would be imminent.

A key ingredient in the computation of initial stiffness matrix is the use of squared terms in the strain metric; recall that these terms are usually truncated for small strains. For more details on the initial stress stiffness matrix, see Cook [14, p. 429]

Details on Beam, Plate, and Shell Theory
Timoshenko [36] provides a discussion of plate and shell theory as does Love [37]. Plate and Shell elements are considered in Cook [14], and in more detail in Ashwell [11] and Morely [32]. Ashwell also considers other curved structures, such as arches.

4.7 A Truss Element via Matrix Transformation

Key Concept: It is sometimes convenient to develop an element stiffness matrix in a local coordinate system, then transform this locally defined element into one that can be used in the global coordinate system.

Some structures have stiffness properties that depend upon both the orientation of the structure, and how the structure is loaded. For instance, in a cantilever beam, loads acting normal to the neutral surface are resisted by the bending stiffness of the structure while loads applied axially are resisted by the membrane stiffness. Alternately, stiffness in a pinned-truss member is defined only along the local longitudinal axis of the member, such that only loads resolved into this direction will be resisted.

How does one identify when a local coordinate system might be advantageous in the development of a certain type of finite element? When the number of normal stress components that an element computes is fewer than the number of coordinates axes used to define the element space, a local coordinate system might be useful. For instance, if an element that computes only uniaxial stress were used in two- or three-dimensional space, it would be a candidate for a local coordinate system. Consider the example of the suspension bridge from Chapter 1, shown again in Figure 4.49.

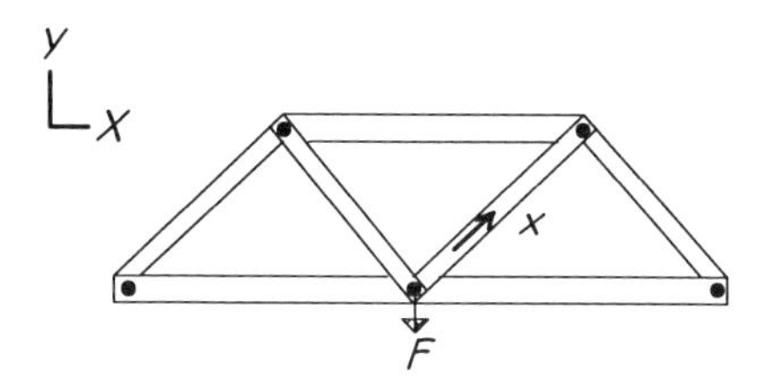

Figure 4.49 A Structure to Be Analyzed with Two-Dimensional Truss Elements

Each pinned member is presumed to be in a state of uniaxial stress.

Although loads can be easily applied to the truss structure in either of the global *X*- or *Y*-coordinate directions, the resulting loads are resisted only along the axial direction of each truss. The stiffness in each truss element is therefore defined in a local coordinate system, *x*, that is aligned along the length of each member. Since the element computes only one normal stress component (along the axis of the element) but the element is used in two-dimensional space, it is convenient to use a local coordinate system. Upper case variables are used to denote the global coordinate directions while lower case is used to denote the local directions.

A structure represented by a truss idealization, such as the structure shown in Figure 4.49, can be modeled with elements that are essentially the same as the rod element from Chapter 2.

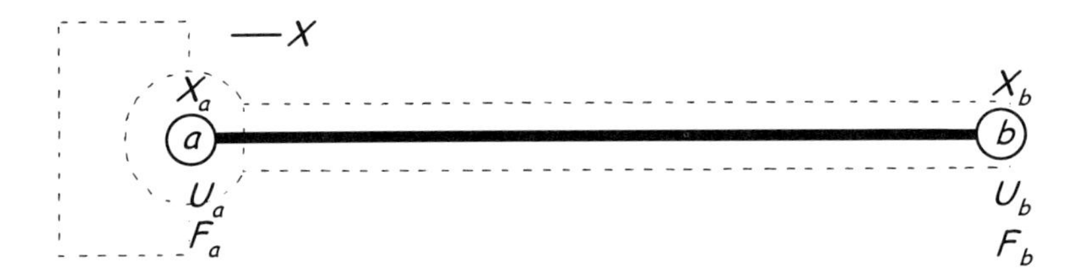

Figure 4.50 Truss Element in 2-D Space

Recall from Chapter 2 the rod element in one-dimensional space, as shown again in Figure 4.50. To be used in a two-dimensional truss structure, the one-dimensional rod element is modified to account for loads in two global coordinate directions. The global loads are resolved into a force along the longitudinal axis of the element. In this manner, the one-dimensional rod element is transformed into a *two-dimensional truss* element.

While both the truss and rod elements compute uniaxial stress only, the truss element accounts for externally applied loads (and displacement boundary conditions) that may have components in two coordinate directions.

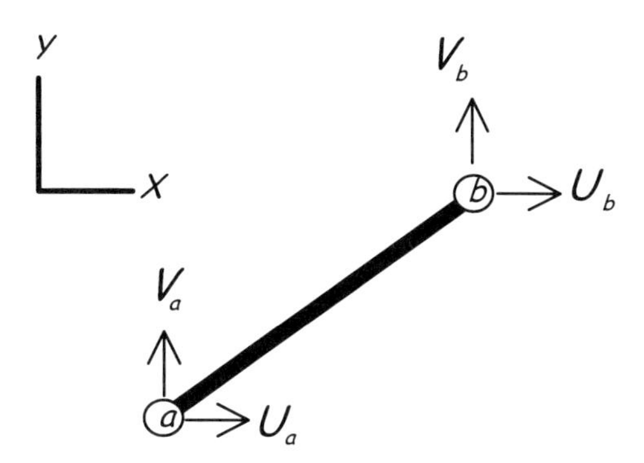

Figure 4.51 Truss Element in Global Space

The truss element will have the same uniaxial stress characteristic as the rod element, but will be used in two-dimensional space, as depicted in Figure 4.51. A global coordinate system will be used to define external forces and displacements, but stress will be computed only in the local x-axis direction, as shown in Figure 4.52. With displacements defined in the global system, the essential boundary conditions will also be imposed in the global directions.

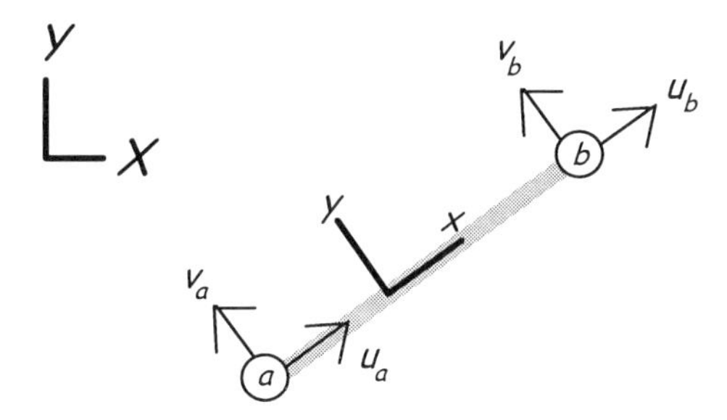

Figure 4.52 Local Coordinate System for a Truss Element

Some means to convert the global displacements and forces into the local system must be developed, since, as is, the element in Figure 4.52 resists loads only along the local x-coordinate direction. Although not shown, force variables, defined in terms of the local coordinate system, are also present at each node. To begin the development of a truss element in two-dimensional space, recall the equilibrium equations for the rod element ${}^e\underline{\underline{K}}\,{}^e\underline{D} = {}^e\underline{F}$, as described in Chapter 2:

$$\frac{EA}{H}\begin{bmatrix} 1 & -1 \\ -1 & 1 \end{bmatrix}\begin{Bmatrix} U_a \\ U_b \end{Bmatrix} = \begin{Bmatrix} F_a \\ F_b \end{Bmatrix} \tag{4.7.1}$$

In the local coordinate system, the equilibrium equations would appear as:

$$\frac{EA}{H}\begin{bmatrix} 1 & -1 \\ -1 & 1 \end{bmatrix}\begin{Bmatrix} u_a \\ u_b \end{Bmatrix} = \begin{Bmatrix} f_{xa} \\ f_{xb} \end{Bmatrix} \tag{4.7.2}$$

The above can be expanded to include the displacement and force variables in the y-coordinate direction, such that the equilibrium equations in local coordinates, ${}^e\underline{\underline{k}}\,{}^e\underline{d} = {}^e\underline{f}$, appear as:

$$\frac{EA}{H}\begin{bmatrix} 1 & 0 & -1 & 0 \\ 0 & 0 & 0 & 0 \\ -1 & 0 & 1 & 0 \\ 0 & 0 & 0 & 0 \end{bmatrix}\begin{Bmatrix} u_a \\ v_a \\ u_b \\ v_b \end{Bmatrix} = \begin{Bmatrix} f_{xa} \\ f_{ya} \\ f_{xb} \\ f_{yb} \end{Bmatrix} \tag{4.7.3}$$

Notice the use of the lowercase variables in (4.7.3), and also notice that there exists no stiffness in the y-coordinate direction, as indicated by the fact that Rows 2 and 4 contain only zeros.

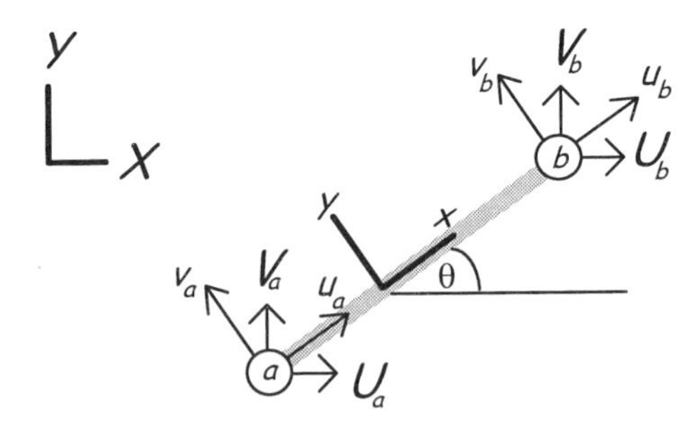

Figure 4.53 Global and Local Displacement Variables

Consider both local and global displacement variables as defined in Figure 4.53. Global displacement variables can be transformed into local directions using trigonometric relationships. The task is to resolve the global displacement variables at the nodes into local displacement variables. First consider the transformation of both global displacement variables (U and V) at Node b into a single displacement variable defined in the local x-coordinate direction:

$$u_b = U_b \cos(\theta) + V_b \cos(90 - \theta) \tag{4.7.4}$$

Using trigonometric relationships, the above is alternately defined as:

$$u_b = U_b \cos(\theta) + V_b \sin(\theta) \tag{4.7.5}$$

Likewise, the remaining local displacement variables are defined in terms of global variables:

$$\begin{aligned} u_a &= U_a \cos(\theta) + V_a \sin(\theta) \\ v_a &= -U_a \sin(\theta) + V_a \cos(\theta) \\ v_b &= -U_b \sin(\theta) + V_b \cos(\theta) \end{aligned} \tag{4.7.6}$$

In matrix form, the transformation of displacement variables from global to local coordinates takes the form:

$$\underset{\underset{{}^e\underline{d}}{\Uparrow}}{\begin{Bmatrix} u_a \\ v_a \\ u_b \\ v_b \end{Bmatrix}} = \underset{\underset{\underline{T}}{\Uparrow}}{\begin{bmatrix} c & s & 0 & 0 \\ -s & c & 0 & 0 \\ 0 & 0 & c & s \\ 0 & 0 & -s & c \end{bmatrix}} \underset{\underset{{}^e\underline{D}}{\Uparrow}}{\begin{Bmatrix} U_a \\ V_a \\ U_b \\ V_b \end{Bmatrix}} \tag{4.7.7}$$

For clarity, the sine and cosine terms in (4.7.7) are denoted as:

$$c = \cos(\theta) \qquad\qquad s = \sin(\theta)$$

Substituting (4.7.7) into (4.7.3):

$$\frac{EA}{H}\begin{bmatrix} 1 & 0 & -1 & 0 \\ 0 & 0 & 0 & 0 \\ -1 & 0 & 1 & 0 \\ 0 & 0 & 0 & 0 \end{bmatrix}\begin{bmatrix} c & s & 0 & 0 \\ -s & c & 0 & 0 \\ 0 & 0 & c & s \\ 0 & 0 & -s & c \end{bmatrix}\begin{Bmatrix} U_a \\ V_a \\ U_b \\ V_b \end{Bmatrix} = \begin{Bmatrix} f_{xa} \\ f_{ya} \\ f_{xb} \\ f_{yb} \end{Bmatrix} \tag{4.7.8}$$

Instead of using the global coordinate system for displacement and a local coordinate system for forces, as is the case in (4.7.8), the local force vector can also be transformed:

$$\underset{\underline{\underline{^e f}}}{\begin{Bmatrix} f_{xa} \\ f_{ya} \\ f_{xb} \\ f_{yb} \end{Bmatrix}} = \underset{\underline{\underline{T}}}{\begin{bmatrix} c & s & 0 & 0 \\ -s & c & 0 & 0 \\ 0 & 0 & c & s \\ 0 & 0 & -s & c \end{bmatrix}} \underset{\underline{\underline{^e F}}}{\begin{Bmatrix} F_{Xa} \\ F_{Ya} \\ F_{Xb} \\ F_{Yb} \end{Bmatrix}} \tag{4.7.9}$$

Using (4.7.9) in the right hand side of (4.7.8):

$$\frac{EA}{H}\begin{bmatrix} 1 & 0 & -1 & 0 \\ 0 & 0 & 0 & 0 \\ -1 & 0 & 1 & 0 \\ 0 & 0 & 0 & 0 \end{bmatrix}\begin{bmatrix} c & s & 0 & 0 \\ -s & c & 0 & 0 \\ 0 & 0 & c & s \\ 0 & 0 & -s & c \end{bmatrix}\begin{Bmatrix} U_a \\ V_a \\ U_b \\ V_b \end{Bmatrix} = \begin{bmatrix} c & s & 0 & 0 \\ -s & c & 0 & 0 \\ 0 & 0 & c & s \\ 0 & 0 & -s & c \end{bmatrix}\begin{Bmatrix} F_{Xa} \\ F_{Ya} \\ F_{Xb} \\ F_{Yb} \end{Bmatrix} \tag{4.7.10}$$

The equation for the above is expressed as:

$$\underline{\underline{^e k T D}} = \underline{\underline{T F}} \tag{4.7.11}$$

Where:

$$\underline{\underline{^e k}} = \frac{EA}{H}\begin{bmatrix} 1 & 0 & -1 & 0 \\ 0 & 0 & 0 & 0 \\ -1 & 0 & 1 & 0 \\ 0 & 0 & 0 & 0 \end{bmatrix} \qquad \underline{\underline{T}} = \begin{bmatrix} c & s & 0 & 0 \\ -s & c & 0 & 0 \\ 0 & 0 & c & s \\ 0 & 0 & -s & c \end{bmatrix} \tag{4.7.12}$$

As it stands, (4.7.10) provides the desired means of using global force and displacement variables while allowing the element to resist loads resolved into the local x-coordinate axis only. However, there is a more convenient way to accomplish the same, using a slightly different computational procedure.

Instead of operating on both the global displacement and force vectors, using the transformation matrix, it is easier to simply transform the local stiffness matrix. Multiplying both sides of (4.7.11) by the inverse of the transformation matrix, and using the identity matrix:

$$\left(\underline{\underline{T}}\right)^{-1} \underline{\underline{^e k T D}} = \underline{\underline{F}} \tag{4.7.13}$$

Since the transformation matrix is orthogonal, its inverse is equal to its transpose (Bathe [13, p. 35]), such that (4.7.13) is more easily expressed as:

$$\underline{\underline{T}}^T \, \underline{\underline{^e k T D}} = \underline{\underline{F}} \tag{4.7.14}$$

In general, an element stiffness matrix may be conveniently established in a local coordinate system, then operated on by a transformation matrix to render a stiffness matrix that can be used with global force and displacement variables. Referring to (4.7.14), stiffness matrix transformation is expressed as:

$$\underline{\underline{^e K}} = \underline{\underline{T}}^T \, \underline{\underline{^e k T}} \tag{4.7.15}$$

For the specific case of the 2-node line element for truss applications, the global stiffness matrix as computed by (4.7.15) appears as:

$$\underline{\underline{^e K}} = \begin{bmatrix} c & -s & 0 & 0 \\ s & c & 0 & 0 \\ 0 & 0 & c & -s \\ 0 & 0 & s & c \end{bmatrix} \frac{EA}{H} \begin{bmatrix} 1 & 0 & -1 & 0 \\ 0 & 0 & 0 & 0 \\ -1 & 0 & 1 & 0 \\ 0 & 0 & 0 & 0 \end{bmatrix} \begin{bmatrix} c & s & 0 & 0 \\ -s & c & 0 & 0 \\ 0 & 0 & c & s \\ 0 & 0 & -s & c \end{bmatrix} \tag{4.7.16}$$

Forces and displacements defined in the global coordinate system can be used with the stiffness matrix above, while the loads are resisted in the local x-coordinate direction only. It is important to note that both the displacement interpolation function and the stress vector are expressed in the local coordinate system:

$$\tilde{u}(x) = N_1 u_a + N_2 u_b \qquad N_1 = \frac{x_b - x}{H} \qquad N_2 = \frac{x - x_a}{H}$$

$$\underline{\underline{\tau}} = \{\tau_{xx}\} \qquad \tau_{xx} = E\frac{d\tilde{u}}{dx}$$

The element associated with the equations above is shown in Figure 4.54.

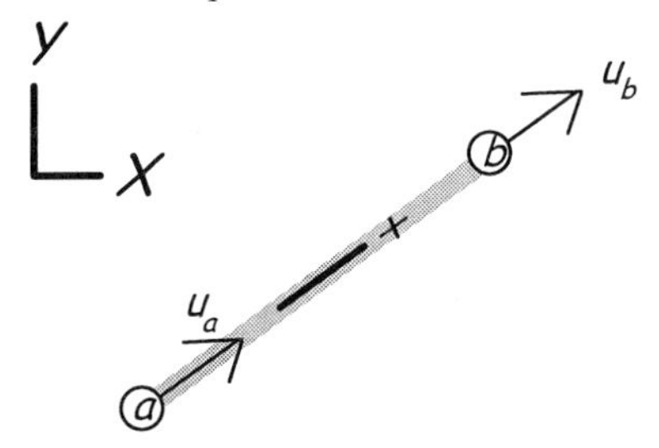

Figure 4.54 A 2-Node Line Element for 2-D Truss Applications

The element equilibrium equations are, as always:

$$\underline{\underline{{}^eK}}\,\underline{\underline{{}^eD}} = \underline{\underline{{}^eF}} \tag{4.7.17}$$

The element stiffness matrix used in (4.7.17) is given by (4.7.16); the displacement and force vectors are as shown on the right side of (4.7.7) and (4.7.9), respectively. After the global displacements are computed using (4.7.17), the local displacements can be found using (4.7.7); values of stress and strain in the local coordinate system can then be computed.

Summary

Several different types of elements have been considered in this chapter. However, there are many more elements for structural analysis, each with its own particular characteristics. The purpose of Chapter 4 was not to review every type of element available for structural and stress analysis, but to give the reader a brief look at some of the more common elements that are currently used, or were used in the past.

Although the elements in Chapter 4 have been considered separately, recall from Section 2.3 that finite elements are joined by adding together the corresponding stiffness from each element to form the global stiffness matrix.

References

1. MacNeal, R.H., *Finite Elements: Their Design And Performance*, Marcel Dekker, Inc., N.Y., 1994
2. Ugural, A.C., Fenster, S.K., *Advanced Strength And Applied Elasticity*, Elsevier, N.Y., 1987
3. Dvorkin, E.N., Bathe, K., "A continuum mechanics based four-node shell element for general non-linear analysis," Eng. Comput., Vol. 1, p. 78, 1984
4. Melosh, R.J., Lobitz, D.W., "On a Numerical Sufficiency Test for Monotonic Convergence of Finite Element Models," AIAA Journal, Vol. 12, No. 5, p. 675, 1975
5. Mikhilin, S.G., "The Problem Of The Minimum Of A Quadratic Functional," Holden Day, San Francisco, 1965
6. Oliveira, E.R. Arantes, "Theoretical Foundations Of The Finite Element Method," Int. J. Solid Structures, Vol. 4, pp. 929–952, 1968
7. Felippa, C.A., Clough, R.W., "The Finite Element Method In Solid Mechanics," in Proceedings of SIAM-AMS, Vol. 2, American Mathematics Society, 1970
8. Verma, A., Melosh, R.J., "Numerical Tests For Assessing Finite Element Model Convergence," Int. J. Numer. Methods In Engineering, pp. 843–857, 1987
9. Melosh, R.J., "Principles For Design of Finite Element Meshes," in "State-of-the-Art Surveys on Finite Element Analysis Technology," ASME, Chapter 3, 1983
10. Key, S.W., "A Convergence Investigation of the Direct Stiffness Method," PhD Thesis, University Of Washington, 1966
11. Ashwell, D.G., Gallagher, R.H., *Finite Elements for Thin Shells and Curved Members*, John Wiley & Sons, N.Y., 1976
12. Haisler, W.E., Stricklin, J.A., "Rigid Body Displacements of Curved Elements in the Analysis of Shells by the Matrix Displacement Method," AIAA Journal, Vol. 5, No. 8, pp. 1525–1527, 1967

13. Bathe, K., *Finite Element Procedures In Engineering Analysis*, Prentice-Hall, Inc., Englewood Cliffs, New Jersey, 1982
14. Cook, R.D., Malkus, D.S., and Plesha, M.E., *Concepts and Applications of Finite Element Analysis*, 3rd Edition, John Wiley & Sons, N.Y., 1989
15. MSC, "Practical Finite Element Modeling And Techniques Using MSC/NASTRAN," NA/1/000/PMSN, The MacNeal-Schwendler Corporation, Los Angles
16. Wilson, E.L., Taylor, R.L., Doherty, W.P., and Ghaboussi, J., "Incompatible Displacement Models," Numerical and Computer Methods in Strcutural Mechanics, Academica Press, N.Y., pp. 43–57, 1973
17. Simo, J.C., Rifai, M.S., "A Class of Assumed Strain Methods and the Method of Incompatible Modes," International Journal for Numerical Methods in Engineering, Vol. 29, pp. 1595–1638, 1990
18. Irons, B.M., "Numerical Integration Applied to Finite Element Methods," Conference on the Use of Digital Computers in Structural Engineering, University of Newcastle, England, 1966
19. Bazeley, G.P., Cheung, Y.K., Irons, B.M., and Zienkiewicz, O.C., "Triangular Elements in Bending: Conforming and Nonconforming Solutions," Proceeding of the Conference on Matrix Methods in Structural Mechanics, Air Force Institute of Technologoy, Wright Patterson Air Force Base, Ohio, pp. 547–576, 1965
20. Irons, B.M., Razzaque, A., "Experience With The Patch Test for Convergence of Finite Elements," in "The Mathematical Foundations of the Finite Element Method with Appilcations to Partial Differential Equations," Aziz, A.K., ed., Academic Press, pp. 557–587, 1972
21. MacNeal, R.H., Harder, R.L., "A Proposed Standard Set of Problems to Test Finite Element Accuracy," Finite Element Analysis & Design, Vol. 1, pp. 3–20, 1985
22. Beardsley, M.C., ed., *The European Philosophers From Descartes To Nietzsche*, Modern Library Edition, Random House, Inc., N.Y., 1992
23. Huebner, K.H., *The Finite Element Method For Engineers*, John Wiley & Sons, N.Y., 1982
24. Fisher, R.C., Ziebur, A.D., *Calculus and Analytic Geometry*, Prentice-Hall, Englewood Cliffs, NJ, 1975
25. Selby, S.M., ed., *CRC Standard Mathematical Tables*, Twentieth Edition, The Chemical Rubber Co., Cleveland, OH, 1972
26. Hoole, S.R.H., *Computer-Aided Analysis And Design Of Electromagnetic Devices*, Elsevier Science Publishing Co., 1989
27. Higdon, A., *Mechanics of Materials*, 4th Edition, John Wiley & Sons, N.Y., 1985

28. Thomas, D.L., Wilson, J.M., Wilson, R.R., "Timoshenko Beam Finite Elements," Journal of Sound and Vibration, Vol. 31, No. 3, p. 315–330, 1973
29. Huang, T.C., "The Effect of Rotatory Inertia and of Shear Deformation on the Frequency and Normal Mode Equations of Uniform Beams With Simple End Conditions," Journal of Applied Mechanics, p. 579–583, 1961
30. Timoshenko, S.P., *Theory of Elasticity*, McGraw-Hill, N.Y., 1987
31. Young, W.C., *Roark's Formulas for Stress & Strain*, 6th Edition, McGraw-Hill, N.Y., 1989
32. Morley, L.S.D., Morris, A.J., "Conflict Between Finite Elements and Shell Theory," Royal Aircraft Establishment, Controller HMSO, London, 1978
33. Koiter, W.T., "A Consistent First Approximation in the General Theory of Thin Elastic Shells," in Proceedings of the Symposium on The Theory Of Thin Elastic Shells, North Holland Publishing Company, Amsterdam, pp. 12–33, 1960
34. Schwab, C., "Mesh Design for Structural Plates," in ESRD Technical Brief, Engineering Software Research & Development, Inc., St. Louis, 1995
35. Cheung, Y.K., King, I.P., and Zienkiewicz, O.C., "Slab Bridges With Arbitrary Shape and Support Condition - A General Method of Analysis Based on Finite Elements," Proc. Inst. Civ. Eng., Vol. 40, pp. 9–36, 1968
36. Timoshenko, S.P., *Theory of Plates and Shells,* McGraw-Hill, Inc., N.Y., reissued 1987
37. Love, A.E.H., *A Treatise On The Mathematical Theory Of Elasticity*, 4th Edition, Dover Publications, N.Y., 1987

Chapter 5

Parametric Elements

"One of the more persistent delusions of mankind is that some sections of the human race are morally better or worse than others."
—Betrand Russell, on the "superior virtue" of the oppressed

5.1 Introduction to Parametric Elements

Key Concept: Parametric elements are helpful in maintaining compatibility, deriving shape functions, and performing numerical integration.

There are several reasons why parametric elements are used in most commercially available finite element software packages. Some of the advantages that will be covered in this chapter are:

- Aiding the process of shape function derivation
- Allowing elements of arbitrary shape and orientation to maintain compatibility
- Aiding in numerical integration

Shape Function Generation and Parametric Elements
There are several ways to develop shape functions for finite elements:

- Perform mathematical procedures
- Apply a known type of interpolation
- Inspection

The mathematical approach to developing shape functions has been used in this text since it is more straightforward, albeit more tedious; we will continue to use the mathematical approach.

As shown in Chapter 4, shape functions for elements with more than three nodal variables are difficult to generate using the mathematical approach, at least for the general case. To circumvent this problem, the shape functions for some elements are not expressed explicitly, as illustrated by the 3-node plate element from Chapter 4. In such a case, the shape functions are left in matrix form, and the element stiffness matrix is generated using additional matrix operations. The increase in computational time for additional matrix manipulations is not considered overly burdensome; the loss of accuracy during the inversion operation may be of more concern.

Another way to deal with shape functions for more complicated elements is to choose spatial coordinates such that matrix manipulation is greatly simplified. This technique was used for the 4-node tetrahedral element in Chapter 4. Although the choice of spatial coordinates allowed the shape functions of the tetrahedral element to be expressed in simple terms, one cannot, as a general rule, ensure that the coordinates of all nodes within a finite element model have simple coordinates. One can, however, define a *local* coordinate system for each element's spatial coordinates, generate shape functions in simple terms, and then later provide a transformation between global and local entities. This is, in essence, the technique used to develop parametric elements, where the key feature is a local coordinate system using *natural coordinates*. In many cases, using natural coordinates simplifies the process of generating shape functions to such an extent that shape functions can be established by inspection.

Natural Coordinates

The key to parametric elements is a local coordinate system in which points are defined using natural coordinates. The natural coordinates at a parametric element's nodes have a magnitude of either zero or unity, as illustrated by Figure 5.1, below.

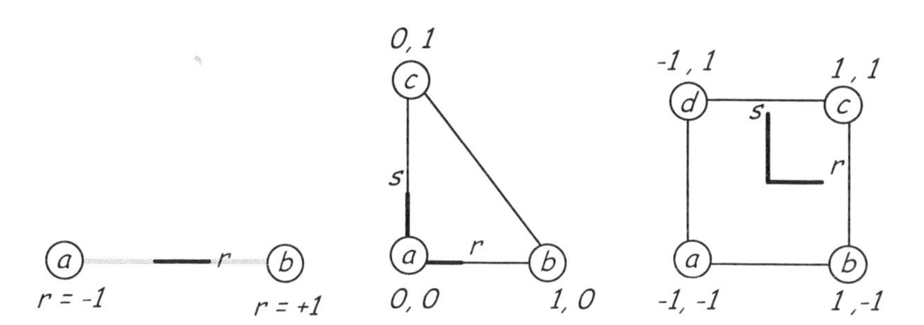

Figure 5.1 Natural Coordinates

Although only three elements are shown, other parametric elements, using similar natural coordinate systems, are commonly used. Parametric elements can be established for all the elements shown in Table 4.1, if desired. It will be shown that the natural coordinate system is used for the spatial coordinates of the element only—the force and displacement vectors are referenced to either the global coordinate system or a local coordinate system defined by the analyst.

A Simple Parametric Element

To illustrate the parametric principle, the 2-node rod element that was developed in Chapter 2 will be considered in parametric form. This is done for illustrative purposes; the 2-node rod element is simple enough so that there is little to profit from parameterization.

An element in parametric space is called a *parent element.* The parent element might be considered an "ideal element" that exists only as a model for "real elements" of the same type. The parent element is ideal because it:

- Has spatial coordinates that allow shape functions to be developed easily
- Represents a non-distorted element
- Facilitates numerical integration

Although the simple 2-node parametric rod element can be used to illustrate some basic principles, the nature of all of the items above will not be revealed until more complex elements are considered.

The "real" elements exist in global space and their spatial coordinates reflect the true size and shape of the element. In contrast, the parent element for a given type of element always has the same coordinates and is always the same size. Observe Figure 5.2, below.

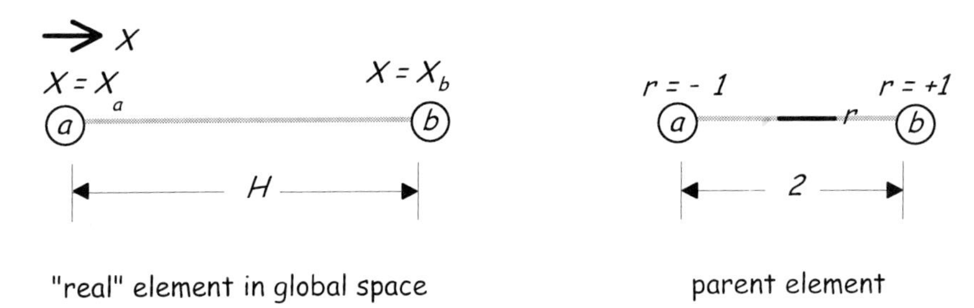

Figure 5.2 A 2-Node Rod Element, Global and Parametric Space

Another way of thinking of a parametric element is that it exists in global space, with each point on the element corresponding to a unique point on the parent element. Hence, the global coordinates can be superimposed or "mapped" to the parent element, as illustrated in Figure 5.3.

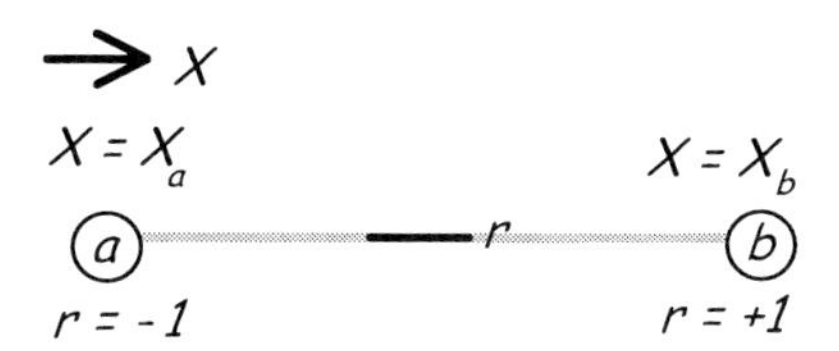

Figure 5.3 Global Element Superimposed on Parent Element

To map the global element to the parent, coordinate values of the element in global space are associated with (*constrained to*) respective coordinate values on the parent. In the case of the 2-node rod element, this means that X-coordinate values are constrained to certain values of r on the parent element.

Parametric mapping in finite elements can be either linear or non-linear, depending upon how many nodes the element has. Since the rod element currently under consideration has only two nodes, mapping is limited to a linear function, for instance:

$$X(r) = a_1 + a_2 r \tag{5.1.1}$$

Analogous to the process used to establish displacement interpolation functions, the mapping function is constrained to certain values at the nodes of the element. Since the left end of the global element maps to the left end of the parent element, we have, using (5.1.1):

$$X(-1) = a_1 + a_2(-1) = X_a \tag{5.1.2}$$

Likewise, the right end of the global element maps to the right end of the parent:

$$X(1) = a_1 + a_2(1) = X_b \tag{5.1.3}$$

The two equations above are expressed as:

$$\begin{aligned} a_1 + a_2(-1) &= X_a \\ a_1 + a_2(1) &= X_b \end{aligned} \tag{5.1.4}$$

In matrix form, (5.1.4) may be expressed as:

$$\begin{bmatrix} 1 & -1 \\ 1 & 1 \end{bmatrix} \begin{Bmatrix} a_1 \\ a_2 \end{Bmatrix} = \begin{Bmatrix} X_a \\ X_b \end{Bmatrix} \tag{5.1.5}$$

The system of equations above is easily solved for the unknown parameters:

$$a_1 = \frac{X_b + X_a}{2} \qquad a_2 = X_b - \frac{X_b + X_a}{2} \tag{5.1.6}$$

Substituting (5.1.6) into (5.1.1) and rearranging:

$$X(r) = \frac{1}{2}(1-r)X_a + \frac{1}{2}(1+r)X_b \tag{5.1.7}$$

Hence, the parametric shape functions for *any* 2-node rod element are:

$$N_1 = \frac{1}{2}(1-r) \qquad N_2 = \frac{1}{2}(1+r) \tag{5.1.8}$$

The term *shape function* comes from the parametric mapping process. It implies that the actual, global dimensions (and the associated shape) of an element are controlled by functions of parametric variables. In the present case, the global X-variable is controlled by the parameter r. The manner in which the shape of an element is controlled by parametric variables will become more clear when two-dimensional parametric elements are considered.

Displacement Interpolation for the 2-Node, Isoparametric Rod Element

One might notice that the shape functions for parametric mapping have the same characteristics required for the displacement interpolation functions. Figure 5.4 reviews the requirements for a one-dimensional displacement interpolation function with respect to parametric coordinates.

Figure 5.4 Displacement Interpolation

The parametric displacement interpolation associated with the element in Figure 5.4 can be expressed as:

$$\tilde{U}(r) = \frac{1}{2}(1-r)U_a + \frac{1}{2}(1+r)U_b \tag{5.1.9}$$

The equation above is formed by replacing the X values in (5.1.7) with U values. One can verify the validity of (5.1.9) quite simply: setting $r=-1$, $\tilde{U}=U_a$; with $r = 1$, $\tilde{U}=U_b$.

It is possible to use one class of shape functions for displacement interpolation and another class for coordinate mapping. For example, a 3-node rod element could use quadratic shape functions for mapping while using linear shape functions for displacement interpolation, or vice versa. However, this text will consider only elements that use the same parametric shape functions for both displacement and mapping functions, i.e., *isoparametric* elements.

Stiffness Matrix for a 2-Node, Isoparametric Rod Element

The parametric shape functions from (5.1.8) will be used to compute the B-matrix, from which the element stiffness matrix for a 2-node rod element can be generated. Recalling the equation to compute element stiffness, (3.5.12):

$$ {}^{e}\underline{\underline{K}} = \int_V \underline{B}^T \underline{\underline{E}}\,\underline{B}\, dV \tag{5.1.10}$$

The material matrix for the rod element is simply equal to the elastic modulus. Assuming that the modulus is constant along the length of the element, (5.1.10) may be expressed as:

$$ {}^{e}\underline{\underline{K}} = E\int_V \underline{B}^T \underline{B}\, dV \tag{5.1.11}$$

Now consider the strain-displacement matrix. The strain vector for the rod element contains just one component, normal strain in the axial direction:

$$ \underline{\varepsilon} = \varepsilon_{XX} \approx \frac{d\tilde{U}}{dX} \tag{5.1.12}$$

Substituting the approximate expression for displacement, (5.1.9), into (5.1.12):

$$ \varepsilon_{XX} \approx \frac{d\tilde{U}}{dX} = \frac{d}{dX}\left(\frac{1}{2}(1-r)U_a + \frac{1}{2}(1+r)U_b\right) \tag{5.1.13}$$

Notice that in (5.1.13) we wish to compute a derivative with respect to X, but this variable does not exist on the right hand side of the equation. A logical approach would be to employ the chain rule:

$$\frac{d\tilde{U}}{dX} = \frac{d\tilde{U}}{dr}\frac{dr}{dX} \tag{5.1.14}$$

The second factor on the right hand side of (5.1.14) requires the computation of the derivative of r with respect to X, which means an expression for the function $r(X)$ must exist. Although (5.1.7) provides an expression for $X(r)$, no explicit expression for the inverse function, $r(X)$, exists. This function could be established by inverting (5.1.7), however, the process of inverting functions is often a difficult task. Alternately, one could use the chain rule, in a backwards sort of way. First, the derivative of displacement with respect to r is computed:

$$\frac{d\tilde{U}}{dr} = \frac{dX}{dr}\frac{d\tilde{U}}{dX} \tag{5.1.15}$$

Second, manipulating (5.1.15), the derivative that describes normal strain is expressed as:

$$\frac{d\tilde{U}}{dX} = \left(\frac{dX}{dr}\right)^{-1}\frac{d\tilde{U}}{dr} \tag{5.1.16}$$

Notice that both derivatives on the right hand side of (5.1.16) can easily be computed without inverting a function. Instead, the derivative of (5.1.7) with respect to r is computed, and the result is inverted. That is:

$$\frac{dX}{dr} = \frac{d}{dr}\left(\frac{1}{2}(1-r)X_a + \frac{1}{2}(1+r)X_b\right) = \frac{X_b - X_a}{2} \tag{5.1.17}$$

from which:

$$\left(\frac{dX}{dr}\right)^{-1} = \frac{2}{X_b - X_a} \tag{5.1.18}$$

Again, this procedure is used because it is typically easier to compute a derivative and then invert the result than it is to invert a function and then take the derivative.

Notice that (5.1.17) computes the derivative of the mapping function, X, with respect to the parametric variable, r. This derivative is known as the *Jacobian*. For the 2-node rod element, the Jacobian is defined as:

$$J \equiv \frac{dX}{dr} \tag{5.1.19}$$

One way to view the Jacobian is that it "scales" parent space to render global space:

$$dX = \underset{\substack{\Uparrow \\ \textit{scale factor}}}{J}\, dr$$

The equation above states that incremental changes in distance along the parent element are multiplied by a scaling factor to yield corresponding changes along the global element. The Jacobian is also used to provide a measure of how much an isoparametric element is distorted. This topic will be discussed in more depth in Chapter 7.

To compute normal strain, we substitute (5.1.19) into (5.1.16):

$$\frac{d\tilde{U}}{dX} = \left(\frac{1}{J}\right)\frac{d\tilde{U}}{dr} \tag{5.1.20}$$

The inverse of the Jacobian serves to transform derivatives in the local coordinate system into derivatives in the global coordinate system. In (5.1.20), the derivative of $\tilde{U}$ with respect to r is transformed into a derivative with respect to X, via the Jacobian inverse. In one-dimensional problems, the Jacobian is a scalar, while in two- and three-dimensional problems, the Jacobian is a matrix. Substituting (5.1.9) into (5.1.20):

$$\frac{d\tilde{U}}{dX} = \left(\frac{1}{J}\right)\frac{d}{dr}\left(\frac{1}{2}(1-r)U_a + \frac{1}{2}(1+r)U_b\right) \tag{5.1.21}$$

or:

$$\frac{d\tilde{U}}{dX} = \left(\frac{1}{J}\right)\left(\frac{1}{2}(U_b - U_a)\right) \tag{5.1.22}$$

Since strain is expressed in terms of a strain displacement matrix and nodal displacement variables, (5.1.22) can be arranged as:

$$\underline{\underline{\varepsilon}} \approx \frac{d\tilde{U}}{dX} = \underline{\underline{B}}\,{}^{e}\underline{\underline{D}} = \left(\frac{1}{J}\right)\frac{1}{2}\begin{bmatrix}-1 & 1\end{bmatrix}\begin{Bmatrix}U_a \\ U_b\end{Bmatrix} \tag{5.1.23}$$

Hence, the strain-displacement matrix for this element is expressed as:

$$\underline{\underline{B}} = \left(\frac{1}{2J}\right)\begin{bmatrix}-1 & 1\end{bmatrix} \tag{5.1.24}$$

With the nodal displacement vector:

$${}^{e}\underline{\underline{D}} = \begin{Bmatrix}U_a \\ U_b\end{Bmatrix}$$

Substituting (5.1.24) into (5.1.11):

$${}^{e}\underline{\underline{K}} = E\int_V \left(\frac{1}{2J}\right)\begin{Bmatrix}-1 \\ 1\end{Bmatrix}\left(\frac{1}{2J}\right)\begin{bmatrix}-1 & 1\end{bmatrix} dV \tag{5.1.25}$$

Recall that all operations associated with the global element, including the integration process, are performed within the domain of the parent element. Since the parent element always has the same parametric coordinates, the limits of integration for any isoparametric element (of the same type) always have the same values, therefore (5.1.25) appears as:

$${}^{e}\underline{\underline{K}} = E\int_{r_a}^{r_b} \left(\frac{1}{2J}\right)\begin{Bmatrix}-1 \\ 1\end{Bmatrix}\left(\frac{1}{2J}\right)\begin{bmatrix}-1 & 1\end{bmatrix} dV \tag{5.1.26}$$

Since the limits of integration are on the r domain, the differential volume must be expressed in terms of r, also. Note that the volume of the element is expressed as:

$$V = AX$$

Assuming that the cross sectional area is constant, the differential volume can be expressed as:

$$dV = A\,dX \tag{5.1.27}$$

Referring to (5.1.19), it is apparent that $dX=Jdr$, such that the equation for differential volume can be expressed as:

$$dV = A\,dX = A\,Jdr \tag{5.1.28}$$

Using (5.1.28) in (5.1.26):

$$\underline{\underline{{}^{e}K}} = E\int_{r_a}^{r_b}\left(\frac{1}{2J}\right)\begin{Bmatrix}-1\\1\end{Bmatrix}\left(\frac{1}{2J}\right)\begin{bmatrix}-1 & 1\end{bmatrix}A\,Jdr \tag{5.1.29}$$

Rearranging (5.1.29) and substituting in the limits of integration:

$$\underline{\underline{{}^{e}K}} = EA\int_{-1}^{1}\left(\frac{1}{4J}\right)\begin{bmatrix}1 & -1\\-1 & 1\end{bmatrix}dr \tag{5.1.30}$$

Since the terms in the integrand are all constant, they can be withdrawn from the integral:

$$\underline{\underline{{}^{e}K}} = \left(\frac{EA}{4J}\right)\begin{bmatrix}1 & -1\\-1 & 1\end{bmatrix}\int_{-1}^{1}dr$$

Performing the integration above and simplifying:

$$\underline{\underline{{}^{e}K}} = \frac{EA}{2J}\begin{bmatrix}1 & -1\\-1 & 1\end{bmatrix} \tag{5.1.31}$$

Referring to (5.1.19) and (5.1.17):

$$J = \frac{X_b - X_a}{2} \tag{5.1.32}$$

It may be obvious that for a 2-node line element used for rod applications, the Jacobian is simply one-half the length of the element (in global space):

$$J = \frac{H}{2} \tag{5.1.33}$$

Substituting (5.1.33) into (5.1.31):

$$\underline{\underline{^eK}} = \frac{EA}{H}\begin{bmatrix} 1 & -1 \\ -1 & 1 \end{bmatrix} \tag{5.1.34}$$

Comparing (5.1.34) with the stiffness matrix given by (2.2.22) in Chapter 2, it is seen that they are, of course, identical.

For a 2-node rod element, there does not appear to be much advantage to using parametric elements. The real advantage of parametric elements becomes apparent in the more complicated elements that are designed for use in two- and three-dimensional problems.

5.2 Compatibility and Parametric Elements

Key Concept: Parametric elements aid in the process of generating shape functions while also providing a means of assuring inter-element compatibility.

In Section 5.1, it was stated that parametric elements allow shape functions to be generated more easily and also allow elements of various shapes and orientations to maintain compatibility. Using a 2-node rod element, the non-parametric shape functions are already quite simple, and since line elements join only at adjacent nodes, compatibility is easily maintained. However, both of these issues are more important in two and three-dimensional elements. Elementary issues related to shape function generation and compatibility will be considered in this section.

Orientation and Non-Parametric Elements

The 3-node triangular surface element for plane stress is not very robust, since it is limited in the modes of displacement that it can represent. Recall that the linear terms in this element's displacement assumptions can, at best, render constant strain. Hence, if the element is used in an area where strain is changing rapidly, many elements of this type will be needed to characterize the phenomenon.

To improve the performance of h-type finite elements, more boundary nodes are typically added.[1] The 4-node quadrilateral surface element may be considered to be an improvement upon the 3-node triangular surface element in that an additional node is employed, thereby allowing an additional term in the displacement assumption. As a result of the additional term, the 4-node element is considerably more robust than the 3-node triangular surface element. However,

---

[1] Some h-type elements, such as the 9-node quadrilateral surface element, gain increased accuracy by adding a 9th node at the interior of the element. Elements that are p-type increase their accuracy without explicitly adding nodes.

the 4-node element does have some inherent deficiencies, such as lack of compatibility, depending upon how it is oriented with respect to the global coordinate axis. Also, a 4-node, quadrilateral surface element is not as adept in areas of curved geometry when compared to the 3-node surface triangle.

To illustrate the compatibility issue associated with arbitrarily orientated elements, consider the non-parametric, 4-node, surface element shown in Figure 5.5. Notice that the edges of the element do not align with the global axes. Recall that to be compatible, displacement on any one edge of the element must be a function of nodal variables on that edge only. The following endeavors to show that some elements, for instance 4-node surface elements, do not exhibit compatibility when used in an arbitrary orientation.

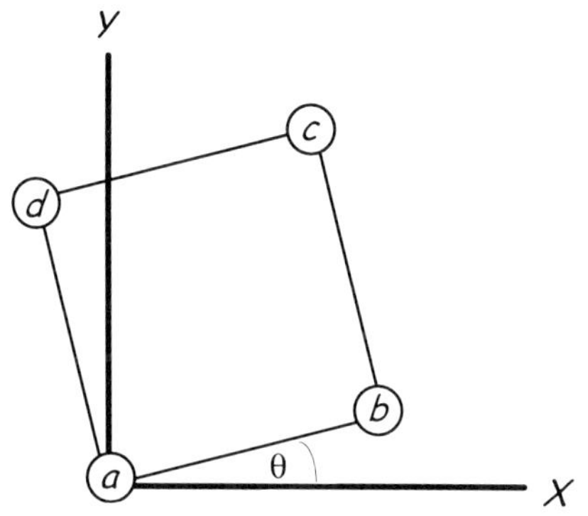

Figure 5.5 A 4-Node Surface Element, Arbitrary Orientation

The element from Figure 5.5 is shown again in Figure 5.6, this time illustrating that at each node there is one nodal variable related to displacement in the global *X*-coordinate direction:

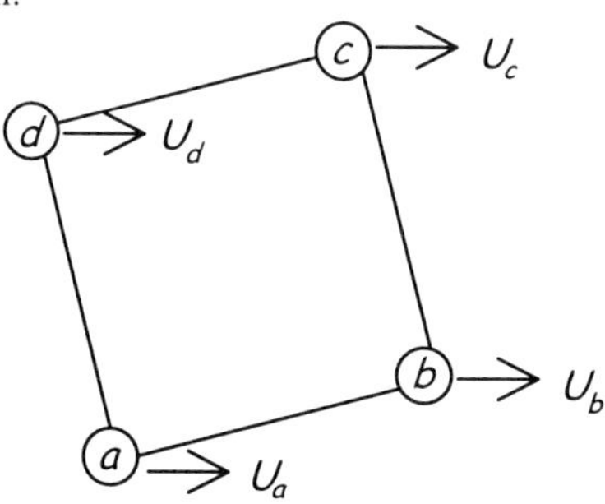

Figure 5.6 A 4-Node Surface Element, Nodal Displacement Variables

When used for plane stress or plane strain, the element in Figure 5.6 would also have nodal displacement variables in the Y-coordinate direction. However, this is not of concern at present.

With the four nodal displacement variables shown in Figure 5.6, a four-term displacement assumption can be employed. Referring to the triangle of Pascal, a four-term displacement assumption in two variables, that maintains geometric isotropy, takes the form:

$$\tilde{U}(X,Y) = a_1 + a_2 X + a_3 Y + a_4 XY \tag{5.2.1}$$

Since the element is not oriented with the global coordinate system, it is difficult to assess the compatibility of the element. Therefore, a local coordinate system that is aligned with the element edges is specified, as illustrated in Figure 5.7. Now, what is the expression for displacement along the a-b edge of the element in terms of the local x-coordinate?

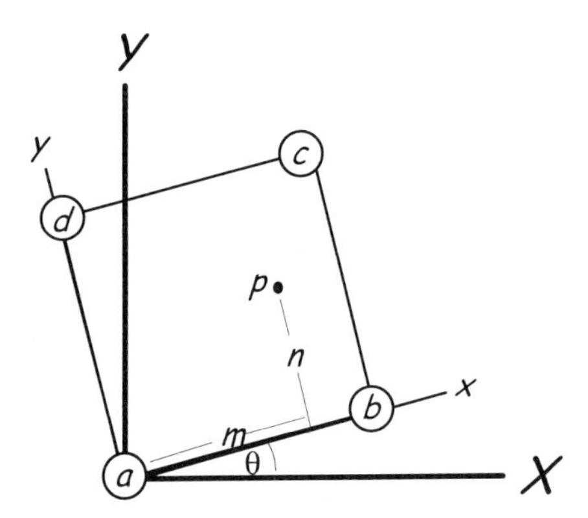

Figure 5.7 A 4-Node Surface Element, Arbitrary Point

To investigate, the displacement assumption given by (5.2.1) is modified to be expressed in terms of the local coordinate system. An arbitrary point p, defined in the local coordinate system by coordinates m and n, will have global coordinates that are determined by first projecting the lengths m and n onto the X-axis, then also projecting them onto the Y-axis. Projecting m and n onto the X-axis:

$$X = m\cos\theta + n\cos(90+\theta)$$

Projecting m and n onto the Y-axis:

$$Y = m\cos(90-\theta) + n\cos\theta$$

Using trigonometric relationships, the two equations above are more easily expressed as:

$$X = m\cos\theta - n\sin\theta \qquad Y = m\sin\theta + n\cos\theta \tag{5.2.2}$$

In general, any point in the local coordinate system, (x,y), will have the global coordinates:

$$X = x\cos\theta - y\sin\theta \qquad Y = x\sin\theta + y\cos\theta \tag{5.2.3}$$

Substituting (5.2.3) into (5.2.1):

$$\begin{aligned} \tilde{U}(x,y) = a_1 + a_2(x\cos\theta - y\sin\theta) + a_3(x\sin\theta + y\cos\theta) \\ + a_4(x\cos\theta - y\sin\theta)(x\sin\theta + y\cos\theta) \end{aligned} \tag{5.2.4}$$

Notice that when the local coordinate system aligns with the global system, angle theta in the equation above is zero, and the expression reverts back to the same form as (5.2.1):

$$\tilde{U}(x,y) = a_1 + a_2x + a_3y + a_4xy$$

Using algebra and rearranging (5.2.4):

$$\tilde{U}(x,y) = a_1 + b_1x + b_2y + b_3x^2 + b_4y^2 + b_5xy \tag{5.2.5}$$

The new constants in the equation above, b_i, are functions of the sine and cosine terms. Now, referring to Figure 5.7, it is seen that along Edge *a-b*, the local y-coordinate is zero therefore (5.2.5) becomes:

$$\tilde{U}(x,0) = a_1 + b_1x + b_3x^2$$

Note that the equation above, representing displacement along the *a-b* edge of the element, is a quadratic function of x. Three nodal variables related to $\tilde{U}$, along Edge *a-b,* are therefore required to allow this function to be compatible with adjacent elements. However, there are only two nodal variables along Edge *a-b,* therefore, $\tilde{U}$ can at most be a linear function. If the displacement assumption of an element is initially defined in terms of global variables, such as (5.2.1), the compatibility of the displacement assumption will depend upon its orientation, at least for the 4-node surface element.

Incompatibility Due to Element Orientation and Displacement Assumption

The element in Figure 5.7 is depicted as a rotated square. This shape was chosen to show that even when an element is well formed, i.e., not distorted, the element is incompatible. Incompatibility of displacement occurs in straight-sided non-isoparametric elements when two factors are present:

1. The element boundaries do not align with the global coordinates
2. The element's displacement assumption is not a complete polynomial

Surface elements with four and eight nodes have incomplete polynomials for displacement assumptions, hence, they are incompatible, in general. In contrast, straight-sided triangular surface elements, both 3- and 6-node versions, do not have C^0 compatibility problems because they employ a complete polynomial displacement assumptions. This issue is directly extended to three-dimensional elements: 4- and 10-node, straight-sided tetrahedral volume elements, having complete polynomial displacement assumptions, are compatible. Straight-sided 8- and 20-node hexahedral volume elements, having incomplete polynomial displacement assumptions, are not compatible.

What about curved sided elements? All curved sided, *non-parametric* elements are incompatible, as suggested by MacNeal [1, p. 99].

Isoparametric Element Compatibility

The discussion of incompatibility above concerns *non-parametric elements.* One reason for using parametric elements is that they can assist in maintaining compatibility. The key to C^0 compatibility in parametric elements is that their displacement interpolation functions are defined, at the outset, with respect to a local coordinate system (the natural coordinate system), and this coordinate system follows the boundaries of the element. Hence, it can be shown that the orientation of an isoparametric element with respect to the global coordinate system does not affect the element's ability to be compatible.

In global space, isoparametric elements may be distorted or even have curved boundaries, but whatever the element's configuration, the displacement interpolation function is defined *in terms of the natural coordinate variables.* Hence, regardless of the shape of the element, or its orientation with respect to the global axes, the displacement assumption can always be compatible on inter-element boundaries assuming that geometric isotropy is maintained and a suitable number of nodes are used on each edge. Again, compatibility of the displacement interpolation function requires that along any element edge, displacement is a function of only the nodal variables which are associated with that edge.

While compatibility is *one* requirement for monotonic convergence, maintaining compatibility alone does not ensure monotonic convergence. Furthermore, increased accuracy does not necessarily result from elements simply because they are compatible.

To illustrate how isoparametric elements can maintain compatibility, even with curved sides, consider the 8-node quadrilateral isoparametric surface element illustrated by Figure 5.8. The isoparametric coordinates are superimposed on the global element. The shape and size of the element in global space is defined by the global coordinates, X_i, Y_i, at each node; for clarity, only the global displacement variables at Node a are shown. Two things to notice: the displacement variables align with the global coordinate system (not the natural), and on any edge, there exist three displacement variables related to $\tilde{U}$, and three related to $\tilde{V}$. Referring to the triangle of Pascal, a geometrically isotropic displacement assumption for this element takes the form:

$$\tilde{U}(r,s) = a_1 + a_2 r + a_3 s + a_4 r^2 + a_5 rs + a_6 s^2 + a_7 r^2 s + a_8 rs^2 \tag{5.2.6}$$

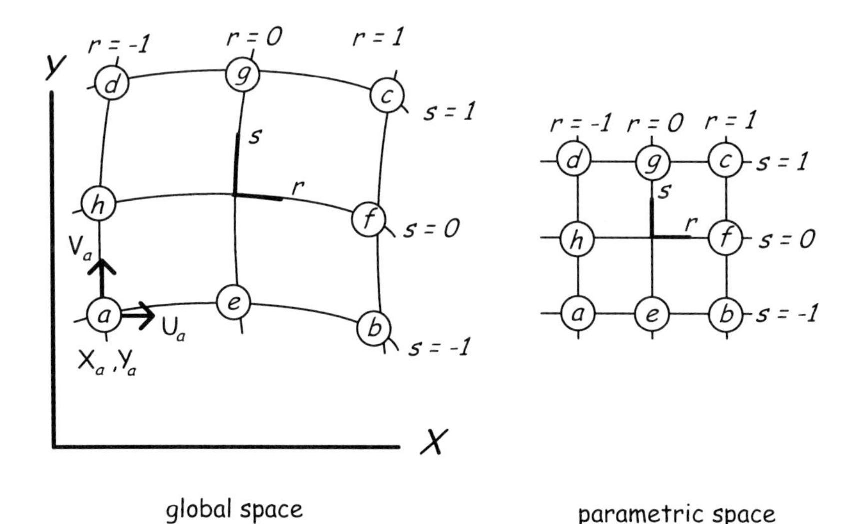

Figure 5.8 An 8-Node Isoparametric Surface Element

Now, consider a typical element edge, say the *a-e-b* edge; for this edge, the s coordinate is equal to −1. Substituting this value for s into (5.2.6):

$$\tilde{U}(r,-1) = a_1 + a_2 r + a_3(-1) + a_4 r^2 + a_5 r(-1) + a_6(-1)^2 + a_7 r^2(-1) + a_8 r(-1)^2 \qquad (5.2.7)$$

Simplifying (5.2.7):

$$\tilde{U}(r,-1) = b_1 + b_2 r + b_3 r^2 \qquad (5.2.8)$$

Notice that (5.2.8) is a quadratic function and, since there are three nodal variables related to $\tilde{U}$ along Edge *a-e-b*, there are enough variables to eliminate the three parameters, b_1, b_2, and b_3. The same characteristic is displayed on any edge, hence, this parametric element is C^0 compatible.

One advantage of isoparametric elements is that they maintain C^0 compatibility irrespective of the element's orientation, boundary shape (straight or curved) or completeness of the element's displacement assumption. If an isoparametric element does not employ a complete polynomial, its displacement assumption must be at least geometrically isotropic. An isoparametric element possesses the desirable characteristic of inter-element compatibility because its natural coordinate system conforms to the shape of the element's boundaries. Using the same procedure just illustrated, it is a simple exercise to show that the 4-node isoparametric surface element is also C^0 compatible because its displacement assumption is also defined with respect to a natural coordinate system that aligns with the element boundaries.

Summary of Compatibility Issues

Many of the elements used in today's general purpose finite element software packages are isoparametric. The following compares C^0 non-parametric elements with isoparametric elements of the same type, and assumes that polynomial displacement assumptions are employed.

Compatibility in Non-Parametric Elements with Straight Sides

If a complete polynomial displacement assumption is used in an element with *straight sides*, the element is C^0 compatible, regardless of the order of the displacement assumption. For example, 3- and 6-node triangular surface elements, along with 4- and 10-node tetrahedral volume elements, with straight sides, exhibit this behavior.

If an element with an incomplete displacement assumption is used, for instance a 4- or 8-node surface element, the element will be incompatible, even with straight sides, unless by chance the element boundaries align with the global coordinate axes. This implies that 4- and 8-node surface elements, to be compatible, must be rectangular and have edges that align with the global axes. Hexahedral volume elements ("bricks") with eight and 20 nodes, like the 4- and 8-node surface elements, also have incomplete polynomial displacement assumptions and therefore exhibit incompatible behavior, in general.

Compatibility in Non-Parametric Elements with Curved Sides
All non-parametric elements with curved sides are incompatible, regardless of the completeness of the displacement assumption or element orientation.

Compatibility and Isoparametric Elements
All isoparametric elements that use geometrically isotropic displacement assumptions are compatible for C^0 applications, regardless of their orientation or boundary curvature. Furthermore, it can be shown that if the parent element is compatible, its "offspring" in global space is also. One can examine a parent element in parametric space and determine if, on a given edge, displacement is a function of variables on the given edge only. If the parent element is compatible, the global element will be also.

Although isoparametric elements aid in allowing elements to maintain compatibility, higher order isoparametric elements with curved boundaries typically have difficulty representing quadratic displacement.

Representing Displacement in Isoparametric Elements
Isoparametric elements encounter difficulties when attempting to represent higher order displacement modes. The following summarizes how isoparametric elements represent (interpolate) displacement, and how this representation influences the element's performance.

Representing Linear Displacement
All properly designed isoparametric elements can correctly represent linear displacement.

Representing Quadratic Displacement
Isoparametric elements often have trouble representing quadratic displacement. However, with side nodes added, an isoparametric element can correctly represent quadratic displacement if the:

- Displacement assumption contains the correct quadratic terms
- Side nodes are placed mid-side
- Element boundaries are straight

In other words, the global element we seek is one which contains the correct quadratic terms and has the same shape as the parent element. How does one determine the "correct quadratic terms"? The most straightforward way is to first identify what terms are required for a complete, *linear* polynomial mapping assumption for an element of the same characteristic shape. If a complete polynomial cannot be used, a geometrically isotropic polynomial should be used instead. For example, consider that a quadratic displacement assumption is desired for an isoparametric, triangular surface element. First, the mapping assumption for a linear triangular surface element is examined, and the following three basis functions are identified:

$$1, r, s$$

The correct quadratic terms are found by forming the product of the basis functions and ignoring duplicate terms:

$$(1, r, s)(1, r, s) = 1, r, s, r, r^2, rs, s, sr, s^2 \tag{5.2.9}$$

Hence, the correct terms to be used in the displacement assumption for this element are:

$$1, r, s, r^2, rs, s^2 \tag{5.2.10}$$

To properly represent quadratic displacement in an isoparametric triangular surface element, the element must have straight sides, edge nodes placed at mid-side, and employ six nodes to accommodate the six basis functions in (5.2.10).

The procedure of establishing the "correct quadratic terms" above is derived from the fact that the basis functions used for quadratic displacement interpolation are contained in the squared terms of a complete, geometrically isotropic, *linear mapping assumption*; see MacNeal [1, p. 108]. For 6-node triangular and 10-node tetrahedral elements, it turns out that using a complete quadratic polynomial displacement assumption renders the "correct quadratic terms." However, 8-node quadrilateral surface elements and 20-node hexahedral volume elements are not quite as simple, since these elements cannot employ a complete quadratic polynomial. This leads to the question of what terms should be included in the displacement assumption to provide the "correct quadratic terms."

To determine the "correct quadratic terms" for a quadrilateral surface element, we again employ the procedure above. The first step is to determine what terms are required for a geometrically isotropic, *linear* mapping assumption for a quadrilateral element. Pascal's triangle suggests that four terms are required:

$$1, r, s, rs$$

Following the same procedure as before, we compute the squared terms from the basis functions, then disregard the duplicates, rendering the following *nine* terms:

$$1, r, s, r^2, rs, s^2, r^2 s, rs^2, r^2 s^2$$

To properly represent quadratic displacement in an *isoparametric* quadrilateral surface element, the element must have straight sides, with edge nodes placed at mid-side, and employ *nine* nodes to accommodate the nine basis functions above. The global element that meets these requirements is illustrated in the figure below:

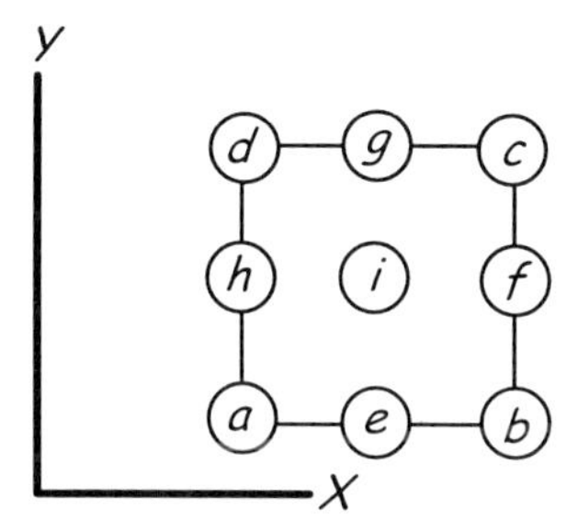

Although the displacement assumption for the 8-node quadrilateral, (5.2.6), is geometrically isotropic, an element using this displacement assumption will not properly represent quadratic displacement. This is true even with the aforementioned restrictions placed upon geometry, since only eight of the nine "correct" terms can be used. The 20-node hexahedral volume element also has too few nodes to include the "correct quadratic terms." Hence, neither of these elements can properly represent quadratic displacement even with straight edges and mid-side nodes.

The Consequences of Improperly Interpolating Quadratic Displacement

Some elements, Kirchhoff plate and Euler-Bernoulli beam elements for example, compute strain using second order derivatives. If isoparametric elements with curved boundaries and quadratic displacement assumptions are used for this type of element, constant strain will not be properly represented. The reason being that strain is computed using derivatives of displacement but the interpolation function that defines displacement within the element is not correct. In elements that compute strains using first order derivatives, linear strain will not be properly represented when quadratic displacement is improperly interpolated. Elements that do not interpolate displacement properly will typically not pass the patch test.

Accuracy and Isoparametric Elements

Parent elements for isoparametric elements that interpolate quadratic displacement always have the same characteristics: straight sides, and the side nodes placed at mid-side. To correctly interpolate quadratic displacement, the element in global space must have the same characteristics, such that the global element appears as simply a scaled version of the parent. In general, the more the global element deviates from the characteristics of the parent, the less accurate it is. Deviation of a global element from the parent element's characteristic shape is termed *distortion*. Distortion also renders an element more susceptible to other problems, such as locking.

Some elements are more sensitive than others to distortion. The topic of compatibility and properly representing strain states is well illustrated by MacNeal [1, p. 108]. Element distortion will be considered in the following section and in Chapter 7.

5.3 A 4-Node Isoparametric Surface Element

Key Concept: A 4-node surface element is somewhat more robust than a 3-node element of the same type, since it includes an additional term in the displacement assumption.

The three node triangular surface element has too few terms in its displacement assumption to exhibit good performance. While the six-node triangular surface element performs well (by virtue of its three additional nodes), models using these elements often end up with too many nodes where they are not needed, with the net effect of greatly increasing the computational cost of the analysis. The 4-node surface element is a very popular element because it represents a good compromise between accuracy and computational costs. A 4-node, isoparametric surface element is considered below. This element is one of a group of elements called "serendipity" elements; Ball [2].

4-Node, Isoparametric Surface Element, C^0 Applications—Shape Functions

Isoparametric elements of a given type always have the same parent in parametric space. In global space, an element assumes a shape which is specified by its global coordinates. As suggested by Figure 5.8, parent elements of a given type always appear the same, having straight and equal sides, even if higher order displacement assumptions are used and the element has curved boundaries in global space. Observe the 4-node isoparametric surface element in Figure 5.9; although global displacement variables exist at each node, for clarity, only those at Node a are shown.

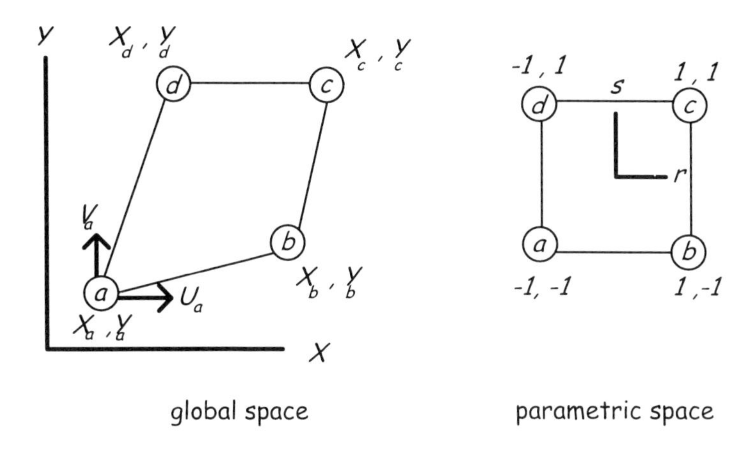

Figure 5.9 A 4-Node Isoparametric Surface Element

Again, we can visualize a parametric element in global space with the parametric coordinates superimposed, as shown in Figure 5.10.

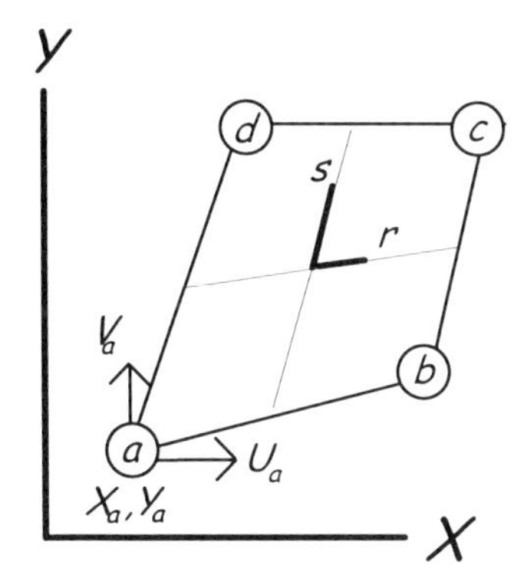

Figure 5.10 Isoparametric Element, Natural Coordinates Superimposed

As with the 2-node, isoparametric line element, the formulation for the 4-node, isoparametric element begins with a mapping function. A four-term mapping function can be used with the 4-node element:

$$X(r,s) = a_1 + a_2 r + a_3 s + a_4 rs \tag{5.3.1}$$

To map the global element to the parent, the mapping function from (5.3.1) is constrained to the nodal coordinate values:

$$\begin{aligned}
X(-1,-1) &= a_1 + a_2(-1) + a_3(-1) + a_4(-1)(-1) = X_a \\
X(\ 1,-1) &= a_1 + a_2(\ 1) + a_3(-1) + a_4(\ 1)(-1) = X_b \\
X(\ 1,\ 1) &= a_1 + a_2(\ 1) + a_3(\ 1) + a_4(\ 1)(\ 1) = X_c \\
X(-1,\ 1) &= a_1 + a_2(-1) + a_3(\ 1) + a_4(-1)(\ 1) = X_d
\end{aligned} \tag{5.3.2}$$

In matrix form, (5.3.2) is expressed as:

$$\begin{bmatrix} 1 & -1 & -1 & 1 \\ 1 & 1 & -1 & -1 \\ 1 & 1 & 1 & 1 \\ 1 & -1 & 1 & -1 \end{bmatrix} \begin{Bmatrix} a_1 \\ a_2 \\ a_3 \\ a_4 \end{Bmatrix} = \begin{Bmatrix} X_a \\ X_b \\ X_c \\ X_d \end{Bmatrix} \tag{5.3.3}$$

With some manipulation (or symbolic software) the equations above will render:

$$\begin{aligned}
a_1 &= \frac{1}{4}(\ X_a + X_b + X_c + X_d) \\
a_2 &= \frac{1}{4}(-X_a + X_b + X_c - X_d) \\
a_3 &= \frac{1}{4}(-X_a - X_b + X_c + X_d) \\
a_4 &= \frac{1}{4}(\ X_a - X_b + X_c - X_d)
\end{aligned} \tag{5.3.4}$$

Substituting (5.3.4) into (5.3.1) and rearranging:

$$X(r,s) = \left(\frac{1}{4}(1-r)(1-s)\right)X_a + \left(\frac{1}{4}(1+r)(1-s)\right)X_b + \left(\frac{1}{4}(1+r)(1+s)\right)X_c + \left(\frac{1}{4}(1-r)(1+s)\right)X_d$$

Hence, the shape functions for the 4-node, isoparametric surface element are:

$$N_1 = \frac{1}{4}(1-r)(1-s) \qquad N_2 = \frac{1}{4}(1+r)(1-s)$$
$$N_3 = \frac{1}{4}(1+r)(1+s) \qquad N_4 = \frac{1}{4}(1-r)(1+s) \tag{5.3.5}$$

The parent element for a quadrilateral is always square and straight-sided, with each side two units long. The shape functions control the *shape* of the element in global space, while the global coordinate variables at the nodes control the *size*. The mapping function for the Y-coordinate values can be found in the same manner, yielding:

$$Y(r,s) = N_1 Y_a + N_2 Y_b + N_3 Y_c + N_4 Y_d \tag{5.3.6}$$

Isoparametric elements use the same shape functions for mapping and displacement interpolation, hence, the displacement interpolation functions for the 4-node, isoparametric surface element take the form:

$$\begin{aligned} \tilde{U}(r,s) &= N_1 U_a + N_2 U_b + N_3 U_c + N_4 U_d \\ \tilde{V}(r,s) &= N_1 V_a + N_2 V_b + N_3 V_c + N_4 V_d \end{aligned} \tag{5.3.7}$$

B-Matrix for a 4-Node, Isoparametric, Plane Stress Element

To compute the element stiffness matrix for the 4-node, isoparametric element, a B-matrix must be specified. Recalling the strain terms for the plane stress idealization:

$$\underline{\underline{\varepsilon}} = \begin{Bmatrix} \varepsilon_{XX} \\ \varepsilon_{YY} \\ \gamma_{XY} \end{Bmatrix} \tag{5.3.8}$$

Beginning with the first strain term:

$$\varepsilon_{XX} \approx \frac{\partial \tilde{U}}{\partial X} \tag{5.3.9}$$

To compute normal strain in the X-coordinate direction, the interpolation function for $\tilde{U}$, given in (5.3.7), is substituted into (5.3.9):

$$\frac{\partial \tilde{U}}{\partial X} = \frac{\partial}{\partial X}\left(N_1 U_a + N_2 U_b + N_3 U_c + N_4 U_d\right) \tag{5.3.10}$$

Note that the same difficulty arises here as with the 2-node parametric rod element: a derivative with respect to X is desired but the shape functions, (5.3.5), are expressed in terms of parametric variables, r and s this case. Analogous to the 2-node isoparametric rod example, to compute the derivative of $\tilde{U}$ with respect to X the chain rule is employed in the following manner:

$$\frac{\partial \tilde{U}}{\partial r} = \frac{\partial X}{\partial r}\frac{\partial \tilde{U}}{\partial X} + \frac{\partial Y}{\partial r}\frac{\partial \tilde{U}}{\partial Y} \tag{5.3.11}$$

As compared to the one-dimensional rod element, (5.3.11) illustrates that obtaining the derivative of $\tilde{U}$ with respect to X requires a little more work in the two-dimensional case, since the shape functions contain two variables. Notice that the chain rule requires partial derivatives, and that the partial for $\tilde{U}$ with respect to r contains two terms, since in global space $\tilde{U}$ is a function of both the X- and Y-coordinate directions. In addition to (5.3.11), another equation is required to fully express normal strain:

$$\frac{\partial \tilde{U}}{\partial s} = \frac{\partial X}{\partial s}\frac{\partial \tilde{U}}{\partial X} + \frac{\partial Y}{\partial s}\frac{\partial \tilde{U}}{\partial Y} \tag{5.3.12}$$

In matrix form, (5.3.11) and (5.3.12) are expressed as:

$$\begin{Bmatrix} \dfrac{\partial \tilde{U}}{\partial r} \\ \dfrac{\partial \tilde{U}}{\partial s} \end{Bmatrix} = \begin{bmatrix} \dfrac{\partial X}{\partial r} & \dfrac{\partial Y}{\partial r} \\ \dfrac{\partial X}{\partial s} & \dfrac{\partial Y}{\partial s} \end{bmatrix} \begin{Bmatrix} \dfrac{\partial \tilde{U}}{\partial X} \\ \dfrac{\partial \tilde{U}}{\partial Y} \end{Bmatrix} \tag{5.3.13}$$

To obtain an expression for $\partial \tilde{U}/\partial X$, each side of (5.3.13) is pre-multiplied by the inverse of the square matrix:

$$\begin{bmatrix} \frac{\partial X}{\partial r} & \frac{\partial Y}{\partial r} \\ \frac{\partial X}{\partial s} & \frac{\partial Y}{\partial s} \end{bmatrix}^{-1} \begin{Bmatrix} \frac{\partial \tilde{U}}{\partial r} \\ \frac{\partial \tilde{U}}{\partial s} \end{Bmatrix} = \begin{bmatrix} \frac{\partial X}{\partial r} & \frac{\partial Y}{\partial r} \\ \frac{\partial X}{\partial s} & \frac{\partial Y}{\partial s} \end{bmatrix}^{-1} \begin{bmatrix} \frac{\partial X}{\partial r} & \frac{\partial Y}{\partial r} \\ \frac{\partial X}{\partial s} & \frac{\partial Y}{\partial s} \end{bmatrix} \begin{Bmatrix} \frac{\partial \tilde{U}}{\partial X} \\ \frac{\partial \tilde{U}}{\partial Y} \end{Bmatrix} \tag{5.3.14}$$

Recalling that a matrix multiplied by its inverse yields the identity matrix, (5.3.14) can be simplified and then rearranged to render:

$$\begin{Bmatrix} \frac{\partial \tilde{U}}{\partial X} \\ \frac{\partial \tilde{U}}{\partial Y} \end{Bmatrix} = \begin{bmatrix} \frac{\partial X}{\partial r} & \frac{\partial Y}{\partial r} \\ \frac{\partial X}{\partial s} & \frac{\partial Y}{\partial s} \end{bmatrix}^{-1} \begin{Bmatrix} \frac{\partial \tilde{U}}{\partial r} \\ \frac{\partial \tilde{U}}{\partial s} \end{Bmatrix} \tag{5.3.15}$$

The transformation between derivatives with respect to *global coordinates* and derivatives with respect to *local coordinates* is again achieved through the inverse of the *Jacobian*.

For two- and three-dimensional elements the Jacobian transformation takes the form of a matrix instead of a scalar, as was the case for the 2-node isoparametric line element for rod applications. The Jacobian matrix for elements in two dimensional space is a *2 by 2* matrix, as shown below:

$$\underline{\underline{J}} = \begin{bmatrix} J_{11} & J_{12} \\ J_{21} & J_{22} \end{bmatrix} = \begin{bmatrix} \frac{\partial X}{\partial r} & \frac{\partial Y}{\partial r} \\ \frac{\partial X}{\partial s} & \frac{\partial Y}{\partial s} \end{bmatrix} \tag{5.3.16}$$

Using the notation given by (5.3.16), it is convenient to express (5.3.15) as:

$$\begin{Bmatrix} \frac{\partial \tilde{U}}{\partial X} \\ \frac{\partial \tilde{U}}{\partial Y} \end{Bmatrix} = \underline{\underline{J}}^{-1} \begin{Bmatrix} \frac{\partial \tilde{U}}{\partial r} \\ \frac{\partial \tilde{U}}{\partial s} \end{Bmatrix} \tag{5.3.17}$$

Using elementary matrix manipulations, the inverse of (5.3.16) is expressed as:

$$\underline{\underline{J}}^{-1} = \frac{1}{\det \underline{\underline{J}}} \begin{bmatrix} J_{22} & -J_{12} \\ -J_{21} & J_{11} \end{bmatrix}$$

Using the above in (5.3.17):

$$\begin{Bmatrix} \frac{\partial \tilde{U}}{\partial X} \\ \frac{\partial \tilde{U}}{\partial Y} \end{Bmatrix} = \frac{1}{\det \underline{\underline{J}}} \begin{bmatrix} J_{22} & -J_{12} \\ -J_{21} & J_{11} \end{bmatrix} \begin{Bmatrix} \frac{\partial \tilde{U}}{\partial r} \\ \frac{\partial \tilde{U}}{\partial s} \end{Bmatrix}$$

Performing the matrix algebra above:

$$\frac{\partial \tilde{U}}{\partial X} = \frac{1}{\det \underline{\underline{J}}}\left(J_{22}\frac{\partial \tilde{U}}{\partial r} - J_{12}\frac{\partial \tilde{U}}{\partial s}\right) \qquad \frac{\partial \tilde{U}}{\partial Y} = \frac{1}{\det \underline{\underline{J}}}\left(-J_{21}\frac{\partial \tilde{U}}{\partial r} + J_{11}\frac{\partial \tilde{U}}{\partial s}\right) \tag{5.3.18}$$

Likewise, for derivatives of $\tilde{V}$:

$$\begin{Bmatrix} \frac{\partial \tilde{V}}{\partial r} \\ \frac{\partial \tilde{V}}{\partial s} \end{Bmatrix} = \begin{bmatrix} \frac{\partial X}{\partial r} & \frac{\partial Y}{\partial r} \\ \frac{\partial X}{\partial s} & \frac{\partial Y}{\partial s} \end{bmatrix} \begin{Bmatrix} \frac{\partial \tilde{V}}{\partial X} \\ \frac{\partial \tilde{V}}{\partial Y} \end{Bmatrix} \quad \Rightarrow \quad \begin{Bmatrix} \frac{\partial \tilde{V}}{\partial X} \\ \frac{\partial \tilde{V}}{\partial Y} \end{Bmatrix} = \underline{\underline{J}}^{-1} \begin{Bmatrix} \frac{\partial \tilde{V}}{\partial r} \\ \frac{\partial \tilde{V}}{\partial s} \end{Bmatrix}$$

Performing the same steps used for derivatives of $\tilde{U}$, derivatives of $\tilde{V}$ are:

$$\frac{\partial \tilde{V}}{\partial X} = \frac{1}{\det \underline{\underline{J}}}\left(J_{22}\frac{\partial \tilde{V}}{\partial r} - J_{12}\frac{\partial \tilde{V}}{\partial s}\right) \qquad \frac{\partial \tilde{V}}{\partial Y} = \frac{1}{\det \underline{\underline{J}}}\left(-J_{21}\frac{\partial \tilde{V}}{\partial r} + J_{11}\frac{\partial \tilde{V}}{\partial s}\right) \tag{5.3.19}$$

The derivatives shown in (5.3.18) and (5.3.19) are used to express the strain vector in terms of a strain-displacement matrix, using the usual form:

$$\underline{\underline{\varepsilon}} = \underline{\underline{B}}\,{}^{e}\underline{\underline{D}}$$

Recall that the strain vector for plane stress contains three terms:

$$\underline{\underline{\varepsilon}} = \begin{Bmatrix} \varepsilon_{XX} \\ \varepsilon_{YY} \\ \gamma_{XY} \end{Bmatrix} \approx \begin{Bmatrix} \frac{\partial \tilde{U}}{\partial X} \\ \frac{\partial \tilde{V}}{\partial Y} \\ \frac{\partial \tilde{U}}{\partial Y} + \frac{\partial \tilde{V}}{\partial X} \end{Bmatrix} \tag{5.3.20}$$

Using (5.3.18) and (5.3.19), the right hand side of (5.3.20) can be expressed in terms of the strain-displacement matrix:

$$\begin{Bmatrix} \dfrac{\partial \tilde{U}}{\partial X} \\ \dfrac{\partial \tilde{V}}{\partial Y} \\ \dfrac{\partial \tilde{U}}{\partial Y} + \dfrac{\partial \tilde{V}}{\partial X} \end{Bmatrix} = \underline{\underline{B}}\, {}^{e}\underline{\underline{D}} \tag{5.3.21}$$

With:

$$\underline{\underline{B}} = \begin{bmatrix} B_{11} & 0 & B_{13} & 0 & B_{15} & 0 & B_{17} & 0 \\ 0 & B_{22} & 0 & B_{24} & 0 & B_{26} & 0 & B_{28} \\ B_{22} & B_{11} & B_{24} & B_{13} & B_{26} & B_{15} & B_{28} & B_{17} \end{bmatrix} \qquad {}^{e}\underline{\underline{D}} \equiv \begin{Bmatrix} U_a \\ V_a \\ U_b \\ V_b \\ \vdots \\ U_d \\ V_d \end{Bmatrix} \tag{5.3.22}$$

The terms in the strain-displacement matrix are expressed as:

$$\begin{aligned}
B_{11} &= \frac{1}{\det \underline{\underline{J}}}\left(J_{22}N_{1,r} - J_{12}N_{1,s}\right) & B_{22} &= \frac{1}{\det \underline{\underline{J}}}\left(-J_{21}N_{1,r} + J_{11}N_{1,s}\right) \\
B_{13} &= \frac{1}{\det \underline{\underline{J}}}\left(J_{22}N_{2,r} - J_{12}N_{2,s}\right) & B_{24} &= \frac{1}{\det \underline{\underline{J}}}\left(-J_{21}N_{2,r} + J_{11}N_{2,s}\right) \\
B_{15} &= \frac{1}{\det \underline{\underline{J}}}\left(J_{22}N_{3,r} - J_{12}N_{3,s}\right) & B_{26} &= \frac{1}{\det \underline{\underline{J}}}\left(-J_{21}N_{3,r} + J_{11}N_{3,s}\right) \\
B_{17} &= \frac{1}{\det \underline{\underline{J}}}\left(J_{22}N_{4,r} - J_{12}N_{4,s}\right) & B_{28} &= \frac{1}{\det \underline{\underline{J}}}\left(-J_{21}N_{4,r} + J_{11}N_{4,s}\right)
\end{aligned} \tag{5.3.23}$$

In (5.3.23), it is convenient to use the following notation for shape function derivatives:

$$N_{1,r} \equiv \frac{\partial N_1}{\partial r} \qquad N_{1,s} \equiv \frac{\partial N_1}{\partial s} \tag{5.3.24}$$

In addition, referring to (5.3.16), the determinant of the Jacobian shown in (5.3.23) is computed as:

$$\det \underline{\underline{J}} = J_{11}J_{22} - J_{12}J_{21} \tag{5.3.25}$$

Deriving the terms in (5.3.23) is a somewhat tedious task and is therefore shown in Appendix C. Although many terms are involved in the strain-displacement matrix, the computations are elementary. To gain an appreciation of the terms that are contained in the *B*-matrix, consider the first term:

$$B_{11} = \frac{1}{\det \underline{\underline{J}}}\left(J_{22}N_{1,r} - J_{12}N_{1,s}\right) \tag{5.3.26}$$

Referring to (5.3.16), the Jacobian component J_{22} is expressed as:

$$J_{22} = \frac{\partial Y}{\partial s} \tag{5.3.27}$$

Recalling the mapping function for the *Y*-coordinate:

$$Y(r,s) = N_1Y_a + N_2Y_b + N_3Y_c + N_4Y_d \tag{5.3.28}$$

Substituting (5.3.28) into (5.3.27):

$$J_{22} = \frac{\partial}{\partial s}\left(N_1Y_a + N_2Y_b + N_3Y_c + N_4Y_d\right) \tag{5.3.29}$$

Performing the differentiation above:

$$J_{22} = N_{1,s}Y_a + N_{2,s}Y_b + N_{3,s}Y_c + N_{4,s}Y_d \tag{5.3.30}$$

The J_{12} term in (5.3.26) is handled in a similar manner:

$$J_{12} = \frac{\partial Y}{\partial r} \tag{5.3.31}$$

Hence:

$$J_{12} = N_{1,r}Y_a + N_{2,r}Y_b + N_{3,r}Y_c + N_{4,r}Y_d \tag{5.3.32}$$

Using (5.3.30) and (5.3.32) in (5.3.26):

$$B_{11} = \frac{1}{\det \underline{\underline{J}}}\Big[\big(N_{1,s}Y_a + N_{2,s}Y_b + N_{3,s}Y_c + N_{4,s}Y_d\big)N_{1,r} - \big(N_{1,r}Y_a + N_{2,r}Y_b + N_{3,r}Y_c + N_{4,r}Y_d\big)N_{1,s}\Big] \tag{5.3.33}$$

The determinant of the J-matrix in (5.3.33) is computed using (5.3.25), (5.3.16), and derivatives of the shape functions, shown below, which are derived from (5.3.5). The derivatives computed in (5.3.34) can be substituted into (5.3.33) to obtain a function of the parametric variables r and s.

$$\begin{aligned} N_{1,s} &= \left(\frac{-1}{4}(1-r)\right) & N_{2,s} &= \left(\frac{-1}{4}(1+r)\right) \\ N_{3,s} &= \left(\frac{1}{4}(1+r)\right) & N_{4,s} &= \left(\frac{1}{4}(1-r)\right) \\ N_{1,r} &= \left(-\frac{1}{4}(1-s)\right) & N_{2,r} &= \left(\frac{1}{4}(1-s)\right) \\ N_{3,r} &= \left(\frac{1}{4}(1+s)\right) & N_{4,r} &= \left(-\frac{1}{4}(1+s)\right) \end{aligned} \tag{5.3.34}$$

Strain Representation in the 4-Node, Plane Stress Isoparametric Element

It is important to note that the variables r and s will appear in the B_{11} term (and in the other terms of the B-matrix as well) suggesting that strain need not be constant within this 4-node isoparametric element for plane stress. Hence, the 4-node surface element is somewhat more robust than the 3-node surface element when used for C^0 applications, since it can represent not only constant strain but some modes of linearly varying strain as well. The fact that the 4-node surface element can represent linearly varying strain may seem somewhat peculiar. Since the displacement is linear, and strain is defined in terms of first order derivatives, should not strain be constant? The answer is of course no.

Observe again the shape functions given by (5.3.5). They are not simply linear functions but are instead *bi-linear*, meaning linear in both r and s. Thus when partial derivatives are computed the resulting functions are still linear, but linear in only one variable, r or s, as shown in (5.3.34). So, we see that the bi-linear nature of the shape functions combined with the partial differentiation process renders a strain displacement matrix in terms of both r and s, as suggested by substituting (5.3.34) into (5.3.33). This contrasts with the 3-node triangular surface element for C^0 applications, which also has a linear displacement assumption. However, the displacement interpolation for the 3-node element is not *bi-linear*, hence, the strain-displacement matrix contains only constants. The 3-node triangular surface element will be considered in the next section.

The Jacobian Determinant and Strain-Displacement Matrix

If the determinant of the Jacobian is constant, each term in the strain-displacement matrix is simply expressed as a linear polynomial function of r and s. However, what happens if the determinant of the Jacobian is not constant but is instead a polynomial? In such a case each term of the strain-displacement matrix, for instance (5.3.33), is then described in terms of *a ratio of two polynomials*, and a ratio of two polynomials is not a polynomial, in general. Therefore, only when the determinant of the Jacobian is constant can strain in the 4-node element vary in a truly polynomial fashion.

Element Distortion and the Jacobian Determinant

It will be shown in Section 5.7 [Equation 5.7.25] that the Jacobian terms for the 4-node isoparametric element are constant only when certain combinations of the element's nodal coordinates sum to zero. Generally, the Jacobian terms (hence the Jacobian determinant) will not be constant for the 4-node isoparametric element, and this characteristic prevails in most other types of isoparametric elements as well.

There are two problems associated with a Jacobian determinant that is not constant. The first problem is that certain types of element distortion can cause the Jacobian determinant to vary in such a way that the strain distribution is influenced more by the shape of the element than by the shape of the structure! For instance, if element distortion causes the Jacobian determinant in (5.3.33) to assume a very small value in a certain area of the element, strain (hence stress) can exhibit asymptotic behavior. If this asymptotic behavior occurs in an area of the structure where a stress raiser is present, it may be difficult to determine if the predicted stress values are due to the stress raiser (a real affect) or due to problems with the Jacobian determinant (an artificial affect). This concept is well illustrated by Tenchev [3] and Haggenmacher [4].

The other problem brought about by a non-constant Jacobian determinant is that error is introduced when the element stiffness matrix is computed. Error is introduced because the integrand that is used to compute the element stiffness matrix contains strain-displacement matrix terms that cannot be expressed as a polynomial when the Jacobian is non-constant.[2] This issue will be considered again in Section 5.7 where the topic of numerical integration for isoparametric

---

[2] It is theoretically possible to contrive a surface element such that the thickness is made to vary in the same manner as the Jacobian determinant, such that the two effects cancel each other. This would allow the integrand to be expressed as a polynomial, even with a non-constant Jacobian determinant.

elements is discussed. Several types of element distortion are illustrated in Chapter 7.

Computing the Strain-Displacement Matrix

Considering the effort required to compute B_{11}, as given by (5.3.33), it is apparent that forming the entire B-matrix using hand calculations would be quite a task. However, this task is easily accomplished using a few lines of computer code. In Section 5.7, a FORTRAN computer program to calculate the B-matrix for a 4-node isoparametric surface element for plane stress applications is presented.

Computing the Element Stiffness Matrix

To form the element stiffness matrix, the B-matrix from (5.3.22) is substituted into the element stiffness matrix formula, (3.5.12):

$$
{}^{e}\underline{\underline{K}} = \int_{r_1}^{r_2}\int_{s_1}^{s_2} \begin{bmatrix} B_{11} & 0 & B_{22} \\ 0 & B_{22} & B_{11} \\ B_{13} & 0 & B_{24} \\ 0 & B_{24} & B_{13} \\ B_{15} & 0 & B_{26} \\ 0 & B_{26} & B_{15} \\ B_{17} & 0 & B_{28} \\ 0 & B_{28} & B_{17} \end{bmatrix} \underline{\underline{E}} \begin{bmatrix} B_{11} & 0 & B_{13} & \cdots & 0 \\ 0 & B_{22} & 0 & \cdots & B_{28} \\ B_{22} & B_{11} & B_{24} & \cdots & B_{17} \end{bmatrix} t\left(\det \underline{\underline{J}}\right) ds\, dr
$$

$$
\underline{\underline{E}} = \frac{E}{1-\nu^2} \begin{bmatrix} 1 & \nu & 0 \\ \nu & 1 & 0 \\ 0 & 0 & \dfrac{1-\nu}{2} \end{bmatrix}
$$

Performing the matrix multiplication and subsequent integration indicated above by manual methods is impractical, as with nearly all parametric elements. Since for the general case closed-form integration cannot be used, the element stiffness matrices are computed using numerical integration, a topic that is briefly considered in Section 5.7. Notice that the integration is performed on the *r-s* domain, as suggested by the limits on the integral above. The transformation from the global *X-Y* domain to the parametric *r-s* domain is shown in Appendix C.

Irons [5] and Taig [6] consider early work with isoparametric elements while a detailed review of interpolation functions for the 4-node quadrilateral element is found in Ball [2].

A 3-Node, Isoparametric, Triangular Surface Element

In the next section, a 3-node isoparametric element is developed using the same procedures that have been used so far. It should be noted that this element can also be generated by simply "collapsing" a 4-node, isoparametric surface element. In short, the collapsed element is formed by simply making two nodes of the 4-node element coincident. This technique can also be used for other isoparametric elements. For instance, an 8-node hexahedral volume element can be collapsed into a three-dimensional wedge element. Bathe [7, p. 220] provides a good illustration of this procedure.

5.4 A 3-Node Isoparametric Surface Element

Key Concept: The 3-node triangular surface element is typically avoided due to its poor performance; however, it does have the advantage of representing curved geometry more easily than the 4-node quadrilateral surface element.

When surface elements are required in a finite element model, analysts will often use a mesh of predominately 4-node quadrilaterals, while allowing (perhaps) a limited number of 3-node triangular elements. Even though the 3-node triangular surface element is not recommended in general, it is sometimes employed on a limited basis, since it can more easily represent curved geometry. In addition, "automatic" meshing algorithms are more easily designed for triangular elements than for quadrilaterals.

This section will briefly consider the 3-node, isoparametric surface element for C^0 applications. The 3-node, isoparametric element is developed in nearly the same manner as the 4-node isoparametric element that was considered in the previous section.

3-Node, Isoparametric Surface Element, C^0 Applications—Shape Functions

Observe the 3-node isoparametric surface element in Figure 5.11. As before, global displacement variables exist at each node, however, for clarity, only those at Node a are shown. As with the other isoparametric elements, the formulation for the 3-node element begins with a mapping function. For a 3-node element, a

three-term mapping function can be used:

$$X(r,s) = a_1 + a_2 r + a_3 s \tag{5.4.1}$$

To map the global element to the parent, the mapping function is constrained to the nodal values of the coordinates:

$$
\begin{aligned}
X(0,\ 0) &= a_1 + a_2(0) + a_3(0) = X_a \\
X(1,\ 0) &= a_1 + a_2(1) + a_3(0) = X_b \\
X(0,\ 1) &= a_1 + a_2(0) + a_3(1) = X_c
\end{aligned}
\tag{5.4.2}
$$

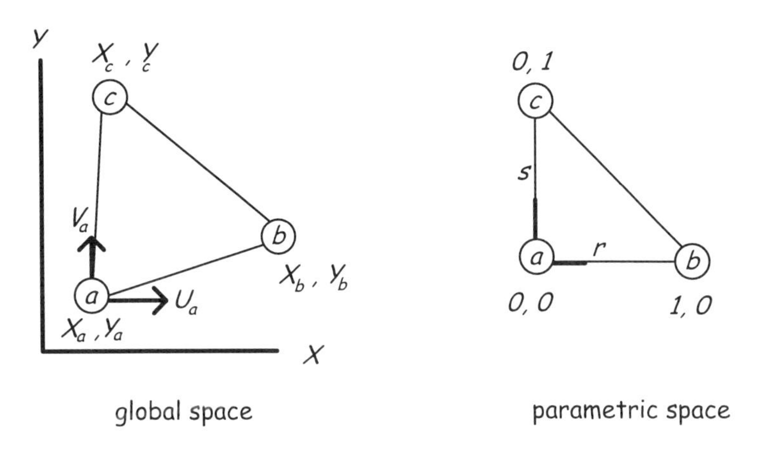

Figure 5.11 A 3-Node Isoparametric Surface Element

In matrix form, the equations above are expressed as:

$$
\begin{bmatrix} 1 & 0 & 0 \\ 1 & 1 & 0 \\ 1 & 0 & 1 \end{bmatrix}
\begin{Bmatrix} a_1 \\ a_2 \\ a_3 \end{Bmatrix} =
\begin{Bmatrix} X_a \\ X_b \\ X_c \end{Bmatrix}
\tag{5.4.3}
$$

With some manipulation, (5.4.3) renders:

$$
\begin{aligned}
a_1 &= X_a \\
a_2 &= X_b - X_a \\
a_3 &= X_c - X_a
\end{aligned}
\tag{5.4.4}
$$

Substituting (5.4.4) into (5.4.1) and rearranging:

$$
X(r,s) = (1 - r - s)X_a + (r)X_b + (s)X_c
\tag{5.4.5}
$$

Hence, the shape functions for the 3-node, isoparametric surface element are:

$$N_1 = (1 - r - s) \qquad N_2 = r \qquad N_3 = s \tag{5.4.6}$$

Again, the shape functions control the shape of the element in global space; the parent always has the same shape and orientation with respect to the natural coordinate system.

The mapping function for the Y-coordinate values can be found in the same manner as shown above, yielding:

$$Y(r,s) = N_1 Y_a + N_2 Y_b + N_3 Y_c \tag{5.4.7}$$

Using the shape functions from (5.4.6) for displacement interpolation:

$$\begin{aligned} \tilde{U}(r,s) &= N_1 U_a + N_2 U_b + N_3 U_c \\ \tilde{V}(r,s) &= N_1 V_a + N_2 V_b + N_3 V_c \end{aligned} \tag{5.4.8}$$

Equations 5.4.8 can be expressed in matrix form:

$$\begin{Bmatrix} \tilde{U}(r,s) \\ \tilde{V}(r,s) \end{Bmatrix} = \underline{\underline{N}}\, {}^{e}\underline{\underline{D}}$$

With:

$$\underline{\underline{N}} = \begin{bmatrix} N_1 & 0 & N_2 & 0 & N_3 & 0 \\ 0 & N_1 & 0 & N_2 & 0 & N_3 \end{bmatrix} \qquad {}^{e}\underline{\underline{D}} = \begin{Bmatrix} U_a \\ V_a \\ U_b \\ V_b \\ U_c \\ V_c \end{Bmatrix} \tag{5.4.9}$$

The shape function matrix will be used again in Chapter 6. With the shape functions and the associated displacement interpolation functions identified, the strain-displacement matrix can now be computed.

B-Matrix for a 3-Node, Isoparametric, Plane Stress Element
Recalling the strain terms for the plane stress idealization:

$$\underline{\underline{\varepsilon}} = \begin{Bmatrix} \varepsilon_{XX} \\ \varepsilon_{YY} \\ \gamma_{XY} \end{Bmatrix}$$

Beginning with the first strain term:

$$\varepsilon_{XX} \approx \frac{\partial \tilde{U}}{\partial X} \tag{5.4.10}$$

To compute normal strain in the X-coordinate direction, the interpolation function for $\tilde{U}$ from (5.4.8) is substituted into (5.4.10):

$$\frac{\partial \tilde{U}}{\partial X} = \frac{\partial}{\partial X}\left(N_1 U_a + N_2 U_b + N_3 U_c\right)$$

The chain rule is employed in the following manner:

$$\frac{\partial \tilde{U}}{\partial r} = \frac{\partial X}{\partial r}\frac{\partial \tilde{U}}{\partial X} + \frac{\partial Y}{\partial r}\frac{\partial \tilde{U}}{\partial Y} \tag{5.4.11}$$

Again, two equations are required to fully express normal strain in the X-coordinate direction; (5.4.11) is one equation, the other equation being:

$$\frac{\partial \tilde{U}}{\partial s} = \frac{\partial X}{\partial s}\frac{\partial \tilde{U}}{\partial X} + \frac{\partial Y}{\partial s}\frac{\partial \tilde{U}}{\partial Y} \tag{5.4.12}$$

In matrix form, (5.4.11) and (5.4.12) are expressed as:

$$\begin{Bmatrix} \dfrac{\partial \tilde{U}}{\partial r} \\ \dfrac{\partial \tilde{U}}{\partial s} \end{Bmatrix} = \begin{bmatrix} \dfrac{\partial X}{\partial r} & \dfrac{\partial Y}{\partial r} \\ \dfrac{\partial X}{\partial s} & \dfrac{\partial Y}{\partial s} \end{bmatrix} \begin{Bmatrix} \dfrac{\partial \tilde{U}}{\partial X} \\ \dfrac{\partial \tilde{U}}{\partial Y} \end{Bmatrix} \tag{5.4.13}$$

To obtain an expression for the strain terms, each side of (5.4.13) is pre-multiplied by the inverse of the square (Jacobian) matrix:

$$\begin{bmatrix} \frac{\partial X}{\partial r} & \frac{\partial Y}{\partial r} \\ \frac{\partial X}{\partial s} & \frac{\partial Y}{\partial s} \end{bmatrix}^{-1} \begin{Bmatrix} \frac{\partial \tilde{U}}{\partial r} \\ \frac{\partial \tilde{U}}{\partial s} \end{Bmatrix} = \begin{bmatrix} \frac{\partial X}{\partial r} & \frac{\partial Y}{\partial r} \\ \frac{\partial X}{\partial s} & \frac{\partial Y}{\partial s} \end{bmatrix}^{-1} \begin{bmatrix} \frac{\partial X}{\partial r} & \frac{\partial Y}{\partial r} \\ \frac{\partial X}{\partial s} & \frac{\partial Y}{\partial s} \end{bmatrix} \begin{Bmatrix} \frac{\partial \tilde{U}}{\partial X} \\ \frac{\partial \tilde{U}}{\partial Y} \end{Bmatrix} \tag{5.4.14}$$

Using matrix algebra and rearranging (5.4.14):

$$\begin{Bmatrix} \frac{\partial \tilde{U}}{\partial X} \\ \frac{\partial \tilde{U}}{\partial Y} \end{Bmatrix} = \begin{bmatrix} \frac{\partial X}{\partial r} & \frac{\partial Y}{\partial r} \\ \frac{\partial X}{\partial s} & \frac{\partial Y}{\partial s} \end{bmatrix}^{-1} \begin{Bmatrix} \frac{\partial \tilde{U}}{\partial r} \\ \frac{\partial \tilde{U}}{\partial s} \end{Bmatrix} \tag{5.4.15}$$

The strain terms now appear on the left hand side of (5.4.15), as desired. Note again that, just as in the case of the 4-node isoparametric surface element, the inverse of the Jacobian transforms derivatives from the local coordinate system into derivatives expressed in the global system. The Jacobian in (5.4.15) is the same as that which evolved when the 4-node quadrilateral element was developed:

$$\underline{\underline{J}} = \begin{bmatrix} J_{11} & J_{12} \\ J_{21} & J_{22} \end{bmatrix} = \begin{bmatrix} \frac{\partial X}{\partial r} & \frac{\partial Y}{\partial r} \\ \frac{\partial X}{\partial s} & \frac{\partial Y}{\partial s} \end{bmatrix} \tag{5.4.16}$$

Using (5.4.16), it is convenient to express (5.4.15) as:

$$\begin{Bmatrix} \frac{\partial \tilde{U}}{\partial X} \\ \frac{\partial \tilde{U}}{\partial Y} \end{Bmatrix} = \underline{\underline{J}}^{-1} \begin{Bmatrix} \frac{\partial \tilde{U}}{\partial r} \\ \frac{\partial \tilde{U}}{\partial s} \end{Bmatrix} \tag{5.4.17}$$

Likewise, for derivatives of $\tilde{V}$:

$$\begin{Bmatrix} \frac{\partial \tilde{V}}{\partial r} \\ \frac{\partial \tilde{V}}{\partial s} \end{Bmatrix} = \begin{bmatrix} \frac{\partial X}{\partial r} & \frac{\partial Y}{\partial r} \\ \frac{\partial X}{\partial s} & \frac{\partial Y}{\partial s} \end{bmatrix} \begin{Bmatrix} \frac{\partial \tilde{V}}{\partial X} \\ \frac{\partial \tilde{V}}{\partial Y} \end{Bmatrix} \Rightarrow \begin{Bmatrix} \frac{\partial \tilde{V}}{\partial X} \\ \frac{\partial \tilde{V}}{\partial Y} \end{Bmatrix} = \underline{\underline{J}}^{-1} \begin{Bmatrix} \frac{\partial \tilde{V}}{\partial r} \\ \frac{\partial \tilde{V}}{\partial s} \end{Bmatrix} \tag{5.4.18}$$

The next step is to express the strain vector in terms of a strain-displacement matrix:

$$\underline{\underline{\varepsilon}} = \underline{\underline{B}}\, {}^e\underline{\underline{D}} \tag{5.4.19}$$

Recall that the strain vector for plane stress contains three components:

$$\underline{\underline{\varepsilon}} = \begin{Bmatrix} \varepsilon_{XX} \\ \varepsilon_{YY} \\ \gamma_{XY} \end{Bmatrix} \approx \begin{Bmatrix} \dfrac{\partial \tilde{U}}{\partial X} \\ \dfrac{\partial \tilde{V}}{\partial X} \\ \dfrac{\partial \tilde{U}}{\partial Y} + \dfrac{\partial \tilde{V}}{\partial X} \end{Bmatrix} \tag{5.4.20}$$

Using (5.4.16), (5.4.17), and (5.4.18), the right hand side of (5.4.20) can be expressed in terms of the strain-displacement matrix:

$$\begin{Bmatrix} \dfrac{\partial \tilde{U}}{\partial X} \\ \dfrac{\partial \tilde{V}}{\partial X} \\ \dfrac{\partial \tilde{U}}{\partial Y} + \dfrac{\partial \tilde{V}}{\partial X} \end{Bmatrix} = \underline{\underline{B}}\, {}^e\underline{\underline{D}} \tag{5.4.21}$$

The strain-displacement matrix in (5.4.21) is defined as:

$$\underline{\underline{B}} = \begin{bmatrix} B_{11} & 0 & B_{13} & 0 & B_{15} & 0 \\ 0 & B_{22} & 0 & B_{24} & 0 & B_{26} \\ B_{22} & B_{11} & B_{24} & B_{13} & B_{26} & B_{15} \end{bmatrix} \qquad {}^e\underline{\underline{D}} = \begin{Bmatrix} U_a \\ V_a \\ U_b \\ V_b \\ U_c \\ V_c \end{Bmatrix} \tag{5.4.22}$$

Since the 3-node, triangular surface element has a total of six nodal DOF's, the *B*-matrix has six columns. The terms in the strain-displacement matrix are expressed as:

$$
\begin{aligned}
B_{11} &= \frac{1}{\det \underline{\underline{J}}}\left(J_{22}N_{1,r} - J_{12}N_{1,s}\right) & B_{22} &= \frac{1}{\det \underline{\underline{J}}}\left(-J_{21}N_{1,r} + J_{11}N_{1,s}\right) \\
B_{13} &= \frac{1}{\det \underline{\underline{J}}}\left(J_{22}N_{2,r} - J_{12}N_{2,s}\right) & B_{24} &= \frac{1}{\det \underline{\underline{J}}}\left(-J_{21}N_{2,r} + J_{11}N_{2,s}\right) \\
B_{15} &= \frac{1}{\det \underline{\underline{J}}}\left(J_{22}N_{3,r} - J_{12}N_{3,s}\right) & B_{26} &= \frac{1}{\det \underline{\underline{J}}}\left(-J_{21}N_{3,r} + J_{11}N_{3,s}\right)
\end{aligned}
\tag{5.4.23}
$$

The determinant of the Jacobian matrix is given by (5.3.25). Deriving the terms in (5.4.23) is less tedious than in the case of the 4-node surface element, although considerable algebra must still be employed. The terms in (5.4.23) are derived using a process similar to that used for the 4-node surface element shown in Appendix C. Although many terms are involved in the strain-displacement matrix for this 3-node surface element, note that the derivatives of the shape functions are simply constants, as can be shown by differentiating the shape functions in (5.4.6).

To gain an appreciation of the terms that are contained in the B-matrix, consider the first term, B_{11}:

$$
B_{11} = \frac{1}{\det \underline{\underline{J}}}\left(J_{22}N_{1,r} - J_{12}N_{1,s}\right) \tag{5.4.24}
$$

Referring to (5.4.16):

$$
J_{22} = \frac{\partial Y}{\partial s} \tag{5.4.25}
$$

Recalling the mapping function for the Y-coordinate:

$$
Y(r,s) = N_1 Y_a + N_2 Y_b + N_3 Y_c \tag{5.4.26}
$$

Substituting (5.4.26) into (5.4.25):

$$
J_{22} = \frac{\partial}{\partial s}\left(N_1 Y_a + N_2 Y_b + N_3 Y_c\right) \tag{5.4.27}
$$

Or, using the notation for partial derivatives:

$$
J_{22} = N_{1,s} Y_a + N_{2,s} Y_b + N_{3,s} Y_c \tag{5.4.28}
$$

Similarly:

$$J_{12} = \frac{\partial}{\partial r}\left(N_1 Y_a + N_2 Y_b + N_3 Y_c\right) \tag{5.4.29}$$

Or:

$$J_{12} = N_{1,r} Y_a + N_{2,r} Y_b + N_{3,r} Y_c \tag{5.4.30}$$

Substituting (5.4.28) and (5.4.30) into (5.4.24):

$$B_{11} = \frac{1}{\det \underline{\underline{J}}}\Big[\left(N_{1,s} Y_a + N_{2,s} Y_b + N_{3,s} Y_c\right) N_{1,s} - \left(N_{1,r} Y_a + N_{2,r} Y_b + N_{3,r} Y_c\right) N_{1,r}\Big] \tag{5.4.31}$$

The derivatives of the shape functions for this element are formed by differentiating (5.4.6):

$$\begin{aligned} N_{1,s} &= (-1) \qquad & N_{2,s} &= (0) \qquad & N_{3,s} &= (\ 1) \\ N_{1,r} &= (-1) \qquad & N_{2,r} &= (1) \qquad & N_{3,r} &= (0) \end{aligned} \tag{5.4.32}$$

The determinant of the Jacobian is the same as computed by (5.3.25):

$$\det \underline{\underline{J}} = J_{11} J_{22} - J_{12} J_{21} \tag{5.4.33}$$

Substituting the shape function derivatives (5.4.32) into (5.4.31):

$$B_{11} = \frac{1}{\det \underline{\underline{J}}}\left[(Y_c - Y_a)(-1) - (Y_b - Y_a)(-1)\right] \tag{5.4.34}$$

Notice that only constants appear in the B_{11} term above. It can be shown that the rest of the B-matrix terms are also constant. By substituting (5.4.32) into (5.4.28) and (5.4.30), it is seen that these two Jacobian terms are constant; the other two terms of the Jacobian matrix are also constant. Since all the terms in the Jacobian are constant, so is the determinant, as computed by (5.4.33). With a constant Jacobian determinant and constants in all of the B-matrix terms, the strain-displacement matrix for the 3-node isoparametric can only represent constant strain. In contrast, the strain-displacement matrix for the 4-node isoparametric element can represent constant strain, and some modes of linearly varying strain as well.

Computing the Stiffness Matrix for a 3-Node Surface Element

To form the element stiffness matrix, the *B*-matrix from (5.4.22) is substituted into the element stiffness matrix formula, (3.5.12):

$$ {}^{e}\underline{\underline{K}} = \int_{r_1}^{r_2}\int \begin{bmatrix} B_{11} & 0 & B_{22} \\ 0 & B_{22} & B_{11} \\ B_{13} & 0 & B_{24} \\ 0 & B_{24} & B_{13} \\ B_{15} & 0 & B_{26} \\ 0 & B_{26} & B_{15} \end{bmatrix} \underline{\underline{E}} \begin{bmatrix} B_{11} & 0 & B_{13} & 0 & B_{15} & 0 \\ 0 & B_{22} & 0 & B_{24} & 0 & B_{26} \\ B_{22} & B_{11} & B_{24} & B_{13} & B_{26} & B_{15} \end{bmatrix} t\left(\det \underline{\underline{J}}\right) ds\, dr $$

$$ \underline{\underline{E}} = \frac{E}{1-v^2}\begin{bmatrix} 1 & v & 0 \\ v & 1 & 0 \\ 0 & 0 & \dfrac{1-v}{2} \end{bmatrix} $$

Since only constants are present in the *B*- and *E*-matrices, and also in the determinant of the Jacobian, the stiffness matrix can be computed with these factors removed from within the integral:

$$ {}^{e}\underline{\underline{K}} = \begin{bmatrix} B_{11} & 0 & B_{22} \\ 0 & B_{22} & B_{11} \\ B_{13} & 0 & B_{24} \\ 0 & B_{24} & B_{13} \\ B_{15} & 0 & B_{26} \\ 0 & B_{26} & B_{15} \end{bmatrix} \underline{\underline{E}} \begin{bmatrix} B_{11} & 0 & B_{13} & 0 & B_{15} & 0 \\ 0 & B_{22} & 0 & B_{24} & 0 & B_{26} \\ B_{22} & B_{11} & B_{24} & B_{13} & B_{26} & B_{15} \end{bmatrix} t\left(\det \underline{\underline{J}}\right) \int_{r_1}^{r_2}\int ds\, dr $$

As with all isoparametric elements, the integration used to compute the stiffness matrix is performed on the local domain, the *r-s* domain in this case. The transformation from the global domain to the *r-s* domain follows the same procedure as that for the 4-node surface element, shown in Appendix C.

The double integral in the stiffness matrix formula above represents the surface area of the parent element. Since the parent element always has the same dimensions, as illustrated in Figure 5.11, the area of the triangle is always equal to *1/2 bh*, i.e., $A = 1/2$. Hence, the expression for the stiffness matrix above is reduced to:

$$\underline{\underline{{}^{e}K}} = \begin{bmatrix} B_{11} & 0 & B_{22} \\ 0 & B_{22} & B_{11} \\ B_{13} & 0 & B_{24} \\ 0 & B_{24} & B_{13} \\ B_{15} & 0 & B_{26} \\ 0 & B_{26} & B_{15} \end{bmatrix} \underline{\underline{E}} \begin{bmatrix} B_{11} & 0 & B_{13} & 0 & B_{15} & 0 \\ 0 & B_{22} & 0 & B_{24} & 0 & B_{26} \\ B_{22} & B_{11} & B_{24} & B_{13} & B_{26} & B_{15} \end{bmatrix} t \frac{\left(\det \underline{\underline{J}}\right)}{2}$$

The area calculation above may be accomplished more formally:

$$A = \int_{r_1}^{r_2}\int ds\, dr = \int_{r_1}^{r_2} s\, dr \tag{5.4.35}$$

Since the height of the triangular parent element varies with r, the expression $s=(1-r)$ is required to compute the area of the element. Using this relationship in (5.4.35), and substituting in the limits of integration:

$$A = \int_{r_1}^{r_2} s\, dr = \int_{0}^{1} (1-r)dr = \frac{1}{2}$$

The integration required to compute the stiffness matrix for the 3-node, C^0 surface element is trivial, and a closed-form solution is easily employed. This element differs from the 4-node, quadrilateral surface element, where the B-matrix, hence the integrand, contains functions of r and s. Since the 3-node, non-parametric element has simple shape functions, maintains compatibility regardless of orientation, and does not require numerical integration to compute its stiffness matrix, there is not much incentive to utilize this element in isoparametric form.

5.5 An 8-Node Isoparametric Volume Element

Key Concept: The most simple volume element, the 4-node tetrahedral, is a poor performer, since it is too stiff and can grossly under-predict stress. The 8-node, hexahedral (six-sided) volume element is more complex but provides a good compromise between the poor performing 4-node tetrahedral and more "computationally expensive" elements, such as the 10-node tetrahedral or the 20-node hexahedral.

An analogy can be drawn between 3- and 4-node surface elements and 4- and 8-node volume elements. The 4-node, quadrilateral surface element requires somewhat more computational effort to generate a stiffness matrix than the 3-node, triangular surface element, hence, the 4-node element is termed "more expensive." While the 3-node surface element is less expensive, it is typically avoided due to its poor performance. The 4-nodc surface element, while more expensive, is considered worth the cost because of its increased accuracy.

Now, consider the 4-node tetrahedral volume element. Like the 3-node triangular surface element, the 4-node tet is "cheap" but performs poorly. It is therefore avoided, while the 8-node hexahedral, albeit more expensive, is considered a better choice because of its improved accuracy.

As discussed in Chapter 4, the 4-node tetrahedral element is the most simple volume element available. Although not shown explicitly, the strain-displacement matrix for the 4-node tetrahedral contains only constants. It is therefore possible to generate the stiffness matrix for the tetrahedral element using closed-form methods. Any volume element more complex than the 4-node tetrahedral will require numerical integration either as a matter of practicality or in most cases, necessity. Hence, the 8-node, hexahedral volume element is formulated as an isoparametric element to facilitate numerical integration while easing the task of shape function generation and allowing compatibility in any orientation.

8-Node, Isoparametric Volume Element, C^0 Applications—Shape Functions

An 8-node, isoparametric, volume element is depicted in Figure 5.12, with global displacement variables shown only at Node *h*, for clarity.

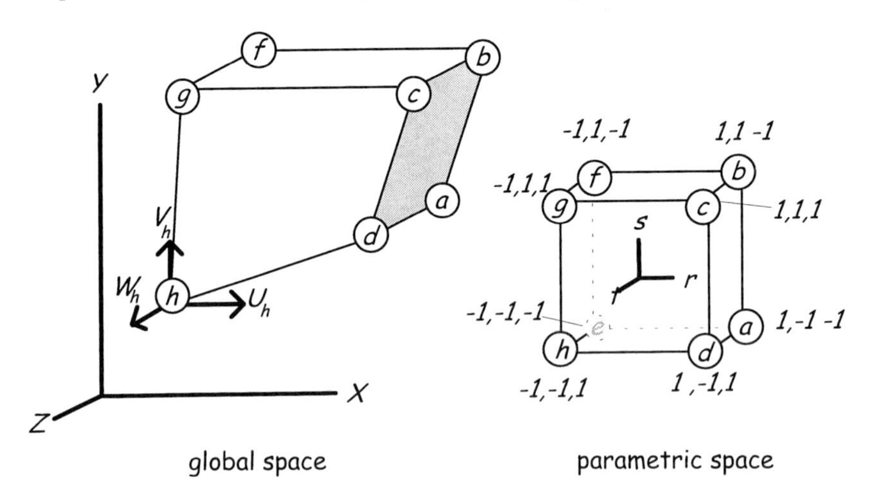

Figure 5.12 An 8-Node Isoparametric Volume Element

As with the other isoparametric elements, the formulation for the 8-node volume element will begin with the mapping function. An eight-term mapping function (Reddy [8, p. 410]) can be used with the 8-node element:

$$X(r,s,t) = a_1 + a_2 r + a_3 s + a_4 t + a_5 rs + a_6 rt + a_7 st + a_8 rst \tag{5.5.1}$$

To map the global element to the parent, the mapping function is constrained to the nodal values of the coordinates:

$$\begin{aligned}
X(1,-1,-1) &= a_1 + a_2(1) + a_3(-1) + a_4(-1) + \cdots + a_8(1)(-1)(-1) = X_a \\
X(1,\ 1,-1) &= a_1 + a_2(1) + a_3(1) + a_4(-1) + \cdots + a_8(1)(1)(-1) = X_b \\
&\vdots \\
X(-1,-1,\ 1) &= a_1 + a_2(-1) + a_3(-1) + a_4(1) + \cdots + a_8(-1)(-1)(1) = X_h
\end{aligned}$$

In matrix form, the above is expressed as:

$$\begin{bmatrix} 1 & 1 & -1 & -1 & \cdots & 1 \\ 1 & 1 & 1 & -1 & \cdots & -1 \\ & & & \vdots & & \\ 1 & -1 & -1 & 1 & \cdots & 1 \end{bmatrix} \begin{Bmatrix} a_1 \\ a_2 \\ \vdots \\ a_8 \end{Bmatrix} = \begin{Bmatrix} X_a \\ X_b \\ \vdots \\ X_h \end{Bmatrix} \tag{5.5.2}$$

Following the same procedure, the unknown a_i's in (5.5.2) are determined, substituted into (5.5.1), and the resulting equation rearranged to yield an interpolation function for *X(r,s,t)*, with the eight shape functions below:

$$\begin{aligned}
N_1 &= \frac{1}{8}(1+r)(1-s)(1-t) & N_2 &= \frac{1}{8}(1+r)(1+s)(1-t) \\
N_3 &= \frac{1}{8}(1+r)(1+s)(1+t) & N_4 &= \frac{1}{8}(1+r)(1-s)(1+t) \\
N_5 &= \frac{1}{8}(1-r)(1-s)(1-t) & N_6 &= \frac{1}{8}(1-r)(1+s)(1-t) \\
N_7 &= \frac{1}{8}(1-r)(1+s)(1+t) & N_8 &= \frac{1}{8}(1-r)(1-s)(1+t)
\end{aligned} \tag{5.5.3}$$

The mapping function for the *Y*- and *Z*-coordinate values can be found in the same manner, yielding:

$$Y(r,s,t) = N_1 Y_a + N_2 Y_b + N_3 Y_c + N_4 Y_d + N_5 Y_e + N_6 Y_f + N_7 Y_g + N_8 Y_h$$
$$Z(r,s,t) = N_1 Z_a + N_2 Z_b + N_3 Z_c + N_4 Z_d + N_5 Z_e + N_6 Z_f + N_7 Z_g + N_8 Z_h$$

Using the same shape functions (above) for displacement, interpolation functions for the 8-node, isoparametric volume element take the form:

$$\begin{aligned}
\tilde{U}(r,s,t) &= N_1 U_a + N_2 U_b + N_3 U_c + N_4 U_d + N_5 U_e + N_6 U_f + N_7 U_g + N_8 U_h \\
\tilde{V}(r,s,t) &= N_1 V_a + N_2 V_b + N_3 V_c + N_4 V_d + N_5 V_e + N_6 V_f + N_7 V_g + N_8 V_h \\
\tilde{Z}(r,s,t) &= N_1 Z_a + N_2 Z_b + N_3 Z_c + N_4 Z_d + N_5 Z_e + N_6 Z_f + N_7 Z_g + N_8 Z_h
\end{aligned} \tag{5.5.4}$$

B-Matrix for an 8-Node, Isoparametric, Three-Dimensional Stress Element
To compute the element stiffness matrix for the 8-node, isoparametric volume element, a B-matrix must be specified. Recalling the strain terms for three dimensional stress:

$$\underline{\underline{\varepsilon}} = \begin{Bmatrix} \varepsilon_{XX} \\ \varepsilon_{YY} \\ \varepsilon_{ZZ} \\ \gamma_{XY} \\ \gamma_{YZ} \\ \gamma_{ZX} \end{Bmatrix} \approx \begin{Bmatrix} \dfrac{\partial \tilde{U}}{\partial X} \\ \dfrac{\partial \tilde{V}}{\partial Y} \\ \dfrac{\partial \tilde{W}}{\partial Z} \\ \dfrac{\partial \tilde{U}}{\partial Y} + \dfrac{\partial \tilde{V}}{\partial X} \\ \dfrac{\partial \tilde{V}}{\partial Z} + \dfrac{\partial \tilde{W}}{\partial Y} \\ \dfrac{\partial \tilde{W}}{\partial X} + \dfrac{\partial \tilde{U}}{\partial Z} \end{Bmatrix}$$

Beginning with the first strain term above:

$$\varepsilon_{XX} \approx \frac{\partial \tilde{U}}{\partial X} \tag{5.5.5}$$

To compute normal strain in the X-coordinate direction, the eight-term interpolation function for $\tilde{U}$ from (5.5.4) is substituted into (5.5.5):

$$\frac{\partial \tilde{U}}{\partial X} = \frac{\partial}{\partial X}\left(N_1 X_a + N_2 X_b + N_3 X_c + N_4 X_d + N_5 X_e + N_6 X_f + N_7 X_g + N_8 X_h\right) \tag{5.5.6}$$

The derivative in (5.5.6) is with respect to X while the shape functions are expressed in terms of the free variables r, s, and t. To compute the derivative of $\tilde{U}$ with respect to X, the chain rule is employed in the following manner:

$$\frac{\partial \tilde{U}}{\partial r} = \frac{\partial X}{\partial r}\frac{\partial \tilde{U}}{\partial X} + \frac{\partial Y}{\partial r}\frac{\partial \tilde{U}}{\partial Y} + \frac{\partial Z}{\partial r}\frac{\partial \tilde{U}}{\partial Z} \tag{5.5.7}$$

In addition to (5.5.7), two other equations are required to fully express the normal strain terms associated with the U variable:

$$\begin{aligned} \frac{\partial \tilde{U}}{\partial s} &= \frac{\partial X}{\partial s}\frac{\partial \tilde{U}}{\partial X} + \frac{\partial Y}{\partial s}\frac{\partial \tilde{U}}{\partial Y} + \frac{\partial Z}{\partial s}\frac{\partial \tilde{U}}{\partial Z} \\ \frac{\partial \tilde{U}}{\partial t} &= \frac{\partial X}{\partial t}\frac{\partial \tilde{U}}{\partial X} + \frac{\partial Y}{\partial t}\frac{\partial \tilde{U}}{\partial Y} + \frac{\partial Z}{\partial t}\frac{\partial \tilde{U}}{\partial Z} \end{aligned} \tag{5.5.8}$$

In matrix form, (5.5.7) and (5.5.8) are expressed as:

$$\begin{Bmatrix} \dfrac{\partial \tilde{U}}{\partial r} \\ \dfrac{\partial \tilde{U}}{\partial s} \\ \dfrac{\partial \tilde{U}}{\partial t} \end{Bmatrix} = \begin{bmatrix} \dfrac{\partial X}{\partial r} & \dfrac{\partial Y}{\partial r} & \dfrac{\partial Z}{\partial r} \\ \dfrac{\partial X}{\partial s} & \dfrac{\partial Y}{\partial s} & \dfrac{\partial Z}{\partial s} \\ \dfrac{\partial X}{\partial t} & \dfrac{\partial Y}{\partial t} & \dfrac{\partial Z}{\partial t} \end{bmatrix} \begin{Bmatrix} \dfrac{\partial \tilde{U}}{\partial X} \\ \dfrac{\partial \tilde{U}}{\partial Y} \\ \dfrac{\partial \tilde{U}}{\partial Z} \end{Bmatrix} \tag{5.5.9}$$

To obtain an expression for $\partial \tilde{U} / \partial X$, each side of (5.5.9) is pre-multiplied by the inverse of the square (Jacobian) matrix:

$$\begin{bmatrix} \frac{\partial X}{\partial r} & \frac{\partial Y}{\partial r} & \frac{\partial Z}{\partial r} \\ \frac{\partial X}{\partial s} & \frac{\partial Y}{\partial s} & \frac{\partial Z}{\partial s} \\ \frac{\partial X}{\partial t} & \frac{\partial Y}{\partial t} & \frac{\partial Z}{\partial t} \end{bmatrix}^{-1} \begin{Bmatrix} \frac{\partial \tilde{U}}{\partial r} \\ \frac{\partial \tilde{U}}{\partial s} \\ \frac{\partial \tilde{U}}{\partial t} \end{Bmatrix} = \begin{bmatrix} \frac{\partial X}{\partial r} & \frac{\partial Y}{\partial r} & \frac{\partial Z}{\partial r} \\ \frac{\partial X}{\partial s} & \frac{\partial Y}{\partial s} & \frac{\partial Z}{\partial s} \\ \frac{\partial X}{\partial t} & \frac{\partial Y}{\partial t} & \frac{\partial Z}{\partial t} \end{bmatrix}^{-1} \begin{bmatrix} \frac{\partial X}{\partial r} & \frac{\partial Y}{\partial r} & \frac{\partial Z}{\partial r} \\ \frac{\partial X}{\partial s} & \frac{\partial Y}{\partial s} & \frac{\partial Z}{\partial s} \\ \frac{\partial X}{\partial t} & \frac{\partial Y}{\partial t} & \frac{\partial Z}{\partial t} \end{bmatrix} \begin{Bmatrix} \frac{\partial \tilde{U}}{\partial X} \\ \frac{\partial \tilde{U}}{\partial Y} \\ \frac{\partial \tilde{U}}{\partial Z} \end{Bmatrix}$$

Using the identity matrix and rearranging the equation above:

$$\begin{Bmatrix} \frac{\partial \tilde{U}}{\partial X} \\ \frac{\partial \tilde{U}}{\partial Y} \\ \frac{\partial \tilde{U}}{\partial Z} \end{Bmatrix} = \begin{bmatrix} \frac{\partial X}{\partial r} & \frac{\partial Y}{\partial r} & \frac{\partial Z}{\partial r} \\ \frac{\partial X}{\partial s} & \frac{\partial Y}{\partial s} & \frac{\partial Z}{\partial s} \\ \frac{\partial X}{\partial t} & \frac{\partial Y}{\partial t} & \frac{\partial Z}{\partial t} \end{bmatrix}^{-1} \begin{Bmatrix} \frac{\partial \tilde{U}}{\partial r} \\ \frac{\partial \tilde{U}}{\partial s} \\ \frac{\partial \tilde{U}}{\partial t} \end{Bmatrix} \tag{5.5.10}$$

As with the other examples, the transformation between derivatives with respect to global coordinates and derivatives with respect to local coordinates is performed via the inverse of the *Jacobian*. Note that the Jacobian matrix for elements in three-dimensional space is a 3 *by 3* matrix, as shown in (5.5.11):

$$\underline{\underline{J}} = \begin{bmatrix} J_{11} & J_{12} & J_{13} \\ J_{21} & J_{22} & J_{23} \\ J_{31} & J_{32} & J_{33} \end{bmatrix} = \begin{bmatrix} \frac{\partial X}{\partial r} & \frac{\partial Y}{\partial r} & \frac{\partial Z}{\partial r} \\ \frac{\partial X}{\partial s} & \frac{\partial Y}{\partial s} & \frac{\partial Z}{\partial s} \\ \frac{\partial X}{\partial t} & \frac{\partial Y}{\partial t} & \frac{\partial Z}{\partial t} \end{bmatrix} \tag{5.5.11}$$

Using (5.5.11), it is convenient to express (5.5.10) as:

$$\begin{Bmatrix} \frac{\partial \tilde{U}}{\partial X} \\ \frac{\partial \tilde{U}}{\partial Y} \\ \frac{\partial \tilde{U}}{\partial Z} \end{Bmatrix} = \underline{\underline{J}}^{-1} \begin{Bmatrix} \frac{\partial \tilde{U}}{\partial r} \\ \frac{\partial \tilde{U}}{\partial s} \\ \frac{\partial \tilde{U}}{\partial t} \end{Bmatrix} \tag{5.5.12}$$

Likewise, for derivatives of $\tilde{V}$ and $\tilde{W}$:

$$\left\{\begin{matrix}\dfrac{\partial\tilde{V}}{\partial r}\\ \dfrac{\partial\tilde{V}}{\partial s}\\ \dfrac{\partial\tilde{V}}{\partial t}\end{matrix}\right\}=\begin{bmatrix}\dfrac{\partial X}{\partial r} & \dfrac{\partial Y}{\partial r} & \dfrac{\partial Z}{\partial r}\\ \dfrac{\partial X}{\partial s} & \dfrac{\partial Y}{\partial s} & \dfrac{\partial Z}{\partial s}\\ \dfrac{\partial X}{\partial t} & \dfrac{\partial Y}{\partial t} & \dfrac{\partial Z}{\partial t}\end{bmatrix}\left\{\begin{matrix}\dfrac{\partial\tilde{V}}{\partial X}\\ \dfrac{\partial\tilde{V}}{\partial Y}\\ \dfrac{\partial\tilde{V}}{\partial Z}\end{matrix}\right\}\Rightarrow\left\{\begin{matrix}\dfrac{\partial\tilde{V}}{\partial X}\\ \dfrac{\partial\tilde{V}}{\partial Y}\\ \dfrac{\partial\tilde{V}}{\partial Z}\end{matrix}\right\}=\underline{\underline{J}}^{-1}\left\{\begin{matrix}\dfrac{\partial\tilde{V}}{\partial r}\\ \dfrac{\partial\tilde{V}}{\partial s}\\ \dfrac{\partial\tilde{V}}{\partial t}\end{matrix}\right\} \tag{5.5.13}$$

$$\left\{\begin{matrix}\dfrac{\partial\tilde{W}}{\partial r}\\ \dfrac{\partial\tilde{W}}{\partial s}\\ \dfrac{\partial\tilde{W}}{\partial t}\end{matrix}\right\}=\begin{bmatrix}\dfrac{\partial X}{\partial r} & \dfrac{\partial Y}{\partial r} & \dfrac{\partial Z}{\partial r}\\ \dfrac{\partial X}{\partial s} & \dfrac{\partial Y}{\partial s} & \dfrac{\partial Z}{\partial s}\\ \dfrac{\partial X}{\partial t} & \dfrac{\partial Y}{\partial t} & \dfrac{\partial Z}{\partial t}\end{bmatrix}\left\{\begin{matrix}\dfrac{\partial\tilde{W}}{\partial X}\\ \dfrac{\partial\tilde{W}}{\partial Y}\\ \dfrac{\partial\tilde{W}}{\partial Z}\end{matrix}\right\}\Rightarrow\left\{\begin{matrix}\dfrac{\partial\tilde{W}}{\partial X}\\ \dfrac{\partial\tilde{W}}{\partial Y}\\ \dfrac{\partial\tilde{W}}{\partial Z}\end{matrix}\right\}=\underline{\underline{J}}^{-1}\left\{\begin{matrix}\dfrac{\partial\tilde{W}}{\partial r}\\ \dfrac{\partial\tilde{W}}{\partial s}\\ \dfrac{\partial\tilde{W}}{\partial t}\end{matrix}\right\} \tag{5.5.14}$$

The next step is to express the strain vector in terms of a strain-displacement matrix:

$$\underline{\underline{\varepsilon}}=\underline{\underline{B}}\,{}^{e}\underline{\underline{D}}$$

Recall that the strain vector for three-dimensional stress contains the terms:

$$\underline{\underline{\varepsilon}}=\left\{\begin{matrix}\varepsilon_{XX}\\ \varepsilon_{YY}\\ \varepsilon_{ZZ}\\ \gamma_{XY}\\ \gamma_{YZ}\\ \gamma_{ZX}\end{matrix}\right\}\approx\left\{\begin{matrix}\dfrac{\partial\tilde{U}}{\partial X}\\ \dfrac{\partial\tilde{V}}{\partial Y}\\ \dfrac{\partial\tilde{W}}{\partial Z}\\ \dfrac{\partial\tilde{U}}{\partial Y}+\dfrac{\partial\tilde{V}}{\partial X}\\ \dfrac{\partial\tilde{V}}{\partial Z}+\dfrac{\partial\tilde{W}}{\partial Y}\\ \dfrac{\partial\tilde{W}}{\partial X}+\dfrac{\partial\tilde{U}}{\partial Z}\end{matrix}\right\} \tag{5.5.15}$$

Manipulating (5.5.11), (5.5.12), (5.5.13), and (5.5.14), the right hand side of

(5.5.15) can be expressed in terms of the strain-displacement matrix:

$$\left\{ \begin{matrix} \dfrac{\partial \tilde{U}}{\partial X} \\ \dfrac{\partial \tilde{V}}{\partial Y} \\ \dfrac{\partial \tilde{W}}{\partial Z} \\ \dfrac{\partial \tilde{U}}{\partial Y} + \dfrac{\partial \tilde{V}}{\partial X} \\ \dfrac{\partial \tilde{V}}{\partial Z} + \dfrac{\partial \tilde{W}}{\partial Y} \\ \dfrac{\partial \tilde{W}}{\partial X} + \dfrac{\partial \tilde{U}}{\partial Z} \end{matrix} \right\} = \underline{\underline{B}}\, {}^{e}\underline{\underline{D}} \tag{5.5.16}$$

A considerable amount of matrix algebra is required to express the strain-displacement matrix terms in (5.5.16), although the manipulations are essentially the same as those used to generate the other isoparametric elements. In short, to find an expression for the first term in the strain-displacement matrix (5.5.16), the Jacobian matrix given by (5.5.11) is first inverted, then substituted into (5.5.12). After performing the matrix algebra in (5.5.12), the first term in the strain vector of (5.5.16) can be expressed in terms of shape function derivatives and nodal displacement variables, from which the first row in the strain-displacement matrix is expressed. Following the same process, the other terms in the strain-displacement matrix are generated, resulting in a strain-displacement matrix as shown in (5.5.17).

$$\underline{\underline{B}} = \begin{bmatrix} B_{1/1} & 0 & 0 & B_{1/4} & \cdots & 0 \\ 0 & B_{2/2} & 0 & 0 & \cdots & 0 \\ 0 & 0 & B_{3/3} & 0 & \cdots & B_{3/24} \\ B_{4/1} & B_{4/2} & 0 & B_{4/4} & \cdots & 0 \\ 0 & B_{5/2} & B_{5/3} & 0 & \cdots & B_{5/24} \\ B_{6/1} & 0 & B_{6/3} & B_{6/4} & \cdots & B_{6/24} \end{bmatrix} \qquad {}^{e}\underline{\underline{D}} = \left\{ \begin{matrix} U_a \\ V_a \\ W_a \\ U_b \\ V_b \\ W_b \\ \vdots \\ U_h \\ V_h \\ W_h \end{matrix} \right\} \tag{5.5.17}$$

There are a total of 24 nodal displacement variables (*8 nodes times 3 variables per node*), such that the strain-displacement matrix in (5.5.17) contains 24 columns.

Somewhat analogous to the 4-node, isoparametric quadrilateral surface element previously considered, the variables *r, s, and t* appear in the *B*-matrix for the 8-node volume element, suggesting that strain need not be constant within the element. Although the terms of the Jacobian determinant are not shown, they also contain both *r, s,* and *t* variables, in general. The full equations for an 8-node, isoparametric volume element are found in Grandin [9]; the natural coordinate directions used in that reference differ from those used in this text. While forming the *B*-matrix of (5.5.17) would be quite a task using hand calculations, it is easily accomplished using a few lines of computer code. In Section 5.7, a FORTRAN computer program to calculate the *B*-matrix for a 4-node, isoparametric surface element is given. A routine for the 8-node, isoparametric volume element could be developed by modifying the routine for the 4-node element.

Computing the Element Stiffness Matrix

To form the element stiffness matrix, the *B*-matrix from (5.5.17) is substituted into the element stiffness matrix formula, (3.5.12):

$$\underline{\underline{{}^{e}K}} = \int_{r_1}^{r_2}\int_{s_1}^{s_2}\int_{t_1}^{t_2} \underline{\underline{B}}^{T}\,\underline{\underline{E}}\,\underline{\underline{B}}\left(\det \underline{\underline{J}}\right) dt\, ds\, dr$$

$$\underline{\underline{E}} = \begin{bmatrix} 1-\nu & \nu & \iota & 0 & 0 & 0 & \\ \nu & 1-\nu & \nu & 0 & 0 & 0 & \\ \nu & \nu & 1-\nu & 0 & 0 & 0 & \\ 0 & 0 & 0 & \dfrac{1}{2(1+\nu)} & 0 & 0 & 0 \\ 0 & 0 & 0 & 0 & \dfrac{1}{2(1+\nu)} & 0 & \\ 0 & 0 & 0 & 0 & 0 & \dfrac{1}{2(1+\nu)} & \end{bmatrix}$$

Performing the matrix multiplication and subsequent integration by closed-form methods is impractical, as with nearly all parametric elements. The element stiffness matrices are therefore computed using numerical integration, a topic to be covered in Section 5.7. Notice that the integration is performed on the *r-s-t* domain.

5.6 Higher Order C^0 Isoparametric Elements

Key Concept: The parametric elements considered so far have been limited to linear displacement assumptions. When more nodes are added to linear elements, quadratic (or higher) order displacement interpolation is possible.

As shown in the previous section, when more nodes are added to elements, the equations that describe the stiffness matrix become unwieldy. Therefore, higher order parametric elements will be considered only in a cursory manner. The reader is referred to MacNeal [1] and Bathe [7] for details regarding higher order elements. Generally speaking, each lower order C^0 element has a higher order version. Consider the illustration of commonly used elements from Chapter 4, shown again in Figure 5.13.

The elements on the left hand side of Figure 5.13, when used for C^0 applications, employ linear displacement and mapping functions while the ones on the right use quadratic functions. For example, as previously shown, the 2-node line element for rod applications has a two-term displacement assumption of the form:

$$\tilde{U} = a_1 + a_2 r \qquad (5.6.1)$$

By adding another node, a 3-node rod element could employ a three term displacement assumption:

$$\tilde{U} = a_1 + a_2 r + a_3 r^2 \qquad (5.6.2)$$

Similarly, a 3-node, triangular surface element for plane stress (or strain) employs the three-term displacement assumption:

$$\tilde{U} = a_1 + a_2 r + a_3 s \qquad (5.6.3)$$

The shape functions for the above were given by (5.4.6):

$$N_1 = (1 - r - s) \qquad N_2 = r \qquad N_3 = s \qquad (5.6.4)$$

Employing three additional nodes, a 6-node triangular element for the same application could utilize a displacement assumption of the form:

$$\tilde{U} = a_1 + a_2 r + a_3 s + a_4 r^2 + a_5 rs + a_6 s^2 \qquad (5.6.5)$$

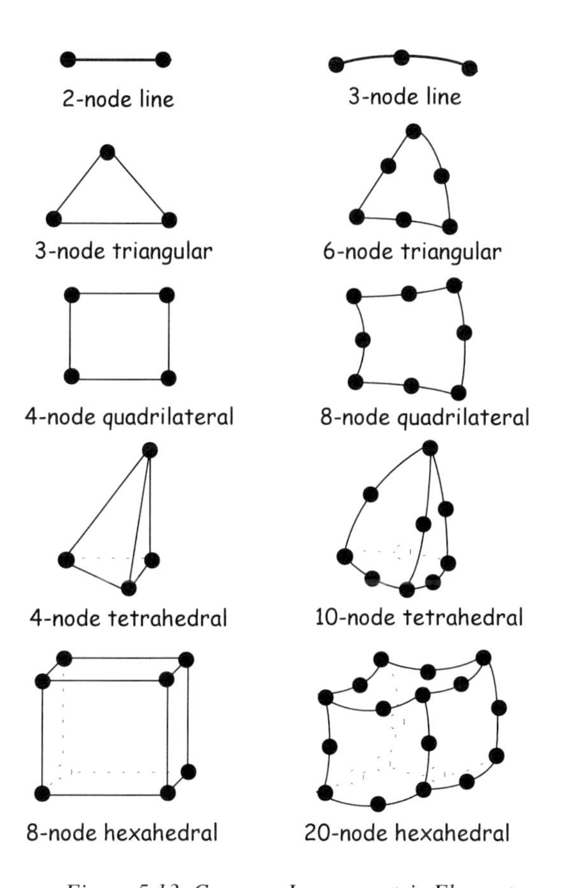

Figure 5.13 Common Isoparametric Elements

Following the usual procedure of constraining the displacement assumption to the nodal variables, the six-term displacement assumption of (5.6.5) can be shown to yield:

$$\tilde{U} = N_1 U_a + N_2 U_b + N_3 U_c + N_4 U_d + N_5 U_e + N_6 U_f \tag{5.6.6}$$

The shape functions in (5.6.6) are:

$$
\begin{aligned}
N_1 &= 2\left(\frac{1}{2} - r - s\right)(1 - r - s) & N_2 &= 2r\left(r - \frac{1}{2}\right) \\
N_3 &= 2s\left(s - \frac{1}{2}\right) & N_4 &= 4r(1 - r - s) \\
N_5 &= 4rs & N_6 &= 4s(1 - r - s)
\end{aligned}
\tag{5.6.7}
$$

The node numbering for the 6-node triangle, consistent with the shape functions above, appears in Figure 5.14.

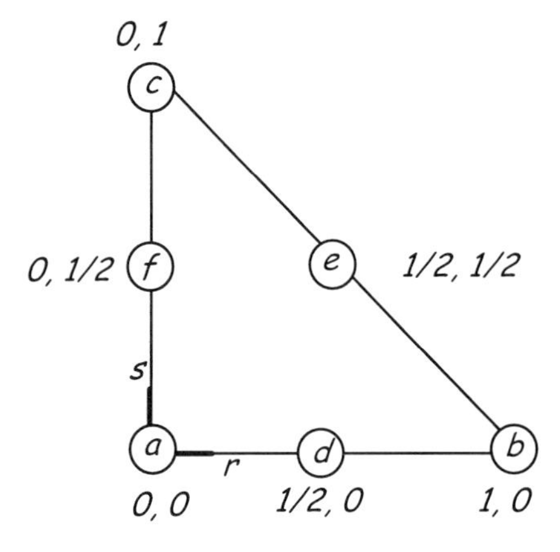

Figure 5.14 A 6-Node, Isoparametric, Triangular Surface Element (Parent)

Consider one last higher order, isoparametric element: an 8-node quadrilateral surface element. With eight nodes present, an eight term polynomial can be employed:

$$\tilde{U}(r,s) = a_1 + a_2 r + a_3 s + a_4 r^2 + a_5 rs + a_6 s^2 + a_7 r^2 s + a_8 rs^2$$

Using the usual process, the displacement assumption above can be expressed as:

$$\tilde{U} = N_1 U_a + N_2 U_b + N_3 U_c + N_4 U_d + N_5 U_e + N_6 U_f + N_7 U_g + N_8 U_h \tag{5.6.8}$$

The shape functions in (5.6.8) are:

$$\begin{aligned}
N_1 &= \frac{1}{4}(1-r)(1-s)(-1-r-s) & N_2 &= \frac{1}{4}(1+r)(1-s)(-1+r-s) \\
N_3 &= \frac{1}{4}(1+r)(1+s)(-1+r+s) & N_4 &= \frac{1}{4}(1-r)(1+s)(-1-r+s) \\
N_5 &= \frac{1}{2}\left(1-r^2\right)(1-s) & N_6 &= \frac{1}{2}\left(1+r^2\right)\left(1-s^2\right) \\
N_7 &= \frac{1}{2}\left(1-r^2\right)(1+s) & N_8 &= \frac{1}{2}(1-r)\left(1-s^2\right)
\end{aligned} \tag{5.6.9}$$

The parent for the 8-node isoparametric surface element appears in Figure 5.15.

As mentioned, there are easier methods to develop shape functions for isoparametric elements. Using the "shape function by inspection" method, shape functions are developed by logical argument rather than by mathematical manipulation; see Cook [10, p. 176] for an example.

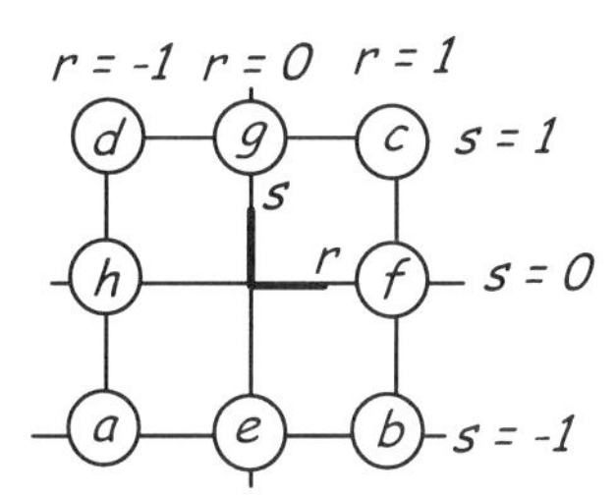

Figure 5.15 8-Node, Isoparametric Quadrilateral Surface Element (Parent)

5.7 Numerical Integration and Isoparametric Elements

Key Concept: Integration is needed to compute the individual finite element stiffness matrices. While closed-form integration is impractical for all but the most simple elements, numerical integration provides a practical alternative.

As shown, the 2-node, isoparametric line element and the 3-node, isoparametric triangular surface element are both simple enough that their element stiffness matrices can be computed using closed-form integration. However, recall that the stiffness matrix integrand for the 4-node, isoparametric quadrilateral is quite complex, making closed-form integration impractical. Except for a limited number of cases, closed-form integration is not possible, because the integrand does not represent a function that has a simple anti-derivative. The same is true for most other isoparametric elements. Therefore, an isoparametric element typically employs numerical integration to compute its stiffness matrix.

Numerical integration is used not only to compute the element stiffness matrix but can also be used to compute the nodal forces that approximate continuous loads, such as surface tractions, body loads, etc. There exist several numerical integration schemes; one commonly used scheme, *Gauss-Legendre* numerical integration, will be briefly considered here. Those totally unfamiliar with numerical integration may choose to review suitable reference before proceeding; Fisher [11]. Recall that a brief example of a crude numerical integration scheme was given in Chapter 1.

Gauss-Legendre Integration

Gauss-Legendre integration requires a three step process. The first step is to transform the domain of the integral under consideration such that the new limits of integration are from -1 to $+1$; these limits, used in one-dimensional problems, allow the integration process to be expressed in a generic form. These same limits are again employed when integration of two- and three-dimensional elements is performed. For triangular surface elements and tetrahedral volume elements, the limits of integration are slightly different, but the principle remains the same.

After transforming the domain, the second step is to transform the original integrand to one that corresponds to the new limits of integration. Lastly, in Step 3, the value of the transformed integrand is computed at one or more Gauss points, and the results summed to obtain the required approximation. It other words, suppose that the following integral is to be evaluated:

$$\int_{X_a}^{X_b} f(X)\, dX$$

The limits of integration are transformed onto -1 to $+1$:

$$\int_{X_a}^{X_b} f(X)\, dX = \int_{-1}^{1} g(r) dr$$

The function $g(r)$ in the equation above results from replacing the X variable in $f(X)$ with a function of the parametric variable, r. The function of the parametric variable evolves from mapping the original domain to a new domain with limits of -1 to $+1$. The integration process above may be familiar; it is typically introduced in undergraduate integral calculus course work, under the topics of *integrating via the method of substitution* and *integration by change of variable*; Fisher [11, p. 334].

The last step in the Gauss-Legendre procedure is to approximate the transformed integral by a sum of terms:

$$\int_{-1}^{1} g(r)dr \approx w_1 g(r_1) + w_2 g(r_2) + \cdots w_n g(r_n) \tag{5.7.1}$$

The values of r_i in (5.7.1) are termed *Gauss points*, and the integral is approximated using n such points. It is at the Gauss points that the function $g(r)$ is evaluated, and then multiplied by a weighting factor, w_i. Table 5.1 lists the first three Gauss points, along with the associated weighting factors, for one-dimensional numerical integration; see Stroud [12] for in-depth details.

Table 5.1 One-Dimensional Gauss Integration Points and Weights

n	r_i	w_i
1	0	2
2	$\sqrt{\frac{1}{3}}$	1
	$-\sqrt{\frac{1}{3}}$	1
3	$\sqrt{\frac{3}{5}}$	$\frac{5}{9}$
	0	$\frac{8}{9}$
	$-\sqrt{\frac{3}{5}}$	$\frac{5}{9}$

The process of numerical integration is best explained using a simple example. Recall Example 2.4, where the average cross sectional area of a non-uniform shaft was computed:

$$\bar{A} = \frac{1}{4}\int_{1}^{5} \pi X^{-2}\, dX = \left(\frac{-\pi}{4}\right)\left(\frac{1}{5} - \frac{1}{1}\right) = 0.2\pi \tag{5.7.2}$$

Notice that the limits of integration are $X_a=1$ and $X_b=5$. The first step in the numerical integration process is to transform (*map*) the limits of integration onto the domain -1 to $+1$:

$$\overline{A} = \int_1^5 \frac{1}{4}\pi X^{-2}\, dX = \int_{-1}^{1} g(r)\, dr \tag{5.7.3}$$

The mapping of the domain can be either linear or non-linear. Assuming that linear mapping is to be employed, a two-term polynomial can be used:

$$X(r) = a_1 + a_2 r \tag{5.7.4}$$

The mapping function relates the limits of integration on X to corresponding values on r. At the lower limit of the integrals in (5.7.3), it is noted that $r = -1$ and $X=1$. Using (5.7.4) to map these points:

$$X(-1) = a_1 + a_2(-1) = 1 \tag{5.7.5}$$

Likewise, at the upper limit of the integrals, $r = +1$ and $X=5$, thus:

$$X(1) = a_1 + a_2(1) = 5 \tag{5.7.6}$$

The two equations above can be solved for the variables a_1 and a_2. The values are then substituted back into (5.7.4), yielding:

$$X(r) = 3 + 2r \tag{5.7.7}$$

It may be apparent that this process is exactly the same as that which was used to create the 2-node, parametric line element for rod applications, from which the generic mapping function was shown to be:

$$X(r) = \frac{1}{2}(1-r)X_a + \frac{1}{2}(1+r)X_b \tag{5.7.8}$$

Substituting $X_a=1$ and $X_b=5$ into (5.7.8), and using a little algebra, it can be shown that for this case the generic form (5.7.8) is equivalent to (5.7.7).

When parametric finite elements are created, using the natural coordinate system, there are several benefits: Shape functions are more easily derived, compatibility is ensured, and the element is transformed onto a domain that facilitates the use of numerical integration.

Continuing with the numerical integration example, we return to (5.7.3). Using linear mapping, the X variable in the integrand of (5.7.3) is expressed using (5.7.7), then, substituting (5.7.7) into (5.7.3):

$$\overline{A} = \int_1^5 \frac{1}{4}\pi\, X^{-2}\, dX = \int_1^5 \frac{1}{4}\pi\,(3+2r)^{-2}\, dX \tag{5.7.9}$$

Since the integrand on the right of (5.7.9) is a function of r, the variable of integration must be replaced with r. Taking the derivative of (5.7.7), it is found that $2dr$ is equivalent to dX:

$$\frac{dX}{dr} = \frac{d}{dr}(3+2r) = 2 \qquad \therefore dX = 2dr \tag{5.7.10}$$

Substituting $2dr$ for dX, and applying the appropriate limits, the right side of (5.7.9) yields:

$$\overline{A} = \int_{-1}^{1} \frac{1}{4}\pi\,(3+2r)^{-2}\, 2dr \tag{5.7.11}$$

Observing both (5.7.3) and (5.7.11), it is apparent that the transformed integrand $g(r)$ is, for this problem:

$$g(r) = \frac{\pi}{4}(3+2r)^{-2}(2) \tag{5.7.12}$$

Note that the closed-form solution[3] to (5.7.11) is:

$$\overline{A} = \int_{-1}^{1} \frac{1}{4}\pi\,(3+2r)^{-2}\, 2dr = 0.2\pi \tag{5.7.13}$$

The closed-form solution to (5.7.11) is of course equivalent to (5.7.2), since (5.7.11) is the same integral, expressed in a different form. However, we wish to employ *numerical* integration. To do so, the integral in (5.7.11) is expressed as a sum of terms, each term evaluated at a Gauss point. As you might imagine, the more Gauss points used, the better the approximation, until convergence is achieved. It can be shown that, using p Gauss points, a polynomial of degree $2p-1$

---

[3] The substitution $Z=3+2r$ is made. Taking the derivative, $dZ=2dr$. Using both, the integral is expressed as: $\overline{A} = \int_{-1}^{1} \frac{\pi}{4}(3+2r)^{-2}\, 2dr = \int_1^5 \frac{\pi}{4}(Z)^{-2}\, dZ = 0.2\pi$.

is exactly integrated. For example, if two Gauss points are used, a third order polynomial can be exactly integrated, after which using more points will have no further benefit. Again, the approximation for one-dimensional Gauss-Legendre integration is stated as:

$$\int_{-1}^{1} g(r)dr \approx w_1 g(r_1) + w_2 g(r_2) + \cdots w_n g(r_n) \tag{5.7.14}$$

Substituting (5.7.12) into (5.7.14), and applying a 2-point approximation:

$$\overline{A} = \int_{-1}^{1} \frac{\pi}{4}(3+2r)^{-2}\, 2dr \approx w_1 \frac{\pi}{2}(3+2r_1)^{-2} + w_2 \frac{\pi}{2}(3+2r_2)^{-2} \tag{5.7.15}$$

Referring to Table 5.1, the weighting factors and values of r_i are substituted into (5.7.15):

$$\overline{A} \approx (1)\frac{\pi}{2}\left(3+2\left(\sqrt{\frac{1}{3}}\right)\right)^{-2} + (1)\frac{\pi}{2}\left(3+2\left(-\sqrt{\frac{1}{3}}\right)\right)^{-2} = 0.176\pi \tag{5.7.16}$$

Although two-point integration can exactly integrate a third order polynomial, note that the integrand in this particular problem is not a polynomial.

Because the closed-form solution is 0.2π, the approximation above may be considered too inaccurate. However, using a 3-point approximation, the solution becomes:

$$\begin{aligned}\overline{A} \approx & \left(\frac{5}{9}\right)\frac{\pi}{2}\left(3+2\left(\sqrt{\frac{3}{5}}\right)\right)^{-2} + \left(\frac{8}{9}\right)\frac{\pi}{2}(3+2(0))^{-2} \\ & + \left(\frac{5}{9}\right)\frac{\pi}{2}\left(3+2\left(-\sqrt{\frac{3}{5}}\right)\right)^{-2} = 0.195\pi\end{aligned} \tag{5.7.17}$$

The approximation given by (5.7.17) compares quite favorably with the correct result. If more accuracy is desired, additional Gauss points can be employed.

It is observed that the lower the integration order, the smaller the value for the approximation. It can be inferred that, when using numerical integration to compute the element stiffness matrix, using too few Gauss points results in a "smaller magnitude" stiffness matrix. This renders an element that is less stiff than an element using the correct number of Gauss points. This fact allows the finite element software developer to mitigate, in some cases, the overly stiff nature of finite elements. An element whose stiffness is computed using this technique is called a *reduced integration* element, and will be considered shortly.

Numerical Integration in Two Dimensions

Although numerical integration in one dimensional problems can be helpful, two- and three-dimensional problems receive the most benefit. For two-dimensional, square domains, the integration process is a direct extension of the one-dimensional numerical procedure just illustrated.

Given an integrand that is a function of two variables, for instance, the integral used to compute the stiffness for the 4-node quadrilateral surface element, it can be shown that the integral may be approximated by a double sum:

$$^{e}\underline{\underline{K}} = \int_{-1}^{1} \int_{-1}^{1} h(r,s)\, dr\, ds \approx \sum_{i=1}^{i=n} \sum_{j=1}^{j=n} w_i w_j\, h(r_i, s_j) \tag{5.7.18}$$

The n value in (5.7.18) is the number of Gauss points in one direction. For approximations in three-dimensional space, a triple sum is used.

Applying Numerical Integration to Quadrilateral Elements

When used for one-dimensional problems, the integral to be evaluated is transformed to the interval −1 to +1. For a quadrilateral domain, the same interval is used but in two directions, as illustrated in Figure 5.16.

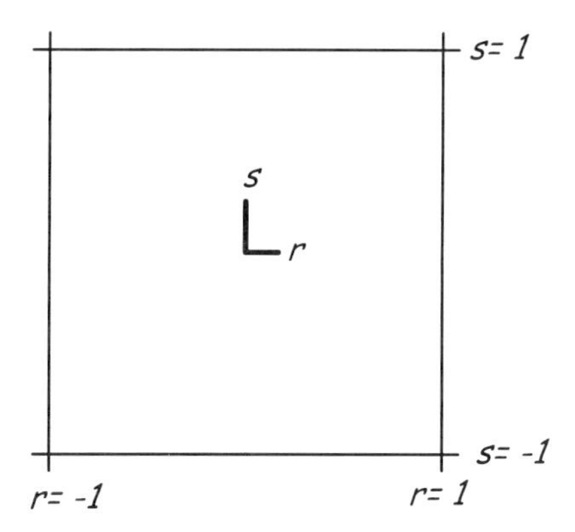

Figure 5.16 Gauss-Legendre Integration Domain for a Rectangular Area

Note that these are the same coordinates chosen as the natural coordinates for the 4-node, isoparametric quadrilateral element.

As in the one-dimensional examples, one can choose the number of integration points to achieve the desired degree of accuracy in two-dimensional problems as well. In the examples that are considered in this text, the number of integration points in one direction are the same as in any other direction. However, this need not always be the case. Some shell elements, for example, may use fewer integration points in the thickness direction than in the membrane directions of the shell. The reason for this is that displacement in these elements often varies linearly in the thickness but, in the membrane directions of the shell, higher order displacement functions may be required to characterize the response. An example illustrating this principle is given in Bathe [7, p. 286].

As illustrated in Figure 5.17, Gaussian integration used on two-dimensional quadrilateral domains is a direct extension of the one-dimensional case.

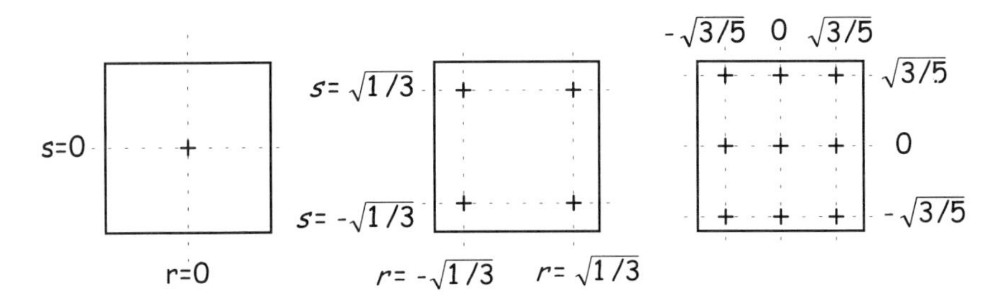

Figure 5.17 Gauss Points: 1 by 1, 2 by 2, and 3 by 3

Consulting Table 5.1, and observing the *2 by 2* integration scheme in Figure 5.17, it is apparent that the same Gauss point values are used for *2 by 2* integration as were used for 2 point integration in the one-dimensional example; the weighting functions are also the same, having a value of unity.

Besides the 4-node quadrilateral element, there are Gauss-Legendre integration schemes for other element shapes, both two- and three-dimensional. For example, when applied to hexagonal domains, such as the 8-node volume element, the integration scheme analogous to the four-point *2 by 2* is an eight-point *2 by 2 by 2* scheme.

The accuracy rules applied in two-dimensional problems are the same as used in one-dimension: using p points in any one direction allows a polynomial of $2p-1$ to be exactly integrated. Hence, while a *2* point scheme in a one-dimensional problem can integrate a polynomial up to order three, a *2 by 2* scheme can integrate a polynomial of *two* variables up to order three; a *2 by 2 by 2* scheme integrates a polynomial of *three* variables up to order three. Consider the example below in which the equations for *2 by 2* Gauss-Legendre integration are given for a 4-node surface element.

Example 5.1—Gauss-Legendre for a 4-Node Plane Stress Element
The stiffness matrix equations are established for a 4-node, isoparametric, quadrilateral, C^0 surface element using *2 by 2* Gauss-Legendre numerical integration. The parent element is shown in Figure 5.18.

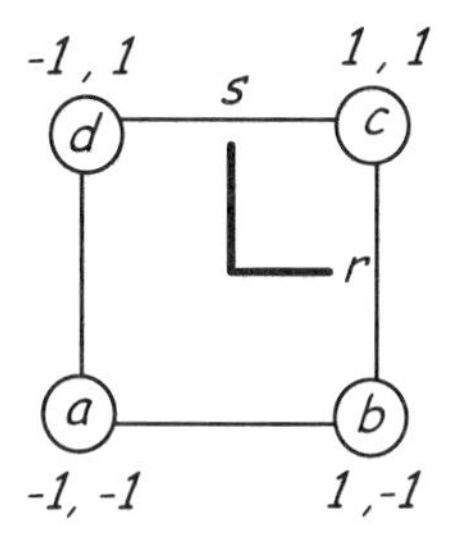

Figure 5.18 A 4-Node Isoparametric Element

Recall from Section 5.3 that the equation for computing the element stiffness for the 4- node, quadrilateral, plane stress element was given as:

$$
{}^{e}\underline{\underline{K}} = \int_{-1}^{1}\int_{-1}^{1} \begin{bmatrix} B_{11} & 0 & B_{22} \\ 0 & B_{22} & B_{11} \\ B_{13} & 0 & B_{24} \\ 0 & B_{24} & B_{13} \\ B_{15} & 0 & B_{26} \\ 0 & B_{26} & B_{15} \\ B_{17} & 0 & B_{28} \\ 0 & B_{28} & B_{17} \end{bmatrix} \underline{\underline{E}} \begin{bmatrix} B_{11} & 0 & B_{13} & \cdots & 0 \\ 0 & B_{22} & 0 & \cdots & B_{28} \\ B_{22} & B_{11} & B_{24} & \cdots & B_{17} \end{bmatrix} t\left(\det \underline{\underline{J}}\right) ds\, dr
$$

Since the element to be integrated is isoparametric, it has already been mapped to the correct domain. Referring to the left hand side of (5.7.18), the integral above will be approximated by a double sum, with *h(r,s)* given as:

$$h(r,s)=\begin{bmatrix} B_{11} & 0 & B_{22} \\ 0 & B_{22} & B_{11} \\ B_{13} & 0 & B_{24} \\ 0 & B_{24} & B_{13} \\ B_{15} & 0 & B_{26} \\ 0 & B_{26} & B_{15} \\ B_{17} & 0 & B_{28} \\ 0 & B_{28} & B_{17} \end{bmatrix} \underline{\underline{E}} \begin{bmatrix} B_{11} & 0 & B_{13} & \cdots & 0 \\ 0 & B_{22} & 0 & \cdots & B_{28} \\ B_{22} & B_{11} & B_{24} & \cdots & B_{17} \end{bmatrix} t\left(\det \underline{\underline{J}}\right) \tag{5.7.19}$$

Substituting (5.7.19) into (5.7.18):

$$\underline{\underline{{}^{e}K}} \approx \sum_{i=1}^{i=n}\sum_{j=1}^{j=n} w_i w_j \begin{bmatrix} B_{11} & 0 & B_{22} \\ 0 & B_{22} & B_{11} \\ B_{13} & 0 & B_{24} \\ 0 & B_{24} & B_{13} \\ B_{15} & 0 & B_{26} \\ 0 & B_{26} & B_{15} \\ B_{17} & 0 & B_{28} \\ 0 & B_{28} & B_{17} \end{bmatrix} \underline{\underline{E}} \begin{bmatrix} B_{11} & 0 & B_{13} & \cdots & 0 \\ 0 & B_{22} & 0 & \cdots & B_{28} \\ B_{22} & B_{11} & B_{24} & \cdots & B_{17} \end{bmatrix} t\left(\det \underline{\underline{J}}\right)$$

Since *2 by 2* Gauss points are employed, *n=2* in the summations above; and, because the weighting functions are equal to unity for *2 by 2* integration, the stiffness formula in this case can be expressed as:

$$\underline{\underline{{}^{e}K}} \approx \sum_{i=1}^{i=2}\sum_{j=1}^{j=2} \begin{bmatrix} B_{11} & 0 & B_{22} \\ 0 & B_{22} & B_{11} \\ B_{13} & 0 & B_{24} \\ 0 & B_{24} & B_{13} \\ B_{15} & 0 & B_{26} \\ 0 & B_{26} & B_{15} \\ B_{17} & 0 & B_{28} \\ 0 & B_{28} & B_{17} \end{bmatrix} \underline{\underline{E}} \begin{bmatrix} B_{11} & 0 & B_{13} & \cdots & 0 \\ 0 & B_{22} & 0 & \cdots & B_{28} \\ B_{22} & B_{11} & B_{24} & \cdots & B_{17} \end{bmatrix} t\left(\det \underline{\underline{J}}\right)$$

Or, in more compact form:

$$\underline{\underline{^{e}K}} \approx \sum_{i=1}^{i=2} \sum_{j=1}^{j=2} \underline{\underline{B}}^T \underline{\underline{E}}\,\underline{\underline{B}}\, t \left(\det \underline{\underline{J}}\right) \tag{5.7.20}$$

Note that (5.7.20) is equivalent to four terms, each term evaluated at one *Gauss* point:

$$\begin{aligned}\underline{\underline{^{e}K}} \approx &\left(\underline{\underline{B}}^T \underline{\underline{E}}\,\underline{\underline{B}}\, t \det \underline{\underline{J}}\right)\Big|_{r_1, s_1} + \left(\underline{\underline{B}}^T \underline{\underline{E}}\,\underline{\underline{B}}\, t \det \underline{\underline{J}}\right)\Big|_{r_1, s_2} + \left(\underline{\underline{B}}^T \underline{\underline{E}}\,\underline{\underline{B}}\, t \det \underline{\underline{J}}\right)\Big|_{r_2, s_1} \\ &+ \left(\underline{\underline{B}}^T \underline{\underline{E}}\,\underline{\underline{B}}\, t \det \underline{\underline{J}}\right)\Big|_{r_2, s_2}\end{aligned} \tag{5.7.21}$$

The Gauss points used in (5.7.21) have the respective values of r and s:

$$r_1 = -\sqrt{\frac{1}{3}} \qquad s_1 = -\sqrt{\frac{1}{3}}$$

$$r_2 = \sqrt{\frac{1}{3}} \qquad s_2 = \sqrt{\frac{1}{3}}$$

Another way to consider (5.7.21) is that the element stiffness matrix consists of four *partial* matrices, with each partial matrix representing the stiffness associated with its respective Gauss point.

Although the equations in (5.7.21) may appear formidable, they are actually quite easy to program on a digital computer. A software routine to compute the elements stiffness matrix is shown in Example 5.2, below. More elegant and powerful routines that accomplish this are given by Cook [10, p. 175] and Bathe [7, p. 295].

Example 5.2—FORTRAN Routine to Compute 4-Node, Plane Stress Stiffness
The equations for computing the stiffness matrix for a 4-node, quadrilateral, C^0 surface element, (5.7.21), are quite simple to program. Figure 5.19 depicts a 4-node, isoparametric surface element in global space. Assume that this element is to be used for a plane stress application.

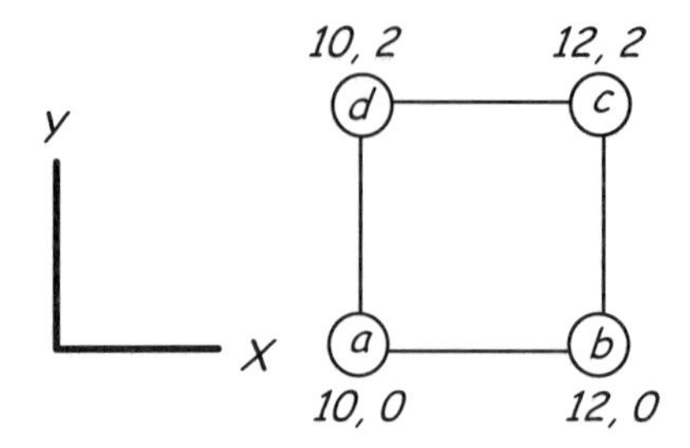

Figure 5.19 A 4-Node Isoparametric Element

To compute the element stiffness, the FORTRAN routine must be supplied with the coordinates of the nodes, a thickness value, and the components of the material matrix. The global coordinates, as input to the FORTRAN routine, are:

$$XA = 10 \quad XB = 12 \quad XC = 12 \quad XD = 10$$
$$YA = 0 \quad YB = 0 \quad YC = 2 \quad YD = 2$$

For simplicity, assume that the elastic modulus is equal to unity, and Poisson's ratio is zero. Hence, the material matrix for this hypothetical element is:

$$\underline{\underline{E}} = \frac{E}{1-\nu^2}\begin{bmatrix} 1 & \nu & 0 \\ \nu & 1 & 0 \\ 0 & 0 & \frac{1-\nu}{2} \end{bmatrix} = \begin{bmatrix} 1 & 0 & 0 \\ 0 & 1 & 0 \\ 0 & 0 & \frac{1}{2} \end{bmatrix}$$

The non-zero components of the matrix above are:

$$E(1,1) = 1 \qquad E(2,2) = 1 \qquad E(3,3) = 0.5$$

Assume that the thickness is equal to unity:

$$THICK = 1$$

The values for the variables above are defined in the software routine, using DATA statements. The routine uses *2 by 2* integration; however, this can be modified by simply adding more Gauss coordinates and changing the counters in the DO 70 and DO 80 statements. The equations used in this routine are derived from equations found at the end of Section 5.3.

```
**********************************************************************
FORTRAN to compute stiffness for 4-node, isoparametric, C^0 surface element
**********************************************************************
      DIMENSION E(3,3),GAUSS_R(2),GAUSS_S(2),B(3,8),EB(3,8),BT_EB(4,8,8)
      DIMENSION PARTSTIF(4,8,8),STIFF(8,8)
      INTEGER GPOINT
      REAL J11,J12,J21,J22
1000  FORMAT(' ',8(F9.5,1X))
* Define global coordinates, the material matrix, and thickness
      DATA XA,XB,XC,XD,YA,YB,YC,YD /10,12,12,10,0,0,2,2/
      DATA E(1,1),E(2,2),E(3,3) /1,1,0.5/
      THICK=1
* Begin loop to evaluate integrand at each of 4 Gauss points
      DO 80 I=1,2
         DO 70 J=1,2
         GPOINT=GPOINT+1
* Define the r and s coordinate values of the Gauss integration points
         DATA GAUSS_R(1), GAUSS_R(2) /-0.5773502692, 0.5773502692/
         DATA GAUSS_S(1), GAUSS_S(2) /-0.5773502692, 0.5773502692/
* Define derivative of shape functions for current Gauss Point
         R=GAUSS_R(I)
         S=GAUSS_S(J)
         C=-0.25
         DR1=C*(1-S)
         DR2=-1*C*(1-S)
         DR3=-1*C*(1+S)
         DR4=C*(1+S)
         DS1=C*(1-R)
         DS2=C*(1+R)
         DS3=-1*C*(1+R)
         DS4=-1*C*(1-R)
* Calc Jacobian components and determinant for current G.P.
         J11=DR1*XA+DR2*XB+DR3*XC+DR4*XD
         J12=DR1*YA+DR2*YB+DR3*YC+DR4*YD
         J21=DS1*XA+DS2*XB+DS3*XC+DS4*XD
         J22=DS1*YA+DS2*YB+DS3*YC+DS4*YD
         DETJ=(J11*J22)-(J12*J21)
* Calc B matrix for the current G.P.
         B(1,1)=(J22*DR1-J12*DS1)/DETJ
         B(1,3)=(J22*DR2-J12*DS2)/DETJ
         B(1,5)=(J22*DR3-J12*DS3)/DETJ
         B(1,7)=(J22*DR4-J12*DS4)/DETJ
         B(2,2)=(-1.*J21*DR1+J11*DS1)/DETJ
```

```
            B(2,4)=(-1.*J21*DR2+J11*DS2)/DETJ
            B(2,6)=(-1.*J21*DR3+J11*DS3)/DETJ
            B(2,8)=(-1.*J21*DR4+J11*DS4)/DETJ
            B(3,1)=B(2,2)
            B(3,2)=B(1,1)
            B(3,3)=B(2,4)
            B(3,4)=B(1,3)
            B(3,5)=B(2,6)
            B(3,6)=B(1,5)
            B(3,7)=B(2,8)
            B(3,8)=B(1,7)
* Calc product EB for the current G.P.
              DO 30 M=1,3
                 DO 20 N=1,8
                 EBSUM=0.0
                    DO 10 P=1,3
                    EB(M,N)=EBSUM+E(M,P)*B(P,N)
                    EBSUM=EB(M,N)
10                  CONTINUE
20               CONTINUE
30            CONTINUE
* Multiply EB by B^T
                 DO 60 M=1,8
                    DO 50 N=1,8
                    BT_EBSUM=0
                       DO 40 P=1,3
                       BT_EB(GPOINT,M,N)=BT_EBSUM+B(P,M)*EB(P,N)
                       BT_EBSUM=BT_EB(GPOINT,M,N)
40                     CONTINUE
* Multiply B^T EB by DETJ and THICK to form the partial stiffness matrix
                       PARTSTIF(GPOINT,M,N)=BT_EB(GPOINT,M,N)*DETJ*THICK
50                  CONTINUE
60               CONTINUE
70            CONTINUE
80         CONTINUE
* Open a file called ESTIFF to write the results
           OPEN(UNIT=1,FILE='ESTIFF',STATUS='UNKNOWN')
* Sum the partial matrices from each Gauss point
              DO 110 J=1,8
                 DO 100 K=1,8
                 STIFFSUM=0
                    DO 90 GCOUNT=1,4
```

```
                 STIFF(J,K)=PARTSTIF(GCOUNT,J,K)+STIFFSUM
                 STIFFSUM=STIFF(J,K)
90               CONTINUE
100           CONTINUE
              WRITE(1,1000)(STIFF(J,K),K=1,8)
110         CONTINUE
            END
```

The output from the FORTRAN program shows that the element stiffness matrix for this element is:

0.500	0.125	-0.250	-0.125	-0.250	-0.125	0.000	0.125
0.125	0.500	0.125	0.000	-0.125	-0.250	-0.125	-0.250
-0.250	0.125	0.500	-0.125	0.000	-0.125	-0.250	0.125
-0.125	0.000	-0.125	0.500	0.125	-0.250	0.125	-0.250
-0.250	-0.125	0.000	0.125	0.500	0.125	-0.250	-0.125
-0.125	-0.250	-0.125	-0.250	0.125	0.500	0.125	0.000
0.000	-0.125	-0.250	0.125	-0.250	0.125	0.500	-0.125
0.125	-0.250	0.125	-0.250	-0.125	0.000	-0.125	0.500

One might see that the subroutine does more work than necessary: since the element stiffness matrix is symmetric, only the top (or bottom) triangle of the matrix need be computed, with the other entries determined by symmetry:

0.500	0.125	-0.250	-0.125	-0.250	-0.125	0.000	0.125
	0.500	0.125	0.000	-0.125	-0.250	-0.125	-0.250
		0.500	-0.125	0.000	-0.125	-0.250	0.125
			0.500	0.125	-0.250	0.125	-0.250
				0.500	0.125	-0.250	-0.125
					0.500	0.125	0.000
		-SYMMETRY-				0.500	-0.125
							0.500

Guidelines for Numerical Integration

When using numerical integration to compute element stiffness, the analyst is sometimes given the choice of using "full integration" or "reduced integration." Before these terms are defined, the nature of the stiffness matrix integrand must be examined.

Constant Jacobian Determinant

Consider the double sum that is used to approximate the stiffness matrix for the 4-

node, isoparametric plane stress element:

$$\underline{\underline{{}^{e}K}} \approx \sum_{i=1}^{i=2}\sum_{j=1}^{j=2} w_i w_j \underline{\underline{B}}^T \underline{\underline{E}}\,\underline{\underline{B}}\, t\left(\det \underline{\underline{J}}\right) \tag{5.7.22}$$

Or, writing out the terms in the matrices:

$$\underline{\underline{{}^{e}K}} \approx \sum_{i=1}^{i=n}\sum_{j=1}^{j=n} w_i w_j \begin{bmatrix} B_{11} & 0 & B_{22} \\ 0 & B_{22} & B_{11} \\ B_{13} & 0 & B_{24} \\ 0 & B_{24} & B_{13} \\ B_{15} & 0 & B_{26} \\ 0 & B_{26} & B_{15} \\ B_{17} & 0 & B_{28} \\ 0 & B_{28} & B_{17} \end{bmatrix} \frac{E}{1-v^2}\begin{bmatrix} 1 & v & 0 \\ v & 1 & 0 \\ 0 & 0 & \frac{1-v}{2} \end{bmatrix}\begin{bmatrix} B_{11} & 0 & B_{13} & \cdots & 0 \\ 0 & B_{22} & 0 & \cdots & B_{28} \\ B_{22} & B_{11} & B_{24} & \cdots & B_{17} \end{bmatrix} t\left(\det \underline{\underline{J}}\right) \tag{5.7.23}$$

The equation above actually represents a system of equations. If each equation in the system is a polynomial, and p Gauss points are used in each coordinate direction, a polynomial with two variables of maximum order $2p-1$ is exactly integrated. However, if the equations are not polynomials, exact integration cannot be achieved, irrespective of the number of Gauss points used. Consider one of the terms in the B-matrix:

$$B_{11} = \frac{1}{\det \underline{\underline{J}}}\Big[\left(N_{1,s}Y_a + N_{2,s}Y_b + N_{3,s}Y_c + N_{4,s}Y_d\right)N_{1,r} - \left(N_{1,r}Y_a + N_{2,r}Y_b + N_{3,r}Y_c + N_{4,r}Y_d\right)N_{1,s}\Big]$$

Referring to derivatives given by (5.3.34), it is seen that B_{11} contains linear terms in r and s; likewise for the rest of the terms in the B-matrix. Thus, when the matrix multiplication $\underline{\underline{B}}^T \underline{\underline{E}}\underline{\underline{B}}$ is performed, the double summation in (5.7.23) will contain equations of the following form:

$$\underline{\underline{{}^{e}K}} = \sum_{i=1}^{i=n}\sum_{j=1}^{j=n} w_i w_j \frac{1}{\left(\det \underline{\underline{J}}\right)^2}\left(\textit{up to 2nd order polynomials}\right) t\left(\det \underline{\underline{J}}\right)$$

Or, canceling the $\det \underline{\underline{J}}$ terms:

$$
{}^{e}\underline{\underline{K}} = \sum_{i=1}^{i=n} \sum_{j=1}^{j=n} w_i w_j \frac{1}{\left(\det \underline{\underline{J}}\right)} \left(up\ to\ 2nd\ order\ polynomials\right) t \tag{5.7.24}
$$

Now, if determinant of the Jacobian matrix and t are constant, the above simply amounts to integrating a second order polynomial. If two points are used in both directions, (5.7.22) is adequately integrated, because a polynomial of up to order three can be exactly integrated. However, if determinant of the Jacobian matrix also contains polynomial terms, numerical integration will not be exact, since the ratio of two polynomials is not, in general, a polynomial. The question is: "When is $\det \underline{\underline{J}}$ constant?" Recalling (5.3.25):

$$
\det \underline{\underline{J}} = J_{11} J_{22} - J_{12} J_{21}
$$

If the individual terms of the Jacobian are constant, then the determinant will also be constant. What causes the individual terms of the determinant to be constant? Consider the first Jacobian term as given by (5.3.16):

$$
J_{11} = \frac{\partial X}{\partial r} = \frac{\partial}{\partial r}\left(N_1 X_a + N_2 X_b + N_3 X_c + N_4 X_d\right)
$$

Substituting the shape functions given by (5.3.5) into the above, and performing the differentiation:

$$
J_{11} = \frac{-1}{4}\left[(1-s)X_a - (1-s)X_b - (1+s)X_c + (1+s)X_d\right]
$$

Rearranging the above:

$$
J_{11} = \frac{-1}{4}\left(X_a - X_b - X_c + X_d + s(-X_a + X_b - X_c + X_d)\right) \tag{5.7.25}
$$

It may be apparent that if the nodal coordinates acting as the coefficient of the s term are combined as shown, and the result is zero, J_{11} is simply a constant; that is:

$$
\begin{aligned}
&if: \\
&(-X_a + X_b - X_c + X_d) = 0
\end{aligned}
$$

then:

$$J_{11} = \frac{-1}{4}\left(X_a - X_b - X_c + X_d\right) = constant$$

There are some element configurations that will result in the Jacobian terms being constant; for example, consider the rectangular element in Figure 5.20, below.

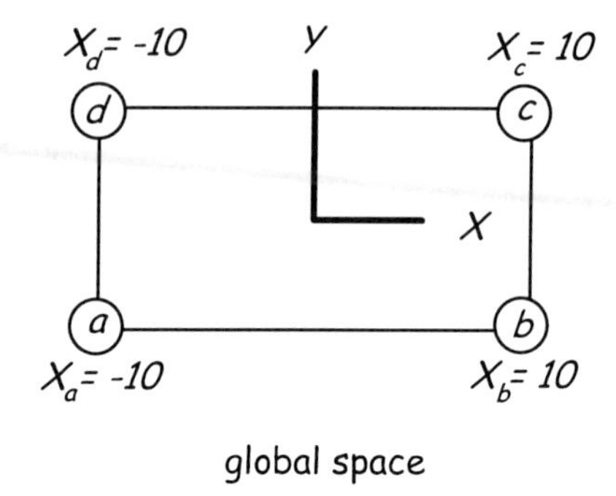

Figure 5.20 A 4-Node Isoparametric Surface Element

Using the nodal coordinates from Figure 5.20 in (5.7.25):

$$J_{11} = \frac{-1}{4}\left(-10 - 10 - 10 - 10 + s\left[-(-10) + (10) - (10) + (-10)\right]\right)$$

Hence:

$$J_{11} = \frac{-1}{4}(-40) = \text{constant}$$

Examining the other terms in the Jacobian matrix, it becomes apparent that certain combinations of element nodal coordinates can result in a constant Jacobian determinant. Typically, if an element in global space is simply a scaled version of the parent element, the Jacobian determinant will be constant. This implies that for higher order elements, the element in global space must have its additional nodes at mid-side, i.e., halfway between the nodes that are at the corners of the element.

If the element in global space deviates from the characteristic of the parent, by virtue of either its shape or the placement of its nodes, the element is considered to be *distorted*. Some types of distortion lead to a non-constant Jacobian determinant. Recalling that the parent element for a 4-node quadrilateral is square, the element shown in Figure 5.20 would be considered distorted, since all of its edges are not of equal length. However, as suggested by computing the J_{11} term given by (5.7.25), this particular type of distortion still renders a constant Jacobian determinant.

Distortion and Numerical Integration

The Jacobian determinant is generally not constant due to element distortion, and numerical integration cannot be performed exactly, regardless of how many Gauss points are used. Some types of element distortion, if severe, can cause the Jacobian determinant to have a significant, detrimental influence upon the numerical integration process. For instance, if element distortion causes the Jacobian determinant to become very small, terms in the stiffness matrix integrand can exhibit asymptotic behavior, making the integration process much less accurate. Element distortion that influences the Jacobian determinant must be held within certain limits if suitable integration accuracy is to be maintained. With a limited amount of distortion, error introduced by inaccuracies in numerical integration can be discounted. Distorted elements will be considered again in Chapter 7.

Full Integration

The term "full integration" of a finite element stiffness matrix refers to the number of Gauss points needed to exactly integrate a given *polynomial*. When the 4-node quadrilateral element is assigned a constant thickness, and has a constant Jacobian determinant, two Gauss points in each direction allows the integrand to be fully integrated. This is because the stiffness matrix integrand of the 4-node element with constant thickness and a constant Jacobian determinant contains up to second order polynomial terms, as shown in (5.7.24). A second order polynomial in two variables is adequately integrated with two Gauss points in each coordinate direction.

In the case of an 8-node quadrilateral, the integrand contains a polynomial with 4th order terms at most, given the same restrictions placed upon thickness and Jacobian determinant mentioned above. Hence, this element is fully integrated using *3 by 3* integration. Table 5.2 lists integration points for these and other elements. It is noted that as the stiffness matrix integrand deviates from a polynomial form, the numerical integration process becomes less accurate.

Reduced Integration

The term "reduced integration" means that integration of the stiffness matrix is being performed using fewer Gauss points than needed to fully integrate a stiffness matrix that contains polynomial terms. Table 5.2 considers typical integration schemes for parametric elements that are used for C^0 applications.

Each element is categorized by the highest order strain state that it can exhibit in its undistorted configuration. This implies that an element can exhibit different order strain states depending upon which component of strain is being considered. Strain states that isoparametric elements exhibit are discussed in MacNeal [1, p. 88].

One advantage of reduced integration is that less computational effort is required to compute the stiffness matrix. Using reduced integration on a 20-node hexahedral (using 8 points instead of 27) can result in a 30% reduction in computational costs; [13, p. 3.2.4-4]. Another advantage of using a lower order integration scheme can result in a less stiff structure. Hence, reduced integration also has the beneficial effect of allowing displacement based finite elements, which are typically overly stiff, to become more flexible. This results in the use of fewer elements to obtain a required level of accuracy.

Table 5.2 Integration Points for C^0 Elements

nodes	shape	characteristic	full	r
constant strain elements				
3	triangular	surface	1	-
4	tetrahedral	volume	1	-
linear strain elements				
4	quad	surface	4	1
6	triangular	surface	3	-
10	tetrahedral	volume	5	4
quadratic strain elements				
8	quad	surface	9	4
8	hexahedral	volume	8	1
cubic strain elements				
9	quad	surface	9	4
20	hexahedral	volume	27	8

Unfortunately, reduced integration can result in the unwanted characteristic called *spurious deformation modes*. Spurious modes of deformation describe the phenomenon where an element using reduced integration exhibits a deformed shape but strain within the element is computed as zero. For example, a 4-node, quadrilateral surface element for plane stress applications using one point integration can exhibit a spurious mode called the hourglass mode, as illustrated

in Figure 5.21. If the elements in Figure 5.21 are integrated only at the center of the element, the hourglass deformation pattern is possible because strain at the center averages to zero and the strain energy of the system is balanced without the presence of a corresponding set of nodal loads.
In contrast, with full integration the non-zero strain values at the four Gauss points would be computed during the numerical integration process, and the strain energy associated with this deformation would have to be accompanied by a corresponding set of nodal loads. There are more precise ways to describe spurious modes of deformation; see MacNeal [1, p. 272] for in-depth details on the topics of reduced integration and spurious modes of deformation. The effect of using reduced integration upon a beam structure is illustrated in Cook [10, p. 561]; in the same reference, page 326, reduced integration of plate elements is considered.

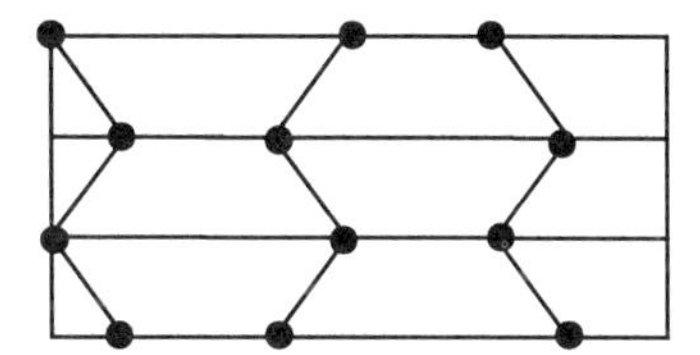

Figure 5.21 A Group of Elements Exhibiting Hourglass Mode

To combat the problem of spurious modes of deformation, some elements using reduced integration may be designed with an *hourglass control* feature. The topic of controlling hourglass patterns is discussed by Kosloff [14], Cook [15], and Jasti [16].

Numerical Integration for Various Elements

Details of numerical integration are considered in Cowper [17], while Bathe [7] provides a convenient table of integration points for triangular and quadrilateral elements. The topic of numerical integration of hexahedral (brick) elements is considered in Irons [18]. In addition to reduced integration, another technique called selective integration also exists. Selective integration allows some terms to be integrated with low level integration (using few points) while other terms are integrated at a higher order (using more points). The interested reader is referred to Pawsey [19] for details on selective integration. The analyst should consult suitable software documentation for details of how, and at what points, a particular element is integrated.

Optimal Stress Sampling Points

Armed with the knowledge of reduced integration, the concept of optimal stress sampling points can be discussed. To begin, a brief overview of some salient points is given.

In general, the computed stress within an element is approximate. With an increasing number of elements, convergence of the predicted stress to the exact is anticipated. To reduce computational costs, the analyst typically attempts to use the fewest number of elements possible while still generating an approximation for stress that is of suitable accuracy. It is possible to compute stress anywhere in the element using the system of equations:

$$\underline{\underline{\tau}} = \underline{\underline{E}}\,\underline{\underline{\varepsilon}} \approx \underline{\underline{E}}\,\underline{\underline{B}}\,{}^{e}\underline{\underline{D}}$$

The above states that stress anywhere within the element can be computed by pre-multiplying the nodal displacement vector by the *B*- and *E*-matrices. If the strain-displacement matrix is constant, as in the 3-node triangular surface element for plane stress applications, a given stress component has the same value anywhere within the domain of the element.

Elements that have variables in the strain-displacement matrix, such as the 4-node quadrilateral surface element, are able to compute differing stress values at various points within the element. The question that arises is: In light of the trade-off between computational costs and accuracy, are there any locations within the element that will produce more accurate stress predictions than others? The answer is yes, there are points within the element where the stress values better represent the stress of the entire element.

In practice, finite element software usually computes the values of stress at discrete locations in the element. Typical locations to compute stress are at the element centroid, at the nodes, or at the Gauss points. However, it has been shown the most accurate prediction of stress occurs when stress is calculated at locations corresponding to the *reduced integration points of the element.* These are the so-called Barlow points, named for the researcher who investigated this phenomenon. The Barlow points are optimal regardless of whether or not the element uses reduced integration. The details of the topic are found in Barlow [20], Barlow [21], Hinton [22] and MacNeal [1, p. 264]. In Tenchev [23] the author suggests that for problems involving stress concentrations, it is better to evaluate the stress directly at the nodes, when using isoparametric elements with quadratic displacement assumptions.

References

1. MacNeal, R.H., *Finite Elements: Their Design And Performance*, Marcel Dekker, Inc., New York, 1994
2. Ball, A.A., "The Interpolation Function of a general Serendipity rectangular Element," Int. J. Num. Meth. Engng., Vol. 15, No. 5, pp. 773–778, 1980
3. Tenchev, R., "Back to Basics: The Stress Distribution in a Finite Element," in *Benchmark*, Alex Ramsay, Editor, NAFEMS, Glasgow, p. 36–37, March, 1994
4. Haggenmacher, G.W., "Diagnostics in Finite Element Analysis," in Proceedings of First Chautauqua on FEM, p. 193–213
5. Irons, B.M., "Engineering Application of Numerical Integration in Stiffness Methods," J. AIAA, No. 14, pp. 2035–2037, 1966
6. Taig, I.C., "Structural Analysis by the Matrix Displacement Method," English Electric Aviation Report, S017, 1961
7. Bathe, K., *Finite Element Procedures In Engineering Analysis*, Prentice-Hall, Inc., Englewood Cliffs, N.J., 1982
8. Reddy, J.N., *An Introduction to the Finite Element Method*, McGraw-Hill Book Company, New York, p. 410, 1984
9. Grandin, H., *Fundamentals of The Finite Element Method*, Macmillan Publishing Company, New York, 1986
10. Cook, R.D., Malkus, D.S., and Plesha, M.E., *Concepts and Applications of Finite Element Analysis*, 3rd Edition, John Wiley & Sons, N.Y., 1989
11. Fisher, R.C., Ziebur, A.D., *Calculus and Analytic Geometry*, 3rd Edition, Prentice-Hall, Inc., Englewood Cliffs, N.J., 1975
12. Stroud, A.H., Secrest, D., *Gaussian Quadrature Formulas*, Prentice-Hall, Englewood Cliffs, N.J., 1966
13. Anonymous, *ABAQUS Theory Manual*, Version 5.5, Hibbitt, Karlsson & Sorensen, Inc., Pawtucket, R.I., 1995
14. Kosloff, D., and Frazier, G.A., "Treatment of Hourglass Patterns in Low Order Finite Element Codes," Number. Analyt. Methods in Geomechanics, 2, pp. 57–72, 1978
15. Cook, R.D., Feng, Z.H., "Control of Spurious Modes in the Nine-Node Quadrilateral Element," Int. J. Num. Meth. Engng., Vol. 18, No. 10, pp. 1576–1580, 1982
16. Jasti, R., "MARC K.5 Reduced Integration Elements With Hourglass Control," in On The MARC Newsletter, pp. 12–14, MARC Analysis Research Corporation, Palo Alto, CA, February, 1995
17. Cowper, G.R., "Gaussian Quadrature Formulas for Triangles," Int. J. Num. Meth. in Eng., Vol. 7, pp. 405–408, 1973
18. Irons, B.M., "Quadrature Rules for Brick-Based Finite Elements," Int. J. Num. Meth. in Eng., Vol. 3, pp. 293–294, 1971

19. Pawsey, S.F., Clough, R.W., "Improved Numerical Integration of Thick Shell Finite Elements," International Journal for Numerical Methods in Engineering, Vol. 3, pp. 575–586, 1971
20. Barlow, J., "Optimal Stress Locations In Finite Element Models," International Journal for Numerical Methods in Engineering, Vol. 10, pp. 243–251, 1976
21. Barlow, J., "Optimal Stress Locations In Finite Element Models," International Journal for Numerical Methods in Engineering, Vol. 28, pp. 1487–1504, 1989
22. Hinton, E., Campbell, J.S., "Local and Global Smoothing of Discontinuous Finite Element Functions Using a Least Squares Method," Int. J. Num. Meth. Engng., Vol. 8, No. 3, pp. 461–480, 1974
23. Tenchev, R., "Accuracy of Stress Recovering and a Criterion for Mesh Refinement in Areas of Stress Concentration," in Finite Element News, Part I, Issue No. 5, October 1991, and Part II, Issue No. 1, February 1992

Chapter 6

Loads and Boundary Conditions

"In the midst of great joy, do not promise anyone anything. In the midst of great anger, do not answer anyone's letter."

—Chinese Proverb

6.1 Introduction to Loads and Boundary Conditions

Key Concept: The finite element method allows complicated loads and boundary conditions to be applied in a simplified manner.

One of the chief strengths of the finite element method is that it allows complicated, continuous loads to be approximated by an appropriate combination of nodal forces. The application of complicated essential boundary conditions is also simplified using approximations based upon individual nodal restraints. Chapter 6 considers the fundamentals of how loads are applied to finite element idealized structures, and how essential boundary conditions are imposed.

6.2 Load Idealization

Key Concept: The finite element method uses the concept of *equivalent nodal forces* to approximate many types of structural loads.

There are several types of loads commonly applied in structural finite element models. Although concentrated loads (point loads) are the only type of loads considered in this text so far, many load types that are encountered in structural problems can be applied using finite element methods. In general, loads for structural problems typically fall into three categories.

Load Types
Structures which are analyzed using the finite element method typically utilize one or more of the following load types:

- Concentrated loads (force units)
- Surface tractions (force per unit area)
- Body loads (force per unit volume)

Body loads, induced by gravity or by imparting an acceleration to a body, will be considered in Section 6.4.

A load that acts over a relatively small portion of a structure may be considered a *concentrated load*, while a load acting upon a large surface area may be considered a *surface traction*. A surface traction is balanced by stress that develops within the structure. For example, consider a cube restrained on the bottom surface with a distributed load applied to the top, as illustrated in Figure 6.1.

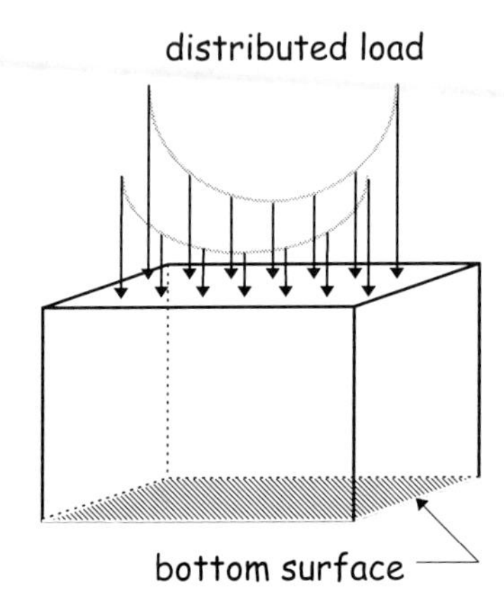

Figure 6.1 Cube with Surface Traction

With the bottom surface of the cube fixed, the distributed load acting upon the surface is in equilibrium with the stress that develops within the cube. Since the load is distributed upon the surface of the structure, and stress results within the body, this loading may be considered a *surface traction*. A surface traction need not be induced by a uniform load distribution. A distributed load that varies along the surface (Figure 6.1) is also considered a surface traction. In addition, a surface traction need not be normal to the surface; a surface traction may act normal, parallel, or at some intermediate angle. Regardless of how a surface traction may be oriented, the traction (by definition) induces some stress within the body to which it is applied.

Equivalent Nodal Forces

Using the finite element method, it is perhaps natural (and correct) to assume that concentrated loads may be represented by forces applied at selected nodes in the model. But how are distributed and body loads represented? Using the finite element method, distributed and body loads are represented by a combination of *equivalent nodal forces*.

There exist two ways to compute the equivalent nodal forces that represent a surface traction or a body load. One method uses the element's shape functions, while a more crude method simply assigns a fractional amount of the total distributed load, acting upon the element, to appropriate element nodes.

If the equivalent nodal forces are computed using the same shape functions that are used in the displacement interpolation function, the resulting forces are termed *consistent nodal forces*. If the nodal forces are computed in some other manner, they are termed *lumped nodal forces*. In either case, the forces must be both statically and kinematically equivalent to the load that is being approximated. Static equivalence requires that the summation of all the equivalent nodal forces and moments must balance the total applied load. Kinematic equivalence will be considered in Section 6.3.

An example of both lumped and consistent nodal forces will be given in the next section. In this chapter, a 3-node, isoparametric surface element is used to illustrate loading principles. However, the same principles apply to many types of isoparametric elements.

Types of Equivalent Nodal Forces

Recall the element equilibrium equations from previous chapters:

$$\underline{\underline{{}^{e}K}}\;\underline{\underline{{}^{e}D}} = \underline{\underline{{}^{e}F}}$$

A stiffness matrix, displacement vector, and force vector can be defined for each element "*e*," as suggested above. In Appendix B, it is shown that the force vector may contain contributions from three different types of loads:

- Concentrated (point) loads
- Surface tractions
- Body loads

In other words, the force vector for each element actually represents the contributions of three constituent vectors:

$$\underline{\underline{{}^{e}F}} = \underline{\underline{{}^{e}F^{C}}} + \underline{\underline{{}^{e}F^{S}}} + \underline{\underline{{}^{e}F^{\beta}}} \tag{6.2.1}$$

where:

$$\underline{\underline{{}^eF}}^C \equiv \text{concentrated forces}$$

$$\underline{\underline{{}^eF}}^S \equiv \text{equivalent nodal forces due to surface tractions}$$

$$\underline{\underline{{}^eF}}^\beta \equiv \text{equivalent nodal forces due to body forces}$$

The three types of loads shown above are used extensively in solid mechanics problems. In addition, forces due to initial stress and strain may be computed. Forces due to initial stress are utilized in the study of structural collapse and the somewhat opposite phenomenon of stress stiffening. Forces due to initial strains may be utilized when studying structures subjected to constrained thermal expansion (or contraction).

While the element force vector (6.2.1) is defined as a combination of three contributing types of forces attributed to a single element, the element force vector is *never actually assembled* as shown. The individual element equilibrium equations that were defined in Chapters 2 and 3 are not used in practice; that is, the individual equilibrium equations are never solved. Recall that the finite element technique first defines all of the individual element stiffness matrices, combines them to form the global stiffness matrix, then solves the *global* system of equilibrium equations, using the *global* force and displacement vectors. Thus, it is really the global force vector that is of interest.

To assemble the global force vector, we first compute, for each element, the equivalent nodal forces due to surface and body loads. The surface and body contributions from all elements are summed, then added to the concentrated forces that are defined at each node in the *assembled model.* Hence, surface and body forces are computed for each individual element, while concentrated forces are defined with respect to the global nodes in the assembled model. The global force vector is typically defined as:

$$\underline{\underline{F}} = \underline{\underline{F}}^C + \sum_{e=1}^{e=nel} \underline{\underline{{}^eF}}^S + \sum_{e=1}^{e=nel} \underline{\underline{{}^eF}}^\beta \tag{6.2.2}$$

6.3 Surface Loads

Key Concept: The finite element method uses the concept of *equivalent nodal forces* to approximate a load applied to the surface of a structure.

The following provides an example of how equivalent nodal forces due to surface tractions may be computed for a 3-node, isoparametric, triangular surface element.

Assume that a surface traction is to be applied to the structure shown in Figure 6.2.

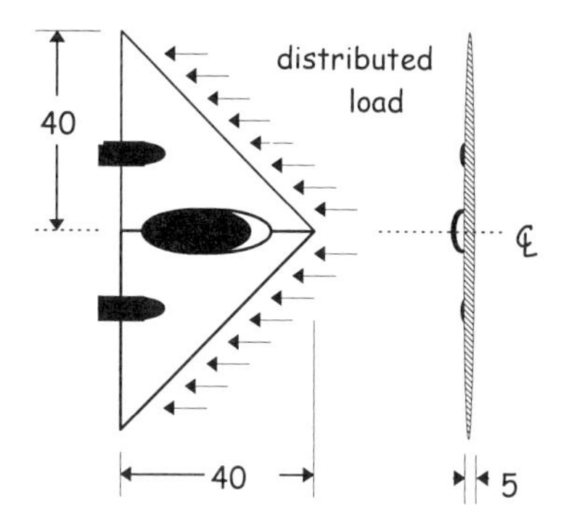

Figure 6.2 Distributed Load on Wing Section

Note that since both the distributed load and the geometry are symmetrical about the centerline, the structure shown in Figure 6.2 may be modeled using mirror symmetry. The crude finite element model illustrated in Figure 6.3 employs mirror symmetry and a single plane stress element.

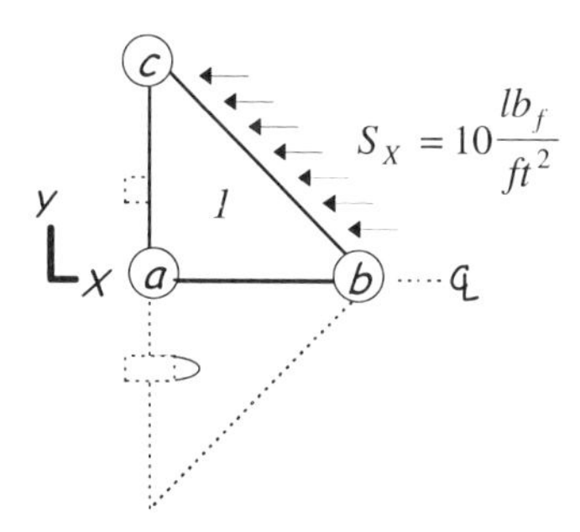

Figure 6.3 Crude Finite Element Model, One Element

The element in Figure 6.3 is labeled Element 1 for future reference. This model, using a single isoparametric element, is entirely too stiff to have any degree of accuracy. In addition, airframe structures may be more appropriately idealized using thin skins, frames, and stringers. The model will suffice to illustrate how equivalent nodal forces due to surface tractions are calculated. The objective is to compute the equivalent nodal forces at Nodes *b* and *c* such that the surface traction on the *b-c* edge of the element is properly represented. The method of computing consistent nodal forces will be illustrated first, while the lumped method will be considered afterwards.

As shown in Appendix B, the equation for equivalent nodal forces due to a surface traction is given as:

$$\underline{\underline{{}^{e}F^{S}}} = \int_{A} \underline{\underline{N}}^{T}\, \underline{\underline{S}}\, dA \tag{6.3.1}$$

$$\underline{\underline{S}} = \begin{Bmatrix} S_X \\ S_Y \\ S_Z \end{Bmatrix} \tag{6.3.2}$$

The denotation $\underline{\underline{{}^{e}F^{S}}}$ in (6.3.1) indicates that the equivalent nodal force vector is due to a surface traction acting upon Element *e,* and the components of the surface traction are contained in the *S*-vector, (6.3.2). The *N*-matrix is the shape function matrix for element *e*.).

The integration procedures considered in the previous chapters were directed at the task of computing element stiffness. As such, the integrals were defined over either the length, surface area, or volume *of an element*. However, considering the integral that defines the equivalent nodal force vector in (6.3.1), the area over which the surface traction is integrated is not the area of the element but, instead, is the area *upon which the surface traction is acting*). The surface traction vector (the *S*-vector) must have units of *force per unit area* since the terms of the *S*-vector, multiplied by the differential *dA* having area units, renders terms with units of force as denoted by the *F* variable on the left hand side of (6.3.1). If the integrals needed to compute the forces are simple, closed-form integration may be used. Otherwise, numerical integration is employed.

For the current problem, a surface traction is applied to Element 1, as illustrated in Figure 6.3. Since a total of six nodal forces are possible for this element, i.e., *X* and *Y* forces at three nodes, the equivalent nodal force vector has six components:

$$\underline{\underline{{}^{1}F^{S}}} = \begin{Bmatrix} {}^{1}F^{S}_{aX} \\ {}^{1}F^{S}_{aY} \\ {}^{1}F^{S}_{bX} \\ {}^{1}F^{S}_{bY} \\ {}^{1}F^{S}_{cX} \\ {}^{1}F^{S}_{cY} \end{Bmatrix} \tag{6.3.3}$$

The surface traction vector for this problem is defined as:

$$\underline{\underline{S}} = \begin{Bmatrix} S_X \\ S_Y \end{Bmatrix} = \begin{Bmatrix} -10 \\ 0 \end{Bmatrix} \tag{6.3.4}$$

The number of components in the surface traction vector is determined by the dimensionality of space in which the finite element problem is posed. For the current problem, using the 3-node triangular element in two-dimensional space, the traction vector need only consider components in the X- and Y-coordinate directions.

Substituting (6.3.4) and (6.3.3) into (6.3.1), the equivalent nodal force vector for this problem is calculated as:

$$\begin{Bmatrix} {}^{1}F^{S}_{aX} \\ {}^{1}F^{S}_{aY} \\ {}^{1}F^{S}_{bX} \\ {}^{1}F^{S}_{bY} \\ {}^{1}F^{S}_{cX} \\ {}^{1}F^{S}_{cY} \end{Bmatrix} = \int_A \underline{\underline{N}}^T \begin{Bmatrix} -10 \\ 0 \end{Bmatrix} dA \tag{6.3.5}$$

Recall from Chapter 5 the shape function matrix for the 3-node, isoparametric, triangular element is defined by (5.4.9):

$$\underline{\underline{N}} = \begin{bmatrix} N_1 & 0 & N_2 & 0 & N_3 & 0 \\ 0 & N_1 & 0 & N_2 & 0 & N_3 \end{bmatrix} \tag{6.3.6}$$

Using the transpose of (6.3.6) in (6.3.5):

$$\begin{Bmatrix} {}^1F_{aX}^S \\ {}^1F_{aY}^S \\ {}^1F_{bX}^S \\ {}^1F_{bY}^S \\ {}^1F_{cX}^S \\ {}^1F_{cY}^S \end{Bmatrix} = \int_A \begin{bmatrix} N_1 & 0 \\ 0 & N_1 \\ N_2 & 0 \\ 0 & N_2 \\ N_3 & 0 \\ 0 & N_3 \end{bmatrix} \begin{Bmatrix} -10 \\ 0 \end{Bmatrix} dA \tag{6.3.7}$$

Performing the matrix computation in (6.3.7):

$$\begin{Bmatrix} {}^1F_{aX}^S \\ {}^1F_{aY}^S \\ {}^1F_{bX}^S \\ {}^1F_{bY}^S \\ {}^1F_{cX}^S \\ {}^1F_{cY}^S \end{Bmatrix} = \int_A \begin{Bmatrix} N_1(-10) \\ 0 \\ N_2(-10) \\ 0 \\ N_3(-10) \\ 0 \end{Bmatrix} dA \tag{6.3.8}$$

Note that because the surface traction in (6.3.4) has non-zero components only in the X-coordinate direction, the equivalent nodal force vector given by (6.3.8) is also of the same form, $F_{ax} = F_{bx} = F_{cx} \neq 0$.

One additional detail needs to be considered before the integration in (6.3.8) is performed. Again observe that the single element model has a traction on the *b-c* edge of the element:

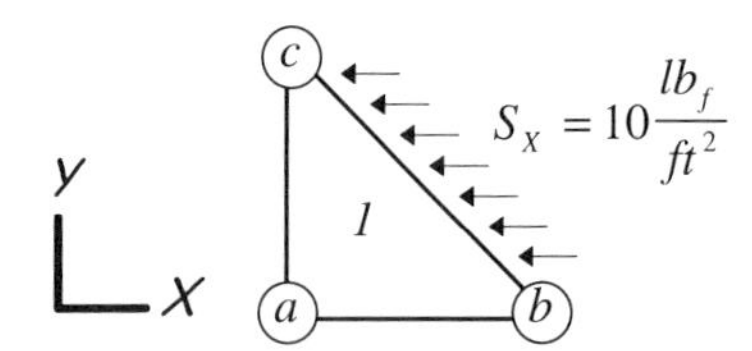

Figure 6.4 Finite Element Representation of Wing

The differential area in the integrand of (6.3.8) refers to the surface area upon which the traction is acting. It is convenient to express this differential in terms of dX and dY, hence, given an element of thickness t, the differential area along the *b-c* side of the element may be expressed as:

$$dA = t\ dl = t\ \sqrt{(dX)^2 + (dY)^2} \tag{6.3.9}$$

The incremental length along Edge *b-c* is denoted as *dl*. However, when using parametric elements, all operations involving integration are performed with respect to the parent element. As such, (6.3.9) should be expressed in terms of differentials *dr* and *ds*. In Figure 6.5, the parent element for the 3-node, isoparametric element used in the present model is considered.

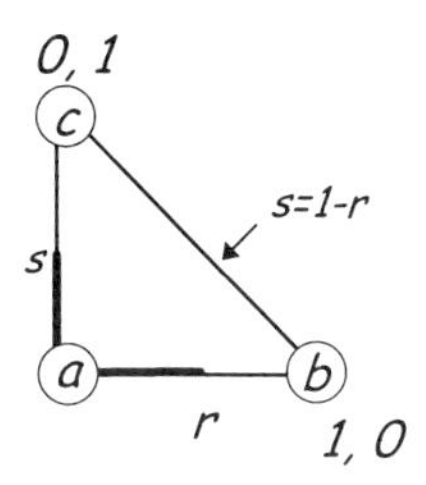

Figure 6.5 Parent Element for 3-Node Triangle

Instead of the differentials dX and dY in (6.3.9), an alternate expression in terms of *dr* and *ds* is required if the integration process is to be performed upon the domain of the parent element. This allows numerical integration to be used, if desired. Knowing that for this particular parametric element, the X and Y variables are mapped with functions of the variables r and s, the differentials dX and dY are expressed as:

$$dX = \frac{\partial X}{\partial r} dr + \frac{\partial X}{\partial s} ds \qquad dY = \frac{\partial Y}{\partial r} dr + \frac{\partial Y}{\partial s} ds \tag{6.3.10}$$

To simplify (6.3.10), it is convenient to replace the s variable with r, so that r is the only parametric variable. Along the *b-c* edge of the parent element:

$$s = 1 - r \tag{6.3.11}$$

Differentiating both sides of (6.3.11), the expression for *ds* is:

$$ds = -dr \tag{6.3.12}$$

Using (6.3.12) in the first equation of (6.3.10):

$$dX = \frac{\partial X}{\partial r}dr + \frac{\partial X}{\partial s}ds = \frac{\partial X}{\partial r}dr - \frac{\partial X}{\partial s}dr \tag{6.3.13}$$

Simplifying the equation above:

$$dX = \left(\frac{\partial X}{\partial r} - \frac{\partial X}{\partial s}\right)dr \tag{6.3.14}$$

Likewise, the differential dY can be expressed in terms of derivatives with respect to parametric variables:

$$dY = \left(\frac{\partial Y}{\partial r} - \frac{\partial Y}{\partial s}\right)dr \tag{6.3.15}$$

In Chapter 5, the mapping function for Y was expressed by (5.4.26):

$$Y = N_1 Y_a + N_2 Y_b + N_3 Y_c \tag{6.3.16}$$

Substituting (6.3.16) into (6.3.15), performing the differentiation, and using the derivatives calculated in (5.4.32), it can easily be shown that (6.3.15) is equivalent to:

$$dY = \left(Y_b - Y_c\right)dr \tag{6.3.17}$$

Through a similar process, it can be shown that, beginning with (6.3.14):

$$dX = \left(X_b - X_c\right)dr \tag{6.3.18}$$

Using (6.3.17) and (6.3.18) in (6.3.9):

$$dA = t\sqrt{(dX)^2 + (dY)^2} = t\sqrt{\left(X_b - X_c\right)^2 + \left(Y_b - Y_c\right)^2}\ dr \tag{6.3.19}$$

Substituting (6.3.19) into (6.3.8):

$$\begin{Bmatrix} {}^1F^S_{aX} \\ {}^1F^S_{aY} \\ {}^1F^S_{bX} \\ {}^1F^S_{bY} \\ {}^1F^S_{cX} \\ {}^1F^S_{cY} \end{Bmatrix} = \int_0^1 \begin{Bmatrix} N_1(-10) \\ 0 \\ N_2(-10) \\ 0 \\ N_3(-10) \\ 0 \end{Bmatrix} t\sqrt{\left(X_b - X_c\right)^2 + \left(Y_b - Y_c\right)^2}\; dr \tag{6.3.20}$$

Considering the first component in the equivalent nodal force vector above:

$${}^1F^s_{aX} = \int_0^1 N_1(-10)t\sqrt{\left(X_b - X_c\right)^2 + \left(Y_b - Y_c\right)^2}\; dr \tag{6.3.21}$$

The preceding superscript on the left hand side of (6.3.21) indicates that this equation applies to Element 1; the "*s*" superscript indicates that the equation is computing equivalent nodal forces due to a *surface* traction; the subscript indicates that the force at Node *a*, in the *X*-coordinate direction, is being calculated.

The shape functions for this element are given by (5.4.6). Using the first shape function, (1-*r*-*s*), in (6.3.21):

$${}^1F^s_{aX} = \int_0^1 (1-r-s)(-10)t\sqrt{\left(X_b - X_c\right)^2 + \left(Y_b - Y_c\right)^2}\; dr$$

Recalling again the expression for *s* given by (6.3.11), and using this relationship in the equation above:

$$\begin{aligned} {}^1F^s_{aX} &= \int_0^1 (1-r-(1-r))(-10)t\sqrt{\left(X_b - X_c\right)^2 + \left(Y_b - Y_c\right)^2}\; dr \\ &= \int_0^1 (0)(-10)t\sqrt{\left(X_b - X_c\right)^2 + \left(Y_b - Y_c\right)^2}\; dr \\ &= 0 \end{aligned} \tag{6.3.22}$$

Hence, the equivalent nodal force at Node *a* is equal to zero. Referring to Figure 6.4, it is perhaps intuitive that a distributed load acting upon the *b*-*c* edge should be expressed in terms of equivalent nodal forces at Nodes *b* and *c* only. Noting that the expression for the second shape function is simply r, the third component in (6.3.20), the equivalent nodal force in the X-coordinate direction at Node b, is:

$$ {}^{1}F_{bX}^{s} = \int_{0}^{1} (r)(-10)\, t\sqrt{\left(X_b - X_c\right)^2 + \left(Y_b - Y_c\right)^2}\, dr \tag{6.3.23} $$

Removing the constants from the integrand of (6.3.23):

$$ \begin{aligned} {}^{1}F_{bX}^{s} &= (-10)t\sqrt{\left(X_b - X_c\right)^2 + \left(Y_b - Y_c\right)^2} \int_{0}^{1} r\, dr \\ &= \frac{(-10)(t)\sqrt{\left(X_b - X_c\right)^2 + \left(Y_b - Y_c\right)^2}}{2} \end{aligned} \tag{6.3.24} $$

In a similar fashion, the fifth component of the equivalent nodal force vector is found to be:

$$ \begin{aligned} {}^{1}F_{cX}^{s} &= (-10)t\sqrt{\left(X_b - X_c\right)^2 + \left(Y_b - Y_c\right)^2} \int_{0}^{1} s\, dr \\ &= (-10)t\sqrt{\left(X_b - X_c\right)^2 + \left(Y_b - Y_c\right)^2} \int_{0}^{1} (1-r)\, dr \\ &= \frac{(-10)(t)\sqrt{\left(X_b - X_c\right)^2 + \left(Y_b - Y_c\right)^2}}{2} \end{aligned} \tag{6.3.25} $$

What (6.3.24) and (6.3.25) suggest is that to compute the nodal equivalent, the magnitude of the uniform surface traction is multiplied by the area upon which it acts, then the product is divided by two, i.e.:

$$ {}^{1}F_{cX}^{s} = \frac{(traction)(thickness)(length\ of\ \ side\ b-c)}{2} \tag{6.3.26} $$

In other words, (6.3.24) and (6.3.25) reveal that equal portions of the force resulting from the uniform surface traction are assigned to each node. For linear elements with uniform surface tractions, the total force resulting from a surface traction acting upon the surface is simply divided into two portions, with one portion assigned to each node. This would not be the case if the traction were non-uniform along the edge, or if a non-linear element, for instance a 6-node triangle, were used. A 6-node triangular element has three nodes along each edge. Intuitively, one might guess that one-third of the total load should be assigned to each node in such a case. However, this is not true of consistent nodal forces. More on this issue later.

Substituting the known values for the nodal coordinates, thickness, and traction into (6.3.24), the equivalent nodal force at Node b, due to a uniform surface traction on Edge b-c, is:

$$^{1}F_{bX}^{s} = \frac{(-10)\,(5)\,\sqrt{(40)^2+(40)^2}}{2} = -1414 \tag{6.3.27}$$

Using (6.3.25), the same value is computed for the nodal force at Node c, such that equal loads are applied as shown in Figure 6.6.

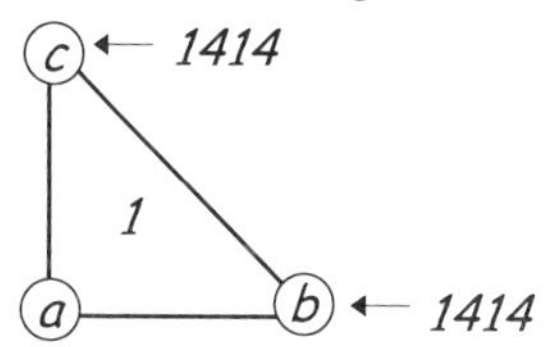

Figure 6.6 Consistent, Equivalent Nodal Forces

The results above likely seem reasonable. It is interesting to note, however, what happens when the surface traction is distributed over more than one element. For example, consider the same problem as illustrated in Figure 6.2, except that four elements are used.

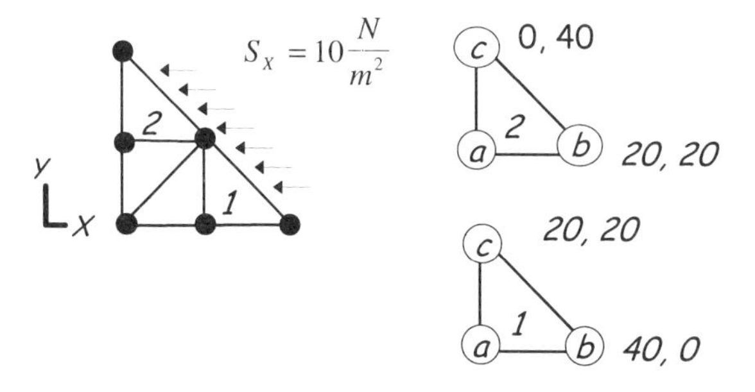

Figure 6.7 Crude Finite Element Model of Wing Section, Four Elements

The model above, although using a few more elements than the previous model, would still not be expected to render accurate results. As mentioned, airframe structures typically require more elaborate modeling techniques than simply using plane stress elements.

Using (6.3.24) to compute the equivalent nodal force for Element 1, Node b:

$$\begin{aligned} {}^{1}F_{bX}^{s} &= (-10)t\sqrt{\left({}^{1}X_{b}-{}^{1}X_{c}\right)^{2}+\left({}^{1}Y_{b}-{}^{1}Y_{c}\right)^{2}}\int_{0}^{1} r\,dr \\ &= (-10)(5)\left(\sqrt{(40-20)^{2}+(0-20)^{2}}\right)\left(\frac{1}{2}\right) \\ &= -707 \end{aligned} \tag{6.3.28}$$

Likewise, for Element 1, Node c, from (6.3.25):

$$\begin{aligned} {}^{1}F_{cX}^{s} &= (-10)(t)\sqrt{\left({}^{1}X_{b}-{}^{1}X_{c}\right)^{2}+\left({}^{1}Y_{b}-{}^{1}Y_{c}\right)^{2}}\int_{0}^{1} s\,dr \\ &= (-10)(t)\sqrt{\left({}^{1}X_{b}-{}^{1}X_{c}\right)^{2}+\left({}^{1}Y_{b}-{}^{1}Y_{c}\right)^{2}}\int_{0}^{1} (1-r)\,dr \\ &= (-10)(5)\left(\sqrt{(40-20)^{2}+(0-20)^{2}}\right)\left(\frac{1}{2}\right) \\ &= -707 \end{aligned} \tag{6.3.29}$$

Following the same procedure for Element 2, Node b:

$$\begin{aligned} {}^{2}F_{bX}^{s} &= (-10)t\sqrt{\left({}^{2}X_{b}-{}^{2}X_{c}\right)^{2}+\left({}^{2}Y_{b}-{}^{2}Y_{c}\right)^{2}}\int_{0}^{1} r\,dr \\ &= (-10)(5)\left(\sqrt{(20-0)^{2}+(20-40)^{2}}\right)\left(\frac{1}{2}\right) \\ &= -707 \end{aligned} \tag{6.3.30}$$

And the same result will be found for Element 2, Node c. The two element force vectors are then defined as:

$$\underline{\underline{{}^1F^S}} = \begin{Bmatrix} {}^1F^S_{aX} \\ {}^1F^S_{aY} \\ {}^1F^S_{bX} \\ {}^1F^S_{bY} \\ {}^1F^S_{cX} \\ {}^1F^S_{cY} \end{Bmatrix} = \begin{Bmatrix} 0 \\ 0 \\ -707 \\ 0 \\ -707 \\ 0 \end{Bmatrix} \qquad \underline{\underline{{}^2F^S}} = \begin{Bmatrix} {}^2F^S_{aX} \\ {}^2F^S_{aY} \\ {}^2F^S_{bX} \\ {}^2F^S_{bY} \\ {}^2F^S_{cX} \\ {}^2F^S_{cY} \end{Bmatrix} = \begin{Bmatrix} 0 \\ 0 \\ -707 \\ 0 \\ -707 \\ 0 \end{Bmatrix}$$

To add the two vectors, a process similar to adding element stiffness matrices needs to be performed using (6.2.2). In such a case, each nodal degree of freedom in the finite element model is assigned a global identification number and the connectivity table is consulted to specify the relationship between global and local degrees of freedom. We will forgo this detailed process, and simply note that local Node c from Element 1 is coincident with local Node b of Element 2, hence, the contributions from these nodes would be added when forming the global force vector using (6.2.2).

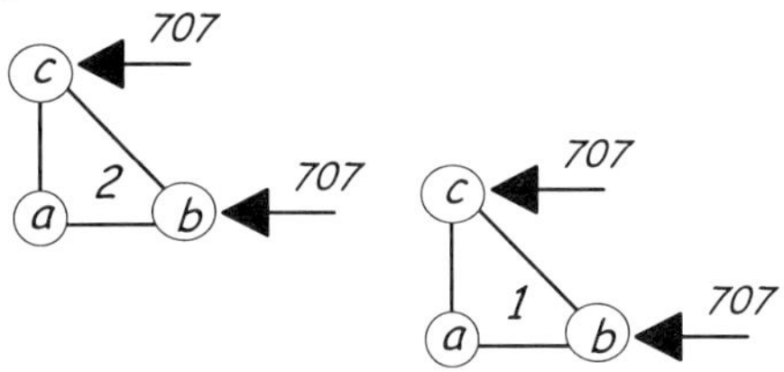

When the global force vector is computed for this problem, the result appears as in Figure 6.8.).

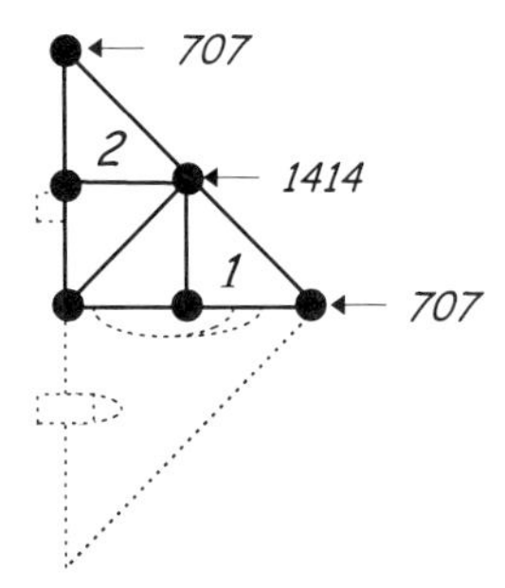

Figure 6.8 Correctly Adding Consistent, Equivalent Nodal Forces

Notice that the sum of all three equivalent nodal forces is equal to –2828, just as in the single element example. The equivalent force at each node is –707, and because two coincident nodes are present on the leading edge where Elements 1 and 2 meet, the correct equivalent nodal force there is the sum of the equivalent nodal forces from each element. This summation process is specified by (6.2.2).

The manner in which the forces are distributed in Figure 6.8 may appear strange. A more "intuitive" set of equivalent nodal forces might be computed by simply assigning one-third of the total force to each node. In other words, the total force on the leading edge of the wing due to the surface traction is first determined:

$$\begin{aligned} \textit{total load} &= (\textit{traction})(\textit{length of leading edge})(\textit{thickness}) \\ &= (-10)(5)\left(\sqrt{(40)^2+(40)^2}\right) \approx -2828 \end{aligned}$$

Next, the total load is divided by three, since there are three nodes on the leading edge:

$$\textit{intuitive equivalent nodal force} = \frac{-2828}{3} \approx -943$$

Hence, using the method above, the equivalent nodal forces would appear as shown in Figure 6.9.

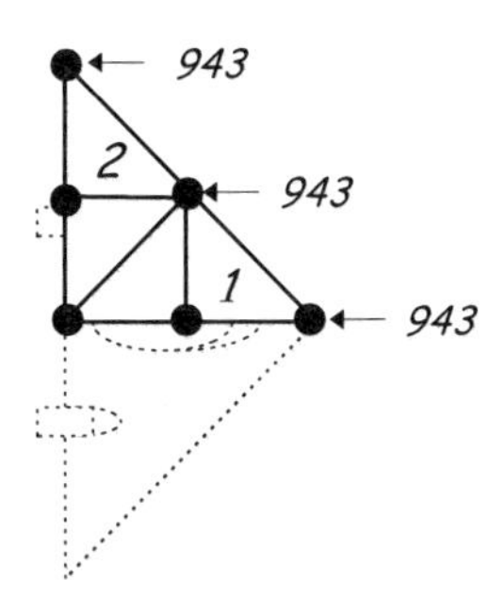

Figure 6.9 Incorrect Nodal Forces

In either model, the total load has the same magnitude, 2828 lb_f. In addition, computing moments about any point renders the same results in either case.

Hence, the loading schemes are *statically equivalent*, if one considers the structure as a non-deformable (rigid) body. However, when structures are considered to be deformable, the stress distribution using the two methods will appear strikingly different. This is because a deformable body must use loads that are *kinematically* equivalent, meaning that loads must be applied with regard to the stiffness associated with a particular nodal degree of freedom. In Figure 6.8, the two end nodes have a stiffness of one-half that of the node that shares Elements 1 and 2. If all three nodes were given a force magnitude of 943, the two end nodes would displace twice as much as the shared node; this behavior is not consistent with a uniform load applied on the leading edge of the wing. Kinematically equivalent loads are further discussed in [1, pp. 36, 258].

Lumped, Equivalent Nodal Forces for Surface Tractions

The consistent loading method described above is sometimes replaced by a more simplistic approach called *lumped* equivalent nodal forces. Using the lumped method for the problem illustrated in Figure 6.2, one first computes the total load on the edge of the *element*:

$$\textit{Force on element edge} = \int_A S_X \, dA$$

In the one-element example from Figure 6.3, the computation of the force on the element edge amounts to:

$$\begin{aligned}\textit{Force on element edge} &= \int_0^1 -10t\sqrt{\left(X_b - X_c\right)^2 + \left(Y_b - Y_c\right)^2}\, dr \\ &= -2828\end{aligned}$$

After the total force is computed, it is fractionally applied to each node. If, in the example above, one chooses to use ½ of the total at each node, the lumped method would render the same results as the consistent method.

The fractional amount to be applied to each node is not always intuitive. For example, if the magnitude of the surface traction varies along the element edge, the consistent nodal load approach might be needed to compute the correct fraction at each node. Also, when using higher order elements, the magnitude of the nodal forces in the equivalent nodal force vector for a single element may be difficult to intuit, even if a uniform surface traction is applied along the edge. For example, using 8-node quadrilateral elements, having three nodes per edge as depicted in Figure 6.10, the consistent approach to equivalent nodal forces would *not* prescribe the use of one-third the total force at each node, even though this might seem to be the correct amount, based upon intuition.

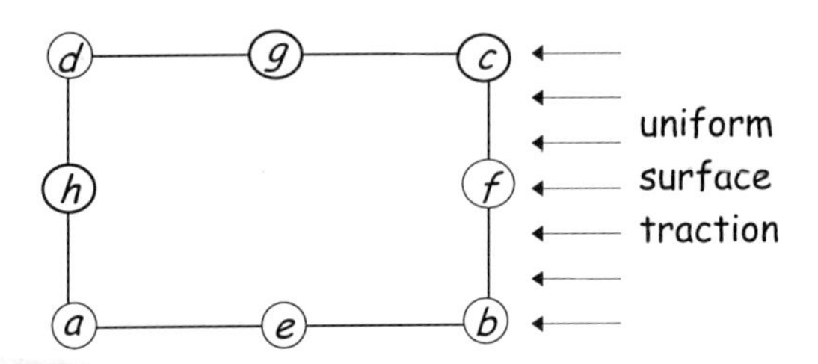

Figure 6.10 An 8-Node, Isoparametric, Quadrilateral Surface Element

The reason that one-third of the total force would not be prescribed at each node in the case of the 8-node quadrilateral is that this element uses second order shape functions, such that displacement is not a linear function along the edges. Hence, when the consistent, equivalent nodal force vector is computed, non-linear functions are integrated in the equation for equivalent nodal forces, (6.3.1):

$$\underline{\underline{{}^{e}F^{S}}} = \int_{A} \underline{\underline{N}}^{T} \; \underline{\underline{S}} \, dA$$

non-linear functions

The result of integrating the equation above is that the mid-side node will have a force magnitude that differs from the two corner nodes. In other words, because the displacement is not linearly interpolated along the edge, the consistent, kinematically equivalent nodal forces are not evenly distributed along the edge. For two good examples that illustrate this principle, see Bathe [2, p. 164, 216] and [3, p. 1–59].

6.4 Body Loads

Key Concept: It is often necessary to impose loads that act on an entire body, as opposed to a surface; these loads are termed *body loads*. A load due to gravity is an example of a body load.

As discussed, surface tractions are approximated by equivalent nodal forces. This section will briefly consider *body loads,* which are also approximated by an equivalent nodal force vector. Two common load cases make use of body loads: forces in a static body under gravitational load, and forces in an accelerating body.

To illustrate how equivalent nodal forces due to gravity are computed, consider a clamped structural beam, suspended under its own weight, as illustrated in Figure 6.11.

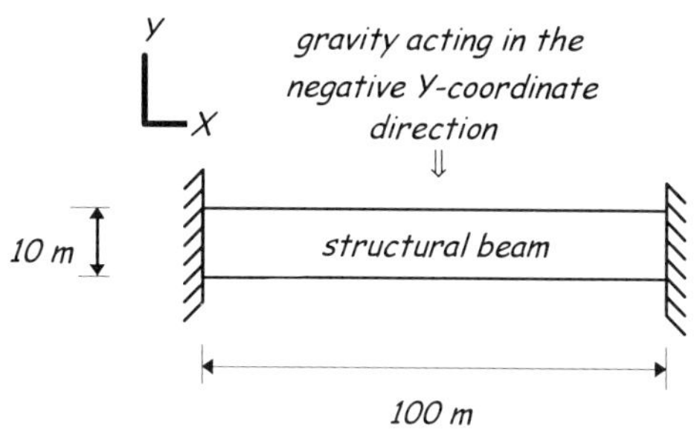

Figure 6.11 Structural Beam with Gravitational Loading

Assume that the beam is $1m$ wide (i.e., the Z-coordinate dimension is 1 meter). The beam can be modeled using 3-node, isoparametric surface elements for plane stress applications, as shown in Figure 6.12.

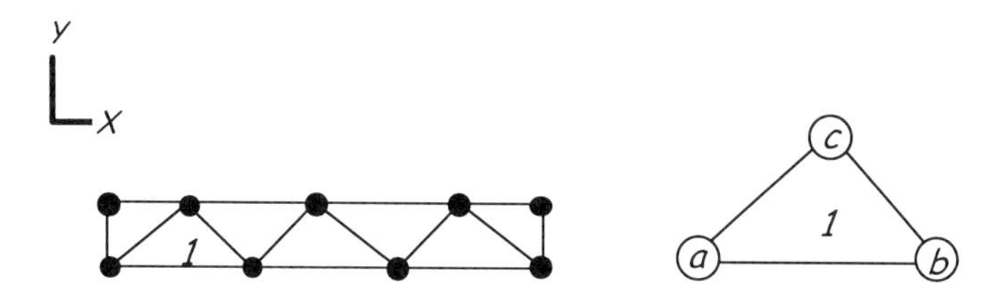

Figure 6.12 Crude Finite Element Model of Beam

The structure is modeled as one longitudinal slice of the beam, using a surface mesh boundary. The plane stress idealization provides a good approximation for beams that are not too wide, because stress in the width direction (Z) can be assumed negligible. Although the type of element chosen for this analysis is appropriate, the mesh depicted in Figure 6.12 is far too coarse to accurately predict stress or strain, especially with the poor performing 3-node triangular element.

Consistent, equivalent nodal forces that approximate a body load are calculated in a manner similar to that used to compute equivalent nodal forces due to surface tractions. For example, consider Element 1 from the crude beam model in Figure 6.12. The equation for computing the body force vector for Element *e*, as given in Appendix B, is:

$$\underline{\underline{{}^{e}F^{\beta}}} = \int_{v} \underline{\underline{N}}^{T}\, \underline{\underline{\beta}}\, dV \qquad \underline{\underline{\beta}} = \begin{Bmatrix} \beta_X \\ \beta_Y \\ \beta_Z \end{Bmatrix} \tag{6.4.1}$$

The N-matrix represents the matrix of shape functions, just like in the surface element example, while the β-matrix represents the body force vector, with each component of the vector expressed in terms of *force per unit volume*. The integral is evaluated over the entire volume of the element. For the 3-node triangular surface element with constant thickness, (6.4.1) is expressed as:

$$\underline{\underline{{}^{e}F^{\beta}}} = \int_{A} \underline{\underline{N}}^{T}\, \underline{\underline{\beta}}\, t\, dA$$

Note that for the uniform thickness surface element, the differential volume in (6.4.1), dV, can be replaced by the $t\,dA$, as shown above. However, recall from Chapter 5 that all integration processes related to isoparametric elements take place within the domain of the parent element. To express the integral above with respect to the parametric domain, the following equation is used:

$$\underline{\underline{{}^{e}F^{\beta}}} = \int_{0}^{1}\int \underline{\underline{N}}^{T}\, \underline{\underline{\beta}}\, t \left(\det \underline{\underline{J}}\right) ds\, dr \tag{6.4.2}$$

Notice that the determinant of the Jacobian matrix (multiplied by $ds\ dr$) is used in place of dA, in accordance with the procedure to transform the global domain to the local (parametric) domain, as discussed in Appendix C.

Analogous to the equivalent nodal force vector due to surface loads given by (6.3.3), the 3-node, isoparametric, triangular surface element in this problem has an equivalent nodal force vector due to body loads which takes the form:

$$\underline{\underline{{}^{e}F^{\beta}}} = \begin{Bmatrix} {}^{1}F^{\beta}_{aX} \\ {}^{1}F^{\beta}_{aY} \\ {}^{1}F^{\beta}_{bX} \\ {}^{1}F^{\beta}_{bY} \\ {}^{1}F^{\beta}_{cX} \\ {}^{1}F^{\beta}_{cY} \end{Bmatrix} \tag{6.4.3}$$

In the current example, the body force vector (the β–vector) contains only one non-zero component, namely, the body force due to gravity:

$$\underline{\underline{\beta}} = \begin{Bmatrix} \beta_X \\ \beta_Y \end{Bmatrix} = \begin{Bmatrix} 0 \\ -g\rho \end{Bmatrix} \tag{6.4.4}$$

The mass density in the above is defined by ρ, while acceleration due to gravity is defined by g. Using (6.4.3) and (6.4.4) in (6.4.2):

$$\begin{Bmatrix} {}^{1}F^{\beta}_{aX} \\ {}^{1}F^{\beta}_{aY} \\ {}^{1}F^{\beta}_{bX} \\ {}^{1}F^{\beta}_{bY} \\ {}^{1}F^{\beta}_{cX} \\ {}^{1}F^{\beta}_{cY} \end{Bmatrix} = \int_0^1 \int \underline{\underline{N}}^T \begin{Bmatrix} 0 \\ -g\rho \end{Bmatrix} t\left(\det \underline{\underline{J}}\right) ds\, dr \tag{6.4.5}$$

The shape function matrix to be used in (6.4.5) was previously given by (6.3.6), when the surface load example was considered. Substituting (6.3.6) into (6.4.5):

$$\begin{Bmatrix} {}^{1}F^{\beta}_{aX} \\ {}^{1}F^{\beta}_{aY} \\ {}^{1}F^{\beta}_{bX} \\ {}^{1}F^{\beta}_{bY} \\ {}^{1}F^{\beta}_{cX} \\ {}^{1}F^{\beta}_{cY} \end{Bmatrix} = \int_0^1 \int \begin{Bmatrix} N_1 & 0 \\ 0 & N_1 \\ N_2 & 0 \\ 0 & N_2 \\ N_3 & 0 \\ 0 & N_3 \end{Bmatrix} \begin{Bmatrix} 0 \\ -g\rho \end{Bmatrix} t\left(\det \underline{\underline{J}}\right) ds\, dr \tag{6.4.6}$$

Performing the matrix multiplication in (6.4.6):

$$\begin{Bmatrix} {}^1F_{aX}^{\beta} \\ {}^1F_{aY}^{\beta} \\ {}^1F_{bX}^{\beta} \\ {}^1F_{bY}^{\beta} \\ {}^1F_{cX}^{\beta} \\ {}^1F_{cY}^{\beta} \end{Bmatrix} = \int_0^1 \int \begin{Bmatrix} 0 \\ N_1(-g\rho) \\ 0 \\ N_2(-g\rho) \\ 0 \\ N_3(-g\rho) \end{Bmatrix} t\left(\det \underline{\underline{J}}\right) ds\,dr \tag{6.4.7}$$

Notice that all of the components of the equivalent nodal force vector associated with the X-coordinate direction are zero, since the body force vector only contains components in the Y-coordinate direction.

To compute the first non-zero term in (6.4.7), the following integral needs to be evaluated:

$${}^1F_{aY}^{\beta} = \int_0^1 \int N_1(-g\rho)\, t\left(\det \underline{\underline{J}}\right) ds\,dr \tag{6.4.8}$$

Recalling the expression for the first shape function, given by (5.4.6), and substituting into (6.4.8):

$${}^1F_{aY}^{\beta} = \int_0^1 \int (1-r-s)(-g\rho)\, t\left(\det \underline{\underline{J}}\right) ds\,dr \tag{6.4.9}$$

Removing all of the constants from the integrand in (6.4.9):

$${}^1F_{aY}^{\beta} = (-g\rho)\, t\left(\det \underline{\underline{J}}\right) \int_0^1 \int (1-r-s)\,ds\,dr \tag{6.4.10}$$

Performing the integration with respect to s in (6.4.10):

$${}^1F_{aY}^{\beta} = (-g\rho)\, t\left(\det \underline{\underline{J}}\right) \int_0^1 \left(s - rs - \frac{s^2}{2} \right) dr \tag{6.4.11}$$

Making use of the fact that the height of the parent triangular element, s, is equal to $1-r$, (6.4.11) may be expressed as:

$$ {}^{1}F_{aY}^{\beta} = (-g\rho)\, t\, \left(\det \underline{\underline{J}}\right) \int_{0}^{1} \left((1-r) - r(1-r) - \frac{(1-r)^2}{2} \right) dr \qquad (6.4.12) $$

Or, rearranging the equation above:

$$ {}^{1}F_{aY}^{\beta} = (-g\rho)\, t\, \frac{\left(\det \underline{\underline{J}}\right)}{2} \int_{0}^{1} (1-r)^2\, dr \qquad (6.4.13) $$

Performing the integration in (6.4.13) and applying the limits of integration:

$$ {}^{1}F_{aY}^{\beta} = \left(\frac{1}{3}\right)(-g\rho)\, \frac{t\left(\det \underline{\underline{J}}\right)}{2} \qquad (6.4.14) $$

In Appendix C, it was shown that one-half the Jacobian determinant is equivalent to the surface area of the triangular element. Therefore, t multiplied by one-half the Jacobian determinant is equivalent to the volume of the element:

$$ V = t\left(\det \underline{\underline{J}}\right)/2 $$

Using V to denote the volume of the *element*, the first component of the equivalent nodal force vector due to the body force, Equation 6.4.14, is simply:

$$ {}^{1}F_{aY}^{\beta} = \left(\frac{1}{3}\right)(-g\rho)\, V \qquad (6.4.15) $$

Or, noting that the product of the volume, density, and acceleration due to gravity is equivalent to the portion of the structure's weight that is associated with Element 1, (6.4.15) may be expressed as:

$$ {}^{1}F_{aY}^{\beta} = \left(\frac{1}{3}\right)(-1)(\textit{element weight}) \qquad (6.4.16) $$

Computing the other components of the equivalent nodal force vector reveals that:

$$\begin{Bmatrix} {}^1F^{\beta}_{aX} \\ {}^1F^{\beta}_{aY} \\ {}^1F^{\beta}_{bX} \\ {}^1F^{\beta}_{bY} \\ {}^1F^{\beta}_{cX} \\ {}^1F^{\beta}_{cY} \end{Bmatrix} = (-1)(element\ weight)\begin{Bmatrix} 0 \\ 1/3 \\ 0 \\ 1/3 \\ 0 \\ 1/3 \end{Bmatrix} \tag{6.4.17}$$

Notice that (6.4.17) is the same as using the *lumped method of equivalent nodal forces* with a nodal fractional factor of 1/3.

From the examples given in this text, it should be evident that a vector of nodal forces is used to represent three types of loads used in structural analysis: concentrated, distributed, and body loads.

Initial Stress and Strain

Although not considered in this text, the equivalent nodal force vector concept is also employed to model loads due to initial stress, and/or initial strain. See Cook [4] for more information on the topic of initial stress and strain. Huebner [5, p. 572] discusses how thermal effects are introduced into the finite element equations of equilibrium.

6.5 Imposing Essential Boundary Conditions

Key Concept: When using the finite element method for static, structural problems, a singular system of equations results when all of the element stiffness matrices are combined to form the global stiffness matrix. The essential boundary conditions that are needed to render a non-singular system of equations are imposed through matrix manipulations.

Essential Boundary Conditions

As shown in the previous chapters, the ability for a loaded body to exhibit rigid body motion is eliminated by *restraining* the body, through imposition of essential boundary conditions. Finite element idealized structures may be restrained by imposing homogeneous ("zero magnitude") boundary conditions at specified nodes. Boundary conditions can be imposed upon the global stiffness matrix, or upon the stiffness matrix of each individual element. In this text, boundary conditions will be imposed upon the global stiffness matrix, although the same concepts apply if the boundary conditions are imposed upon the individual element stiffness matrices. The terms "restraint" and homogeneous, essential boundary condition are used interchangeably in this text.

When solving a structural problem using a differential equation, boundary conditions are imposed upon the solution to the differential equation. In doing so, the arbitrary constants in the differential equation are replaced with constants that apply to the specific problem at hand. When using the Ritz method, the natural ("force") boundary conditions are included in the functional, while the essential boundary conditions are imposed upon the displacement assumption (Chapter 1). Using the Ritz method, the essential boundary conditions were imposed at the outset of the procedure when the displacement assumption was made to conform to the prescribed displacement boundary conditions of the structure.

Using the finite element method, boundary conditions are also required. While a differential equation without boundary conditions is simply expressed in terms of arbitrary constants, the finite element system of equations for a static problem, without boundary conditions, results in a singular system of equations. To remove the singularity, essential boundary conditions are imposed through matrix operations that can be performed upon the global system of equilibrium equations. One method of imposing boundary conditions was illustrated in Chapter 2: appropriate rows and columns are simply removed from the global system of equations, rendering a non-singular system of equations. While this worked well in the extremely simple problems considered, in practice, such a method would require a significant amount of work to pull out the equations that are associated with homogeneous boundary conditions, then re-arrange the remaining equations in the form of a square, non-singular matrix. This method may be considered a "matrix partitioning" method. Some commercially available FEA software programs use a procedure based upon the matrix partitioning method shown here.

Methods to Impose Essential Boundary Conditions

Three common methods of imposing essential boundary conditions will be considered:

1. Matrix partitioning
2. Penalty
3. Ones-on-diagonal

Matrix Partitioning

Matrix partitioning is quite straight-forward for select, one-dimensional problems, as illustrated in Chapter 2. This method may be conceived as one which eliminates selected rows and columns of the system of equilibrium equations, then rearranges the remaining equations in non-singular form. Consider a global system of equilibrium equations in partitioned form:

$$\begin{bmatrix} \underline{\underline{K_{11}}} & \underline{\underline{K_{12}}} \\ \underline{\underline{K_{21}}} & \underline{\underline{K_{22}}} \end{bmatrix} \begin{Bmatrix} \underline{\underline{D_k}} \\ \underline{\underline{D_u}} \end{Bmatrix} = \begin{Bmatrix} \underline{\underline{F_u}} \\ \underline{\underline{F_k}} \end{Bmatrix} \tag{6.5.1}$$

Notice that each element (component) of the matrices above is actually a matrix, not merely a scalar entry. The subscript k indicates known values of displacement or force, while the u subscript denotes unknown values. The simplicity of (6.5.1) may belie the effort involved in obtaining this type of partitioned form—in most finite element models many matrix manipulations would be required to obtain such a conveniently ordered system. Performing the matrix algebra in (6.5.1), the bottom row produces:

$$\underline{\underline{K_{21} D_k}} + \underline{\underline{K_{22} D_u}} = \underline{\underline{F_k}} \tag{6.5.2}$$

The components of the stiffness matrices in (6.5.2) are computed using the techniques discussed in the previous chapters, hence, all of the K-matrices in (6.5.1) are known. By definition, the vectors D_k and F_k are known, hence (6.5.2) can be solved for the unknown displacement vector by matrix manipulations:

$$\underline{\underline{D_u}} = \left(\underline{\underline{K_{22}}}\right)^{-1} \left(\underline{\underline{F_k}} - \underline{\underline{K_{21} D_k}}\right) \tag{6.5.3}$$

To clarify the meaning of (6.5.3), consider that all known values of displacement (either zero or non-zero values) are included in the vector D_k. For instance, whenever a displacement boundary condition is imposed, the value of displacement at a particular node, in a particular direction, is known. All known forces are contained in vector F_k. Recall that when no external force is specified at a node, a value of zero is assumed, and these zero values are also included in vector F_k. As a result, in a large model, F_k can often contain many zeros. While the inverse of matrix K_{22} is indicated, the system of equations in (6.5.3) is typically solved using a Gaussian elimination scheme, not by formal matrix inversion.

All of the matrices on the right hand side of (6.5.3) are known, and the unknown nodal displacements, vector D_u, can be determined, since the right hand side represents a non-singular system of equations. After computing vector D_u via

(6.5.3), all displacements in the structure are now defined, and we turn our attention to the first row of equations from (6.5.1):

$$\underline{\underline{F_u}} = \underline{\underline{K_{11}D_k}} + \underline{\underline{K_{22}D_u}} \quad \text{(now known)} \tag{6.5.4}$$

The equation above is used to compute the unknown, non-zero nodal forces, called *reaction forces,* which are typically induced at nodes having a prescribed displacement boundary condition. In practice, (6.5.4) is often not computed, since the analyst often has no interest in knowing the reaction forces.

In a small, one-dimensional system of equations, the partitioning method is easily employed, as will be illustrated in Example 6.1, below.

Example 6.1—Imposing Boundary Conditions by Partitioning

The finite element model from Example 2.5 is shown again in Figure 6.13, below. Recall the equilibrium equations for this problem, consisting of three unknown displacements, as originally given by (2.4.19):

$$3.14(10)^6 \begin{bmatrix} 5 & -5 & 0 \\ -5 & 6 & -1 \\ 0 & -1 & 1 \end{bmatrix} \begin{Bmatrix} U_1 \\ U_2 \\ U_3 \end{Bmatrix} = \begin{Bmatrix} F_1 \\ 0 \\ P \end{Bmatrix} \tag{6.5.5}$$

The system of equations above is partitioned in a form suggested by (6.5.1):

$$3.14(10)^6 \left[\begin{array}{c|cc} 5 & -5 & 0 \\ \hline -5 & 6 & -1 \\ 0 & -1 & 1 \end{array}\right] \left\{\begin{array}{c} U_1 \\ \hline U_2 \\ U_3 \end{array}\right\} = \left\{\begin{array}{c} F_1 \\ \hline 0 \\ P \end{array}\right\} \tag{6.5.6}$$

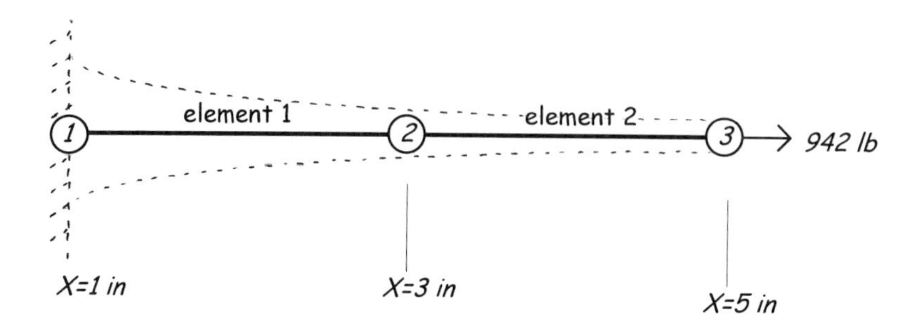

connectivity table

element	node a	node b
1	1	2
2	2	3

Figure 6.13 Non-Uniform Shaft Model from Chapter 2

Referring to (6.5.1) and (6.5.6), the following are defined:

$$\underline{\underline{K_{11}}} = 3.14(10)^6[5] \qquad \underline{\underline{K_{12}}} = 3.14(10)^6\begin{bmatrix} -5 & 0 \end{bmatrix} \qquad \underline{\underline{K_{21}}} = 3.14(10)^6\begin{Bmatrix} -5 \\ 0 \end{Bmatrix}$$

$$\underline{\underline{K_{22}}} = 3.14(10)^6\begin{bmatrix} 6 & -1 \\ -1 & 1 \end{bmatrix} \qquad \underline{\underline{D_k}} = \{U_1\} \qquad \underline{\underline{D_u}} = \begin{Bmatrix} U_2 \\ U_3 \end{Bmatrix} \tag{6.5.7}$$

$$\underline{\underline{F_u}} = \{F_1\} \qquad \underline{\underline{F_k}} = \begin{Bmatrix} 0 \\ P \end{Bmatrix}$$

The non-uniform shaft is fixed on the left end, such that the displacement at Node 1 is zero, hence:

$$\underline{\underline{D_k}} = \{U_1\} = \{0\} \tag{6.5.8}$$

Substituting (6.5.8) into (6.5.3):

$$\underline{\underline{D_u}} = \left(\underline{\underline{K_{22}}}\right)^{-1}\left(\underline{\underline{F_k}} - \underline{\underline{K_{21}}}\{0\}\right) = \left(\underline{\underline{K_{22}}}\right)^{-1}\underline{\underline{F_k}} \tag{6.5.9}$$

Taking note of the definitions in (6.5.7), the above may be expressed as:

$$\underline{\underline{D_u}} = \left(3.14(10)^6\begin{bmatrix} 6 & -1 \\ -1 & 1 \end{bmatrix}\right)^{-1}\begin{Bmatrix} 0 \\ P \end{Bmatrix} \tag{6.5.10}$$

Inverting the *2 by 2* square matrix in (6.5.10):

$$\underline{\underline{D_u}} = \frac{1}{5\left(3.14(10)^6\right)}\begin{bmatrix} 1 & 1 \\ 1 & 6 \end{bmatrix}\begin{Bmatrix} 0 \\ P \end{Bmatrix} \tag{6.5.11}$$

Performing the matrix operations in (6.5.11):

$$\underline{\underline{D_u}} = \frac{1}{5\left(3.14(10)^6\right)}\begin{Bmatrix} (1)(0)+(1)(P) \\ (1)(0)+(6)(P) \end{Bmatrix} = \frac{1}{5\left(3.14(10)^6\right)}\begin{Bmatrix} P \\ 6P \end{Bmatrix} \tag{6.5.12}$$

Recalling that the applied load in this problem, P, is equal to 942, Equation 6.5.12 renders:

$$\underline{\underline{D_u}} = \frac{1}{5\left(3.14(10)^6\right)}\begin{Bmatrix} 942 \\ 5652 \end{Bmatrix} = \begin{Bmatrix} 6(10)^{-5} \\ 36(10)^{-5} \end{Bmatrix} \tag{6.5.13}$$

Noting the definition of the displacement vector given in (6.5.7), another way of stating (6.5.13) is:

$$\underline{\underline{D_u}} = \begin{Bmatrix} U_2 \\ U_3 \end{Bmatrix} = \begin{Bmatrix} 6(10)^{-5} \\ 36(10)^{-5} \end{Bmatrix} \qquad \therefore \quad U_2 = 6(10)^{-5} \qquad U_3 = 36(10)^{-5} \tag{6.5.14}$$

Notice that these results are exactly the same as those found in Chapter 2. This occurs because the same technique (partitioning) was used in Chapter 2, albeit tacitly.

Although the *2 by 2* matrix in (6.5.10) was inverted using a closed form procedure, rendering (6.5.11), a Gaussian elimination procedure is typically utilized to solve the system of equations. In other words, although (6.5.9) shows that $\underline{\underline{K}}_{22}$ is inverted, (6.5.15) would be solved instead of (6.5.9):

$$\underbrace{\underline{\underline{K}}_{22}}_{\underline{\underline{K}}}\ \underbrace{\underline{\underline{D}}_{u}}_{\underline{\underline{D}}} = \underbrace{\left(\underline{\underline{F}}_{k} - \underline{\underline{K}}_{21}\,\underline{\underline{D}}_{k}\right)}_{\underline{\underline{F}}} \tag{6.5.15}$$

The equation above provides a solution to (6.5.9) in "*KD=F*" form, which avails itself to a Gaussian elimination procedure.

One advantage of using the partitioning method of imposing boundary conditions is that the order of the square matrix is reduced. The present example began with a *3 by 3* matrix and was reduced to *2 by 2*. However, the example given does not illustrate the amount of matrix manipulation that is required in any practical problem. Consider that in other problems, the rows and columns that need to be removed from the matrix will not typically be located at the extremities of the square matrix. Hence, rows and columns are removed from internal positions within the matrix, then the reduced matrix reformulated in compacted form to render a non-singular matrix.

The partitioning method has an additional advantage of being "exact," in that no other approximations are necessary for the matrix solution. Of course, round-off error is still introduced. For more information on this approach, see SDRC Masters Series online help under the topic of *The Finite Element Method, Degree of Freedom Families* [6, p. 4].

Penalty Method of Imposing Boundary Conditions

The penalty method is not exact because it introduces an approximation into the solution of the equilibrium equations via arbitrary constants. The penalty method has the advantage of being much easier to implement than the partitioning method.

Consider that the analyst wishes to impose a boundary condition, either zero or non-zero in value, at a given node i. The value of the displacement boundary condition will be represented by the variable a, as shown in (6.5.16).

$$D_i = a \tag{6.5.16}$$

To effect this restraint upon the global system of equations requires two steps:

- The stiffness element K_{ii} is multiplied by a large number, yielding a new term, K_{ii}^*.

- In the force vector, the component F_i is replaced with the value aK_{ii}^*.

Example 6.2—Imposing Boundary Conditions Using Penalty Method
The penalty method will be illustrated below, where the global stiffness matrix from Example 2.5 is considered again. Recalling the system of equations that needs to be solved, (2.4.19):

$$3.14(10)^6 \begin{bmatrix} 5 & -5 & 0 \\ -5 & 6 & -1 \\ 0 & -1 & 1 \end{bmatrix} \begin{Bmatrix} U_1 \\ U_2 \\ U_3 \end{Bmatrix} = \begin{Bmatrix} F_1 \\ 0 \\ P \end{Bmatrix} \tag{6.5.17}$$

The imposed boundary condition is:

$$U_1 = 0$$

Since a boundary condition is prescribed at Node 1, i=1, and we need to multiply K_{11} by a large number, say $1.0(10)^{16}$. In practice, components of the K-matrix can be examined to determine the magnitude of the largest value that appears. Then, a number many orders of magnitude larger is chosen as a *multiplier*. Operating on (6.5.17):

$$3.14(10)^6 \begin{bmatrix} 5(10)^{16} & -5 & 0 \\ -5 & 6 & -1 \\ 0 & -1 & 1 \end{bmatrix} \begin{Bmatrix} U_1 \\ U_2 \\ U_3 \end{Bmatrix} = \begin{Bmatrix} (a)5(10)^{16} \\ 0 \\ P \end{Bmatrix} \tag{6.5.18}$$

In the present example, a=0, hence, the first entry in the force vector on the right hand side of (6.5.18) is replaced by zero, and the system of equations is expressed as:

$$3.14(10)^6 \begin{bmatrix} 5(10)^{16} & -5 & 0 \\ -5 & 6 & -1 \\ 0 & -1 & 1 \end{bmatrix} \begin{Bmatrix} U_1 \\ U_2 \\ U_3 \end{Bmatrix} = \begin{Bmatrix} 0 \\ 0 \\ P \end{Bmatrix} \tag{6.5.19}$$

The square matrix in (6.5.19) is now non-singular, and Gaussian elimination can be employed. The resulting nodal displacements are found to be:

$$U_1 \approx 0 \qquad U_2 = 6.00(10)^{-5} \qquad U_3 = 36.0(10)^{-5} \tag{6.5.20}$$

These values are essentially the same as those found using the partitioning method of imposing boundary conditions (Example 6.1) although the smallness of U_1 is controlled by the magnitude chosen for the multiplier factor.

The Penalty Method of imposing essential boundary conditions is approximate, since a "zero value" of displacement is generated by dividing one number by another number that is relatively large. One disadvantage of using the penalty method is that it increases the maximum Eigenvalue of the matrix, which tends to decrease the accuracy of the results; Bathe [2, p. 485]

Reaction forces may be computed using any row of the *original* set of equilibrium equations, (6.5.17), that has a reaction force associated with it. Notice that the penalty method does not provide a reduction in the order of the stiffness matrix.

The Ones-On-Diagonal Method of Imposing Boundary Conditions
Assume that the analyst wishes to impose the boundary condition:

$$D_i = a$$

- All terms in the stiffness matrix in Row i are set to zero
- The component of the stiffness matrix K_{ii} is given the value of unity
- The value of the vector F_i is set to a

The Ones-on-Diagonal method is considered in the example below.

Example 6.3—Boundary Conditions Using Ones-On-Diagonal
Consider once again the equilibrium equations from Example 2.5:

$$3.14(10)^6 \begin{bmatrix} 5 & -5 & 0 \\ -5 & 6 & -1 \\ 0 & -1 & 1 \end{bmatrix} \begin{Bmatrix} U_1 \\ U_2 \\ U_3 \end{Bmatrix} = \begin{Bmatrix} F_1 \\ 0 \\ P \end{Bmatrix} \tag{6.5.21}$$

The boundary condition that needs to be imposed is:

$$U_1 = 0 \qquad \therefore a = 0$$

Since a boundary condition is prescribed at Node 1, i=1, and all terms in Row 1 of the stiffness matrix are set to zero:

$$3.14(10)^6\begin{bmatrix} 0 & 0 & 0 \\ -5 & 6 & -1 \\ 0 & -1 & 1 \end{bmatrix} \tag{6.5.22}$$

The stiffness matrix component K_{11} is set to unity:

$$3.14(10)^6\begin{bmatrix} 1 & 0 & 0 \\ -5 & 6 & -1 \\ 0 & -1 & 1 \end{bmatrix} \tag{6.5.23}$$

Substituting (6.5.23) for the original stiffness matrix in (6.5.21), and setting F_1 equal to a:

$$3.14(10)^6\begin{bmatrix} 1 & 0 & 0 \\ -5 & 6 & -1 \\ 0 & -1 & 1 \end{bmatrix}\begin{Bmatrix} U_1 \\ U_2 \\ U_3 \end{Bmatrix} = \begin{Bmatrix} a \\ 0 \\ P \end{Bmatrix}$$

Since the boundary condition in this problem is homogeneous, $a=0$, and the above becomes:

$$3.14(10)^6\begin{bmatrix} 1 & 0 & 0 \\ -5 & 6 & -1 \\ 0 & -1 & 1 \end{bmatrix}\begin{Bmatrix} U_1 \\ U_2 \\ U_3 \end{Bmatrix} = \begin{Bmatrix} 0 \\ 0 \\ P \end{Bmatrix} \tag{6.5.24}$$

Solving the *3 by 3* system of equations above renders the same answers as in Example 2.5:

$$U_1 = 0 \qquad U_2 = 6.00(10)^{-5} \qquad U_3 = 36.0(10)^{-5} \tag{6.5.25}$$

The *Ones-On-Diagonal* method does not reduce the size of the original matrix, while it does tend to add many useless zeros. Some solution procedures can be invoked to minimize the computational expense of many zeros (see Chapter 2). In addition to adding many zeros, the stiffness matrix that evolves from this method no longer has the property of symmetry; the square matrix in (6.5.24) illustrates this fact. The loss of symmetry can be a problem if one wishes to utilize a solution procedure that makes use of matrix symmetry, such as schemes that only compute a triangular portion of the stiffness matrix. Another disadvantage of the *Ones-On-Diagonal* method is that may also increase the maximum matrix bandwidth; Meyer [7]. Reaction forces are calculated using the respective row from the original stiffness matrix, (6.5.21) in this example, after all the nodal displacements are computed.

Reaction Forces and Matrix Ill-Conditioning

One measure of matrix ill-conditioning may be computed using the reaction forces. The difference between the summation of the reaction forces and the summation of the applied (i.e., external) loads reflects the degree of ill-conditioning. This difference, called the residual, is computed as:

$$\text{Residual} = \sum \textit{External Loads} - \sum \textit{Reaction Forces}$$

If the mathematics performed when solving for the unknown displacements is done exactly, and no other loads are applied, the residual should be zero; Rigby [8].

Skew Boundary Conditions and Loads

The last topic in this chapter is that of skew boundary conditions. In some finite element models, the analyst may wish to impose an essential boundary condition in a direction that is not aligned with any of the global coordinate axes. In addition, one may need to define concentrated forces or equivalent nodal forces in an off-axis direction.

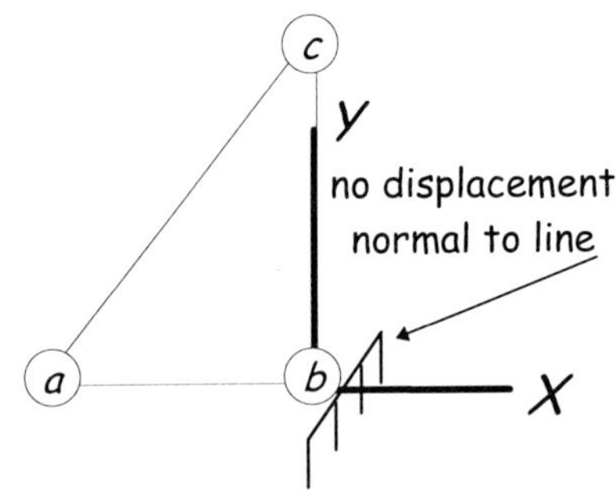

Figure 6.14 Skew Boundary Conditions

For a simple example of a skew boundary condition, consider the 3-node surface element for plane stress applications shown in Figure 6.14.

As indicated, displacement at Node b is to be restrained in a direction that does not align itself with either coordinate axis. A method used in such cases is to first establish a local coordinate system that contains one axis aligned with the direction in which displacement is to be restrained, with the other axis orthogonal to this direction. A local coordinate system corresponding to the problem in Figure 6.14 is shown in Figure 6.15. With this coordinate system, the desired skew boundary conditions shown in Figure 6.14 can be imposed by restraining the displacement in the y-coordinate direction at Node b.

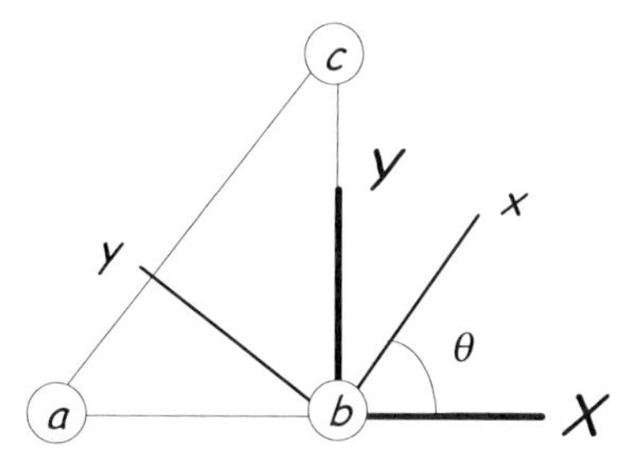

Figure 6.15 Local Coordinate System

Note the distinction between upper and lower case variables in Figure 6.15: the upper case variables correspond with the global system while the lower case variables correspond to the local coordinate system.

Assume that the stiffness matrix for the element shown in Figure 6.15 has been computed in terms of the global coordinate system, with the element equilibrium equations:

$$\underline{\underline{^eK}}\ \underline{\underline{^eD}} = \underline{\underline{^eF}} \tag{6.5.26}$$

or:

$$\underline{\underline{^eK}} \begin{Bmatrix} U_a \\ V_a \\ U_b \\ V_b \\ U_c \\ V_c \end{Bmatrix} = \begin{Bmatrix} F_{Xa} \\ F_{Ya} \\ F_{Xb} \\ F_{Yb} \\ F_{Xc} \\ F_{Yc} \end{Bmatrix} \tag{6.5.27}$$

However, at Node b we wish to have nodal displacement vectors in directions associated with the local coordinate system, hence, we endeavor to resolve the global displacements at Node b into local components. Assume that the line of skew boundary conditions is oriented at angle θ, relative to the global X-axis:

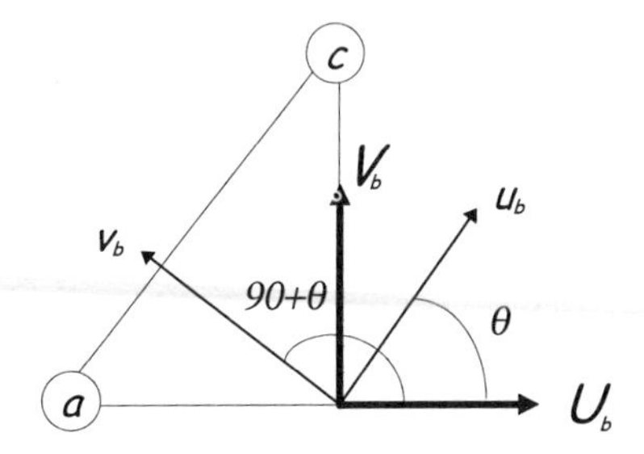

Figure 6.16 Local Coordinate System Details

To impose the boundary conditions in the local coordinate system, we must remove the two displacement components aligned with the global system, U_b and V_b, and replace them with displacement components aligned with the local coordinate system, u_b and v_b. One can then prescribe $v_b=0$ to effect the desired boundary condition.

To begin, displacement from u_b and v_b is resolved into the global X direction using the cosine law; the resulting displacement is denoted as U_b:

$$U_b = u_b \cos\theta + v_b \cos(90+\theta) \tag{6.5.28}$$

Or, using trigonometric relationships, (6.5.28) becomes:

$$U_b = u_b \cos\theta - v_b \sin\theta \tag{6.5.29}$$

Likewise, displacement in the global Y-direction can be expressed in terms of local components:

$$V_b = u_b \cos(90 - \theta) + v_b \cos(\theta) \tag{6.5.30}$$

Using trigonometric relationships, (6.5.30) is expressed as:

$$V_b = u_b \sin(\theta) + v_b \cos(\theta) \tag{6.5.31}$$

The global displacement vector can now be expressed in terms of the local displacement components at Node b, the global coordinates at Nodes a and c, and a transformation matrix as shown in (6.5.32). The c and s terms in (6.5.32) are shorthand for cosine θ and sine θ, respectively:

$$\underline{\underline{{}^{e}D}} = \begin{Bmatrix} U_a \\ V_a \\ U_b \\ V_b \\ U_c \\ V_c \end{Bmatrix} = \begin{bmatrix} 1 & 0 & 0 & 0 & 0 & 0 \\ 0 & 1 & 0 & 0 & 0 & 0 \\ 0 & 0 & c & -s & 0 & 0 \\ 0 & 0 & s & c & 0 & 0 \\ 0 & 0 & 0 & 0 & 1 & 0 \\ 0 & 0 & 0 & 0 & 0 & 1 \end{bmatrix} \begin{Bmatrix} U_a \\ V_a \\ u_b \\ v_b \\ U_c \\ V_c \end{Bmatrix} \tag{6.5.32}$$

$\Uparrow$
transformation matrix

In short, the global displacement vector can be expressed in terms of a transformation vector, T, and a displacement vector that has both global and local coordinates:

$$\underline{\underline{{}^{e}D}} = \underline{\underline{T}}\,\underline{\underline{{}^{e}D^{*}}} \tag{6.5.33}$$

The vector with both global and local displacement values, $\underline{\underline{{}^{e}D^{*}}}$, may be called a mixed vector. The transformation matrix and mixed nodal displacement vector, for this problem, are defined as:

$$\underline{\underline{T}} = \begin{bmatrix} 1 & 0 & 0 & 0 & 0 & 0 \\ 0 & 1 & 0 & 0 & 0 & 0 \\ 0 & 0 & c & -s & 0 & 0 \\ 0 & 0 & s & c & 0 & 0 \\ 0 & 0 & 0 & 0 & 1 & 0 \\ 0 & 0 & 0 & 0 & 0 & 1 \end{bmatrix} \qquad \underline{\underline{{}^{e}D^{*}}} = \begin{Bmatrix} U_a \\ V_a \\ u_b \\ v_b \\ U_c \\ V_c \end{Bmatrix} \tag{6.5.34}$$

The nodal forces can also be expressed with respect to the local coordinate system, as shown in Figure 6.17.

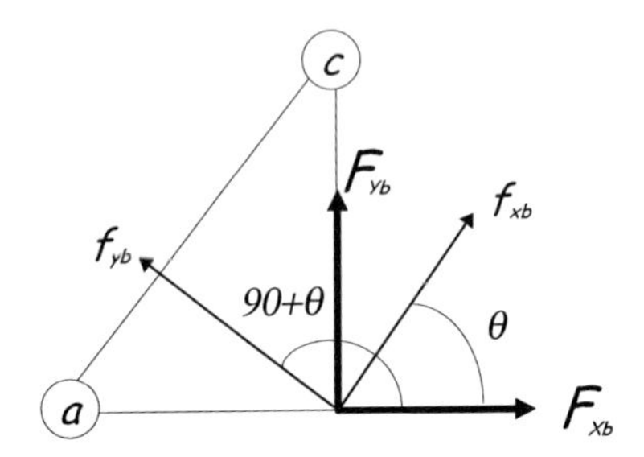

Figure 6.17 Forces in Local Coordinate System

Perhaps one can see that the same transformation process can be followed for the force vector:

$$\underline{\underline{{}^eF}} = \underline{\underline{T}}\,\underline{\underline{{}^eF^*}} \tag{6.5.35}$$

where:

$$\underline{\underline{{}^eF^*}} = \begin{Bmatrix} F_{aX} \\ F_{aY} \\ f_{bx} \\ f_{by} \\ F_{cX} \\ F_{cY} \end{Bmatrix} \tag{6.5.36}$$

The equilibrium equations for the element, (6.5.26), can be expressed in terms of the modified displacement and force vectors:

$$\underline{\underline{{}^eK}}\;\underline{\underline{T}}\,\underline{\underline{{}^eD^*}} = \underline{\underline{T}}\,\underline{\underline{{}^eF^*}} \tag{6.5.37}$$

Or, using the transformation matrix defined in (6.5.34) along with the nodal and force vectors defined above, (6.5.37) can be expressed as:

$$\underline{\underline{{}^eK}}\begin{bmatrix}1&0&0&0&0&0\\0&1&0&0&0&0\\0&0&c&-s&0&0\\0&0&s&c&0&0\\0&0&0&0&1&0\\0&0&0&0&0&1\end{bmatrix}\begin{Bmatrix}U_a\\V_a\\u_b\\v_b\\U_c\\V_c\end{Bmatrix}=\begin{bmatrix}1&0&0&0&0&0\\0&1&0&0&0&0\\0&0&c&-s&0&0\\0&0&s&c&0&0\\0&0&0&0&1&0\\0&0&0&0&0&1\end{bmatrix}\begin{Bmatrix}F_{aX}\\F_{aY}\\f_{bx}\\f_{by}\\F_{cX}\\F_{cY}\end{Bmatrix} \tag{6.5.38}$$

A more compact and effective way of expressing (6.5.37) can be achieved by using matrix algebra. Multiplying both sides of (6.5.37) by the inverse of the transformation matrix:

$$\left(\underline{\underline{T}}\right)^{-1}\underline{\underline{{}^eK}}\;\underline{\underline{T}}\;\underline{\underline{{}^eD^*}}=\left(\underline{\underline{T}}\right)^{-1}\underline{\underline{T}}\;\underline{\underline{{}^eF^*}} \tag{6.5.39}$$

Multiplying the transformation matrix by its inverse yields the identity matrix which simplifies the right hand side of (6.5.39):

$$\left(\underline{\underline{T}}\right)^{-1}\underline{\underline{{}^eK}}\;\underline{\underline{T}}\;\underline{\underline{{}^eD^*}}=\underline{\underline{{}^eF^*}} \tag{6.5.40}$$

Since the transformation matrix is orthogonal, it can be shown that the inverse and the transpose are identical (Bathe [2, p. 35]):

$$\left(\underline{\underline{T}}\right)^{-1}=\left(\underline{\underline{T}}\right)^{T} \tag{6.5.41}$$

Using this relationship in (6.5.40):

$$\left(\underline{\underline{T}}\right)^{T}\underline{\underline{{}^eK}}\;\underline{\underline{T}}\;\underline{\underline{{}^eD^*}}=\underline{\underline{{}^eF^*}} \tag{6.5.42}$$

The expression for the equilibrium equations in (6.5.42) is preferable to that of (6.5.40) since it is easier to compute the transpose of a matrix than the inverse, generally speaking. The expression in (6.5.42) may also be expressed as:

$$\underline{\underline{{}^eK^*}}\;\underline{\underline{{}^eD^*}}=\underline{\underline{{}^eF^*}} \qquad \text{where: } \underline{\underline{{}^eK^*}}\equiv\left(\underline{\underline{T}}\right)^{T}\underline{\underline{{}^eK}}\;\underline{\underline{T}} \tag{6.5.43}$$

The expressions for $\underline{\underline{{}^eD^*}}$ and $\underline{\underline{{}^eF^*}}$ are defined by (6.5.34) and (6.5.36). The stiffness matrix in (6.5.43) is transformed so that its components are all referenced to the global coordinate system. When transformed matrices are all referenced to the same coordinate system, they can be combined. Of course, the stiffness matrix of any other element that also uses Node b would have to be transformed before being combined to form the global stiffness matrix. Notice that using this approach to imposing skew displacement boundary conditions, one is also able to impose nodal forces at skew angles. (An essential boundary condition and force, acting in the same coordinate direction, cannot be applied at the same node, however.)

Before stress values are computed for an element with transformed nodes, the vector ${}^{e}D^{*}$ (the mixed vector) must be transformed back into the original, non-transformed displacement vector, via (6.5.33). Stress can then be computed using:

$$\underline{\underline{\tau}} = \underline{\underline{B}}\,{}^{e}\underline{\underline{D}} \qquad (6.5.44)$$

This concludes the section on loads and boundary conditions. Although a 3-node triangular element was used in the examples, the principles remain the same for other isoparametric elements.

References

1. Anonymous, *A Finite Element Primer*, Department of Trade and Industry, National Engineering Laboratory, Glasgow, U.K., p. 94, 1987
2. Bathe, K., Finite Element Procedures In Engineering Analysis, Prentice-Hall, Inc., Englewood Cliffs, N.J., 1982
3. Anonymous, "Practical Finite Element Modeling And Techniques Using MSC/NASTRAN," NA/1/000/PMSN, The MacNeal-Schwendler Corporation, Los Angeles, 1990
4. Cook, R.D., *Concepts And Applications Of Finite Element Analysis*, John Wiley & Sons, N.Y., 1989
5. Huebner, K.H., *The Finite Element Method For Engineers*, John Wiley & Sons, N.Y., 1982
6. Structural Dynamics Research Corporation, Milford, Ohio, 1985
7. Meyer, C., "Special Problems Related to Linear Equation Solvers," J. of Structural Division, ASCE, July, 1973
8. Rigby, F.N., "Back to Basics-2. Ill-Conditioning in Finite Element Analysis," in *BenchMark*, E.R. Robertson, Editor, NAFEMS, Glasgow, December, 1992

Chapter 7

Practical Considerations and Applications

"Works of art are of an infinite loneliness..."

—R.M. Rilke

7.1 Finite Element Procedures

Key Concept: A checklist or procedure is highly recommended when performing finite element analysis.

This section presents a procedural outline to assist the analyst in organizing finite element projects. The outline should be considered as thought starter, as opposed to an all-inclusive list.

An Outline for Finite Element Projects
In keeping with Section 1.6, the analysis process is considered in three categories: stating the problem in engineering terms, developing a mechanical idealization, and producing a finite element solution. The procedure and associated discussion given in this section targets analysis work performed in conjunction with product design and development, as opposed to engineering research. However, many of the items listed in the procedure would apply in either case.

A Procedure for Finite Element Projects

I. State Problem in Engineering Terms
- Ask the "Engineering Question"
- Pose the problem
- Draw a rough sketch

II. Mechanical Idealization
- Linearity and localized stress
- Boundary Condition Assumptions
- Stress-strain assumptions

- Geometric simplification/symmetry
- Material assumptions
- Loading assumption
- Update the rough sketch

III. Finite Element Solution
- Create mesh boundary and associated finite element geometry
- Choose element type and read element documentation
- Model data sheet
- Input model data
- Model check
- Invoke finite element solver
- Evaluate results
- Re-run with finer mesh for convergence check
- Verify accuracy of results

State Problem in Engineering Terms

There appears to be a significant amount of computer aided engineering analysis that eventually turns out to be of little or no value. This phenomenon is more prevalent in companies that have recently jumped on the CAE bandwagon, as opposed to, for instance, the aircraft companies who have been using computer aided modeling techniques since the early days of the digital computer era.[1] There is no single reason for this waste, however, it appears that a common factor is that problems to which analysis efforts are directed are often not properly specified. In the interest of producing an effective analysis, some recommendations regarding analysis project management are given.

State Problem in Engineering Terms—Ask the "Engineering Question"

It may be helpful to begin an analysis project with an informal meeting of interested parties to determine the question that the analysis is intended to answer. This is not the same thing as stating the purpose of the analysis. Stating the question that one hopes to answer often provides more direction for an analysis than stating a purpose. The latter seems to lend itself to vagueness. For example, consider a hypothetical "widget bracket," for which the stated purpose of an analysis might be "Optimize the design of the widget bracket." Notice that this statement provides little insight as to what is to actually be done. In contrast, consider the *engineering question*: "What is the minimum cross section that can be used for a widget bracket if the bracket is to withstand a static load of 500 lb_f."

[1] Aeronautical engineers at Boeing were responsible (in part) for an early paper on the use of "finite elements"; see Turner [1].

Of course, poorly posed questions are of not of much value either, for instance, the question "What is a good design for the widget bracket?" does little to convey how the results of the analysis will be used. One should try to pose the engineering question using specific terms, while including some of the design variables. It does little good if the results of an analysis suggest changing design parameters that must remain fixed. For instance, without any suggestion of what the design variables of the widget bracket are, the results of an analysis may recommend the use of a shorter bracket. This recommendation is of little value if the length of the bracket is something that cannot be changed.

The purpose of the analysis logically follows from the engineering question. Write the engineering question down, preferably on some type of analysis request document. Often, several parties (designers, engineers, and analysts) are involved in any one analysis project. Do all parties agree that answering the stated engineering question is the purpose of the analysis? Keep in mind that the more narrow the scope of the analysis, the more likely it is to be successful and timely. If there are several engineering questions to be answered, break the project into individual analyses, keep each individual analysis simple, and document the details.

State Problem in Engineering Terms—Pose the Problem

Posing the problem begins by asking the question, "What criteria (or criterion) will be used to answer the Engineering Question." For the widget bracket example, it is assumed that low carbon steel will be used, and that "bracket failure" will be defined as the onset of gross yielding. This information, along with the other details discovered during the initial discussion of the analysis problem, might render the following problem statement:

> What is the magnitude of the maximum von Mises stress that occurs in the widget bracket when a static, monotonic, uniformly distributed load is applied to the top surface of the bracket? The left face of the bracket is assumed to be rigidly fixed, and the material is assumed to be linear, elastic, and isotropic in the range of loads applied. The uniformly distributed load is to result in a total force of 500 lb_f with the resultant force in the *Z*-coordinate direction, as illustrated on the attached sketch.

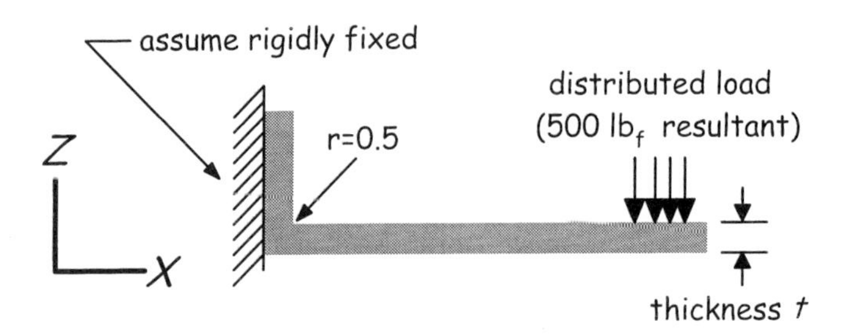

Figure 7.1 Rough Sketch of Widget Bracket

Mechanical Idealization

With the problem posed in engineering terms, the process of mechanical idealization can begin. Linearity, boundary conditions, stress-strain assumptions, geometric simplification, assumptions regarding material behavior and loading are reviewed.

Mechanical Idealization—Linearity and Localized Stress

In an actual structure there may exist regions of *localized stress* (or *stress concentration*) due to stress raisers. The level of stress in the localized areas may be very much higher than the nominal stress which acts upon a larger region of the structure. As the loading upon the structure is increased, the material near a stress raiser may yield and significantly deform, which can cause the stress to be limited. For example, say a small sharp corner is formed during the fabrication of a part. Upon initial loading, the magnitude of stress in the sharp corner may be many times that of the nominal stress, and quickly reach the yield strength of the material. If the material is ductile, and the load continues to increase, large deformation can occur in the vicinity of the sharp corner, perhaps changing the once sharp corner into a more rounded one. The rounded corner now has a much lower stress concentrating effect, and the stress in the localized region takes on a value more in line with the nominal value. In other cases, material near a stress raiser does not yield, and the level of stress remains well above the nominal stress. In either case, a small region of localized stress may have little affect upon the structure, depending upon the type of material and loading. In the case of a ductile material subjected to monotonic loading, some stress concentrations can be ignored.

There are cases when localized yielding affects the distribution of nominal stress. For instance, if localized yielding takes place in the vicinity of a component's mounting area, the nominal stress may be affected, since the primary load path can be altered. If high values of localized stress and associated large deformation is anticipated, a non-linear analysis should be considered.

In other cases localized stress is present but a non-linear analysis is not needed. However, the magnitude of the localized stress will be taken into account to determine if the design is adequate. For instance, structures subjected to cyclic loads, or structures composed of brittle homogeneous materials subjected to any type of loading, often need to be considered in terms of localized stress and associated crack initiation. Several examples of typical stress concentrations are given in Higdon [2, p. 62], while Timoshenko [3, p. 159] illustrates stress concentrations using *photoelastic stress analysis*. A chapter in Juvinall [4] is devoted to the topic of stress concentration, while Pilkey [5] provides an exhaustive study of the subject, with many illustrations and graphs.

Mechanical Idealization—Choice of Stress/Strain Assumption

The novice analyst typically asks "How do I identify the applicable stress/strain assumption for a given structure and loading?" This is a difficult question to answer. One aspect of the art of finite element analysis is in choosing the best stress/strain assumption for a given problem; knowing which is best typically comes with experience in analyzing stress and structural problems. A few guidelines will be offered.

The easiest stress assumption to identify is uniaxial stress. In practice, the uniaxial assumption is typically applied in the case of truss structures, such as bridges that employ pinned trusses in their construction, if bending is not induced in the member. Commercial buildings may also used pinned trusses in their construction, as might some broadcast towers if the reinforcing members are pinned. A pinned truss allows the applied loads to be carried through axial deformation instead of bending deformation, the advantage being that members are typically stronger in membrane tension than in bending (one must be certain that the joint remains truly pinned throughout the life of the structure).

The next most readily identifiable assumption is perhaps the axisymmetric solid, where both the structure and the loading is symmetrical about an axis of revolution[2]. Another stress assumption that is fairly easy to recognize is that of beam bending. Beams may also be used in truss-type structures, however, beams are used when the joints are fixed instead of pinned, such that bending is induced. To employ a beam element, for instance a 3-node line element for beam bending, one typically assumes that the cross section is uniform. If a beam with a non-uniform cross section is used, some other element type might be necessary. Although the beam cross section typically needs to be uniform along the length, the cross section need not be prismatic, for instance, box, T- or I-sections may be

---

[2] Some finite element software allows the use of axisymmetric solid elements when the loading is not symmetrical but able to be described by a Fourier series expansion.

modeled. In such cases the analyst needs to be aware of the assumptions that pertain to the type of section being used; for details see [6], White [7] and Young [8].

The plane stress case may be somewhat more difficult to identify, perhaps because it has more potential applications. A plate subjected to in-plane forces is often used as an example of plane stress. Plane stress may also be used in the case of narrow beams in either bending or extension; beams subjected to combined bending and torsion would not be well represented by plane stress, however. The fundamental presumption is that the load is applied in-plane and stress in one coordinate direction is zero.

The plane strain idealization appears to be less utilized than plane stress. However, water retention dams and perhaps railway rails subjected to pressure to from passing locomotives might be candidates for a plane strain idealization. Some rolling operations may be considered plane strain; Yu [9]. Plane strain presumes that the displacement, and associated strain, is zero in one coordinate direction.

The presumption of plate bending stress and the use of the associated plate bending elements appears to be a thing of the past. With the advent of robust, easy to use shell elements, three choices exist for modeling plate bending: plate elements, shell elements, or three-dimensional brick elements. Until recently, Kirchhoff plate elements were the best choice for thin plate bending since they were less expensive and typically performed better than any of the alternatives. However, as computers have become more powerful, and shell elements more robust, the use of plate elements has declined. Analysts today often employ shell elements even when there is relatively little membrane deformation, such as in the case of plate bending.

Mechanical Idealization—Geometric Simplification

As mentioned in Chapter 1, a substantial reduction in computational expense is often achieved through geometric simplification, using symmetry arguments. A common type of symmetry is *axisymmetry*, where the geometry of the structure can be represented by a surface, theoretically revolved about an axis. Another type of symmetry commonly used is that of mirror symmetry. Mirror symmetry can be utilized when the geometry of a structure and the applied loads exhibit symmetry about some axis, or pair of axes. Both of these cases were considered in Chapter 1; other types of symmetry arguments can also be imposed.

After all of the idealization details are reviewed, the rough sketch that was generated when the problem was initially posed should be updated to reflect the idealization assumptions. If a closed-form approximation for stress or displacement is to be performed, it should be included with the updated sketch.

Finite Element Solution

After the idealization process has been completed, the analyst may begin work on the finite element model. It is entirely possible that after the mechanical idealization is discussed, the engineering problem may need to be re-posed, since the initial problem statement may not lend itself to an idealization that can be cast into finite element form. Assuming that the idealization *does* lend itself to finite element analysis, the next step is to consider the mesh boundary geometry.

If the mesh boundary is complicated, the analyst may enlist the help of a designer to construct the mesh boundary geometry. Once the mesh boundary geometry is available, the analyst is able to begin meshing. Details associated with creating and verifying finite element models are considered in Section 7.2.

Finite Element Solution—Keeping Things Straight

In some cases, an analyst may endeavor for many weeks (or even months) to produce a single, large finite element model. Alternately, an analyst may work on a smaller model, performing many variations in loading, boundary conditions, element type, model size, etc. In such cases, it is quite easy to loose track of which model produced which results, etc. In this regard, a Model Data Sheet, as shown in Figure 7.2, may be helpful.

MODEL DATA SHEET

Date ______________

Analysis Name ________________________________

Part Number ________________________________

Model Part Name ________________________________

Pre-processor ________________________________

Solver ________________________________

Element Type(s) ________________________________

Engineering question:

____

____

____

Notes:

____

____

____

Sketch

Figure 7.2 Model Data Sheet

7.2 Elementary Modeling Techniques

Key Concept: Finite element models span a broad range of purpose and complexity. However, some elementary modeling techniques are applicable to many different finite element models that are used to solve solid mechanics problems.

This section addresses some of the more elementary issues associated with finite element modeling techniques, for instance, what element to use. One of the first questions a novice analyst might ask is "Why not just use volume elements for every application?" This question was briefly addressed in Chapter 4; a few additional words will be offered here.

The Use of Three-Dimensional Elements
Theoretically, three-dimensional volume elements could be used for all applications, (even plate bending or shell stress) although this approach would often be inefficient. The inefficiency is a result of three factors:

- Element distortion
- Effort involved in creating and processing three-dimensional elements
- Element performance

Use of 3-D Elements—Distortion
As will be discussed in Section 7.3, elements may display undesirable characteristics when they are distorted, and one measure of distortion is aspect ratio, as shown in Section 7.3, Figure 7.12. If three-dimensional elements are used to model very thin structures, many small elements would be needed since the length and width of the elements must not be much greater than the thickness. Hence, a better choice of element in such cases might be surface elements, where the thickness (specified by a mathematical constant) would not be subjected to the same restrictions as the thickness of three-dimensional elements. Only the length and the width of the surface element would need to conform to aspect ratio restrictions, allowing far fewer elements to be used.

Use of 3-D Elements—Effort to Create and Process Elements
Another reason why three-dimensional elements are not always efficient is that they require more effort to create and process. Creating three-dimensional mesh boundaries is often a daunting task, while surface boundaries are typically handled more easily. In addition, models with many three-dimensional elements can contain many nodal DOF's, requiring significant computer resources to store and solve.

Use of 3-D Elements—Element Performance
In some modes of deformation, particularly bending, surface elements (plates and shells) typically outperform volume elements such as the 8-node brick. Some sources state that three-dimensional elements should be avoided for bending applications unless surface elements simply cannot be used.

In short, three-dimensional elements are not always the most efficient. The analyst should therefore choose among several candidates to obtain an element that will perform best in a given situation.

The General Class of Element Follows from the Idealization
Assume that the procedure for finite element projects listed in Section 7.1 has been followed for the hypothetical widget bracket: An engineering question is stated, the problem posed, an idealization established, and the mesh boundary constructed. From the idealization, the analyst knows the general class of element(s) that will be employed: Since the widget bracket in Figure 7.1 calls for a surface representation and the presumption of plate bending, a plate (or shell) element will be employed. The issue now is what particular plate or shell element is to be employed.

Which Element to Use?
The novice analyst might be somewhat bewildered with the vast array of elements available in commercial finite element software today: Which element is best?

Table 7.1 Common Element Usage

stress/strain assumption	element geometry	common element
uniaxial stress	line	2-node line
axisymmetric solid	surface	4-node quadrilateral
beam bending	line	3-node line
plane stress	surface	4-node quadrilateral
plane strain	surface	4-node quadrilateral
plate bending/ shell	surface	4-node quadrilateral
3-d stress	volume	8-node hexahedral (* 10-node Tet)

The choice of the best element depends upon several factors. However, once the analyst chooses a stress/strain assumption to characterize the structure under investigation, the choice of which element to use becomes somewhat easier, as suggested by Table 7.1. Referring to the table, the use of a 4-node quadrilateral element is commonly used for both plate and shell applications. True "plate elements" are typically not often used anymore.

Recall from Chapter 4 that shell elements can be used to model plate applications, since shell theory considers both bending and membrane affects. The use of the 10-node tetrahedral element for three-dimensional stress analysis has become increasingly popular in recent years, although it might not be fair to call it a commonly used element at this time. In fact, calling any of the elements in Table 7.1 "common" may elicit protest from some individuals. Indeed, even if universal agreement on what constitutes "best" or "common" elements could be reached, the ever changing nature of finite element technology would surely render the chosen elements obsolete before too long. Consider the elements listed in the table above as those which are currently popular, or have enjoyed popularity in the recent past. Documentation from the finite element software vendor should be consulted to identify recommended elements; for instance, see [10, p. 1–1] for elements that MSC/NASTRAN[3] calls common, or [11] for a similar list of elements used with MARC[4] finite element software. Factors that influence the choice of which particular element to use are now discussed.

Factors that Influence the Choice of Element

During the idealization process, a stress/strain assumption is identified, and this assumption determines the general class of element that is to be used for a given analysis. For instance, if a plane stress idealization is chosen, the mesh boundary and associated elements will be surface type. Alternately, if the idealization is three-dimensional stress, volume boundaries and elements will be used. The choice of the *particular* element is influenced by:

- Element accuracy
- The size of the model, in terms of the total number of elements
- Complexity of the mesh boundary/automeshing
- Linearity
- Element distortion

---

[3] MSC/NASTRAN®, The MacNeal-Schwendler Corporation, 815 Colorado Boulevard, Los Angeles, CA.

[4] MARC®, MARC Analysis Research Corporation, 260 Sheridan Avenue, Palo Alto, CA.

There exists some disagreement as to which elements are "the best." The author is of the opinion that there is no "best" element—the choice depends upon the circumstances. The discussion that follows should be considered as general background, as opposed to rules for choosing elements.

Choice of Element—Element Accuracy

While it may be argued about which elements are best, there seems to be some agreement upon what constitutes "undesirable" elements. Typically, both the 3-node triangular element, and its three-dimensional analogue, the 4-node tetrahedral, are avoided. Both can only characterize constant stress, and in areas of high stress gradients, a considerable number of these elements would be required to maintain suitable accuracy. A limited number of 3-node triangular surface elements may be tolerated in a mesh of 4-node quadrilateral surface elements. The reason for utilizing 3-node triangular elements is that they are handy in regions of curved geometry, and also in regions where mesh transitions take place. An example of mesh transition using 3-node triangular elements is shown in Figure 7.6.

Choice of Element—Size of Model

All of the elements listed in Table 7.1 are linear elements. Why use linear elements, if quadratic elements are typically more effective, due to their higher order displacement assumptions? Using higher order elements, the model often ends up with too many nodes, and hence, too many degrees of freedom, substantially increasing the resources needed to process the model. If the finite element model under consideration does not require too many elements, higher order elements may indeed be the best choice, since the total number of DOF's would be somewhat limited. However, in models that require a moderate to large number of elements, it often happens that the number of nodal DOF's exceeds the number that can be efficiently handled. This may seem surprising, because today's engineering workstations are very fast (mainframe computers more so, perhaps) and can store vast amounts of data.[5] Still, large finite element models can require a relatively large amount of solver time, either in terms of raw CPU time, or time waiting in the queue before the solver is invoked.

If only one solution is needed for a given analysis, solver time may be less significant than when a number of computer runs are to be executed. Analysts may often run an analysis at least two times, the first time with a coarse mesh, and the second run with a refined mesh, to check for convergence. In other cases, the model may be run tens or even hundreds of times to evaluate different mesh

[5] It would appear that a large number of small to moderate size finite element models are now being solved on engineering workstations. Twenty years ago, nearly all of FEA solvers utilized mainframe computers. Workstations, as we know them today, did not exist.

schemes, different loads, different boundary conditions, slightly different geometry, etc. In addition to requiring a considerable amount of computer RAM, large finite element models can also consume a very large amount of hard disk space for both storage of the model, and for "scratch space" that is needed during the solution procedure. Scratch space of 500 Mb is not uncommon in moderate size models. Finally, post-processing large models often becomes a difficult task. For instance, rendering a color contour plot of a finite element model containing 100,000 three-dimensional elements (and the associated nodes) may take a significant amount of time. In short, as a practical matter, models with many nodal DOF's become difficult to process.

Using higher order elements, it is often the case that many of the nodal DOF's are wasted, since they are used in places where the stress is relatively constant, in which case a few linear elements (with far fewer nodal DOF's) would suffice. The enterprising analyst may therefore choose to use both higher order and lower order elements in the same model, using higher order elements in areas of the model that will experience large stress gradients, and lower order elements in areas of constant stress. Figure 7.3 depicts an 8-node quadrilateral element joined to a 4-node quadrilateral using constraint equations.

Constraint equations that involve nodal displacement variables at several nodes are often termed "multipoint constraints." By requiring Node 2 to displace in a manner such that it is always co-linear with Nodes 1 and 3, the quadratic displacement capability of the 8-node element is reduced to a linear function on one edge, such that compatibility is maintained between the elements. Other types of elements can also be joined in this manner or by somewhat similar techniques; see [10], Zemitis [12], and Feld [13] for more on joining elements of dissimilar type.

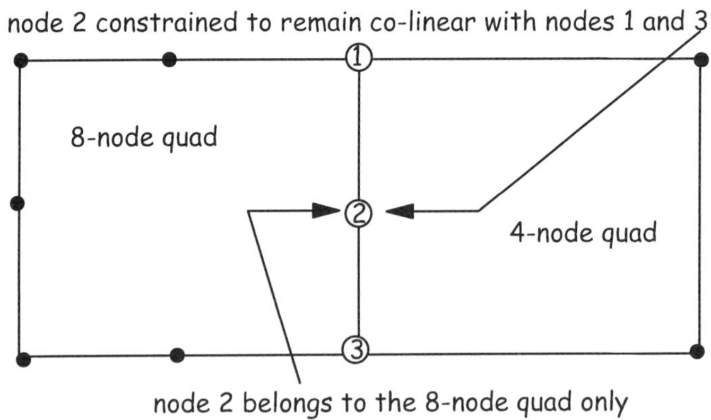

Figure 7.3 Joining an 8-Node Quadrilateral to a 4-Node

The technique of constraining the mid-side node of a higher order element to be co-linear with the other two nodes is an effective means of joining higher order elements to lower order elements, but it may not be efficient, especially if the model is to be re-meshed several times. With many thousand elements, specifying the constraint equations for each element pair may prove tedious, while at the same time introducing another potential source of error. Automatic imposition of such constraints is now possible in some finite element software.

Finite element code developers and academicians typically advance the use of higher order elements, while analysts working "in the trenches" will often opt for lower order elements due to practical limitations. Melosh [14] states that the use of higher order elements is not necessarily more efficient than the use of lower order elements—one must also consider the particular mesh, the level of accuracy required, etc.

In summary, analysts building large to moderate finite element models requiring surface elements often prefer 4-node quadrilateral surface elements (with mesh grading) over 8- and 9-node quadrilaterals. Similarly, 8-node hexahedral volume elements ("bricks") may be preferred over 20-node bricks. However, for tetrahedral elements, the pattern of common element usage changes. While a mesh of lower order quadrilateral or hexahedral elements can perform adequately, a mesh of 4-node tetrahedral volume elements would not be recommended, regardless of the size of the model. If tetrahedral elements are to be used, the 10-node tetrahedral, having a complete quadratic displacement assumption, is the lowest order tetrahedral that should be used. As mentioned, the 4-node tetrahedral is too stiff to be used with any level of confidence.

Choice of Element—Complexity of Mesh Boundary/Meshing Routines

A truly automatic finite element meshing routine would be able to create any class of finite element (line, surface, or volume) within a specified mesh boundary of the respective type by simply specifying a desired element size and issuing a single command. Furthermore, the elements generated within the boundary would be of good quality (no excessive distortion) regardless of the complexity or size of the mesh boundary. An additional requirement for the automatic mesh generator would be for the meshing operation to take place within a reasonable amount of time.

Although the mesh generation technology has improved steadily over the years, there does not appear to be an automatic finite element mesh generator that meets the requirements above. Indeed, until recently, automated meshing dictated the use of 3- or 6-node triangles for surface meshing while requiring either 4- or 10-node tetrahedral elements for volume meshing.

A. Automeshing Two-Dimensional Boundaries: Today, the best finite element pre-processors are equipped with "automeshers" that are fairly robust in generating two-dimensional elements. These meshing routines can create good quality quadrilateral (or triangular) surface elements within moderately complex surface boundaries. Some operator intervention may still be required, particularly when a large, complex boundary is present. In such a case, the analyst may need to first partition the boundary into smaller portions and specify element sizes near complicated geometric features. Some of the two-dimensional automeshers may substitute a limited number of triangular elements in place of quadrilaterals if the mesher cannot produce a mesh entirely of good quality quadrilaterals.

B. Automeshing Three-Dimensional Boundaries: A robust automatic mesher for three-dimensional elements, one that can produce all types of volume elements while maintaining suitable quality under a variety of circumstances, is not available at this time. There do exist fairly robust three-dimensional automatic meshers. However, these meshing routines are limited to *tetrahedral elements*. As with two-dimensional automated meshing routines, some operator intervention is typically required to partition the volume mesh boundary and/or specify local element sizes in areas of difficult geometry. Three-dimensional automeshing routines typically require more intervention than routines for two-dimensional elements.

C. Mapped Meshing of Two- and Three-Dimensional Boundaries: When the analyst desires more control over how the mesh is generated, mapped meshing is used. With mapped meshing, the analyst can choose the number of elements to be produced on each edge of the mesh boundary, and the resulting pattern of elements will be propagated throughout the entire mesh boundary. Using a free mesh, the analyst can also choose the number of element on the edges of the mesh boundary, but the resulting pattern will not typically propagate throughout the entire mesh boundary. Thus, the analyst has less control over the mesh pattern when using a free meshing routine, and more control when using a mapped meshing routine.

Mapped meshing also allows the use of brick elements in volume mesh boundaries. The disadvantage of mapped meshing is that far more mesh boundaries may be required in a model when mapped meshing is used, as compared to free meshing, and the shape of the mapped mesh boundaries need to be more like the shape of the element that is being meshed. For instance, if creating a mesh boundary for brick elements, the boundary often needs to have six faces, just like the elements that will be created within the boundary.[6] This contrasts with the boundaries that are used when automeshing tetrahedral volume elements, where the boundaries can be of far more general shape. The mesh

---

[6] Lately, there are some mapped meshing routines that relax this requirement somewhat.

boundaries used to free mesh tetrahedral elements can be oddly shaped because tetrahedral elements, having four faces, can fill oddly shaped volumes and still maintain an acceptable level of distortion. Brick elements, ideally shaped as cubes, cannot fit into oddly shaped volumes without experiencing a great deal of distortion. Since mesh boundaries for mapped meshes are more restrictive in their geometry, the process of creating mesh boundaries to allow mapped meshing of brick elements can be extremely time consuming when the geometry is complex.

D. Two-Dimensional Meshing Summary: Using current technology, an analyst is often able to utilize automeshing for any type of *two-dimensional* element they desire. For moderate to large size models, the choice is often 4-node quadrilaterals, since they represent a compromise between accuracy and computational expense. For smaller, two-dimensional models, a mesh of 8- or 9-node quadrilaterals may be most effective. If more control of the mesh is required, mapped meshing of two-dimensional elements may be needed.

E. Three-Dimensional Meshing Summary: For small, geometrically non-complicated, three-dimensional models, without bending loads, a mapped mesh of 20-node bricks may be best. When three-dimensional models are of moderate size or larger, but still non-complex, the analyst may choose 8-node brick elements. Again, automeshing cannot be used for brick elements; the analyst must use a mapped meshing routine.[7] However, for non-complex geometry, mapped meshing of bricks is often as easy as the automeshing of tetrahedrals, because few partitions are required in either case. When geometry becomes complex, many more partitions are required for mapped meshing, and the creation of all the required partitions can be extremely time consuming. In such cases, the use of brick elements may not be advisable, unless tetrahedrals simply cannot be used for the type of problem that is being solved.

When a three-dimensional model is geometrically complex, the analyst may resort to automeshing with 10-node tetrahedrals. (Again, the 4-node tetrahedral is generally not recommended under any circumstances.) The advantage of automeshing is that when the geometry is complex, far fewer partitions are needed to produce a meshable volume when compared to mapped meshing. In addition, the 10-node tetrahedral can withstand a considerable amount of distortion, as compared to brick elements, before its accuracy degrades. However, there are some disadvantages commonly associated with 10-node tetrahedrals:

- Too many nodes in model—vast computer resources may be required
- Mesh grading can be difficult
- Difficult to apply specialty elements (contact, gap, and rigid elements)

[7] In some cases, a three-dimensional mesh of brick elements may be created by extruding quadrilateral elements. The resulting mesh is sometimes called a mesh of two and one-half dimensions (i.e., 2 ½-D).

- Convergence checking difficult
- More tetrahedral elements required to represent simple shapes
- Some controversy regarding accuracy when compared to brick elements

Recall from Chapter 4 that to perform convergence checking, the original mesh must be embedded in the refined mesh. However, using standard tetrahedral automeshing routines, the location of all of the original nodes need not be the same when a refined mesh is created. This makes convergence checking difficult. Another drawback to using tetrahedral elements is that to represent simple shapes, more elements and nodes are required than when using brick elements. While a single 8-node brick element can be used to represent a cube, an informal test suggests that *eleven*, 10-node tetrahedrons, with 34 nodes, would be automeshed in the same cube. In addition, there appears to be continuing controversy over how accurate 10-node tetrahedral elements are.

In Russell [15], 10-node tetrahedral elements are compared with shell elements, with the results suggesting the 10-node tetrahedral performs acceptably well, as far as accuracy and computational efficiency is concerned. However, Ramsay [16] compares 10-node tetrahedral elements with brick elements and suggests that the 10-node tet may be inferior in terms of computational efficiency. This is followed by another study, Cifuentes [17], which suggests that Tetrahedral and Hexahedral elements are comparable. At this point in time, the analyst can be reasonably assured that the 10-node tetrahedral element is, at least, a usable element of acceptable accuracy, even if it may not be the most computationally efficient. Regardless of the type of element used, the model should always be checked for convergence using mesh refinement.[8]

One last note on the topic of tetrahedral versus hexahedral elements. If one discounts distortion issues, a single 8-node hexahedral element could outperform a 10-node tetrahedral element, depending upon the strain state that the element is attempting to characterize. MacNeal [18, p. 88] shows that while the 10-node tetrahedral is not able to characterize any quadratic strain states, the 8-node hexahedral can characterized three. Hence, the single 8-node hexahedral has the potential of being more robust if it is used to characterize one of the three quadratic strain states. However, the hexahedral element can only represent nine linear strain states, while the tetrahedral can represent 18. Eighteen strain states are needed to completely define linearly varying strain in three-dimensional problems when polynomial displacement assumptions are used. Hence, while in some modes of deformation a single 8-node hexahedral can outperform a single 10-node tetrahedron, in other modes, the tetrahedron will out perform the

---

[8] One significant advantage of *p*-type elements is that convergence can be checked without re-meshing the model.

hexahedron.[9] A 20-node hexahedron can represent all 18 linear strain states, and in addition, 21 quadratic strain states.

Mesh generation is further discussed by Zienkiewicz [19], Cook [20], and Shephard [21]. The reader interested in three-dimensional meshing routines is encouraged to review works by Shephard.

Choice of Element—Linearity

The analyst needs to consider if the analysis will be linear or non-linear before the particular element is chosen, since some elements do not perform well when certain types of non-linearity are present. For example, higher-order elements have been avoided in the past when contact boundary conditions needed to be modeled. One reason for this is that the mid-side nodes of higher order elements do not behave the same as the corner nodes. Hence, contact models need to distinguish between the types of nodes that are involved in contact. In addition, some higher-order elements may not handle the non-linearity of material *plasticity*, while others are avoided simply because they are computationally expensive. Expensive (higher order) elements become even more burdensome in non-linear problems because the stiffness matrix is often re-computed many times to render a stepwise solution. An introduction to non-linear solution procedures is given in Cook [22, p. 505]

Choice of Element—Distortion

Some elements perform better than others when distorted. Higher order hexahedral elements may perform better when distorted than lower order hexahedrals, for certain types of distortion, as illustrated by Dewhirst [23]. However, recall from Chapter 5 that when higher order elements assume a curved boundary, they become less accurate; MacNeal [18, p. 195]. Ironically, the presence of curved geometry is why one might choose to use higher order elements. Poor performance with curved boundaries is another reason why lower order, 4-node quadrilateral elements for shell applications may be preferable over the higher order shell elements.

The Future of Finite Element Meshing

Although the finite element techniques of today depend upon the ability to idealize a continuous structure using an assemblage of elements that are geometrically simple, research continues in methods that attempt to more fully utilize the geometry that represents the actual structure. One such area of research

---

[9] The inability to characterize necessary strain states in the 8-node hexagonal gives rise to a phenomenon called *locking*, as mentioned in Chapter 5, where the element becomes unrealistically stiff in certain modes of deformation. Software vendors today commonly provide a remedy for this deficiency; MacNeal [15].

considers *domain composition* methods. For three-dimensional models, this means that instead of a three-step process, from solid model to finite elements to continuum equations, domain composition methods attempt to utilize the geometry of the original solid model, and go directly (more or less) to the continuum equations. Domain composition methods are discussed in Cox [24].

Coarse Mesh to Start

Although some conflicting recommendations in literature exist, the analyst may do well to begin with a coarse finite element mesh. Otherwise, one might develop a refined mesh, submit the job to the solver, and find out (many hours later) that some of the required data (such as boundary conditions) was not included in the model. In such a case, the software will often terminate the analysis after hours of wasted computation time. (Other equally important data may also be left out as an oversight.) Another reason for using a coarse mesh is that the process of checking for convergence typically requires a coarse mesh to begin, then finer meshes to test for convergence.

In some cases, a somewhat coarse mesh may provide suitable results if the analysis is concerned only with displacements or modes of vibration. The example of the non-uniform shaft of Chapter 2 suggested that while the stress values were considerably in error, the displacement values were less so. This is because stress is a function of derivatives, which means stress is a function that is at least one order lower than displacement, assuming that polynomials are used for displacement assumptions. Since stress is a function of lower order than displacement, a slower rate of convergence can be expected, other things being equal, as seen by comparing Equations 2.4.24 and 2.4.33.

Mesh Grading

As discussed, one method of improving accuracy while limiting computational expense is to use lower order elements, with mesh grading in areas where stress levels are changing rapidly. Where does stress change rapidly? As mentioned, in regions of stress concentrations, stress can change rapidly from the nominal value to a value that is several times greater. Engineers often have an intuitive sense of where stress raisers are likely: sharp corners, changes in cross section, holes, cutouts, cracks, etc. Again, the topic of stress concentrations is covered in Juvinall [4] and Pilkey [5].

As discussed by Juvinall [4], it may be helpful to visualize stress concentrations using the "force-flow" concept, as illustrated in Figure 7.4. The figure depicts a prismatic bar with notches, subjected to a uniaxial load.

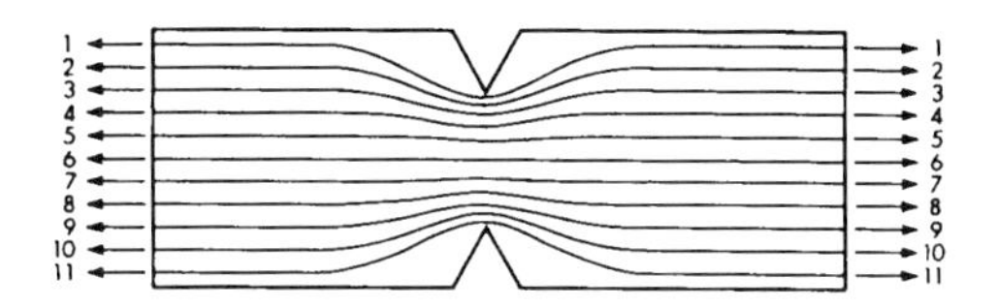

Figure 7.4 The Flow-Force Analogy

Force-flow lines are analogous to fluid streamlines in a non-turbulent fluid flowing through a channel that has the same shape as the structure. As the lines of force attempt to pass through the restriction, they bunch together, resulting in a larger average force per unit area in the region of the restriction. Since normal stress is defined in terms of force per unit area, a higher level of stress occurs in the material near the restriction.

Mesh Grading—Need to Limit the Change in Stiffness
Care must be taken when using mesh grading. As mentioned, it is good practice to avoid placing very stiff elements directly adjacent to very flexible ones, to prevent numerical problems. The stiffness of an element is directly related to its elastic modulus and volume. Smaller elements are more stiff, other things being equal, since for a given load a smaller elements displaces less. It is perhaps intuitive that increasing the elastic modulus increases the stiffness, other things being equal. Hence, the stiffness of an element is directly proportional to elastic modulus and inversely proportional to size, as expressed in (7.2.1).

$$\underline{\underline{K}} \propto \frac{E}{V} \tag{7.2.1}$$

Cook [22, p. 578] recommends that E/V should not change by more than a factor of three across adjacent elements. Using this guideline, the transition pattern in Figure 7.5 would be considered poor because, assuming that the thickness and modulus are the same for both elements, the volume of the smaller appears to differ from volume of the larger by more than a factor of three.

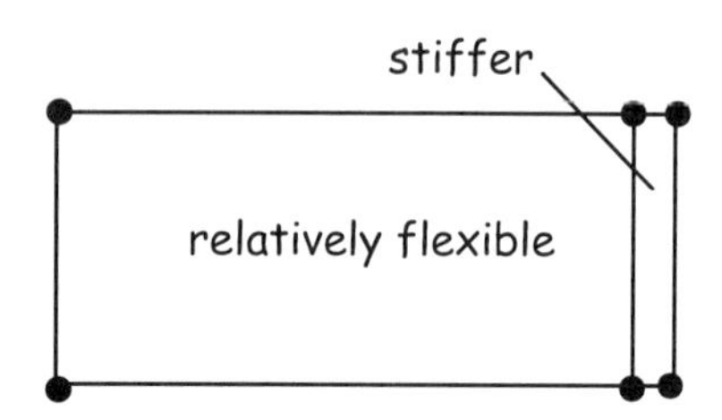

Figure 7.5 Poor Mesh Transition

Mesh Grading—Patterns

To utilize mesh grading, a transition zone is established. There are many transition patterns that would allow the transition from a large, 4-node quadrilateral element to several smaller ones; a few examples are shown in Figure 7.6. The first mesh employs 3-node triangular elements in the transition zone while the second employs 4-node quadrilateral elements only. The second method appears to be more prone to element distortion than the first.

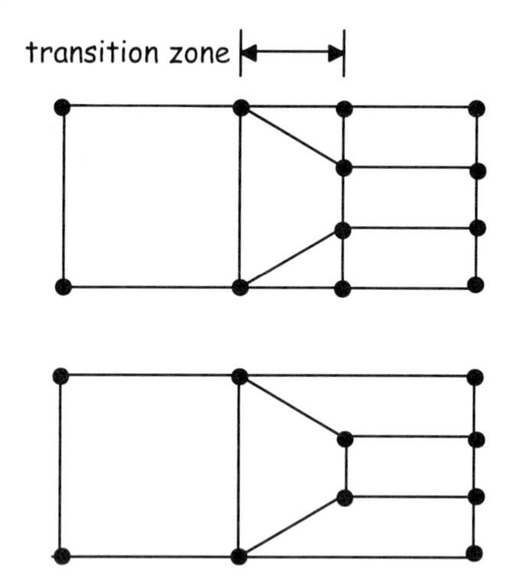

Figure 7.6 Examples of Mesh Grading

Mesh transitions using 8- or 9-node quadrilaterals would follow in an analogous manner. However, one word of caution. Although 8-node quadrilaterals can be adequately connected to 4-node quadrilaterals (Figure 7.3) it is generally unwise to attempt a transition from one type of element to another while at the same time changing the mesh density. This case is illustrated in Figure 7.7. Recognize that the two linear displacement functions generated by the two, 4-node quads, are not compatible with the quadratic displacement function on the edge of the 8-node quadrilateral.

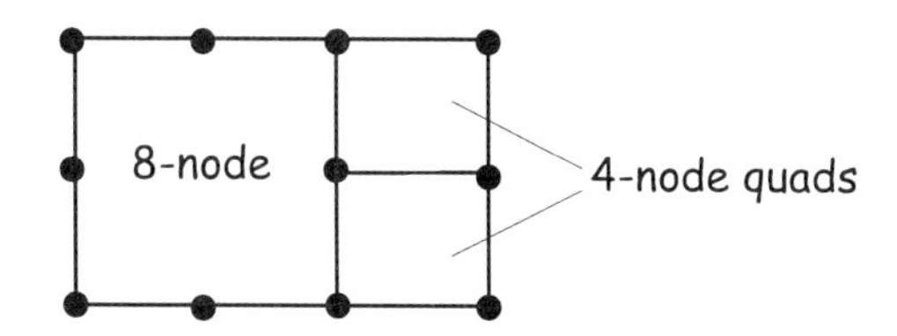

Figure 7.7 Transitions of This Type Are Not Recommended

Another way to accomplish the mesh transition in Figure 7.6 is with multipoint constraints (MPC's), as illustrated in Figure 7.8. In such a case, Nodes 2 and 3 are mathematically *constrained* to remain co-linear with Nodes 1 and 4, regardless of how the structure deforms under load. This particular type of mesh transition should be used with considerable caution.

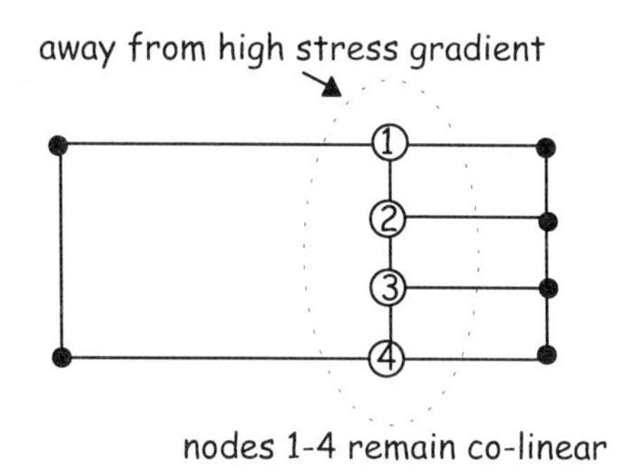

Figure 7.8 A Mesh Transition Using MPC's

In general, the analyst attempts to impose mesh transitions away from regions of high stress gradient. That is, the analyst generates transitions at locations within the model where stress is relatively constant, as opposed to imposing them in an area where stress is changing rapidly. While the mesh transition in Figure 7.6 can be done automatically, the multipoint constraint method in Figure 7.8 is often done manually. However, recent advances in finite element meshing technology allow the imposition of constraints for some types of mesh transitions to be imposed automatically. Mesh transition patterns are illustrated in [25].

Rigid Elements

Some types of multipoint constraints commonly occur, such a rigid constraint between two or more nodes. Consider the case illustrated in Figure 7.9, where Nodes 1 through 3 are, due to some aspect of the idealization, required to remain rigidly connected. For instance, imagine that a rigid plate connects Nodes 1 through 3, and assume that the displacement at Nodes 2 and 3 will be dependent upon Node 1. A multipoint constraint could be imposed upon the three nodes to characterize the rigid plate. To invoke a multipoint constraint, the analyst inputs the node number of the independent node, the node number(s) for the dependent nodes, and the mathematical relationship between them.

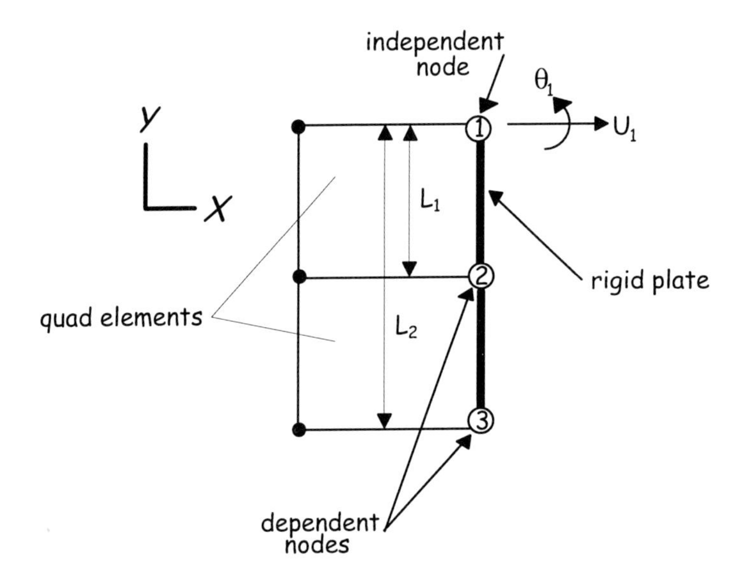

Figure 7.9 Using a Rigid Element

In the present example, the required constraint equations between the three nodes (considering only X-coordinate displacement) would be:

$$U_1 = U_1 + (0)\theta_1$$
$$U_2 = U_1 + L_1\theta_1$$
$$U_3 = U_1 + L_2\theta_1$$

Or, in matrix form:

$$\begin{Bmatrix} U_1 \\ U_2 \\ U_3 \end{Bmatrix} = \begin{bmatrix} 1 & 0 \\ 1 & L_1 \\ 1 & L_2 \end{bmatrix} \begin{Bmatrix} U_1 \\ \theta_1 \end{Bmatrix} \tag{7.2.2}$$

Equation 7.2.2 indicates that displacement at Nodes 2 and 3 is dependent upon the rotation and translation of Node 1. The displacement of Nodes 2 and 3 is *constrained* to the displacement at Node 1, via the constraint equations given by (7.2.2). Similar equations could be written for displacement in the Y-coordinate direction, although the analyst generally has the choice of whether to constrain displacement only in one coordinate direction or in several. Since the need for the type of MPC described above arises often, some code developers have designed "rigid elements" to effect multipoint constraints without the analyst having to write out the constraint equations.

To use a rigid element in the present case, the analyst would first mesh the model as usual, using two 4-node quadrilaterals, and then identify Node 1 as the independent node. Nodes 2 and 3 would then be identified as the dependent nodes, and the analyst would define the X-coordinate direction as the direction in which the rigid element is to constrain. (Typically, many dependent nodes can be related to the independent node.) Up to six directions can be constrained—three translations and three rotations. Using a rigid element, the rest of the constraint process is done automatically; constraint equations such as (7.2.2) are defined by the software, without the analyst's help.

It is important to note that the term "rigid element" may be somewhat misleading. The novice might logically assume that a rigid line element is somehow designed so that its modulus or cross sectional property is so large that the element is essentially rigid relative to any other elements in the mesh. However, recall that juxtaposing very stiff elements with flexible ones is not recommended, due to the associated numerical problems. In addition, simply modifying the stiffness property of a line element to make it rigid would not allow the analyst to choose which coordinate direction is to be constrained.

The advantages of "rigid elements" are that they allow the imposition of common multipoint constraints without having to actually write out the constraint equations. Since they are essentially MPC's, they avoid numerical errors associated with matrix ill-conditioning that very stiff elements would cause. While there exist rigid elements that can be used to invoke common types of multipoint constraints, in some cases the analyst may need to invoke constraint equations manually. In summary, rigid elements are used:

- Instead of assigning an element a very large stiffness property, such as modulus or cross sectional area
- When nodes are to be constrained in one coordinate direction but allowed freedom in another
- As an alternative to distributing loads (or displacements) over several nodes

The Use of Local Coordinate Systems

As discussed in Chapter 6, it is sometimes advantageous to impose loading or displacement boundary conditions in a direction that is not aligned with the global coordinates system. Most finite element pre-processors allow the analyst to choose an arbitrary direction in which to define a coordinate system. For example, consider Figure 7.10, where a plate with a circular hole is allowed to rotate about the hole, but radial displacement is prohibited.

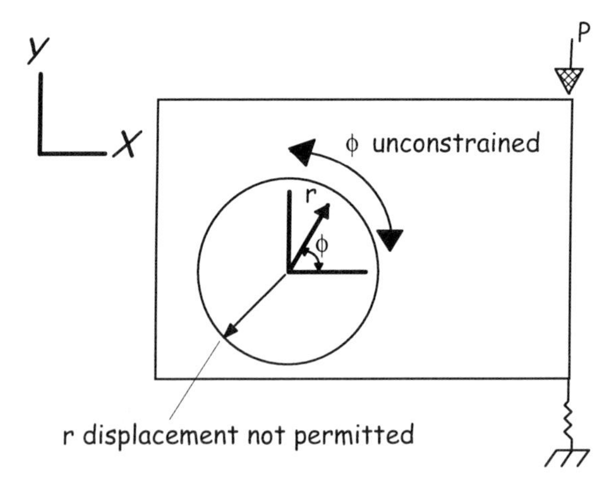

Figure 7.10 A Local Coordinate System

This might be the case if a smooth, fixed shaft were inserted into the hole, with a slip fit between the shaft and plate.

When a local coordinate system is specified, nodes are transformed in the manner described in Chapter 6, and at these transformed nodes the opportunity exists for the user to specify loads or displacements that are not aligned with the global coordinate system. To invoke a local coordinate system, the analyst specifies where the origin is to be located, the coordinate directions, the type of coordinate system (Cartesian, cylindrical, or spherical), and which nodes are to be associated with the local system. Typically, any number of local systems can be specified.

Model Check

Finite element models should always be checked to help guard against fundamental errors, and also guard against the frustration associated with having the solver run for a considerable amount of time, and then abort due to incorrect or missing data. While it is not possible to list all the potential sources of error, a short list of some common error checks is given.

some fundamental pre-analysis checks

- element distortion
- coincident nodes
- coincident elements
- mesh boundaries
- reversed normals
- material properties
- element properties
- boundary conditions

Details of finite element distortion will be considered in the next section. For now, consider the issue of coincident nodes.

Recall from Chapter 4 that overlapping nodes are present when elements are created. Typically, the nodes created within any one mesh boundary are distinct, while coincident nodes typically appear at the interface of two mesh boundaries.

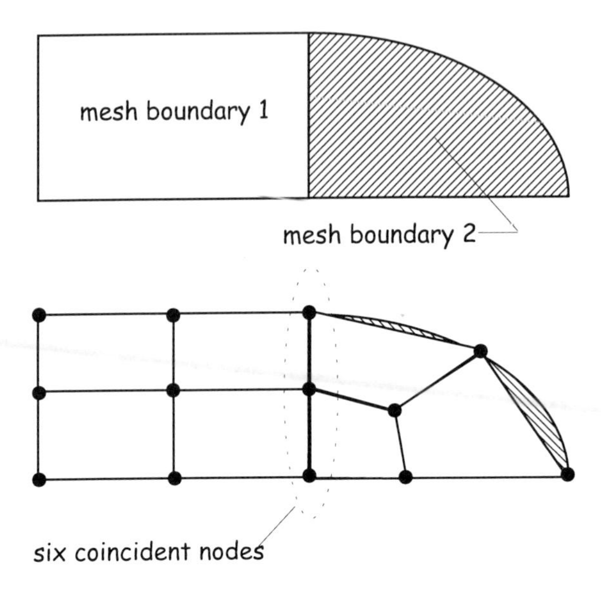

Figure 7.11 Coincident Nodes at Mesh Boundary Interface

Some finite element pre-processors automatically merge coincident nodes, while others require the analyst to manually invoke a node merging routine. If coincident nodes are not merged together, the model may behave as if a crack were present. One check the analyst can use to detect coincident nodes is the "free element edge" check, where all element edges that are not shared with another element are highlighted. When this is done, the boundary between coincident nodes will appear distinct from the rest of the properly connected boundaries.

The problem of *coincident elements* is similar to the problem of coincident nodes: Elements are created and, perhaps by mistake, duplicate elements are created directly over the original ones. This is quite insidious, since the analyst will not be able to visually detect the existence of the duplicate elements, nor will there always be tell-tale signs, such as the stress concentration that would be associated with a "crack" due to improperly merged nodes. With coincident elements, the model can, however, be twice as stiff as it should be. Most pre-processors include some check for coincident elements.

Another problem that can occur is that of "reversed normals" in surface elements. For instance, in shell elements, the "top" surface is distinct from the "bottom." It can happen that the mesh boundaries are so created that an element with a top surface is connected with an element that is turned over, with its "bottom" facing up. This presents problems, for instance, when applying pressure loads: in one case the load will be pushing on the element surface while in the other the load will be pulling. Many pre-processors allow the analyst to display arrows that are normal to the element's surface. If all the arrows do not point in the same direction, the analyst can flip some of the elements over to align all of the elements to have the same normal. This is accomplished by simply redefining the connectivity of the elements that are upside down. Unusual stress patterns can result due to mismatched normals.

Element properties should be checked, since they are easily forgotten; cross sectional properties for beams and trusses, and thickness for plates and shells are often left out of a model. Other difficulties can occur in cases where the software assigns a default value for an element property without the analyst's knowledge. For example, a thickness value of unity is often invoked as a default value when the analyst fails to input the thickness of a structural shell or a plane stress element.

Material properties should be thoroughly checked, especially in the case where several material properties exist within the same model. Some pre-processors allow the analyst to separate elements based upon material properties to determine if certain regions of the model have the correct properties. In addition, some pre-processors check for admissible values of Poisson's ratio (i.e., $0 \le \nu < 0.5$) and consistent values of shear modulus. (Shear modulus is typically considered to be a function of the elastic modulus and Poisson's ratio.)

One caution about material properties: it is common to design, fabricate, and test prototype parts in advance of production tools and processes. However, it is not uncommon for prototype parts to have material properties and geometric features that are not representative of production tooling. This is especially true of cast parts, where sharp corners, porosity, and highly non-isotropic materials properties may be present in the production part, but not in the prototype, or vice-versa. These factors can have a significant impact upon the strength of the part. When evaluating the results of an analysis, the analyst should be aware that the strength of the prototype material may differ substantially from that of the material used for the production part.

Finally, boundary conditions should be checked and re-checked to make sure rigid body motion is restrained in static models. The topic of model check-out procedures is discussed by Zins [26].

7.3 Element Distortion

Key Concept: The shape of an element (in global space) can affect the accuracy of the finite element solution.

In Chapter 5, it was stated that certain types of distortion in finite elements results in a Jacobian determinant that is not constant, which in turn causes the strain-displacement matrix to be expressed as something other than a polynomial. The problems with this are twofold. First, if the strain displacement matrix is heavily influenced by the geometry of the element, it is possible that the strain distribution induced by *element geometry* (an artificial affect) will corrupt the true strain distribution, which is a function of the *structural geometry.*

A second problem with element distortion concerns the numerical integration process that is used to compute the stiffness matrix. Recall from Chapter 5 that the numerical integration schemes used to compute element stiffness matrices are based upon the integration of a polynomial. If the strain-displacement matrix is not a polynomial, due to distortion of the element, the numerical integration process will not be carried out correctly. Some distortion can be tolerated with the level of what is acceptable depending upon the type of element, the loading, and where the element is located in the mesh. In critical areas, less distortion is tolerated, other things being equal.

What Characterizes Distortion in Isoparametric Elements?

When the shape of the element in global space deviates from the parent element (introduced in Chapter 5) the resulting element is said to be distorted. Three considerations should be made in regard to this definition. First, all types of distortion do not have the same affect upon element accuracy and second, some types of elements are more severely influenced by distortion than others. Third, distorted elements may perform well when characterizing certain modes of deformation, while producing very inaccurate results when characterizing other modes. Inaccuracies are often revealed when bending deformation is present.

Measures of Element Distortion

Finite element pre-processors typically compute various measures of element distortion, with the details of the distortion metrics varying somewhat with each software developer. Some basic element distortion concepts will be considered here. The analyst is encouraged to study the documentation for the particular software he is using to gain an understanding of distortion metrics used by the software developer. For examples of suggested distortion limits, see [10, p. 1–15].

The following distortion metrics are commonly used:

- Aspect ratio (or "stretch")
- Skew
- Taper
- Warp
- Normalized Jacobian

The computation of any of the metrics listed above, except the normalized Jacobian, depends upon the type of element. For instance, stretch in triangular surface elements is computed differently than stretch in quadrilateral surface elements. Similarly, because only higher order elements have curved sides, excessive curvature of an edge is not an applicable distortion metric for straight sided elements, nor is warp computed in 3-node triangular elements since the three nodes of the element can only define a flat plane. Warp *is* defined for quadrilateral surface elements, however. (Other elements also make use of the warp metric.)

As shown in Figure 7.12, a quadrilateral surface element is used to illustrate some of the distortion metrics discussed in this chapter; the metrics for a hexahedral volume element would follow in an analogous fashion. Although not shown, the distortion metrics used for triangular surface elements are analogous to those used for tetrahedrals.

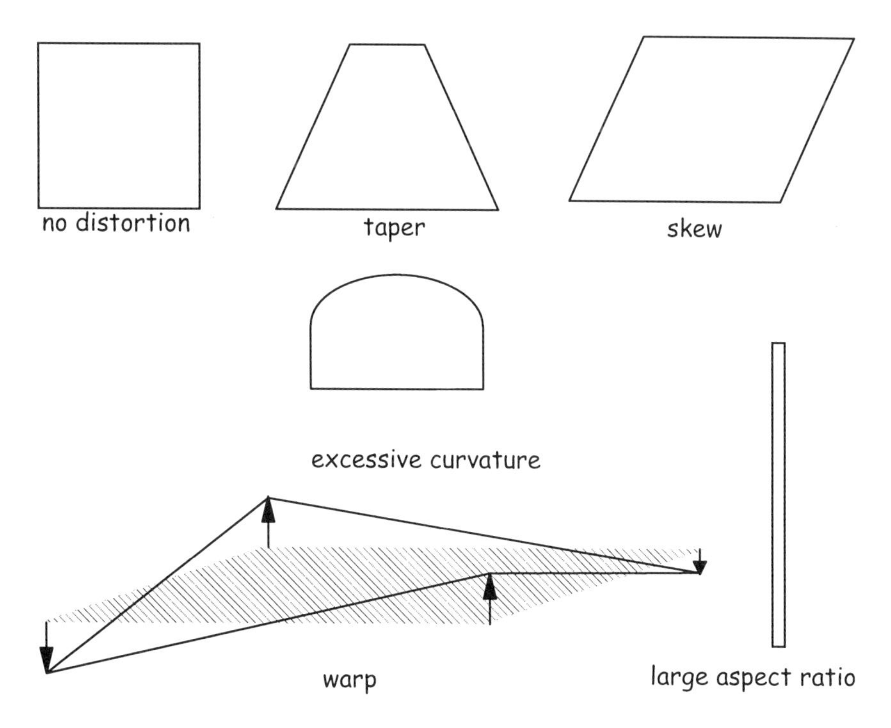

Figure 7.12 Element Distortion

The Normalized Jacobian

The normalized Jacobian (actually, the normalized Jacobian *determinant*) is computed in the same manner, regardless of the particular type of isoparametric element. Recall from Chapter 5 that the Jacobian can be considered a type of scaling factor between the parent element (which is by definition always undistorted) and the global element. The normalized Jacobian is defined as:

$$J_n = \frac{(\det \underline{\underline{J}})V_p}{V_g} \tag{7.3.1}$$

where:

$(\det J) \equiv$ determinant of Jacobain

$V_p \equiv$ volume of parent element

$V_g \equiv$ volume of global element

The determinant of the Jacobian is typically evaluated at each Gauss point, and the minimum of all the values is used as the measure of distortion. A value of unity means that the element is undistorted, while a value of zero means that the element has no volume (the ultimate form of distortion); if J_n is less than zero, the mapping of the element is invalid; this happens when an element is inside-out.

Some distortion metrics predict only certain types of distortion. For example, recall from Chapter 5 that the Jacobian of a 4-node surface element is a constant if the element is rectangular, so, regardless of the aspect ratio, the normalized Jacobian will predict the same value. Thus, the normalized Jacobian alone is not sufficient to predict all types of element distortion.

Many elements now have remedies that allow the element to perform well, even with some distortion (Example 7.1). However, combinations of different types of distortion, for instance large aspect ratio and warping, can still cause poor accuracy. Inaccuracies due to the combination of warping and aspect ratio in three-dimensional elements is illustrated by Dewhirst [23]. Loss of accuracy due to warping in quadrilateral surface elements is discussed by Haftka [27], while Jacobian metrics and associated element distortion are discussed by van Kuffeler [28] and Waagmeester [29]. Robinson [30] discusses tests performed with single elements to determine the effects of distortion. Cook [22, p. 196] shows the affects of element geometry upon a cantilever beam modeled using 4- and 8-node quadrilateral plane stress elements. MacNeal [18, p. 250] illustrates the same, using 20-node hexagonal elements, comparing the differences between rectangular, trapezoidal, and elements that are parallelogram shaped.

Again, the analyst should realize that both the manner in which a particular type of distortion is quantified, and the effects that the distortion has, can vary somewhat with each finite element software package. Two finite elements that are ostensibly identical may be influenced by distortion to differing extents, since their mathematical formulations may be different. Robinson [31] discuss various finite element warning diagnostics and element quality.

Single Element Tests for Accuracy of Distorted Elements

A single element test (Robinson [30]) is one method of examining the accuracy of various types of elements. While it may seem more desirable to use a mesh of elements to check distortion, this often proves difficult because in such cases it is not clear if the loss of accuracy is a result of the distorted elements in the mesh, or a result of the fact that to distort elements, one must move the nodes. When the nodes are moved in a non-uniform stress field, the loss of accuracy may be more due to the fact the nodes are moved away from the area of highest stress. Then why not use a uniform strain field in element distortion tests? The reason is that many elements, even when highly distorted, do not exhibit loss of accuracy in a uniform strain field, while the same distorted element, subjected to bending or torsion, may exhibit severe loss of accuracy. For example, a single, highly distorted plane stress element can render the exact theoretical displacement (or nearly so) when subjected to uniaxial tension. However, this same element, when subjected to bending loads, exhibits accuracy that degrades rapidly with increasing distortion. This behavior is considered in Example 7.1.

Example 7.1—Single Element Test for Aspect Ratio

A tip-loaded cantilever beam is modeled, as depicted in Figure 7.13, with aspect ratios of 10:1, 100:1 and 1000:1; the tip load is adjusted in each case to maintain a constant theoretical value of maximum fiber stress in the beam.

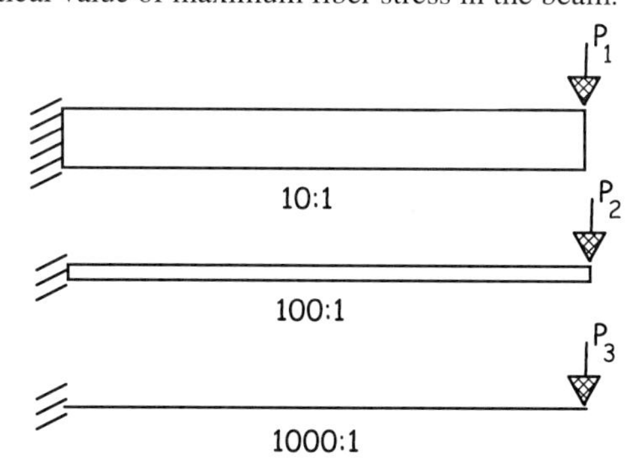

Figure 7.13 Cantilever Beam for Single Element Test

A value for displacement at the tip is computed for each case (using elementary beam theory) and compared with the result of the finite element analysis; the ratio of these two values is termed the "displacement ratio." A single plane stress element, using two different formulations, is evaluated to determine its sensitivity to aspect ratio distortion. The first formulation ("no tricks") is a 4-node quadrilateral, full integration, without any modifications. The other element is also fully integrated but uses a "bubble function." The results of this single element test are plotted in Figure 7.14.

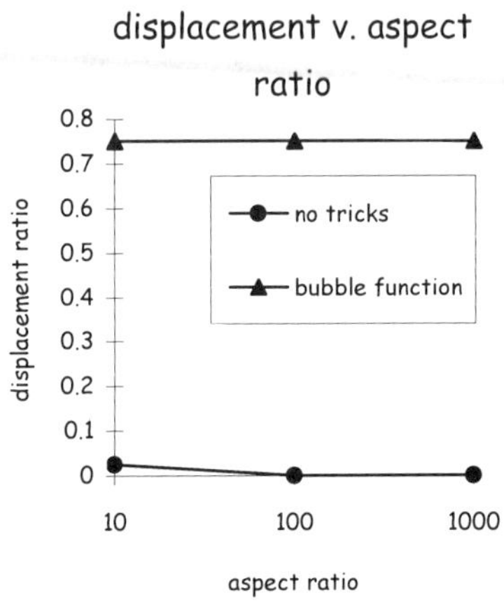

displacement v. aspect ratio

displacement ratio: 0, 0.1, 0.2, 0.3, 0.4, 0.5, 0.6, 0.7, 0.8

no tricks

bubble function

aspect ratio: 10, 100, 1000

Figure 7.14 Single Element Test for Aspect Ratio Sensitivity

Notice that a *single* quadrilateral element, using a bubble function, predicts displacement that is 75 percent of the theoretical. This is a very good result considering that only one element was used. Furthermore, this element does not display any sensitivity to aspect ratio, even for the extreme case where the length to depth ratio is 1000 to 1. The unmodified element, however, suggests that even for a 10:1 aspect ratio, the predicted displacement is less than 5 percent of the theoretical; notice that the performance of the element continues to degrade with increasing aspect ratio. At 1000:1, the FEA predicted value of displacement is $2.5(10)^{-6}$ times the theoretical value. In a uniform strain field, such as that induced by a uniaxial load, both elements predict displacement correctly, even with aspect ratios of 1000:1 or more.

Whenever an element does not possess the ability to deform in the manner required by the loading and boundary conditions, the element will likely behave in an overly stiff manner. When an element grossly under predicts displacement, as in the case of large aspect ratio beam in Figure 7.14, the overly stiff behavior is termed *element locking*.

In the case of cantilever beam bending, deformation of the neutral surface is not simply a linear function. However, in the standard 4-node quadrilateral, only linear displacement can be represented. Hence, this element tends to lock when attempting to represent higher order displacement modes. Complete locking does not occur, because the element uses a shear mode of deformation to represent (albeit poorly) bending deformation. However, as the aspect ratio becomes larger, shear deformation is less able to account for bending deformation.

Notice that the element with the bubble function in Figure 7.14 does not seem to suffer from the locking problem. This is because this element has second order interpolation terms, more closely representing the deformation in the point loaded beam. Since this element does not rely solely upon shear deformation to simulate bending, it experiences little decrease in accuracy as the aspect ratio grows larger (and shear deformation becomes negligible).

While the example above illustrates that the bubble function can greatly improve the performance of the 4-node quadrilateral element, recall from Chapter 5 that the disadvantage of the bubble function (or "incompatible mode") is that the element is rendered incompatible. Incompatible modes are discussed by Cook [22, p. 232]. Remedies for locking have been proposed and are considered in MacNeal [18] and Cook [22].

7.4 Verifying and Interpreting Results

Key Concept: It is sometimes difficult to know whether the results of a given analysis are reasonably correct or substantially in error. The analyst should perform fundamental checks to impart some credibility to the results provided by an analysis.

Even when a finite element analysis is carried our carefully, including model checking and several solutions with different mesh densities to establish convergence, the analyst is often left to ponder if the results are "real." In other words, do the results reflect the exact response of the structure, or are they significantly in error? Furthermore, even when the results can be considered real, the question might then become "What does all this mean?" Several related issues are considered.

Model Validation

Analysts often use the term "validation" in conjunction with the issue of whether or not the results of a finite element analysis bear any relation to physical reality.

It may be best to approach FEA results with a healthy dose of skepticism, only accepting those results that can be supported by independent facts. The author finds it particularly worrisome that, as finite element analysis becomes more widely used, it appears that analysis results are given considerably less scrutiny. (Perhaps this is because the graphical capability of today's post-processors permits very impressive rendering of analysis results, however wrong they may be.)

Post Analysis Checks
As a general rule, computer aided analysis results should never be accepted alone, *prima facie.*

some fundamental post analysis checks

- reasonable displacements
- animation
- closed-form calculation
- physical testing and experience
- reaction forces sum to applied loads
- matrix ill-conditioning
- element stress versus "smoothed"
- peer review

Some additional supporting evidence, such as a closed-form calculation, experimental testing, or substantial experience should always be used to judge if analysis results correlate with the response of the physical structure. Several post-processing checks, listed above, may be helpful in this regard.

Post Analysis Check—Reasonable Displacements
One of the easiest checks for linear, static finite element analyses is to examine the maximum displacement values within a model. Displacement magnitudes on the order of the physical dimensions of the structure (or larger) indicate a problem with either boundary conditions or modulus. Recall that using the linear theory of solid mechanics, only very small deformations are allowed, with strains in the range of a few percent or so.

Post Analysis Check—Animation
Many finite element post-processors allow the analyst to animate the displacement results of an analysis. The animation routine multiplies the computed nodal displacements by some scaling factor to exaggerate the maximum displaced configuration of the model. A series of animation "frames" are computed by

adding, in a stepwise fashion, a portion of the scaled maximum displacement to the undeformed geometry of the structure. When frames are displayed sequentially at a suitable rate, the model appears to deform as it might if the loading were incrementally applied in real time, again, with exaggerated displacements (Figure 7.15).

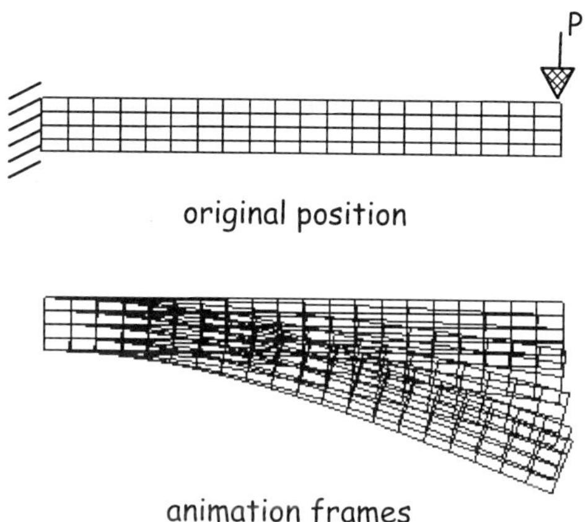

Figure 7.15 Animated Beam Deflection

Animated models often lend an intuitive feel as to the plausibility of deformation patterns in a loaded structure. Gross errors in boundary conditions, load application, or computation can sometimes be identified using animation.

Post Analysis Check—Hand Calculation
Closed-form solutions ("hand calculations") are highly recommended, even if they can only be used to set rough bounds on the solution. The process of performing a closed-form solution not only provides an opportunity to gain some degree of confidence in the finite element results, but may also lend some intuitive feel for the variables used in the analysis.

Post Analysis Check—Testing and Experience
While testing can be very valuable in verifying the accuracy of a finite element model, it may happen that no suitable test specimen exists. As product development cycles shorten, considerable expectation is placed upon having an accurate analytical model long before test parts are available. In other cases, prototypes exist, but the testing necessary to verify the finite element model is cost

prohibitive. (For instance, it is currently not economically feasible to send a space vehicle into orbit simply to verify structural analysis predictions.) In other cases, it may be difficult to design a test that would provide a definitive answer—analysis work in magnetics and acoustics comes to mind in this regard. In still other cases, a prototype and a suitable test may be available but measuring the results becomes difficult—for instance, detecting very small deflections on vibrating or rotating parts often proves challenging. And finally, some testing simply takes too long—fatigue testing can require extended periods of test time. In short, for several reasons, physical testing is not always possible.

When physical testing *is* possible, a simplified test may be performed to help validate a finite element model, even though the test is not required for design verification. The results of a simple test may provide some indication that the model accurately predicts response, at least for a specific type of loading or boundary condition. For example, say that a fatigue test is to be performed, using random vibration to load the structure. Now, instead of performing random excitation to failure, which may require an extensive amount of time, a static load to failure may be applied to verify that the model can predict failure under static loading.

If the finite element model can predict failure under static loading, this may at least provide some check of the static stiffness of the structure. It obviously does not provide information about the dynamic behavior of the structure, since the stiffness of the model could be exact while the mass distribution is grossly in error.[10] Error in the mass distribution of a structure normally causes the dynamic response of the structure to be incorrect, unless some other mitigating factors are also present. The advantage of the static model is that if a static analysis does not correlate with physical testing, the analyst is given some indication that an error exists, and corrective action can be sought.

What types of physical tests are typically employed to validate linear, static finite element models? There are many tests that might be used to validate a model, for instance, force versus displacement, strain gauge testing, photoelastic stress, force to yield, or force to fracture. Of course, a physical test pre-supposes that a representative test specimen is available.

A. Force Versus Displacement Testing: loads are applied to various points on the physical structure and displacements are measured, while the analyst attempts to do the same with the finite element model. Care must be taken that measurements are performed accurately, especially if the displacements are small. Furthermore, one must insure that the loads are distributed over a large enough area if one is measuring global displacement of the structure. That is, using a

[10] One check used in dynamic models is to apply a 1g load to a restrained model and determine if the reaction forces at the restraints sum to the weight of the structure.

small point load could cause a very localized deformation which would not correlate with global measurements.

B. Strain Gauge Testing: Sometimes a structure can be instrumented using strain gauges, and the actual strains under load compared to the strains computed using the finite element method. Although this would appear relatively straightforward, it is often the case that strain gauges are placed in areas where the strain gradient is high. A high strain gradient is characterized by a large change in strain between any two points that are close together. In such a case, the strain recorded by a gauge will be extremely sensitive to slight changes in the location of the gauge. This, coupled with the fact that finite elements tend to smear results to render an average value, has often led to questionable correlation between finite element and strain gauge results. Pinfold [32] discusses the need to have strain gauges placed in areas where the strain field is relatively constant.

C. Photoelastic Stress Testing: This technique is used to produce a stress pattern image on a plastic replica of an actual part. The results of this technique can be used successfully to compare with patterns in a suitably refined finite element mesh. There are limitations to the complexity of the structures that can be represented by the photoelastic technique.

D. Force to Yield, Force to Fracture Testing: Using this technique, a structure is incrementally loaded until either yield or fracture occur. One obvious disadvantage of this method is that it destroys the test specimen. Furthermore, it may be difficult to determine when a structure yields, unless gross yielding occurs. Force to fracture has the advantage that complete fracture is easily recognized. However, while force to fracture may indeed show good correlation with finite element results for the case of brittle fracture, difficulties may occur when attempting to do the same when fracture occurs in a structure composed of a material that would normally be considered ductile. Failure of this sort may show considerable variability in the stress magnitude at fracture due to the size of the specimen, the temperature, the strain rate, and the presence of notches. This issue will be considered again, shortly.

The issues above consider only monotonic loading and isotropic materials. The analyst should be aware that the failure of a structure due to repeated loading can be totally different, and occur in a entirely different location in a structure, when compared to failures that occur due to monotonic loading.

E. A Representative Specimen for Testing: Engineers are sometimes surprised to find that a structure that has passed prototype testing subsequently fails when production samples are tested. There are many reasons for this. For instance, production samples often vary considerably from prototype samples, in both material properties and mechanical properties, such as surface finish. Poor surface finish (e.g., the appearance of cracks at the surface) in stamped parts can have a significant, detrimental affect upon the strength of a structure that is composed of brittle material. Surface finish also influences the strength of a structure that is composed of a ductile material, when the structure is subjected to repeated loading. Cast and injection molded parts can have certain material properties when a production mold is used but may have quite different properties when a prototype mold is used. One reason for this is that a prototype mold may be *gated* differently than a production mold. The gate location can have a significant affect upon directionality of material properties, and also on porosity. There are many more examples of differences between prototype and production parts—care must be taken to assure that test specimens represent the significant attributes of the actual production part.

Another method of validating a finite element model is that of comparing FEA component life predictions with the performance of similar components subjected to typical service loads in the field. An engineer with many years of experience can often tell if a failure predicted by analytical means would occur under actual service conditions. Unfortunately, due to the current trend of corporate downsizing, many of the more experienced engineers are being replaced by engineers who may not have as much intuitive feel for real-world engineering problems.

Post Analysis Check—Reaction Forces

The analyst may find it helpful to evaluate reaction forces at nodes where restraints are prescribed, and determine if they sum to the total force applied to the structure. This is can be particularly handy if the model makes use of surface tractions or body loads. If a structure is assigned a body load that represents gravity, the reaction forces should sum to the weight of the real structure, given that the body load is the only load on the structure.

Analysis Checks—Matrix Conditioning

Recall from Chapter 2 that matrix ill-conditioning can make a finite element model very sensitive to computational errors, such as round-off. Since many computers today are double-precision, matrix conditioning errors are somewhat less of a concern than in the past. However, matrix ill-conditioning may be indicative of a poor idealization. For instance, a very slender beam tied to a single node of a very stiff three-dimensional body is typically a poor representation of an actual structure, and this is reflected by poor matrix conditioning in the model.

Most finite element software routines will compute a matrix conditioning number to alert the analyst of possible problems. In some cases, matrix conditioning is characterized by the ratio of the smallest and largest pivots that evolve during Gaussian elimination. Another indicator of poor matrix conditioning in linear static problems is a large magnitude of the *residual loads*, as discussed in Chapter 6. If all of the mathematics associated with solving the finite element equilibrium equations where carried out exactly, the residual forces would be zero. However, in reality, errors are introduced in each computation of the solution procedure, and ill-conditioned matrices tend to aggravate the error. This error is reflected in a large magnitude of residual force. Matrix ill-conditioning was considered in Chapter 2.

Post Analysis Check—Smoothed Versus Element Stress
As discussed, it is recommended that the analyst perform a convergence check to determined if the mesh density is adequate. One indicator of the need for mesh refinement is large jumps in the magnitude of stress from element to element, assuming that the exact solution is smooth. In most finite element post-processors, the analyst can choose to display smoothed stress values, or values that represent the stress within each individual element. Also, substantial differences between the appearance of the smoothed stress plot, as compared to the element stress plot, may also suggest the need for further mesh refinement.

Post Analysis Check—Peer Review
The analyst is strongly encouraged to have his work checked by another analyst, especially if closed-form calculations have been performed. In some cases, it happens that in the process of explaining the details of an analysis model to another analyst, fundamental assumptions or errors in logic are revealed.

This concludes the section on post analysis checks.

Interpreting Results
If an analysis project has been well planned, the analyst knows at the outset what metric is to be used to make engineering decisions based upon the results of the analysis. The topic of stress metrics, which are used to evaluate analysis results, is briefly considered below.

Which Stress Metric to Use?
If an analysis is performed to determine the stress in a structure, the analyst has the choice of several stress metrics to consider. Most finite element post-processors compute several stress metrics, for example: von Mises, principal, six Cartesian stress components, τ_{xx}, τ_{yy}, ..., Tresca stress, and so on.

The topic of stress metrics and associated failure criteria is a topic worthy of in-depth consideration but is beyond the scope of this text. A few salient points relative to von Mises and principal stress metrics will be provided; however, the reader is warned that the topic of failure modes and stress metrics is a complex subject. Those who wish to use analysis to predict structural failure, even under the "limited" realm of common metals and static, monotonic loading, should consider some study of failure mechanisms, augmented with exposure to practical examples, ideally under the guidance of a mentor with many years of experience.

Using Uniaxial Test Data

Uniaxial tensile tests are relatively cheap and easy to perform, and for many materials these tests provide an adequate description of how stress, strain, yield strength, and fracture are related for a simple state of stress. However, many stress analysis problems of practical interest involve two- or three-dimensional states of stress. Therefore, it would be helpful if it were somehow possible to relate the easily obtainable uniaxial test data to a failure criterion that can also be used for two- and three-dimensional stress states. Both the von Mises and Maximum Principal Stress criteria attempt to do just this.

What is "Failure"?

Perhaps the image of a collapsed structure, or a structural component that has been broken into two or more pieces, comes to mind when the term *failure* is used. However, upon further reflection, one would likely realize that many types of mechanical failure are possible, a few of which are listed below. The following brief discussion will be limited to the topic of failure associated with yielding and fracture in common engineering materials.

types of mechanical failure

- yielding
  (localized, gross)
- fracture
  (brittle, ductile)
- change of dimension
  (wear, corrosion, creep, swelling...)
- excessive deformation
  (elastic, plastic, buckling...)
- stress relaxation

The von Mises Yield Criterion for Ductile Yielding

The von Mises stress criterion is often used, but evidently, sometimes incorrectly. The criterion is known by several other names, for instance, the Distortion Energy criterion; Crandall [33, p. 316] lists several alternate names, and provides a basic introduction; see Higdon [2], also.

The von Mises criterion is best applied (and best understood) when used to predict the onset of yielding in a structure where the material behaves in a ductile fashion. Ductility is characterized by a substantial amount of plastic deformation before final fracture. Collins [34] suggests that fracture may be considered brittle if less than five percent elongation (in two inches) occurs before final fracture; if five or more percent elongation occurs, the fracture can be considered ductile. Uniaxial tensile specimens composed of low carbon steel, copper, or nickel typically exhibit ductile yielding. However, temperature, strain rate, specimen size, or notch sensitivity can cause some *structures* to fail by brittle fracture even though the *material* would normally be considered ductile. More on this later.

Why is the von Mises metric typically a good predictor of the onset of ductile yielding? It has been shown that ductile yielding in metals (and in other materials) is a result of *distortion*. Since the von Mises metric computes the magnitude of stress that tends to distort a body, it is often a good predictor of ductile yielding. To conceptualize distortion, consider a body composed of low carbon steel, in the shape of a cube. If the body is restrained on one face, and a surface traction applied to an opposing face, as illustrated in Figure 7.16, the body distorts. That is, the loaded body no longer has the same shape as it did in its unloaded condition. With sufficient distortion, the body would be expected to yield. In addition to the extensional type of distortion, shear distortion can also occur.

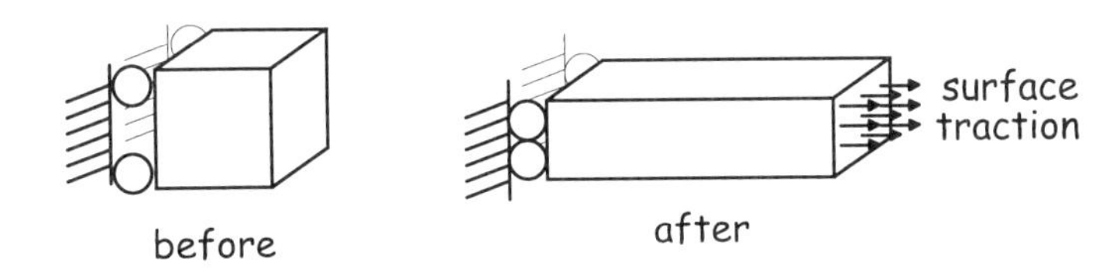

Figure 7.16 Exaggerated Distortion of a Cube Under Uniaxial Load

In contrast to displacements which tend to distort a body, consider a cube again, this time submerged in a fluid under very high pressure, producing a compressive, uniform surface traction normal to all six faces, as illustrated in Figure 7.17.

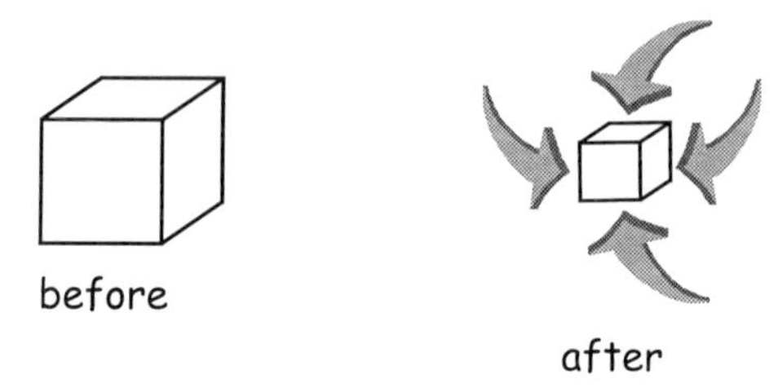

Figure 7.17 Cube Under Hydrostatic Load

Uniform pressure produces hydrostatic stress within the cube, and, regardless of the magnitude of the applied pressure, the cube will always retain its shape, which is to say, no distortion occurs. The reason that the hydrostatic condition does not cause distortion is that the forces acting upon each face are uniform and balanced. Since the cube does not distort, it should not yield; experimental testing supports this assertion; see Bridgman [35]. The same non-distorted result would be anticipated if the cube were subjected to uniform *tensile* traction normal to all six faces.

The illustration above suggests that a stress metric used to predict the onset of ductile yielding would need to predict some finite value of stress when the structure is distorted, but a zero value of stress when the structure is subjected to the hydrostatic condition. To be most useful, the metric would be applicable to one-, two-, and three-dimensional stress states, but would be compared with easily obtainable uniaxial test data. This is exactly how the von Mises metric is contrived, as shown in Crandall [33, p. 316], and given by (7.4.1), below.

$$\tau_{mises} = \sqrt{\frac{1}{2}\left[(\tau_{XX} - \tau_{YY})^2 + (\tau_{YY} - \tau_{ZZ})^2 + (\tau_{ZZ} - \tau_{XX})^2\right] + 3(\tau_{XY}^2 + \tau_{YZ}^2 + \tau_{ZX}^2)} \qquad (7.4.1)$$

The von Mises metric given by (7.4.1) is computed and compared with the stress magnitude at which a tensile test specimen yields in a uniaxial tensile test. If the von Mises stress is greater than the uniaxial yield stress, yielding is considered imminent. In this manner, uniaxial tensile test data is used to predict yielding for one-, two-, and there-dimensional stress states.

It is interesting to note that the von Mises metric is always positive, since distortion is considered neither positive nor negative. Thus, whether a uniaxially loaded bar is compressed or stretched by the same amount, the same value of von Mises stress is computed.

Maximum Principal Stress Criterion for Brittle Fracture

The principal stress metric is often used to predict failure by *cleavage*, in structures composed of brittle material. When cleavage occurs with very little plastic deformation, the fracture is said to be brittle. A structure composed of cast iron would normally be expected to fail by brittle fracture. Brittle fracture, according to the Maximum Principal Stress criterion, (alternately known as the Maximum Normal Stress theory; Juvinall [4], Collins [34]) is directly related to the maximum tensile stress acting normal to a surface.

As mentioned in Chapter 1, a general loading case can induce both normal and shear stress within a structure, with the maximum normal stress magnitude occurring on planes where the shear stress is zero. The planes on which shear stress is zero are called *principal planes*. Mohr's circle is a graphical means that can be used to determine the orientation of principal planes, and the magnitudes of the normal stress on these planes; Crandall [33, p. 222].

The normal stress on a principal plane is termed *principal stress*. The maximum principal stress theory suggests that the largest principal tensile stress is responsible for cleavage associated with brittle fracture. Hence, the maximum principal tensile stress can be computed and compared to the ultimate tensile strength from uniaxial testing. When the maximum principal tensile stress exceeds the value determined by the uniaxial test, fracture is considered imminent. Higdon [2, p. 199] shows interesting illustrations of both brittle and ductile fracture in rods subjected to torsional loads.

Brittle Fracture in Structures with Materials Normally Considered Ductile

It has been observed that structures composed of materials that are normally considered ductile can exhibit brittle fracture, i.e., cleavage without significant plastic deformation. The parameters that control brittle fracture in structures that normally fail in a ductile fashion are:

- The rate at which the material is strained
- Ambient temperature
- Size
- The presence of notches

High strain rates, low temperature, large structures, and the presence of notches can promote brittle fracture in materials normally considered ductile. Because of the complexity and the number of variables involved, accurate prediction of this mode of failure, based upon the magnitude of a given stress metric alone, is generally not possible. In Orowan [36], the author suggests that if the thickness of a specimen composed material normally considered ductile is less than the *critical crack dimension*, brittle fracture will not occur. In the same reference, the author also suggests an equation for computing the critical crack dimension, based upon a modification of the Griffith formula; Griffith [37]. Structures composed of other materials, plastics for instance, can also exhibit a transition from ductile to brittle behavior based upon the same four variables listed above. A comparison of brittle and ductile failure in plastics, attributed to size affects, is given in Trantina [38].

Ductile Fracture
What about ductile fracture? Ductile fracture, separation of the structure into two or more pieces after a significant amount of plastic deformation, is apparently not well understood at this point. Orowan [36, p. 217] suggests that ductile fracture takes place without a "cleavage like process," and illustrates mechanisms for this type of fracture. Although the maximum principal stress criterion typically works well in predicting *brittle* fracture (i.e., cleavage) it does not do as well when used to predict ductile fracture. Ductile fracture may correlate better with the von Mises criterion, since ductile fracture is not thought to occur by cleavage.

Directionally Dependent Strength
Finite element post-processors typically compute principal stress values, and display them just like von Mises stress. However, one can plot both maximum and minimum principal stress. This is helpful for materials that have failure strengths that differ according to whether or not the loading is compressive or tensile. For example, concrete exhibits very high strength in compression but limited strength in tension; see Crandall [33, p. 329] and Collins [34, p. 148]. The von Mises criterion would not be expected to be a good failure criterion in materials that have directionally dependent strength, since the terms compression and tension do not apply.

Figure 7.18 compares the von Mises (distortion energy) criterion with the maximum normal stress theory for several common engineering materials. Example 7.2 considers how stress computed by the von Mises formula compares to the maximum principal stress.

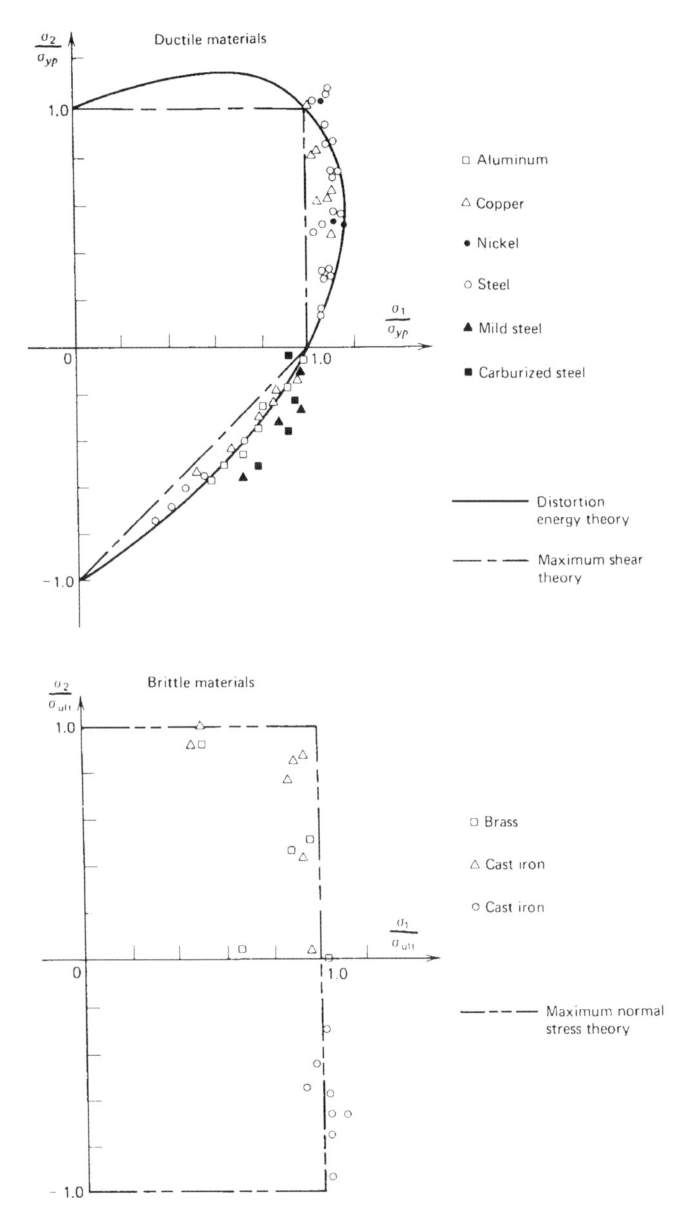

Figure 7.18 Comparison of Failure Theories

Example 7.2—Comparing the von Mises Stress Metric with the Principal Stress Metric to Predict the Onset of Yielding.

A thin plate, with a hypothetical material having a yield strength of 110 psi, is constrained on one end as depicted in Figure 7.19. A uniformly distributed tensile load is placed upon the other end of the plate, inducing a state of uniaxial stress, equal to 100 psi.

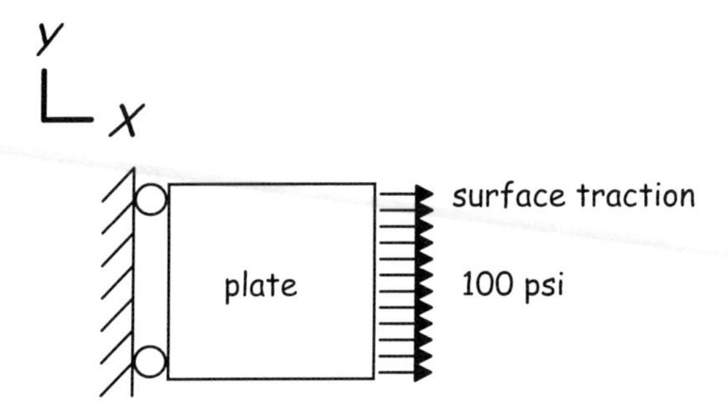

Figure 7.19 Plate Under Uniaxial Load

Using (7.4.1) to compute the von Mises stress, it is noted that the only non-zero stress component is τ_{XX}:

$$\tau_{mises} = \sqrt{\frac{1}{2}\left[(100-0)^2 + (0-0)^2 + (0-100)^2\right] + 3(0)} = 100psi \tag{7.4.2}$$

In this case, the von Mises stress is simply equal to the uniaxial stress, 100 psi. Furthermore, because there is no applied shear stress, the *X*-plane may be considered a principal plane (a plane where shear stress vanishes), hence, the maximum principal stress in this case is also 100 psi. For the case of uniaxial stress, then, there is no difference between the von Mises metric and the maximum principal stress metric; both would not predict yielding since they compute a stress value which is less than the 110 psi yield strength.

Now consider a case of two-dimensional stress. The thin plate from above is now loaded on two edges of the plate, as shown in Figure 7.20. The bottom edge of the plate is constrained in the *Y*-coordinate direction, while the left edge of the plate is constrained in *X*.

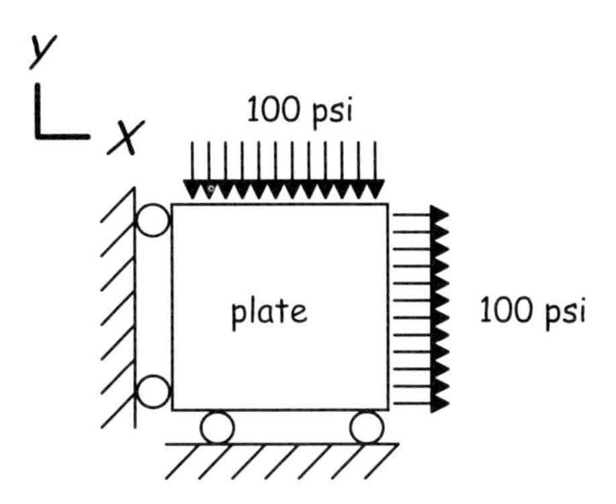

Figure 7.20 Plate Under Bi-axial Load

The maximum principal stress is again 100 psi, for the same reason as before. However, using (7.4.1), the von Mises stress metric now predicts 173 psi:

$$\tau_{mises} = \sqrt{\frac{1}{2}\left[(100+100)^2 + (-100-0)^2 + (0-100)^2\right] + 3(0)} = 173 psi \tag{7.4.3}$$

In this case, von Mises criterion predicts yielding because τ_{mises} is greater than the 110 psi yield stress, but the maximum principal stress, still at 100 psi, does not predict yielding.

Finally, consider the cube from Figure 7.16, and assume that all six exposed surfaces are subjected to a uniform, compressive surface traction of 100 psi, thereby inducing a state of hydrostatic stress in the body. The von Mises stress metric should be zero throughout the body because hydrostatic stress has no effect upon yielding. Computing the von Mises stress for this case:

$$\tau_{mises} = \sqrt{\frac{1}{2}\left[(-100+100)^2 + (-100+100)^2 + (-100+100)^2\right] + 3(0)} = 0 psi \tag{7.4.4}$$

In this loading case, no matter how small the yield strength of the material, the von Mises metric does not predict yielding. However, the principal stress metric is again 100 psi, corresponding to the maximum normal stress on any of the principal faces. (In this case, all the exposed faces could be considered principal, since each is free of shear stress.) Now, if the yield strength of this material were slightly less, say 90 psi instead of 110, the maximum principal strength criterion *would* erroneously predict yielding; as mentioned, yielding is not anticipated for the hydrostatic condition.

The reader is again cautioned that failure prediction is often very complicated, and can have serious consequences in terms of personal safety and expense. Prediction of structural failure is not something recommended for the novice analyst.

A general discussion of failure theories is found in Juvinall [4] and Crandall [33]; metallurgical considerations associated with plastic deformation are found in Van Vlack [39] and Collins [34].

Strain Energy Density

Another metric that finite element post-processors typically compute is strain energy density. Strain energy density can illustrate where most of the deformation is taking place within a structure, and guide the analyst as to the location where the most effective changes can be made to strengthen a structure. For an example of using strain energy in finite element analysis, see [10].

7.5 Applications Hints

Key Concept: There are many places where a finite element analysis can go astray; it may be helpful to consider a few fundamental errors.

Nearly all finite element software vendors provide a collection of sample problems along with their software documentation. The sample problems typically show how finite element analysis is applied to relatively simple problems that have closed-form solutions; see [11] and [40] for instance. There is no need to repeat these types of problems here. Instead, a few common pitfalls will be illustrated, along with some fundamental techniques used in finite element analysis.

Problem 1—Pinned Truss

Problem 1—Description

A pinned truss is designed as illustrated in Figure 7.21. A static load of 3000 lb_f is applied at the top, while the truss is restrained as shown. The engineering question to be answered by this model: "Is cross sectional area of the truss members adequate to prevent failure by gross yielding?"

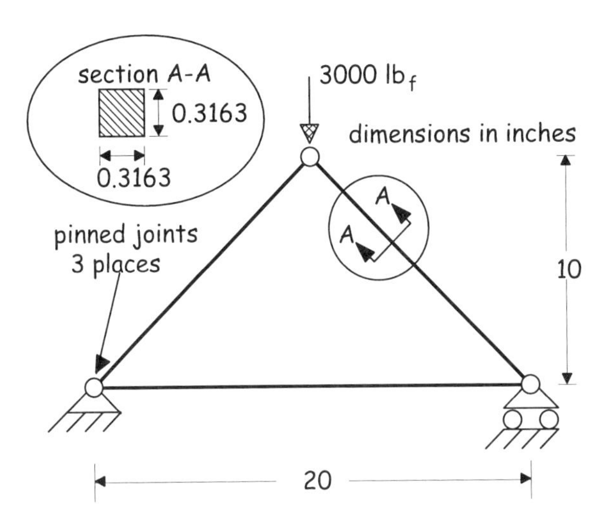

Figure 7.21 A Pinned Truss

The material is AISI C1015 hot rolled steel, having a yield strength of 46,000 psi; Juvinall [4]. The restraint at the left side prevents the truss from translating, but does not prevent rotation. Rotation is prevented by supporting the right side with a *slider mechanism*, which allows translation in the *X*-coordinate direction but prevents translation in *Y*. The load is assumed to be concentrated at the top of the truss.

Problem 1—Finite Element Model
As illustrated in Figure 7.22, three rod elements are used, one for each member of the truss. The rod elements will be used in two-dimensional space, such that each node of the rod element has two degrees of freedom, *U* and *V*. At Node 1, both *U* and *V* are assigned a value of zero, while only *V* is set to zero at Node 2, thereby representing a slider mechanism. A single force is applied at Node 3, in the negative *Y*-coordinate direction. A linear, static solution procedure will be used.

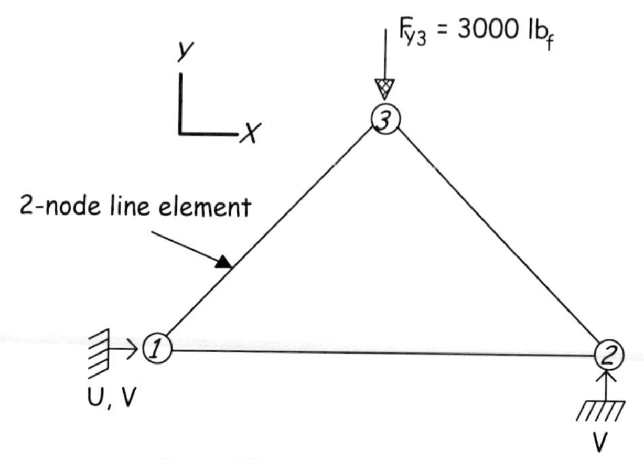

Figure 7.22 Pinned Truss Model

Problem 1—Analysis Results
The results of the analysis are displayed in Figure 7.23.

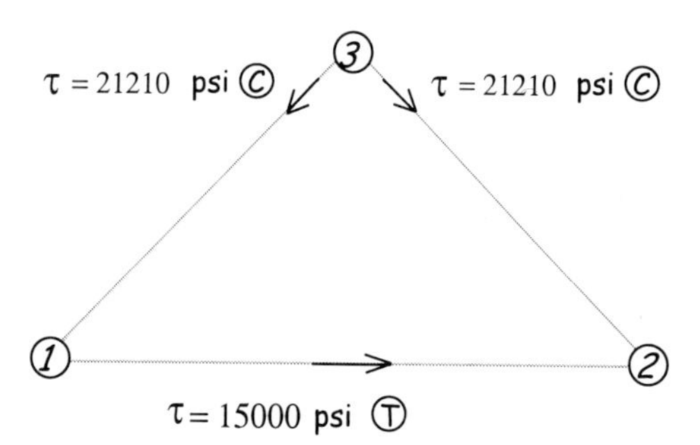

Figure 7.23 Results of Pinned Truss Analysis

The analysis predicts a maximum uniaxial stress of 21210 psi (compression) in the truss, such that even with a factor of safety of two, the members appear to be of large enough cross section, as far as yielding is concerned. However, is yielding the only failure mode of concern in this case?

Since the top members are in compression the alert analyst might realize that buckling is a possibility. To use the Euler buckling equations, a check is made of the slenderness ratio of the two members that are in compression:

$$\frac{L}{c} \overset{?}{>} \sqrt{\frac{2\pi nE}{\tau_y}} \qquad c \equiv \sqrt{\frac{I}{A}} \tag{7.5.1}$$

The quantity on the left hand side of the inequality in (7.5.1), known as the slenderness ratio, must be greater than the quantity on the right if Euler buckling is to be used. Substituting for c, and rearranging (7.5.1):

$$\frac{L^2 A}{I} \overset{?}{>} \frac{2\pi nE}{\tau_y} \tag{7.5.2}$$

$$L = \text{Length}$$
$$A = \text{Cross section}$$
$$\text{I} = \frac{\text{bh}^3}{12}$$
$$E = 30(10)^6\, psi$$
$$\tau_y = 46000 psi$$
$$n = 1$$

The variable n is a factor associated with the end constraints of the member, and in this case is assigned a value of unity; see Shigley [41]. Substituting the known values into (7.5.2):

$$\frac{(14.4)^2(0.1)}{\frac{1}{12}(.3163)(.3163)^3} \overset{?}{>} \frac{2\pi^2(1)30(10)^6}{46000} \tag{7.5.3}$$

Performing the mathematics in (7.5.3):

$$24860 \overset{?}{>} 12873 \tag{7.5.4}$$

Since the left hand side of (7.5.4) is greater than the right, the Euler equation is applicable to this problem, and the critical load is calculated, as shown in (7.5.5):

$$P_{cr} = \frac{\pi^2 EI}{L^2} = \frac{\pi^2 30(10)^6 \left(\frac{1}{12}(.3163)(.3163)^3 \right)}{(14.4)^2} = 1191\, lb_f \tag{7.5.5}$$

Thus, if the load in the members under compression is greater than 1191, collapse is likely. The load is computed in the members using the stress determined by the finite element analysis, along with the cross sectional area:

$$S = \frac{P}{A_0} \qquad \therefore P = SA_0 = 21210 psi\,(0.1) in^2 = 2121\, lb_f \tag{7.5.6}$$

Comparing the computed force in the member, 2121 lb_f, with the critical buckling load, 1191 lb_f, it is noted that the structure is likely to collapse. However, the analysis (as shown) will give the analyst no indication of collapsing!

Problem 1—Summary

It is important to note that when using engineering analysis techniques, *the analyst must understand the physics of the problem* which he is trying to solve. In this case, the structure would have likely collapsed, while the analysis results suggest that the design is adequate. This point cannot be over emphasized—finite element software, or any analysis software available today, does not replace engineering judgment.

Can finite element analysis be used for buckling (and/or collapse) analysis? Yes. Most software codes allow a separate solution procedure to take place if the analyst wishes to invoke this option. However, the analyst must understand when this type of analysis must be invoked.

One additional note on the problem of the pinned truss. The unsuspecting analyst might try to use beam elements in place of the rod elements. Of what consequence would this be? Using beam elements, the idealization would simulate fixed joints, something like having the three vertices of the truss welded together. As such, the state of stress would no longer be characterized by membrane deformation because bending stress would also be induced. Depending upon the particular problem, bending could have a substantial influence upon the results of the analysis.

Problem 2—Beam Bending

Problem 2—Description

A Cantilever beam shown in Figure 7.24 is to be analyzed, and the maximum fiber stress determined. The analyst has several choices of elements to use in this case, since the cross section of the beam is uniform, and the loading simple. One-, two-, or three-dimensional elements could be employed, if desired. A concentrated load is to be applied to the tip of the beam, and the stress state in the vicinity of the load will be ignored.

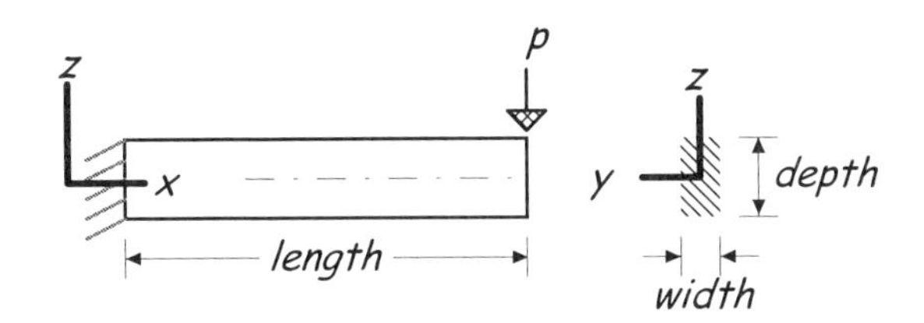

Figure 7.24 Cantilever Beam with Concentrated Load

Problem 2—Finite Element Models

An easy way to represent the beam is to use a single, 2-node line element, as depicted in Figure 7.25. The single element predicts displacement and stress exactly (within beam theory) for the loading and boundary conditions shown; Higdon [2, p. 708].

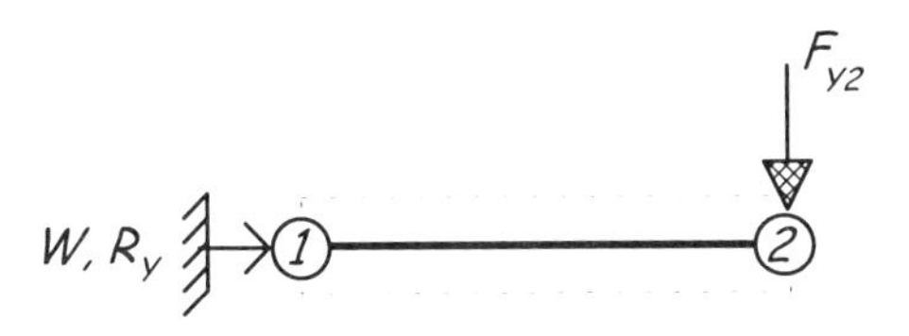

Figure 7.25 Cantilever Beam Using Single Line Element

Another way to model the beam would be to use shell elements, as shown in Figure 7.26.

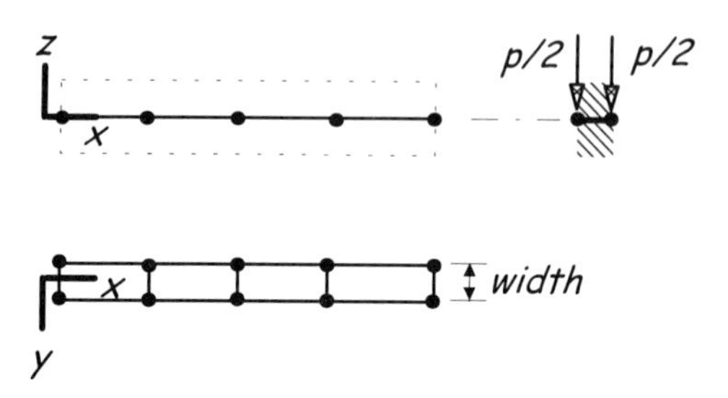

Figure 7.26 Cantilever Beam Using 4-Node Shell Elements

In this case, the length and width of the beam are represented by the physical dimensions of the elements, while the depth is specified by a mathematical constant. A substantial increase in computational expense would be incurred using the shell elements, compared to the single beam element, due to the increase in the number of DOF's associated with the model.

While the line element works very well for a beam with a uniform cross section, it has difficulty representing a beam where the cross section is no longer uniform. For example, the beams shown in Figure 7.27 would not lend themselves to use of line elements.

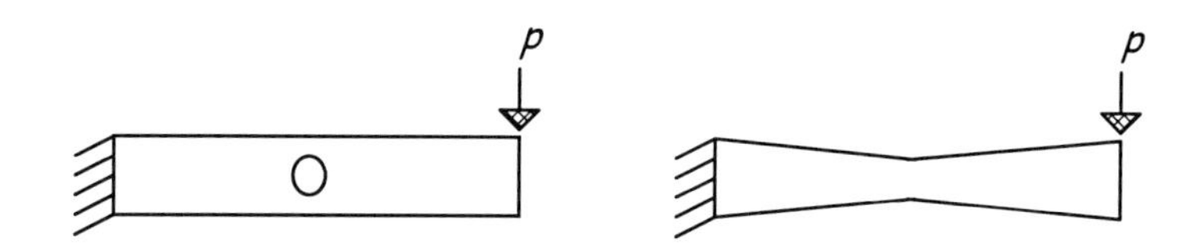

Figure 7.27 Cantilever Beams with Concentrated Load

Either beam in Figure 7.27 would be better modeled using 4-node quadrilateral elements for plane stress applications, assuming that the beam width is uniform. For example, the beam with the central hole is shown with a coarse mesh in Figure 7.28.

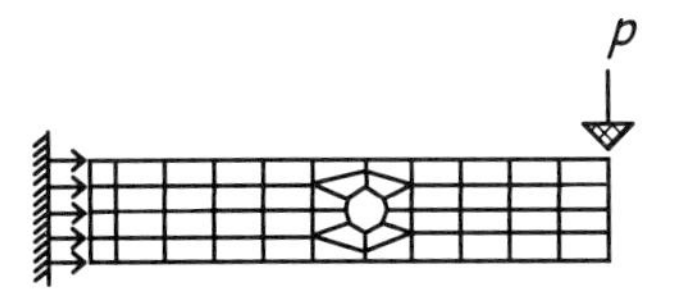

Figure 7.28 Beam with Hole, 4-Node Quad Elements, Coarse Mesh

In the case of a uniform beam, the use of 4-node quadrilateral elements for plane stress applications would require a significant increase in computation to yield the same result as the 2-node line element for beam applications. The beam could also be modeled using volume elements, again with a substantial increase in computational expense over the 2-node line element.

Problem 2—Summary

There often exist several types of elements that can be used to model a given structure, although one particular type of element may provide a far more effective solution. However, perhaps it is evident that making general statements about which element to use for a given type of structure is difficult. Even for a simple application such a cantilever beam, the most effective element type is a function of not only the loading and boundary conditions, but is also a function of the of the geometry of the structure.

Sometimes, the geometry of a structure might suggest the use of two different types of elements. Consider a structure that is thin except for a fairly massive mounting boss, as depicted in Figure 7.29.

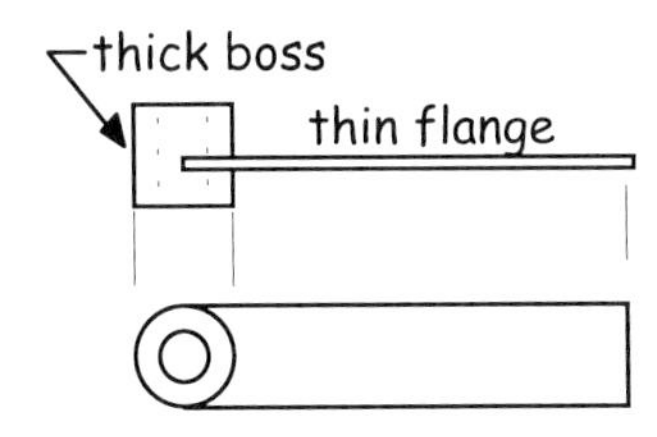

Figure 7.29 Thin Structure with Large Mounting Boss

The thin flange portion would suggest the use of thin shell elements, while the thick boss suggests the use of three-dimensional elements. In cases such as these, the analyst may simply use three-dimensional elements for the entire structure, instead of using shell elements for the flange and three-dimensional elements for the flange, because it is typically easier to continue with one type of element instead of attempting to join volume elements with shell elements. Remember, the volume elements have translational DOF's only, while shell elements have translational and rotational degrees of freedom.

If the stress at the interface of the boss and flange is important, the analyst may choose to use hexahedral volume elements for the boss, and for a portion of the thin flange, and then follow with shell elements for the remainder of the flange, as depicted in Figure 7.30.

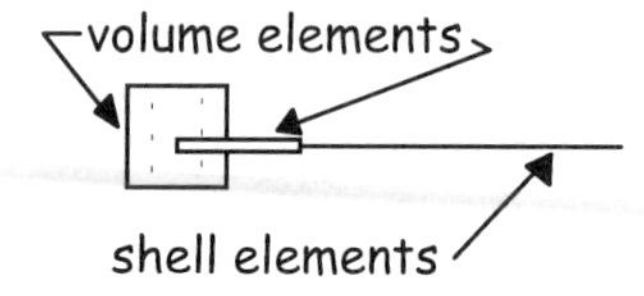

Figure 7.30 Thin Structure with Large Mounting Boss

It is perhaps unwise to make a transition from one type of element to another at a point where accurate stress prediction is needed. The proper connection between shell and volume elements is considered in [13]. It should be noted that one or two tetrahedral elements through the thickness of a thin structure will typically not properly represent bending deformation.

Problem 3—Beam with Roller Support

Problem 3—Description

The cantilever beam from the last example is again examined, but with the addition of a roller support as depicted in Figure 7.31. This problem attempts to illustrate how beam elements, derived using a mechanics of materials approach, differ from elements that are derived using the continuum approach. Recall that using the mechanics of materials approach, the objective is often to determine the load bearing capability or maximum deflection of a structure. As such, stress concentrations are discounted.

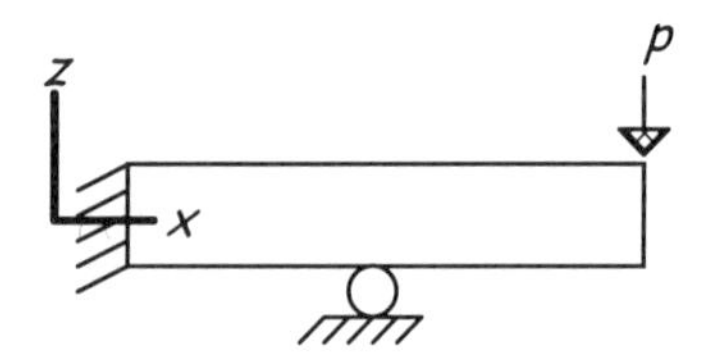

Figure 7.31 Cantilever Beam with Roller Support

Problem 3—Finite Element Models

The first model will use two, 2-node line elements for beam bending applications.

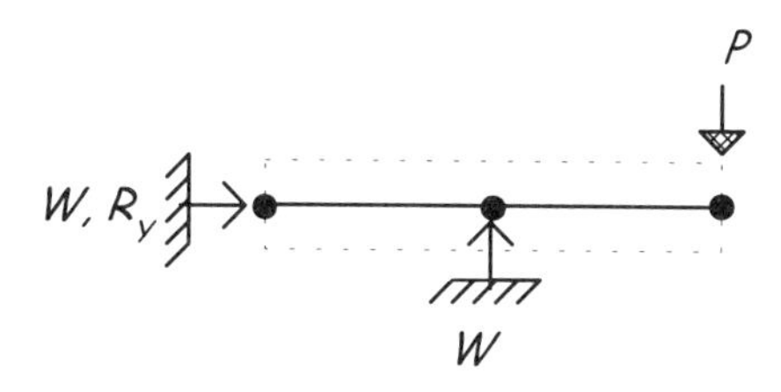

Figure 7.32 Cantilever Beam Using Two Line Elements

Without the roller support, the maximum axial stress computed by the finite element model is 100 psi. With the roller support placed midway along the beam, the stress is drops, to 50 psi, as expected, since the beam is effectively half as long as before. Now, the same idealization is attempted with plane stress elements (Figure 7.33).

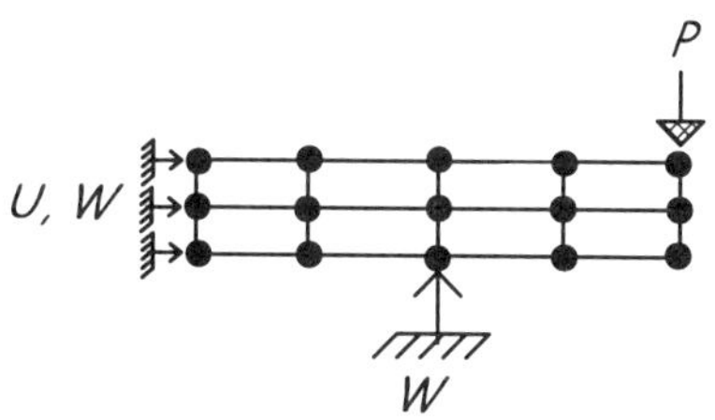

Figure 7.33 Cantilever Beam Using 4-Node Quadrilaterals

The results from the problem idealized as shown in Figure 7.33 are quite interesting. The *W* constraint produces a stress concentration, with the maximum stress limited only by the density of the mesh in the vicinity of the restrained node. That is to say, as the analyst continues to refine the mesh, the peak value of stress will continue to increase, without limit.

The reason that the stress continues to increase is that the stiffness associated with the node continues to decrease as the mesh is refined. This results in continually increasing displacement, and stress behaves in the same fashion; this is indicative of a poor idealization. The same behavior would be anticipated if another type of continuum element were used, for instance, an 8-node hexahedral element.

How would this problem be better idealized? First, consider the physical nature of the real problem. If it is assumed that the roller is essentially rigid with respect to the beam, the beam will experience a significant amount of deformation at the point where it comes into contact with the roller, as the load is increased. The deformation will result in an increased contact area between the roller and the beam.

The analyst has several choices for this problem. The first would be to carry out a nonlinear *contact analysis*, with an incrementally increasing load, until the final value of P is reached. The non-linear analysis would allow a small increment of load to be applied, with the deformation of the beam near the roller computed. The newly deformed geometry would be taken into account, and the software would determine how much more of the beam is in contact with the roller. New boundary conditions at the interface of the beam and roller would be established, the stiffness matrix updated, and the analysis would continue with a new incremental load added to the existing load. This type of analysis is computationally expensive but might more accurately simulate the actual phenomenon. Advances in non-linear finite element software with *automatic contact* routines in recent years allow the analyst to perform such an analysis fairly easily. However, for the present problem, the accuracy gained using a non-linear analysis may be not be needed, especially if the applied loads do not cause significant deformation.

A more crude way to approach the problem is to compute the estimated contact area between the roller and the beam when the beam deforms. This contact area would then be used as a guide in choosing how many nodes to restrain for a given mesh density. See Young [8, p. 647] for equations to compute the contact area between deformable bodies. The accuracy of the solution found using the more crude procedure may be well within the limits engineering accuracy, rendering the more accurate procedure unnecessary.

Problem 3—Summary

Elementary stress and strain problems are often introduced to undergraduate engineers using the mechanics of materials approach. Using this approach, one is typically concerned with determining the load carrying capability (or maximum deflection) of structures such as beams and plates, in contrast to characterizing the state of stress in a three-dimensional body.

Problem 4—Plate with Center Hole, Symmetric Boundary Conditions

Problem 4—Description

A plate loaded in uniaxial tension, Figure 7.34 is to be analyzed. Again, a number of different choices exist as to which type of element might be used.

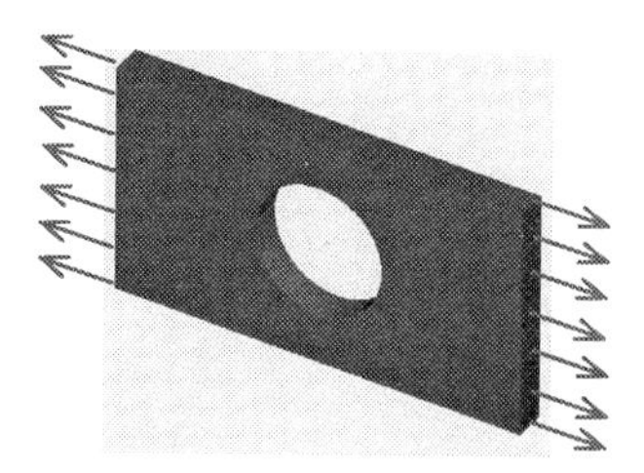

Figure 7.34 Plate with Hole in Uniaxial Tension

Problem 4—Finite Element Model

Several principals are illustrated in this problem. Since the plate, when loaded as illustrated in Figure 7.34, is in a state of plane stress, the 4-node quadrilateral plane stress element is an appropriate choice of element. Figure 7.35 reveals that in this case, both the geometry and loading of the structure are symmetrical. This type of symmetry is known as mirror symmetry, since the structural geometry and loading to the right of the *Y-Z* plane are mirror images of that to the left of the plane. This same type of symmetry is also exhibited to either side of the *X-Z* plane. When two planes of symmetry exist, the condition is sometimes called *quarter symmetry*.

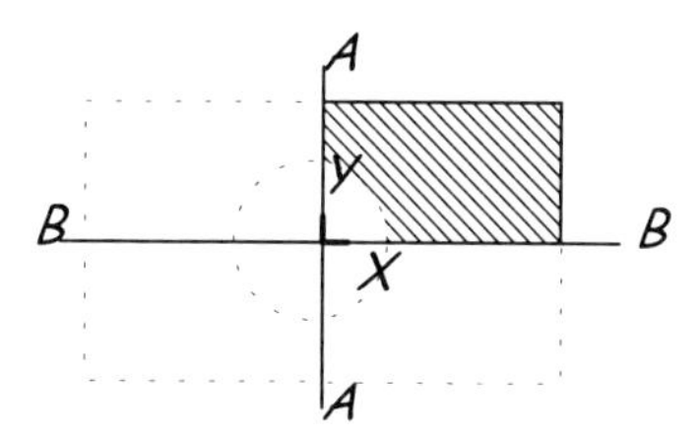

Figure 7.35 Mirror Symmetry About Two Planes

Mirror symmetry is employed when both the geometry of the structure and the loading are symmetrical about a plane. Boundary conditions are then invoked on the nodes that lie on the plane(s) of symmetry to simulate the effects of having the entire structure present.

How are symmetrical planes identified? Typically, by visual inspection one can identify the planes in a given structure about which the geometry is identical. For instance, in Figure 7.35, notice that the geometry is a mirror image about plane cuts *A-A* and *B-B*. Now if the loading to either side of the plane is also a mirror image, mirror symmetry can be employed.

How does one determine what boundary conditions are applicable when using mirror symmetry? One method that seems to work is to imagine that the structure in question is subjected to unconstrained uniform thermal expansion. When this happens, there will exist planes upon which points do not move in a particular coordinate direction, and this suggests what type of boundary condition to apply. For instance, given the plate shown in Figure 7.36, it is seen that regardless of how much the plate expands, points on a vertical plane cut (*A-A*) through the plate will not displace in the *X*-coordinate direction.

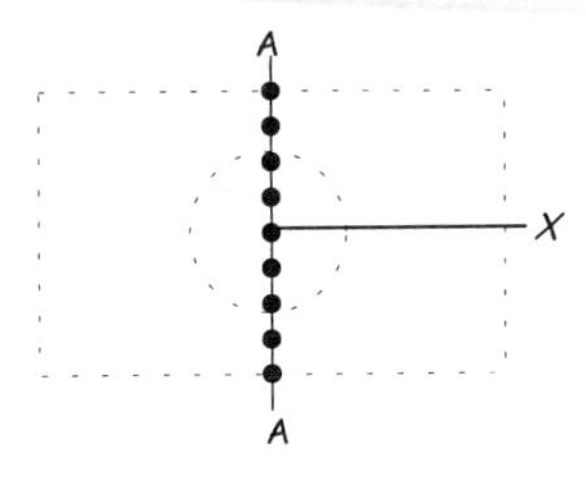

Figure 7.36 Points That Do Not Displace in X

Similarly, with a horizontal plane cut through the center of the plate, points along the *X*-axis will not move in the *Y*-coordinate direction. With this in mind, it is apparent that boundary conditions associated with one-quarter of the structure are characterized by rollers on two edges, where the roller allows displacement in one coordinate direction but not in the orthogonal direction, and shown in Figure 7.37. The rollers indicate that the horizontal edge cannot displace in the *Y*-coordinate direction, while the vertical edge cannot displace in the *X*-coordinate direction.

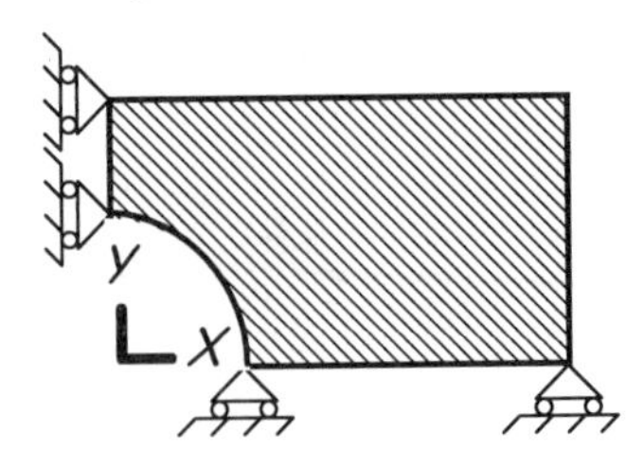

Figure 7.37 Rollers on Edge of Plate Mesh Boundary

A finite element mesh is constructed within the boundary shown in Figure 7.37, using just a few elements, rendering the coarse mesh shown in Figure 7.38.

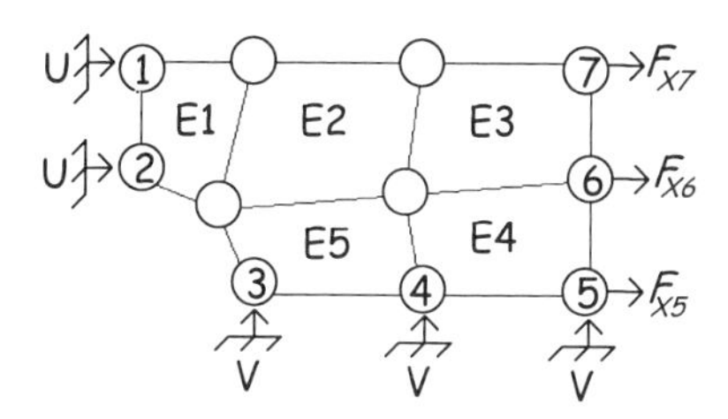

Figure 7.38 Mesh Boundary and Coarse Mesh for Plate with Hole

The boundary conditions invoked in the finite element model above are called symmetrical boundary conditions, and they simulate the effect of having the full geometry. The example of using symmetry given here is perhaps one of the most simple examples of using symmetry to reduce the computational expense of finite element models. Several references related to symmetry were given in Chapter 1.

Problem 4—Summary

Those with minimal finite element modeling experience will typically attempt to exactly replicate, with finite elements, the geometry of the structure which is to be analyzed. Many novice analysts are surprised to find that the practical limits of finite element modeling capability are quickly exceeded. That is, trying to exactly replicate the actual geometry of a complex structure will typically require many more elements than can be employed in a practical fashion. The topic of model size was briefly considered in Section 7.2, under the heading of *Choice of Element—Size of Model.*

In light of the need to limit the size of a finite element model, analysts will typically use various means to reduce the elements that are required for a given idealization. The use of structural symmetry is one means to reduce computational requirements.

Problem 5—Non-Uniform Shaft

Problem 5—Description

The non-uniform shaft with circular cross section, as considered in Chapter 2, is to be analyzed. The displacement at the free end of the shaft is desired.

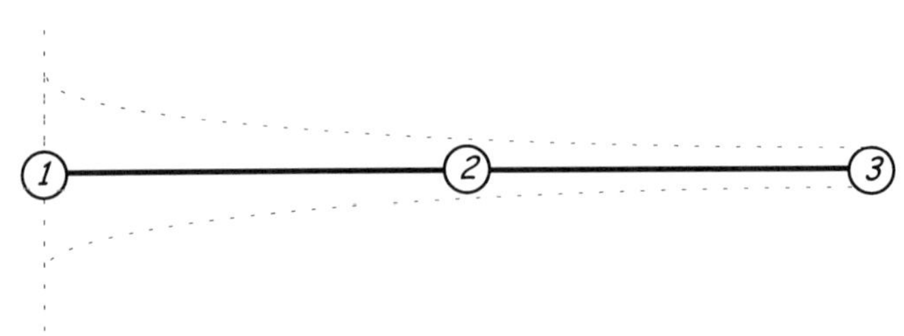

Figure 7.39 Non-Uniform Shaft

Problem 5—Finite Element Model

As illustrated in Chapter 2, this problem may be solved using rod elements with various cross sections. However, another way to solve the problem is to use axisymmetric elements.

Figure 7.40 illustrates a finite element mesh that could be used to model the non-uniform shaft. The mesh boundary is the surface of revolution that, if revolved 360 degrees about the centerline, would render the geometry of the non-uniform shaft. Using this type of modeling technique, a significant reduction in the amount of geometry required for the idealization is achieved.

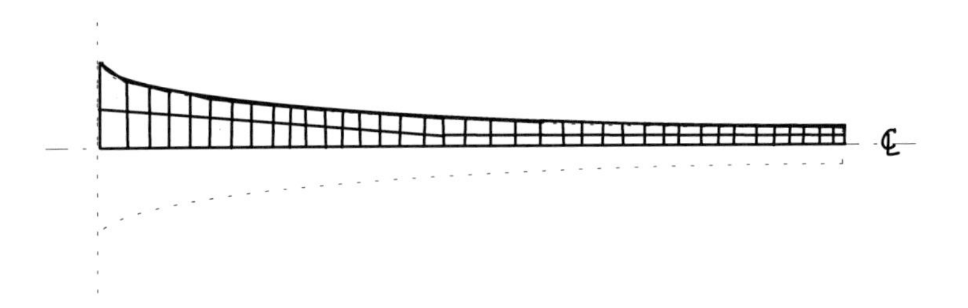

Figure 7.40 Non-Uniform Shaft

The surface elements used for the axisymmetric idealization are formulated to simulate the effects of having the entire geometry present, somewhat in the same way that the surface element for plane stress applications simulates an entire structure.

One word of caution when using the axisymmetric element. Applying a concentrated load to a node of the model actually represents a ring of forces. For example, consider Figure 7.41.

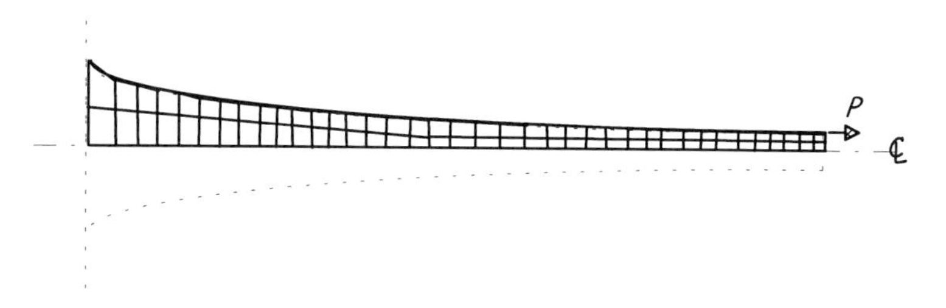

Figure 7.41 Force at Node of Axisymmetric Model

Just as the geometry of the shaft is (theoretically) rotated 360 degrees, so is the force applied to the node, as indicated. It may be best to apply loads in terms of surface tractions, with units of force/area. For instance, if a uniform load of 100 lb_f is to be applied to the right end of the shaft, and the actual surface area of the end of the shaft is 2 in^2, the analyst can specify that a surface traction of 50 psi is to be applied to the edge of the elements on the right end of the shaft. This will result in a total force of 100 lb_f, applied to the end of the shaft.

References

1. Turner, M.J., Clough, R.W., Martin, H.C., Topp, L.J., "Stiffness and Deflection Analysis of Complex Structures," in *Journal of the Aeronautical Sciences*, Vol. 23, No. 9, p. 805–854, 1956
2. Higdon, A., *Mechanics Of Materials*, John Wiley & Sons, N.Y., 1985
3. Timoshenko, S.P., *Theory of Elasticity*, McGraw-Hill, New York, 1987
4. Juvinall, R.C., *Stress, Strain, And Strength*, McGraw-Hill Book Company, N.Y., 1967
5. Pilkey, W.D., Peterson's Stress Concentration Factors, Second Edition, John Wiley & Sons, New York, 1997
6. Anonymous, *Specification for the Design, Fabrication, and Erection of Structural Steel for Building*, American Institute of Steel Construction, 1978
7. White, R.N., Salmon,C.G, (Editiors), *Building Structures Design Handbook*, John Wiley & Sons, Inc., New York, 1987
8. Young, W.C., *Roark's Formulas for Stress & Strain*, 6th Edition, McGraw-Hill, N.Y., 1989

9. Yu, M.T., DuQuesnay, D.L., Topper, T.H., "Fatigue Behavior of a Cold-Rolled SAE Grade 945X SLA Steel," American Society for Testing and Materials, p. 274–280, 1990
10. Anonymous, "Practical Finite Element Modeling And Techniques Using MSC/NASTRAN," NA/1/000/PMSN, The MacNeal-Schwendler Corporation, Los Angeles, 1990
11. Anonymous, *Primer Introductory Textbook*, MARC Analysis Research Corporation, Palo Alto, CA, 1993
12. Zemitis, W.S., Howell, L.L., "Modeling of Axisymmetric Structures with Discontinuous Stiffeners Using 2-D Elements in ANSYS," ANSYS News, Second Edition, Swanson Analysis Systems, Inc., Houston, PA, 1995
13. Feld, O.J., and Soudry, J.G., "Modeling The Interface Between Shell and Solid Elements," Goodyear Aerospace Corporation, Akron, 1983
14. Melosh, R.J., Utku, S., in *State-of-the-Art Surveys on Finite Element Technology*, ASME, Applied Mechanics Division, Chapter 3, 1983
15. Russell, D., Priess, J., "Evaluating Automated Meshing and Element Selection," Machine Design, 21 June 1990
16. Ramsay, A., "Solid FE Modeling—Hex or Tet?," BenchMark, E.R. Robertson, Editor, NAFEMS, Glasgow, p. 12–15, December, 1992
17. Cifuentes, A.O., Kalbag, A., "A Performance Study of Tetrahederal and Hexahedral Elements in 3-D Finite Element Structural Analysis," in Finite Elements in Analysis and Design, p. 313–318, 1992
18. MacNeal, R.H., Finite Elements: Their Design And Performance, Marcel Dekker, Inc., New York, 1994
19. Zienkiewicz, O.C., Phillips, "An Automatic Mesh Generation Scheme for Plane and Curved Surfaces by Isoparametric Coordinates," International Journal for Numerical Methods in Engineering, Vol. 3, p. 519–528, 1971
20. Cook, R.D., Oates, "Mapping Methods for Generating Three-Dimensional Meshes," Computers in Mechanical Engineering, August, p. 67–73, 1982
21. Shephard, M.S., "Approaching the Automatic Generation of Finite Element Meshes," Computers in Mechanical Engineering, p. 49–56, 1983
22. Cook, R.D., *Concepts And Applications Of Finite Element Analysis*, John Wiley & Sons, N.Y., 1989
23. Dewhirst, D.L., "Finite Element Distortion Tests," Technical Report No. Sr-89 -33, Ford Motor Company, Dearborn, 1989
24. Cox, J.J., Charlesworth, W.W., Anderson, D.C., "Domain Composition Methods for Associating Geometric Modeling with Finite Element Modeling," ACM 089791-427-9/91/0006/0443, p. 443–454, 1991
25. Anonymous, *A Finite Element Primer*, Department of Trade and Industry, National Engineering Laboratory, Glasgow, U.K., p. 94, 1987

26. Zins, J.W., "Quality Control of Finite Element Models," Proceedings of the Fifth World Congress on Finite Element Methods, Robinson and Associates, Devon, England, October, 1987
27. Haftka, R.T., Robinson, J.C., "Effect of Out-of-Planeness of Membrane Quadrilateral Finite Elements," AIAA J., Vol. 11, No. 5, p. 742–744, 1973
28. Van Kuffeler, F., Waagmeester, C., and Assendelft, J., "Theoretical Background and Physical Meaning of the Determinant of the Jacobian in Isoparametric Finite Element," Finite Element News, No. 5, October, 1988
29. Waagmeester, C., Assendelft, J., and van Kuffeler, F., "Jacobian Shape Parameters in the Patch Test for 3-D Solid Element," Finite Element News, No. 5, October, 1988
30. Robinson, J., "A Single Element Test," Comp. Meth. App. Mech. Engng., Vol 7., No. 2, p. 191–200, 1976
31. Robinson, J., Haggermacher, G.W., "Element Warning Diagnostics," Finite Element News, Numbers 3, 4 and 5, June, August and October, 1982
32. Pinfold, M., "Validation of the Finite Element Analysis of a Composite Automotive Suspension Arm," BenchMark, E.R. Robertson, Editor, NAFEMS, Glasgow, p. 28–30, December, 1992
33. Crandall, S.H., *An Introduction to the Mechanics of Solids*, Second Edition, McGraw-Hill Book Company, New York, 1972
34. Collins, J.A., *Failure of Materials in Mechanical Design*, John Wiley & Sons, New York, p. 230, 1981
35. Bridgman, P. W., *Studies in Large Plastic Flow and Fracture*, McGraw-Hill Book Company, New York, 1952
36. Orowan, E., "Fracture and Strength of Soilds," in *Reports on Progress in Physics*, The Physical Society, London, Vol. 12, p. 185–232, 1949
37. Griffith, A.A., First International Congress on Applied Mechanics, Delft, p. 55, 1924
38. Trantina, G., "Design Engineering for Performance," GE Plastics, Inc., Schenectady, N.Y., 20 September, 1994
39. Van Vlack, L.H., Elements of Materials Science and Engineering, Third Edition, Addison-Wesley, Reading MA, 1975
40. Anonymous, *ABAQUS/Standard Example Problems Manual*, Hibbitt, Karlsson, & Sorensen, Pawtucket, RI, 1995
41. Shigley, J.E., *Mechanical Engineering Design*, Third Edition, McGraw-Hill, Inc., New York, 1977

Appendix A

A Simple Differential Equation

A differential equation based upon static equilibrium of forces is developed for a prismatic rod in uniaxial tension. Although the differential equation given here considers only axial displacement and one component of stress, the same logic applies to displacement due to normal stress in other coordinate directions as well. As shown in Chapter 2, a *system* of differential equations, (2.1.1), is required to describe stress and displacement for a body of arbitrary shape, subjected to general loading.

Consider the prismatic rod in Figure A.1, restrained on the left face and subjected to an external load, with resultant force passing through the centroids of all the cross sections of the bar.

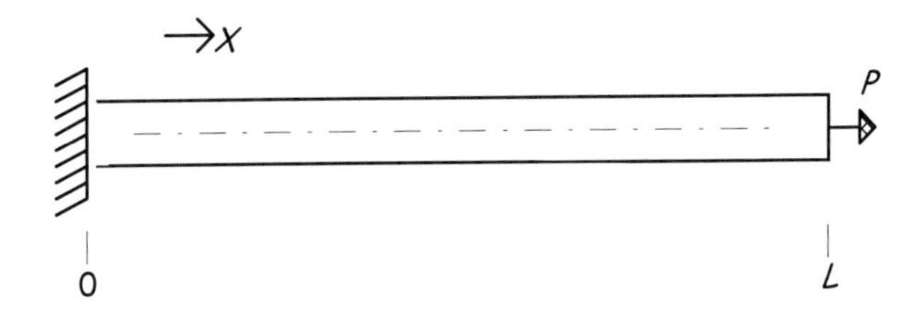

Figure A.1 Rod Under Uniaxial Load

Examining a small (infinitesimal) slice of the rod, normal stress exists on the left and right faces, as depicted in Figure A.2.

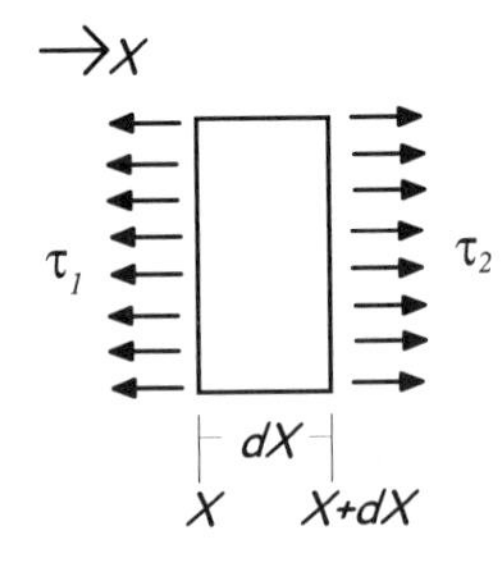

Figure A.2 Normal Stress on a Differential Slice of Rod

The normal stress acting upon the faces can be expressed in terms of force, since normal force is the product of the normal stress and the cross sectional area, as depicted in Figure A.3.

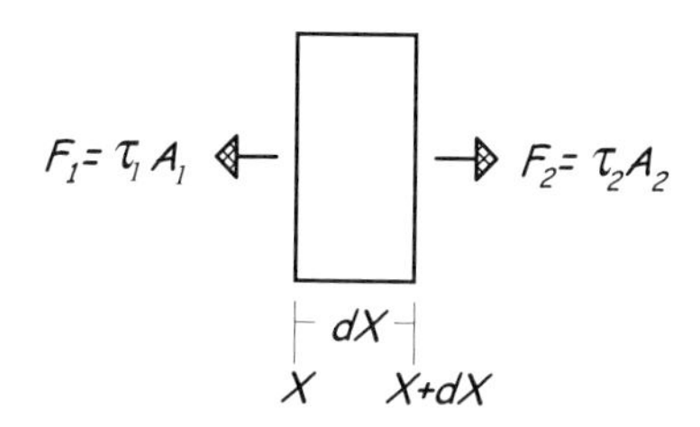

Figure A.3 Differential Slice of Rod

If the slice is not accelerating (i.e., the slice is in static equilibrium) the summation of forces acting upon the slice must be zero:

$$\sum F_X = -F_1 + F_2 = 0 \tag{A.1}$$

Expressing (A.1) in terms of stress instead of force:

$$\sum F_X = -\tau_1 A_1 + \tau_2 A_2 = 0 \tag{A.2}$$

The product of stress and area on the right face can be expressed in terms of stress and area on the left:

$$\tau_2 A_2 = \tau_1 A_1 + \frac{\partial(\tau A)}{\partial X} dX \tag{A.3}$$

Substituting (A.3) into (A.2):

$$-\tau_1 A_1 + \tau_1 A_1 + \frac{\partial(\tau A)}{\partial X} dX = 0 \tag{A.4}$$

Simplifying (A.4):

$$\frac{\partial}{\partial X}(\tau A) = 0 \tag{A.5}$$

The equation above suggests that the product of the uniaxial stress and cross sectional area (i.e., force) is constant on any slice of the rod. Since we wish to generate an expression for displacement, the uniaxial stress term in (A.5) is replaced with the Hooke's law equivalent, which introduces the axial displacement variable into the equation:

$$\frac{\partial}{\partial X}\left(EA\frac{\partial U}{\partial X}\right)=0 \tag{A.6}$$

Assuming that displacement is a function of only one variable, the partial derivatives in (A.6) can be replaced with ordinary derivatives:

$$\frac{d}{dX}\left(EA\frac{dU}{dX}\right)=0 \tag{A.7}$$

The equation above is the second order, ordinary, homogeneous differential equation, as was given by (1.2.1) in Chapter 1. The two boundary conditions associated with (A.7), for this problem, are:

$$U(0)=0 \tag{A.8}$$

$$EA\frac{dU}{dX}\bigg|_{X=L}=P \tag{A.9}$$

The first boundary condition, (A.8), states that displacement on the left face of the rod is zero. The second boundary condition is equivalent to saying that the force on the extreme right face of the rod must be equal to the applied load:

$$F_{X=L}=P \tag{A.10}$$

To arrive at (A.9), (A.10) is manipulated such that the force variable is replaced by the product of stress and cross sectional area:

$$F|_{X=L}=P$$

$$\uparrow$$

$$\tau A$$

$$\therefore\ \tau A|_{X=L}=P \tag{A.11}$$

Using the Hooke's law relationship, the stress at the right face of the rod is expressed as:

$$\tau|_{X=L} = E\frac{dU}{dX}\bigg|_{X=L} \tag{A.12}$$

Using (A.12) in (A.11), the boundary condition of (A.9) is expressed as:

$$EA\frac{dU}{dX}\bigg|_{X=L} = P \tag{A.13}$$

The differential equation given by (A.7), along with boundary conditions (A.8) and (A.9), can be used to generate an expression for displacement in a rod with a static, uniaxial load.

Appendix B

Functionals

Before considering functionals, it may be helpful to review simple, single variable functions. A function assigns a unique value to a variable for a given value of a second variable. For example:

$$y(x) = x^2 \tag{B.1}$$

The function $y(x)$ assigns one value to the variable y for a given value of the variable x.

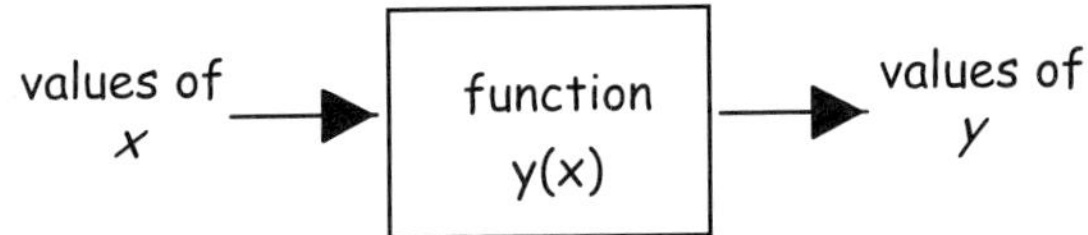

In contrast, a single variable functional assigns a value to a variable for a given *function* of a second variable:

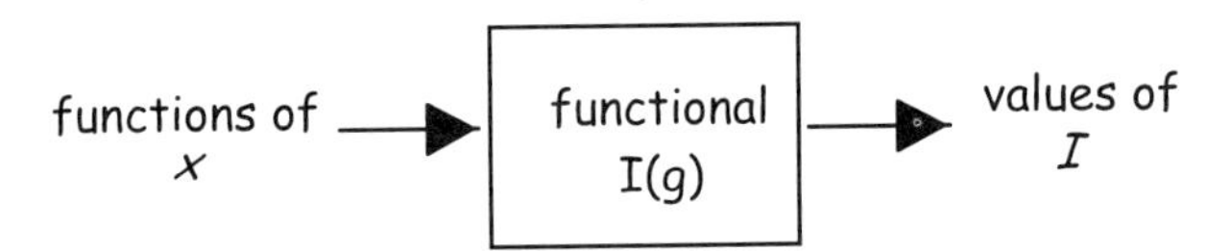

Here, the functional $I(g)$ operates on *functions of* x, and outputs values of I.

Functionals take the form of an integral equation; an example of a functional with a single independent variable is given by (B.2):

$$I(g) = \int_1^3 [g(x)]^2 dx \tag{B.2}$$

For example, if $g(x)=5x$, (B.2) renders:

$$I(g) = \int_1^3 [5x]^2 \, dx = \int_1^3 25x^2 \, dx = 25\frac{x^3}{3}\bigg|_1^3 = \frac{675}{3} \tag{B.3}$$

The functional *I(g)* assigns a value to the variable *I* for any integrable function *g(x)*. A functional may also contain derivative terms, as illustrated by (B.4):

$$J(g) = \int_1^3 \left\{ \left[g(x) \right]^2 + \frac{d}{dx} g(x) \right\} \, dx \tag{B.4}$$

If again *g(x)=5x*, the functional given by (B.4) will render:

$$J(g) = \int_1^3 \left\{ (5x)^2 + \frac{d}{dx} 5x \right\} \, dx = \int_1^3 \left\{ 25x^2 + 5 \right\} \, dx = \frac{675}{3} + 10 \tag{B.5}$$

In general, a single variable functional can be expressed as:

$$I(g) = \int_a^b F\left(g, \frac{dg}{dx}, \frac{d^2 g}{dx^2}, \ldots, \frac{d^n g}{dx^n}, x \right) dx \tag{B.6}$$

Although a functional with a single independent variable was considered in the preceding discussion, functionals are often expressed in terms of several independent variables.

The finite element method employs a functional to generate the equilibrium equations for displacement. A functional is established because it provides a mathematical basis from which a desired equation (or system of equations) can be obtained. In general, functionals may be established irrespective of a physical principle, by manipulating the governing differential equation for a given problem. In other cases, a physical principle, such as minimum total potential energy, provides the motivation to establish a functional. Huebner [1, p. 94] and Reddy [2, p. 544] discuss how differential equations can be manipulated to establish functionals. Bathe [3, p. 97] shows the reverse process: How to manipulate a functional to extract the governing differential equation. Zienkiewicz [4, p. 18] discusses the energy principle in relationship to finite element functionals.

An advantage of using a physical principle to establish a functional for solid mechanics problems is that a single functional, established with respect to a physical principle, can be applied to a whole class of solid mechanics problems. For instance, many linear, static continuum mechanics problems may be solved using the same functional. In contrast, manipulating a differential equation into functional form often lends itself to only one specific problem, or a narrow range of problems. For instance, one can manipulate the differential equation from Appendix A, (A.7), to establish a functional but the functional will apply only to uniaxial stress problems. In addition, the process of manipulating a differential equation to establish a functional also has the disadvantage of requiring the use of variational calculus techniques, which adds to the complexity of the problem.

Another difference between the two approaches of establishing functionals is that developing a functional using a physical principle (such as total potential energy) can lend an intuitive feel for a given type of problem. If both a physical principle and a differential equation are applicable to a given problem, it is possible for the same functional may be arrived at using either approach.

The displacement based finite element method employs a functional representing total potential energy to compute the equilibrium displacement within a structure. An expression for this type of functional, which can be used to solve a broad range of solid mechanics problems, will be now be developed.

Total Potential Energy for a Discrete System

Functionals are used for problems associated with continuous systems. However, before considering a continuous system, it may be helpful to first consider a discrete system to gain a fundamental understanding of energy principles. A simple, discrete mechanical system shown in Figure B.1, below.

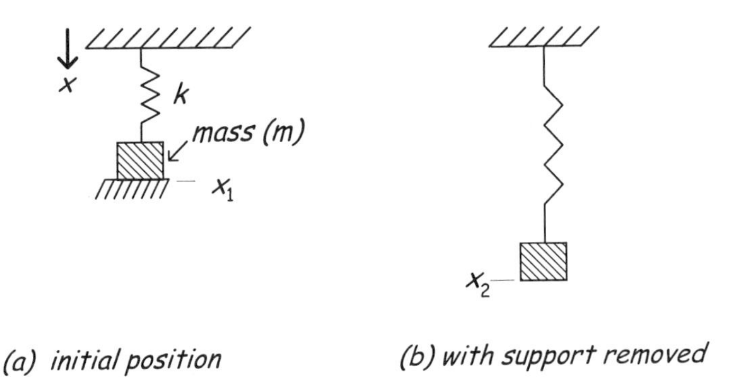

Figure B.1 Energy in a Mass-Spring System

Assume that if the support is removed, the mass will eventually settle at x_2. The total potential energy of the system, defined as the sum of the elastic energy of the spring, E_e, and the potential energy of the load, E_p, is to be examined. Denoting total potential energy of this discrete system as π:

$$\pi = E_e + E_p \tag{B.7}$$

We will examine the energy of the spring and the mass separately. Assume that without the mass attached, we stretch the spring from position x_1 to x_2. The change in the amount of stretch in the spring will be defined as Δu:

$$\Delta u = u_2 - u_1 \tag{B.8}$$

Note that u_i is the amount of stretch when the spring is at a particular value of x_i.

The change in the elastic energy of the spring, when changing the stretch by an amount Δu under magnitude load F, will be considered as equal to the incremental work done on the spring:

$$\Delta E_e = F\Delta u \tag{B.9}$$

The equation above assumes that all of the work energy to stretch the spring is stored as elastic energy. In the limit, as the smaller increments of stretch are considered, the change in elastic energy of the spring is defined as the differential:

$$dE_e = F\,du$$

Integrating both sides of the equation above:

$$E_e = \int F\,du + c_1 \tag{B.10}$$

If the spring is linear, the force can be replaced by a spring constant multiplied by displacement, thus, (B.10) may alternately be expressed as:

$$E_e = \int ku\,du + c_1$$

Or, performing the integration:

$$E_e = \frac{1}{2}ku^2 + c_1 \tag{B.11}$$

The equation above represents the elastic energy of the spring in terms of displacement, a spring constant, and an arbitrary constant.

Now consider the potential energy of the load. We again disconnect the spring from the mass, and then lower the mass from x_1 to x_2. From the first law of thermodynamics, the incremental heat associated with a process is equal to the

sum of three energy terms and the incremental work (Van Wylen [5]) as given by (B.12):

$$\delta Q = dE_i + dE_k + dE_p + \delta W \tag{B.12}$$

For the process of lowering the mass:

$\delta Q \equiv$ incremental change in heat
$dE_i \equiv$ incremental change in internal energy of load
$dE_k \equiv$ incremental change in kinetic energy of load
$dE_p \equiv$ incremental change in potential energy of load
$\delta W \equiv$ incremental change in work of system

We specify that the process of moving the mass from x_1 to x_2 is adiabatic, causes no change in internal energy, nor imparts velocity to the mass, thus:

$$\overset{0}{\cancel{\delta Q}} = \overset{0}{\cancel{dE_i}} + \overset{0}{\cancel{dE_k}} + dE_p + \delta W \tag{B.13}$$

For this process, (B.13) suggests that the incremental change in potential energy of the load is equal to the negative of the incremental work required to lower the mass:

$$dE_p = -\delta W \tag{B.14}$$

The incremental work to lower the mass can be defined as:

$$\delta W = F du \tag{B.15}$$

Substituting (B.15) for the right hand side of (B.14):

$$dE_p = -F du \tag{B.16}$$

Recall that for the process of lowering the mass, the force is presumed constant, and equal to the mass times gravity:

$$F = mg = \text{constant}$$

Noting that F in (B.16) is constant, and integrating both sides of the equation:

$$E_p = -Fu + c_2 \qquad \text{(B.17)}$$

Thus, (B.17) shows that the potential energy of the load is a linear function of displacement. Substituting (B.17) and (B.11) into (B.7):

$$\pi = \frac{1}{2}ku^2 + c_1 - Fu + c_2 \qquad \text{(B.18)}$$

If both the potential energy of the load, E_p, and elastic energy of the spring, E_e, are defined as zero when displacement is zero:

$$c_1 = c_2 = 0 \qquad \text{(B.19)}$$

Taking note of (B.19), the expression for total potential energy given by (B.18) becomes:

$$\pi = \frac{1}{2}ku^2 - Fu \qquad \text{(B.20)}$$

As shown in Section 1.3, (B.20) can be minimized with respect to u to find the equilibrium displacement of the spring/mass system.

Total Potential Energy of a Continuous System

Equation B.20 describes the total potential energy for a discrete system, however, a more helpful equation for stress analysis would consider continuous systems. To develop a expression for total potential energy in a continuous system, let us examine a bar with a uniaxial load, as shown in Figure B.2.

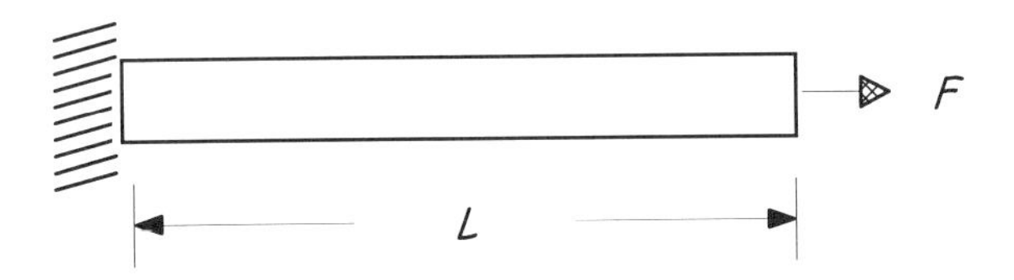

Figure B.2 Bar Under Uniaxial Load

The objective is to generate an expression for total potential energy of the continuous system shown in Figure B.2, where a solid bar with the left end fixed is subjected to a uniaxial load. As before, the expression for total potential energy is composed of terms describing elastic energy and potential energy of the load. The elastic energy in continuous systems can be considered as *strain energy*. For the bar, the strain energy will be denoted as E_ε, such that the total potential energy expressed as:

$$\Pi = E_\varepsilon + E_p \tag{B.21}$$

The upper case pi symbol is used to denote that the total potential energy of a continuous system is being considered.

Observe the term for potential energy of the external load in (B.21). Recall that the potential energy of the load for the discrete mass/spring system was given as:

$$E_p = -F\,u + c_2 \tag{B.22}$$

The potential energy of the continuous bar is nearly the same as for the spring/mass system. The difference is that in a continuous system, displacement is a function, and the load (F) acts at the end of the bar, where $x=L$. Since displacement is a continuous variable, (B.22) is modified to reflect this fact:

$$E_p = -F\,u(L) + c_2 \tag{B.23}$$

The equation above states that the potential energy of the load is a function of the external load and the displacement at the point where the load is applied. Substituting (B.23) into (B.21):

$$\Pi = E_\varepsilon - F\,u(L) + c_2 \tag{B.24}$$

To arrive at an expression for the strain energy of the entire bar, the strain energy in a differential slice of the bar will first be examined.

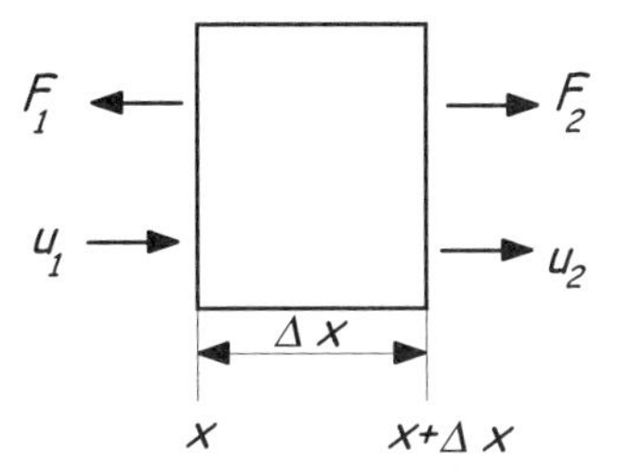

Figure B.3 Slice of Bar Under Uniaxial Load

With the external load applied, the right face of the slice displaces by an amount u_2 while the left faces displaces u_1. Examining the work done on the right face:

$$work\big|_{x+\Delta x} = \int_0^{u_2} F du \tag{B.25}$$

On the left face, the work is negative, since the direction of the force opposes the direction of the displacement:

$$work\big|_{x} = -\int_0^{u_1} F du \tag{B.26}$$

Assuming that a linear relationship between force and displacement prevails when stretching the bar:

$$F = cu \tag{B.27}$$

Note that c in the equation above is a constant. Substituting (B.27) for F in (B.25), and integrating:

$$work\big|_{x+\Delta x} = \frac{1}{2} c_2 u_2^2 \tag{B.28}$$

Likewise, substituting (B.27) into (B.26) and integrating:

$$work\big|_{x} = -\frac{1}{2} c_1 u_1^2 \tag{B.29}$$

Using (B.27), the magnitudes of the forces generated by displacements u_1 and u_2 are:

$$F_1 = c_1 u_1 \qquad F_2 = c_2 u_2 \tag{B.30}$$

Using the definitions of (B.30) in (B.28) and (B.29):

$$work\big|_{x+\Delta x} = \frac{1}{2} F_2 u_2 \tag{B.31}$$

$$work\big|_{x} = -\frac{1}{2} F_1 u_1 \tag{B.32}$$

Assuming that all of the work imparted to the slice is stored as strain energy, the two equations above suggest the energy associated with the small slice can be expressed as the sum of the two work terms:

$$\Delta E = work\big|_{x+\Delta x} + work\big|_{x} \tag{B.33}$$

Substituting (B.31) and (B.32) into (B.33):

$$\Delta E = \frac{1}{2} F_2 u_2 - \frac{1}{2} F_1 u_1 \tag{B.34}$$

To obtain an expression for energy per unit length, both sides of (B.34) are divided by the width of the slice:

$$\frac{\Delta E_\varepsilon}{\Delta x} = \frac{1}{2}\left(\frac{F_2 u_2 - F_1 u_1}{\Delta x}\right) \tag{B.35}$$

The displacement on the right face, u_2, can be expressed in terms of displacement on the left face:

$$u_2 = u_1 + \Delta u \tag{B.36}$$

Substituting (B.36) into (B.35):

$$\frac{\Delta E_\varepsilon}{\Delta x} = \frac{1}{2}\left(\frac{(F_2 - F_1)u_1 + F_2 \Delta u}{\Delta x}\right) \tag{B.37}$$

If the slice is in static equilibrium, the summation of the forces on the slice must sum to zero, i.e., $F_2 + (-F_1) = 0$, thus, (B.37) is equivalent to:

$$\frac{\Delta E_\varepsilon}{\Delta x} = \frac{1}{2}\left(\frac{F_2 \Delta u}{\Delta x}\right) \tag{B.38}$$

As we consider smaller and smaller slices, the thickness approaches zero, and (B.38) becomes:

$$\lim_{x \to 0} \frac{\Delta E_\varepsilon}{\Delta x} = \frac{dE_\varepsilon}{dx} = \frac{1}{2} F \frac{du}{dx} \tag{B.39}$$

Rearranging (B.39):

$$dE_\varepsilon = \frac{1}{2} F \frac{du}{dx} dx \tag{B.40}$$

The force variable in (B.40) can be expressed in terms of normal stress and cross sectional area:

$$\tau_{xx} = \frac{F}{A} \tag{B.41}$$

Rearranging (B.41):

$$F = \tau_{xx} A \tag{B.42}$$

For uniaxial stress, normal stress may also be defined as:

$$\tau_{xx} = E \frac{du}{dx} \tag{B.43}$$

Substituting (B.43) into (B.42):

$$F = EA \frac{du}{dx} \tag{B.44}$$

Substituting (B.44) into (B.40):

$$dE_\varepsilon = \frac{1}{2} EA \left(\frac{du}{dx} \right)^2 dx \tag{B.45}$$

or:

$$E_\varepsilon = \int_0^L \frac{1}{2} EA \left(\frac{du}{dx} \right)^2 dx + c_3 \tag{B.46}$$

Substituting (B.46) into (B.24):

$$\Pi = \int_0^L \frac{1}{2} EA \left(\frac{du}{dx} \right)^2 dx + c_3 - F\,u(L) + c_2 \tag{B.47}$$

If we assume that both strain energy and potential energy of the load are zero when displacement is zero, the arbitrary constants in (B.47) are both zero, therefore:

$$\Pi = \frac{1}{2} \int_0^L EA \left(\frac{du}{dx} \right)^2 dx - F\,u(L) \tag{B.48}$$

The equation above, describing the total potential energy for a simple continuous system (a bar is this case) is the same equation that was given in Chapter 1, (1.3.2). Minimizing this equation with respect to displacement renders equilibrium displacement within the bar.

A Total Potential Energy Functional for Finite Element Problems

The potential energy load term in the functional above considers just one point load, as applied to the end of the bar shown in Figure B.2. However, using the finite element method, many different types of loads can be included in the load potential terms. We will now consider how potential energy, for three types of loads applied when using the finite element method, is expressed.

Recall again that the incremental change in potential energy of the load is a function force and the incremental displacement:

$$dE_p = -F du \tag{B.49}$$

Since we will be considering concentrated, surface, and body load contributions to potential energy, a distinction must be made between them. Thus, the incremental change in potential energy due to a concentrated load will be denoted with a C superscript:

$$dE_p^C = -F du \tag{B.50}$$

When considering forces and displacements in more than one coordinate direction, it is noted that force and displacement variables that are orthogonal to each other have no influence on potential energy. Thus, a more precise way of expressing (B.50) would be to specify the dot product between force and displacement:

$$dE_p^C = -F \bullet dD \tag{B.51}$$

The variables F and D are force and displacement variables, respectively. In a finite element model, there can be many force and displacement components. Hence, for finite element models, (B.51) is equivalent to:

$$dE_p^C = -\underline{\underline{F}}\, d\,{}^e\underline{\underline{D}}^T \tag{B.52}$$

The force vector above contains nodal forces due to concentrated loads, while the displacement vector, as usual, contains finite element nodal displacement variables related to a given element e.

The expression given by (B.52) states that the incremental change in potential energy associated with concentrated loads, for a given element, is a function of the external concentrated load and the incremental change in the nodal displacement vector. The force and displacement vectors used in finite element models are constructed such that their product is equivalent to the dot product. In other words, components of the force and displacement vectors are arranged such that, when multiplied together, each term consists of a force and displacement variable that act in the same coordinate direction. Integrating both sides of (B.52):

$$E_p^C = -\int \underline{\underline{F}}\; d\,{}^e\underline{\underline{D}}^T + c_1 \tag{B.53}$$

Or, performing the integration, considering the nodal forces to be constant:

$$E_p^C = -\,{}^e\underline{\underline{D}}^T\, \underline{\underline{F}} + c_1 \tag{B.54}$$

As before, assume that the potential energy of the load is zero when the displacement is zero. As such, c_1=0, and (B.54) becomes:

$$E_p^C = -\,{}^e\underline{\underline{D}}^T \underline{\underline{F}} \tag{B.55}$$

Since the forces in (B.55) are due to concentrated loads, they will be denoted by with a C superscript:

$$E_p^C = -\,{}^e\underline{\underline{D}}^T \underline{\underline{F}}^C \tag{B.56}$$

Now, consider potential energy due to a surface loads as depicted in Figure B.4, and denoted as E_p^S.

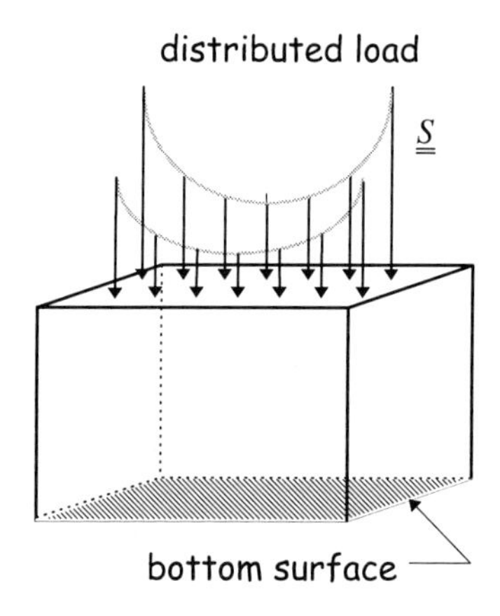

Figure B.4 Surface Load

The potential energy due to surface loads can be expressed in a form similar to (B.53):

$$E_p^S = -\int \underline{\underline{F}}\, d\breve{\underline{\underline{D}}}^T + c_1 \tag{B.57}$$

where:

$$\underline{\underline{\breve{D}}} \equiv \begin{Bmatrix} \tilde{U}(x,y,z) \\ \tilde{V}(x,y,z) \\ \tilde{W}(x,y,z) \end{Bmatrix} \tag{B.58}$$

Since the forces are continuous, the displacement in (B.57) must also be a continuous function, not simply discrete variables (nodal displacements), as in (B.53). Therefore, the displacement vector above contains functions, and is distinguished from the vector of nodal displacements by the *smile overscript.*

The force vector in (B.57) is formed by integrating a surface traction over the area to which it is applied:

$$\underline{\underline{F}} = \int \underline{\underline{S}}\, dA \tag{B.59}$$

Substituting (B.59) into (B.57):

$$E_p^S = -\int \int \underline{\underline{S}}\, dA \; d\underline{\underline{\breve{D}}}^T + c_1 \tag{B.60}$$

If we assume that the surface loads are not a function of displacement, the two integrals in (B.60) can be evaluated separately:

$$E_p^S = -\left(\int \underline{\underline{S}}\, dA\right)\left(\int d\underline{\underline{\breve{D}}}^T\right) + c_1 \tag{B.61}$$

Performing the integration with respect to displacement:

$$E_p^S = -\int \underline{\underline{\breve{D}}}^T \; \underline{\underline{S}}\, dA + c_1 \tag{B.62}$$

The vector of finite element interpolation functions can be expressed in terms of a matrix of shape functions and a vector of nodal displacement variables:

$$\underline{\underline{\breve{D}}} = \begin{Bmatrix} \tilde{U}(x,y,z) \\ \tilde{V}(x,y,z) \\ \tilde{W}(x,y,z) \end{Bmatrix} = \underline{\underline{N}}\; {}^{e}\underline{\underline{D}} \tag{B.63}$$

Substituting (B.63) into (B.62):

$$E_p^S = -\int {}^e\underline{\underline{D}}^T \underline{\underline{N}}^T \underline{\underline{S}}\, dA + c_1 \tag{B.64}$$

Assuming that the potential energy of the load is zero when displacement is zero, (B.64) becomes:

$$E_p^S = -\int {}^e\underline{\underline{D}}^T \underline{\underline{N}}^T \underline{\underline{S}}\, dA \tag{B.65}$$

Since the vector of nodal displacements is not a function of the spatial variables, it can be removed from the integral:

$$E_p^S = -\,{}^e\underline{\underline{D}}^T \int \underline{\underline{N}}^T \underline{\underline{S}}\, dA \tag{B.66}$$

The above represents the potential energy due to surface loads. Likewise, potential energy due to body loads can be expressed as:

$$E_p^\beta = -\,{}^e\underline{\underline{D}}^T \int \underline{\underline{N}}^T \underline{\underline{\beta}}\, dV \tag{B.67}$$

Recall from (B.21) that total potential energy is a function of the strain energy and the potential of the loads:

$$\Pi = E_\varepsilon + E_p \tag{B.68}$$

Using (B.56), (B.66), and (B.67) in (B.68):

$${}^e\tilde{\Pi} = E_\varepsilon - {}^e\underline{\underline{D}}^T \underline{\underline{F^C}} - {}^e\underline{\underline{D}}^T \int \underline{\underline{N}}^T \underline{\underline{S}}\, dA - {}^e\underline{\underline{D}}^T \int {}^T\underline{\underline{N}}\, \underline{\underline{\beta}}\, dV \tag{B.69}$$

Notice that (B.69) gives an approximate expression for total potential energy, and that the expression applies to just one finite element.

Performing the integration over the surface and volume generates equivalent nodal forces, as discussed in Chapter 6, such that (B.69) can alternately be defined as:

$${}^e\tilde{\Pi} = E_\varepsilon - {}^e\underline{\underline{D}}^T\, {}^e\underline{\underline{F^C}} - {}^e\underline{\underline{D}}^T\, {}^e\underline{\underline{F^S}} - {}^e\underline{\underline{D}}^T\, {}^e\underline{\underline{F^\beta}} \tag{B.70}$$

where:

$$\underline{\underline{{}^eF^S}} \equiv \int \underline{\underline{N}}^T \, \underline{\underline{S}}\, dA \qquad \underline{\underline{{}^eF^\beta}} \equiv \int \underline{\underline{N}}^T \, \underline{\underline{\beta}}\, dV \tag{B.71}$$

In Chapter 3, a general expression for strain energy was given by (3.4.8):

$$Strain\ Energy = E_\varepsilon = \frac{1}{2}\int_V \underline{\underline{{}^eD}}^T \, \underline{\underline{B}}^T \, \underline{\underline{E}} \, \left(\underline{\underline{B}}\, \underline{\underline{{}^eD}}\right) dV \tag{B.72}$$

Using (B.72) in (B.70):

$${}^e\tilde{\Pi} = \frac{1}{2}\int_V \underline{\underline{{}^eD}}^T \, \underline{\underline{B}}^T \, \underline{\underline{E}} \, \left(\underline{\underline{B}}\, \underline{\underline{{}^eD}}\right) dV - \underline{\underline{{}^eD}}^T \, \underline{\underline{{}^eF^C}} - \underline{\underline{{}^eD}}^T \, \underline{\underline{{}^eF^S}} - \underline{\underline{{}^eD}}^T \, \underline{\underline{{}^eF^\beta}} \tag{B.73}$$

However, all of the potential terms amount to nodal displacements multiplied by nodal forces, such that the last three terms in (B.73) can be expressed more generally as:

$$-\underline{\underline{{}^eD}}^T \, \underline{\underline{F^C}} - \underline{\underline{{}^eD}}^T \, \underline{\underline{{}^eF^S}} - \underline{\underline{{}^eD}}^T \, \underline{\underline{{}^eF^\beta}} = (-1)\left({}^eD_a \, {}^eF_a + {}^eD_b \, {}^eF_b + \cdots\right) \tag{B.74}$$

The displacement and force terms on the right hand side of (B.74) are general. For a given problem, the can act in any coordinate direction, although the displacement and force, in any one term, must act in the same direction. Using (B.74), the total potential energy for a given element, as given by (B.73), can be expressed as:

$${}^e\tilde{\Pi} = \frac{1}{2}\int_V \underline{\underline{{}^eD}}^T \, \underline{\underline{B}}^T \, \underline{\underline{E}} \, \left(\underline{\underline{B}}\, \underline{\underline{{}^eD}}\right) dV - \left({}^eD_a \, {}^eF_a + {}^eD_b \, {}^eF_b + \cdots\right) \tag{B.75}$$

Equation B.75 is minimized in using the procedure shown in Chapter 3, starting with Equation 3.4.11, to render a general expression for finite element equilibrium equations.

References

1. Huebner, K.H., *The Finite Element Method For Engineers*, John Wiley & Sons, N.Y., 1982

2. Reddy, J.N., *Energy and Variational Methods in Applied Mechanics*, John Wiley & Sons, N.Y., 1984
3. Bathe, K., *Finite Element Procedures In Engineering Analysis*, Prentice-Hall, Inc., Englewood Cliffs, New Jersey, 1982
4. Zienkiewicz, O.C., *The Finite Element Method in Structural and Continuum Mechanics*, McGraw-Hill, N.Y., 1970
5. Van Wylen, G.J., *Fundamentals of Classical Thermodynamics*, John Wiley & Sons, N.Y., 1978

Appendix C

B-Matrix and Domain Transformation

B-Matrix Derivation for a 4-Node, Isoparametric Surface Element

In Chapter 5, a 4-node, isoparametric surface element for C^0 applications was developed, as depicted in Figure C.1.

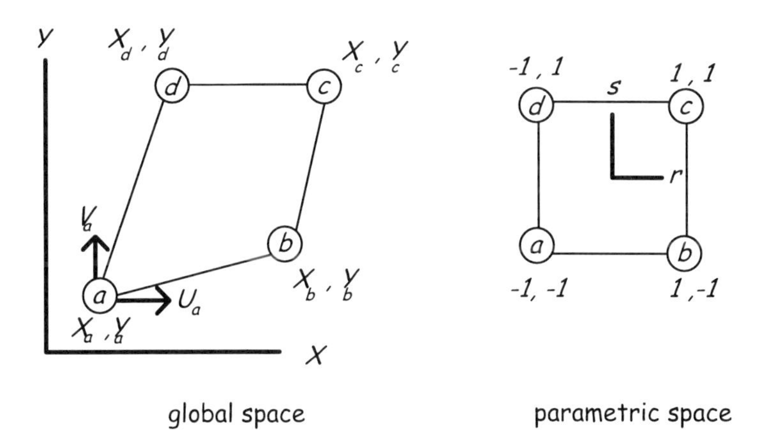

Figure C.1 A 4-Node Isoparametric Surface Element

The terms in the strain-displacement matrix were not derived, since a significant amount of manipulation is required to do so, and the details are not necessarily of general interest. The terms for the *B*-matrix are now considered.

When used for plane stress or plane strain, the 4-node surface element employs a strain vector having three terms:

$$\underset{=}{\varepsilon} = \begin{Bmatrix} \varepsilon_{XX} \\ \varepsilon_{YY} \\ \gamma_{XY} \end{Bmatrix} \tag{C.1}$$

Using the finite element approach, the strain terms are approximated by replacing the exact displacement variables with interpolation functions:

$$\underline{\underline{\varepsilon}} = \begin{Bmatrix} \varepsilon_{XX} \\ \varepsilon_{YY} \\ \gamma_{XY} \end{Bmatrix} \approx \begin{Bmatrix} \dfrac{\partial \tilde{U}}{\partial X} \\ \dfrac{\partial \tilde{V}}{\partial Y} \\ \dfrac{\partial \tilde{U}}{\partial Y} + \dfrac{\partial \tilde{V}}{\partial X} \end{Bmatrix} \tag{C.2}$$

We wish to express the finite element approximation for strain using a strain-displacement matrix and a vector of nodal displacements:

$$\underline{\underline{\varepsilon}} \approx \begin{Bmatrix} \dfrac{\partial \tilde{U}}{\partial X} \\ \dfrac{\partial \tilde{V}}{\partial Y} \\ \dfrac{\partial \tilde{U}}{\partial Y} + \dfrac{\partial \tilde{V}}{\partial X} \end{Bmatrix} = \underline{\underline{B}}^{e} \underline{\underline{D}} \tag{C.3}$$

For the 4-node element under consideration, the vector of nodal displacements contains eight terms, such that equation above can also be written as:

$$\underline{\underline{\varepsilon}} \approx \begin{Bmatrix} \dfrac{\partial \tilde{U}}{\partial X} \\ \dfrac{\partial \tilde{V}}{\partial Y} \\ \dfrac{\partial \tilde{U}}{\partial Y} + \dfrac{\partial \tilde{V}}{\partial X} \end{Bmatrix} = \underline{\underline{B}} \begin{Bmatrix} U_a \\ V_a \\ U_b \\ V_b \\ U_c \\ V_c \\ U_d \\ V_d \end{Bmatrix} \tag{C.4}$$

The objective here is to show how the terms for the B-matrix of (C.4) are derived.

It was shown in Chapter 5 that for the 4-node element under consideration, the derivatives that approximate the strain terms are defined as:

$$\frac{\partial \tilde{U}}{\partial X} = \frac{1}{\det \underline{\underline{J}}} \left(J_{22} \frac{\partial \tilde{U}}{\partial r} - J_{12} \frac{\partial \tilde{U}}{\partial s} \right) \tag{C.5}$$

$$\frac{\partial \tilde{U}}{\partial Y} = \frac{1}{\det \underline{\underline{J}}} \left(-J_{21} \frac{\partial \tilde{U}}{\partial r} + J_{11} \frac{\partial \tilde{U}}{\partial s} \right) \tag{C.6}$$

$$\frac{\partial \tilde{V}}{\partial X} = \frac{1}{\det \underline{\underline{J}}} \left(J_{22} \frac{\partial \tilde{V}}{\partial r} - J_{12} \frac{\partial \tilde{V}}{\partial s} \right) \tag{C.7}$$

$$\frac{\partial \tilde{V}}{\partial Y} = \frac{1}{\det \underline{\underline{J}}} \left(-J_{21} \frac{\partial \tilde{V}}{\partial r} + J_{11} \frac{\partial \tilde{V}}{\partial s} \right) \tag{C.8}$$

Using (C.5) through (C.8) in (C.4):

$$\underline{\underline{\varepsilon}} \approx \left\{ \begin{array}{c} \dfrac{\partial \tilde{U}}{\partial X} \\ \dfrac{\partial \tilde{V}}{\partial Y} \\ \dfrac{\partial \tilde{U}}{\partial Y} + \dfrac{\partial \tilde{V}}{\partial X} \end{array} \right\} = \left\{ \begin{array}{c} \dfrac{1}{\det \underline{\underline{J}}} \left(J_{22} \dfrac{\partial \tilde{U}}{\partial r} - J_{12} \dfrac{\partial \tilde{U}}{\partial s} \right) \\ \dfrac{1}{\det \underline{\underline{J}}} \left(-J_{21} \dfrac{\partial \tilde{V}}{\partial r} + J_{11} \dfrac{\partial \tilde{V}}{\partial s} \right) \\ \dfrac{1}{\det \underline{\underline{J}}} \left(-J_{21} \dfrac{\partial \tilde{U}}{\partial r} + J_{11} \dfrac{\partial \tilde{U}}{\partial s} \right) + \dfrac{1}{\det \underline{\underline{J}}} \left(J_{22} \dfrac{\partial \tilde{V}}{\partial r} - J_{12} \dfrac{\partial \tilde{V}}{\partial s} \right) \end{array} \right\} \tag{C.9}$$

Simplifying (C.9) yields:

$$\underline{\underline{\varepsilon}} \approx \frac{1}{\det \underline{\underline{J}}} \left\{ \begin{array}{c} \left(J_{22} \dfrac{\partial \tilde{U}}{\partial r} - J_{12} \dfrac{\partial \tilde{U}}{\partial s} \right) \\ \left(-J_{21} \dfrac{\partial \tilde{V}}{\partial r} + J_{11} \dfrac{\partial \tilde{V}}{\partial s} \right) \\ \left(-J_{21} \dfrac{\partial \tilde{U}}{\partial r} + J_{11} \dfrac{\partial \tilde{U}}{\partial s} \right) + \left(J_{22} \dfrac{\partial \tilde{V}}{\partial r} - J_{12} \dfrac{\partial \tilde{V}}{\partial s} \right) \end{array} \right\} \tag{C.10}$$

Recalling the displacement interpolation functions for the 4-node element:

$$\tilde{U}(r,s) = N_1U_a + N_2U_b + N_3U_c + N_4U_d$$
$$\tilde{V}(r,s) = N_1V_a + N_2V_b + N_3V_c + N_4V_d \tag{C.11}$$

Computing the partial derivative of $\tilde{U}$ with respect to r:

$$\frac{\partial\tilde{U}}{\partial r} = \frac{\partial}{\partial r}\left(N_1U_a + N_2U_b + N_3U_c + N_4U_d\right) \tag{C.12}$$

Or, taking the derivative of each term:

$$\frac{\partial\tilde{U}}{\partial r} = \left(\frac{\partial N_1}{\partial r}U_a + \frac{\partial N_2}{\partial r}U_b + \frac{\partial N_3}{\partial r}U_c + \frac{\partial N_4}{\partial r}U_d\right) \tag{C.13}$$

It is convenient to use an alternate expression to denote the derivatives of the shape functions:

$$\frac{\partial N_1}{\partial r} \equiv N_{1,r} \tag{C.14}$$

Using the alternate expressions for derivatives, (C.13) is more conveniently expressed as:

$$\frac{\partial\tilde{U}}{\partial r} = \left(N_{1,r}U_a + N_{2,r}U_b + N_{3,r}U_c + N_{4,r}U_d\right) \tag{C.15}$$

Likewise:

$$\frac{\partial\tilde{U}}{\partial s} = \left(N_{1,s}U_a + N_{2,s}U_b + N_{3,s}U_c + N_{4,s}U_d\right) \tag{C.16}$$

Using (C.15) and (C.16) in (C.10):

$$\underline{\underline{\varepsilon}} \approx \frac{1}{\det \underline{\underline{J}}} \left\{ \begin{array}{c} J_{22}\left(N_{1,r}U_a + N_{2,r}U_b + N_{3,r}U_c + N_{4,r}U_d\right) - J_{12}\left(N_{1,s}U_a + N_{2,s}U_b + N_{3,s}U_c + N_{4,s}U_d\right) \\ -J_{21}\dfrac{\partial \tilde{V}}{\partial r} + J_{11}\dfrac{\partial \tilde{V}}{\partial s} \\ \left(-J_{21}\dfrac{\partial \tilde{U}}{\partial r} + J_{11}\dfrac{\partial \tilde{U}}{\partial s}\right) + \left(J_{22}\dfrac{\partial \tilde{V}}{\partial r} - J_{12}\dfrac{\partial \tilde{V}}{\partial s}\right) \end{array} \right\}$$

(C.17)

The first row in (C.17), describing the normal strain in the X-coordinate direction, can be rearranged as:

$$\frac{1}{\det \underline{\underline{J}}}\Big\{\left(J_{22}N_{1,r} - J_{12}N_{1,s}\right)U_a + \left(J_{22}N_{2,r} - J_{12}N_{2,s}\right)U_b + \left(J_{22}N_{3,r} - J_{12}N_{3,s}\right)U_c + \left(J_{22}N_{4,r} - J_{12}N_{4,s}\right)U_d\Big\}$$

(C.18)

Or, in matrix form:

$$\varepsilon_{XX} \approx \frac{1}{\det \underline{\underline{J}}}\begin{bmatrix} J_{22}N_{1,r} - J_{12}N_{1,s} & 0 & J_{22}N_{2,r} - J_{12}N_{2,s} & 0 & J_{22}N_{3,r} - J_{12}N_{3,s} & 0 & J_{22}N_{4,r} - J_{12}N_{4,s} & 0 \end{bmatrix} \begin{Bmatrix} U_a \\ V_a \\ U_b \\ \vdots \\ V_d \end{Bmatrix}$$

(C.19)

Derivatives of $\tilde{V}$ can be computed as were derivatives of $\tilde{U}$:

$$\frac{\partial \tilde{V}}{\partial r} = \left(N_{1,r}V_a + N_{2,r}V_b + N_{3,r}V_c + N_{4,r}V_d\right) \tag{C.20}$$

$$\frac{\partial \tilde{V}}{\partial s} = \left(N_{1,s}V_a + N_{2,s}V_b + N_{3,s}V_c + N_{4,s}V_d\right) \tag{C.21}$$

Using (C.20) and (C.21), the second row of (C.17), describing normal strain in the Y-coordinate direction, can be expressed as:

$$\frac{1}{\det \underline{\underline{J}}}\left\{-J_{21}\left(N_{1,r}V_a + N_{2,r}V_b + N_{3,r}V_c + N_{4,r}V_d\right) + J_{11}\left(N_{1,s}V_a + N_{2,s}V_b + N_{3,s}V_c + N_{4,s}V_d\right)\right\} \tag{C.22}$$

Rearranging the above:

$$\frac{1}{\det \underline{\underline{J}}}\left\{\left(-J_{21}N_{1,r} + J_{11}N_{1,s}\right)V_a + \left(-J_{21}N_{2,r} + J_{11}N_{2,s}\right)V_b + \left(-J_{21}N_{3,r} + J_{11}N_{3,s}\right)V_c + \left(-J_{21}N_{4,r} + J_{11}N_{4,s}\right)V_d\right\} \tag{C.23}$$

In matrix form, (C.23) is expressed as:

$$\varepsilon_{YY} \approx \frac{1}{\det \underline{\underline{J}}}\begin{bmatrix} 0 & -J_{21}N_{1,r} + J_{11}N_{1,s} & 0 & -J_{21}N_{2,r} + J_{11}N_{2,s} & 0 & J_{21}N_{3,r} + J_{11}N_{3,s} & 0 & -J_{21}N_{4,r} + J_{11}N_{4,s} \end{bmatrix} \begin{Bmatrix} U_a \\ V_a \\ U_b \\ \vdots \\ V_d \end{Bmatrix} \tag{C.24}$$

The third row in (C.17) can be set into matrix form using the same procedure that was used for the first two rows. Using (C.19), (C.24), and the third row after it is set into matrix form, the strain vector can be expressed in terms of the strain displacement matrix and the vector of nodal displacements:

$$\frac{1}{\det \underline{\underline{J}}}\begin{bmatrix} (J_{22}N_{1,r} - J_{12}N_{1,s}) & 0 & \cdots & 0 \\ 0 & (-J_{21}N_{1,r} + J_{11}N_{1,s}) & \cdots & (-J_{21}N_{4,r} + J_{11}N_{4,s}) \\ (-J_{21}N_{1,r} + J_{11}N_{1,s}) & (-J_{21}N_{1,r} + J_{11}N_{1,s}) & \cdots & (J_{22}N_{4,r} - J_{12}N_{4,s}) \end{bmatrix} \begin{Bmatrix} U_a \\ V_a \\ U_b \\ \vdots \\ V_d \end{Bmatrix}$$

Which is to say:

$$\underline{\underline{\varepsilon}} \approx \underline{\underline{B}}\,{}^{e}\underline{\underline{D}} = \begin{bmatrix} B_{11} & 0 & B_{13} & 0 & B_{15} & 0 & B_{17} & 0 \\ 0 & B_{22} & 0 & B_{24} & 0 & B_{26} & 0 & B_{28} \\ B_{22} & B_{11} & B_{24} & B_{13} & B_{26} & B_{15} & B_{28} & B_{17} \end{bmatrix} \begin{Bmatrix} U_a \\ V_a \\ U_b \\ \vdots \\ V_d \end{Bmatrix} \tag{C.25}$$

The terms in the strain-displacement matrix are expressed as:

$$\begin{aligned} B_{11} &= \frac{1}{\det \underline{\underline{J}}}\left(J_{22}N_{1,r} - J_{12}N_{1,s}\right) & B_{22} &= \frac{1}{\det \underline{\underline{J}}}\left(-J_{21}N_{1,r} + J_{11}N_{1,s}\right) \\ B_{13} &= \frac{1}{\det \underline{\underline{J}}}\left(J_{22}N_{2,r} - J_{12}N_{2,s}\right) & B_{24} &= \frac{1}{\det \underline{\underline{J}}}\left(-J_{21}N_{2,r} + J_{11}N_{2,s}\right) \\ B_{15} &= \frac{1}{\det \underline{\underline{J}}}\left(J_{22}N_{3,r} - J_{12}N_{3,s}\right) & B_{26} &= \frac{1}{\det \underline{\underline{J}}}\left(-J_{21}N_{3,r} + J_{11}N_{3,s}\right) \\ B_{17} &= \frac{1}{\det \underline{\underline{J}}}\left(J_{22}N_{4,r} - J_{12}N_{4,s}\right) & B_{28} &= \frac{1}{\det \underline{\underline{J}}}\left(-J_{21}N_{4,r} + J_{11}N_{4,s}\right) \end{aligned} \tag{C.26}$$

Domain Transformation for a 4-Node, Isoparametric Surface Element
Using isoparametric elements, the element stiffness matrix is computed within the domain of the parent element, and the element in global space is mapped to the parent. The transformation from global space to parametric space is discussed in the following.

Recall the equation to compute the stiffness matrix for a single element:

$${}^{e}\underline{\underline{K}} = \int_V \underline{\underline{B}}^T \, \underline{\underline{E}}\,\underline{\underline{B}}\, dV \tag{C.27}$$

For surface elements of uniform thickness:

$$dV = t\; dX\; dY \tag{C.28}$$

However, we wish to perform integration within the domain of the parent element, such that the differentials used above are expressed as:

$$dX = \frac{\partial X}{\partial r} dr \left(\hat{i}\right) + \frac{\partial X}{\partial s} ds \left(\hat{j}\right) \qquad dY = \frac{\partial Y}{\partial r} dr \left(\hat{i}\right) + \frac{\partial Y}{\partial s} ds \left(\hat{j}\right) \tag{C.29}$$

We note that both dX and dY are vectors, having direction and magnitude, and they can be expressed as a combination of vector components in parametric space, as suggested by (C.29). The differential area of a rectangle is formed by the cross product of the two vectors (Fisher [1, p. 592]):

$$dA = dX \otimes dY \tag{C.30}$$

Actually, the differential area should be expressed as the *magnitude* of the cross product given in (C.30). However, if the connectivity and geometry of a finite element is valid, the cross product will always be non-negative, such that:

$$|dX \otimes dY| = dX \otimes dY$$

Using (C.29) in (C.30):

$$dA = \left(\frac{\partial X}{\partial r} dr \left(\hat{i}\right) + \frac{\partial X}{\partial s} ds \left(\hat{j}\right) \right) \otimes \left(\frac{\partial Y}{\partial r} dr \left(\hat{i}\right) + \frac{\partial Y}{\partial s} ds \left(\hat{j}\right) \right) \tag{C.31}$$

Forming the cross product above:

$$dA = \frac{\partial X}{\partial r}\frac{\partial Y}{\partial r} drdr \left(\hat{i} \otimes \hat{i}\right) + \frac{\partial X}{\partial r}\frac{\partial Y}{\partial s} drds \left(\hat{i} \otimes \hat{j}\right) + \frac{\partial X}{\partial s}\frac{\partial Y}{\partial r} dsdr \left(\hat{j} \otimes \hat{i}\right) + \frac{\partial X}{\partial s}\frac{\partial Y}{\partial s} dsds \left(\hat{j} \otimes \hat{j}\right) \tag{C.32}$$

Note that:

$$\hat{i} \otimes \hat{i} = 0 \qquad \hat{i} \otimes \hat{j} = k \qquad \hat{j} \otimes \hat{i} = -\hat{k} \qquad \hat{j} \otimes \hat{j} = 0 \tag{C.33}$$

Using (C.33), (C.32) reduces to:

$$dA = \left(\frac{\partial X}{\partial r}\frac{\partial Y}{\partial s} - \frac{\partial X}{\partial s}\frac{\partial Y}{\partial r} \right) dr\, ds \left(\hat{k}\right) \tag{C.34}$$

Recall the Jacobian matrix for the 4-node element:

$$\underline{\underline{J}} = \begin{bmatrix} \dfrac{\partial X}{\partial r} & \dfrac{\partial X}{\partial s} \\ \dfrac{\partial Y}{\partial r} & \dfrac{\partial Y}{\partial s} \end{bmatrix} \tag{C.35}$$

The determinant of the Jacobian is:

$$\det \underline{\underline{J}} = \frac{\partial X}{\partial r}\frac{\partial Y}{\partial s} - \frac{\partial X}{\partial s}\frac{\partial Y}{\partial r} \tag{C.36}$$

Thus, (C.34) is equivalent to:

$$dA = \left(\det \underline{\underline{J}}\right) dr\, ds \left(\hat{k}\right) \tag{C.37}$$

Since the $\hat{k}$ is simply a unit vector that denotes direction, the magnitude of the differential area is:

$$dA = \left(\det \underline{\underline{J}}\right) dr\, ds \tag{C.38}$$

Therefore:

$$dV = t\, dA = t\left(\det \underline{\underline{J}}\right) dr\, ds \tag{C.39}$$

Using (C.39) the stiffness of the element can be expressed with integrals defined on the parametric domain:

$$^{e}\underline{\underline{K}} = \int_{r_1}^{r_2}\int_{s_1}^{s_2} \underline{\underline{B}}^T \underline{\underline{E}}\,\underline{\underline{B}}\, t\left(\det \underline{\underline{J}}\right) dr\, ds \tag{C.40}$$

In summary, what we have shown is that the two dimensional domain of the global element, described by *dX dY*, can be transformed to the parent domain, *dr ds*, via the determinant of the Jacobian matrix. An analogous procedure can be used to transform three-dimensional domains. In one-dimensional problems, the transformation is accomplished simply through a scalar variable; see Chapter 5, (5.1.19).

References

1. Fisher, R.C., Ziebur, A.D., *Calculus and Analytic Geometry*, 3rd Edition, Prentice-Hall, Inc., Englewood Cliffs, N.J., 1975

Index